The IMA Volumes
in Mathematics
and its Applications

Volume 59

Series Editors
Avner Friedman   Willard Miller, Jr.

Institute for Mathematics and
its Applications
IMA

The **Institute for Mathematics and its Applications** was established by a grant from the National Science Foundation to the University of Minnesota in 1982. The IMA seeks to encourage the development and study of fresh mathematical concepts and questions of concern to the other sciences by bringing together mathematicians and scientists from diverse fields in an atmosphere that will stimulate discussion and collaboration.

The IMA Volumes are intended to involve the broader scientific community in this process.

Avner Friedman, Director
Willard Miller, Jr., Associate Director

* * * * * * * * * *

IMA ANNUAL PROGRAMS

1982–1983	**Statistical and Continuum Approaches to Phase Transition**
1983–1984	**Mathematical Models for the Economics of Decentralized Resource Allocation**
1984–1985	**Continuum Physics and Partial Differential Equations**
1985–1986	**Stochastic Differential Equations and Their Applications**
1986–1987	**Scientific Computation**
1987–1988	**Applied Combinatorics**
1988–1989	**Nonlinear Waves**
1989–1990	**Dynamical Systems and Their Applications**
1990–1991	**Phase Transitions and Free Boundaries**
1991–1992	**Applied Linear Algebra**
1992–1993	**Control Theory and its Applications**
1993–1994	**Emerging Applications of Probability**
1994–1995	**Waves and Scattering**
1995–1996	**Mathematical Methods in Material Science**

IMA SUMMER PROGRAMS

1987	**Robotics**
1988	**Signal Processing**
1989	**Robustness, Diagnostics, Computing and Graphics in Statistics**
1990	**Radar and Sonar (June 18 - June 29)**
	New Directions in Time Series Analysis (July 2 - July 27)
1991	**Semiconductors**
1992	**Environmental Studies: Mathematical, Computational, and Statistical Analysis**
1993	**Modeling, Mesh Generation, and Adaptive Numerical Methods for Partial Differential Equations**
1994	**Molecular Biology**

* * * * * * * * * *

SPRINGER LECTURE NOTES FROM THE IMA:

The Mathematics and Physics of Disordered Media

> Editors: Barry Hughes and Barry Ninham
> (Lecture Notes in Math., Volume 1035, 1983)

Orienting Polymers

> Editor: J.L. Ericksen
> (Lecture Notes in Math., Volume 1063, 1984)

New Perspectives in Thermodynamics

> Editor: James Serrin
> (Springer-Verlag, 1986)

Models of Economic Dynamics

> Editor: Hugo Sonnenschein
> (Lecture Notes in Econ., Volume 264, 1986)

W.M. Coughran, Jr.   Julian Cole
Peter Lloyd   Jacob K. White
Editors

Semiconductors
Part II

With 133 Illustrations

Springer-Verlag

New York Berlin Heidelberg London Paris
Tokyo Hong Kong Barcelona Budapest

W.M. Coughran, Jr.
AT&T Bell Laboratories
600 Mountain Ave., Rm. 2T-502
Murray Hill, NJ 07974-0636 USA

Peter Lloyd
AT&T Bell Laboratories
Technology CAD
1247 S. Cedar Crest Blvd.
Allentown, PA 18103-6265 USA

Julian Cole
Department of Mathematical Sciences
Rensselaer Polytechnic Institute
Troy, NY 12180 USA

Jacob K. White
Massachusetts Institute of Technology
Department of Electrical Engineering and
  Computer Science
50 Vassar St., Rm. 36-880
Cambridge, MA 02139 USA

Series Editors:
Avner Friedman
Willard Miller, Jr.
Institute for Mathematics and its
  Applications
University of Minnesota
Minneapolis, MN 55455 USA

Mathematics Subject Classifications (1991): 35-XX, 60-XX, 76-XX, 76P05, 81UXX, 82DXX, 35K57, 47N70, 00A71, 00A72, 81T80, 93A30, 82B40, 82C40, 82C70, 65L60, 65M60, 94CXX, 34D15, 35B25

Library of Congress Cataloging-in-Publication Data
Semiconductors / W.M. Coughran, Jr. ... [et al.].
        p.   cm. — (The IMA volumes in mathematics and its
    applications ; v. 58–59)
      Includes bibliographical references and index.
      ISBN 0-387-94250-5 (v. 1 : alk. paper). — ISBN 0-387-94251-3 (v.
    2 : alk. paper)
        1. Semiconductors—Mathematical models.   2. Semiconductors-
    -Computer simulation.   3. Computer-aided design.   I. Coughran,
    William Marvin.   II. Series.
    TK7871.85.S4693   1994
    621.3815′2—dc20   93-50622
Printed on acid-free paper.

Production managed by Laura Carlson; manufacturing supervised by Jacqui Ashri.
Camera-ready copy prepared by the IMA.
Printed and bound by Edwards Brothers, Inc., Ann Arbor, MI.
Printed in the United States of America.

9 8 7 6 5 4 3 2 1

ISBN 0-387-94251-3 Springer-Verlag New York Berlin Heidelberg
ISBN 3-540-94251-3 Springer-Verlag Berlin Heidelberg New York

The IMA Volumes
in Mathematics and its Applications

Current Volumes:

Forthcoming Volumes:

Systems & Control Theory for Power Systems

Adaptive Control, Filtering and Signal Processing

Discrete Event Systems, Manufacturing, Systems, and Communication Networks

Mathematical Finance

FOREWORD

This IMA Volume in Mathematics and its Applications

SEMICONDUCTORS, PART II

is based on the proceedings of the IMA summer program "Semiconductors." Our goal was to foster interaction in this interdisciplinary field which involves electrical engineers, computer scientists, semiconductor physicists and mathematicians, from both university and industry. In particular, the program was meant to encourage the participation of numerical and mathematical analysts with backgrounds in ordinary and partial differential equations, to help get them involved in the mathematical aspects of semiconductor models and circuits. We are grateful to W.M. Coughran, Jr., Julian Cole, Peter Lloyd, and Jacob White for helping Farouk Odeh organize this activity and trust that the proceedings will provide a fitting memorial to Farouk.

We also take this opportunity to thank those agencies whose financial support made the program possible: the Air Force Office of Scientific Research, the Army Research Office, the National Science Foundation, and the Office of Naval Research.

Avner Friedman

Willard Miller, Jr.

Preface to Part II

Semiconductor and integrated-circuit modeling are an important part of the high-technology "chip" industry, whose high-performance, low-cost microprocessors and high-density memory designs form the basis for supercomputers, engineering workstations, laptop computers, and other modern information appliances. There are a variety of differential equation problems that must be solved to facilitate such modeling.

During July 15–August 9, 1991, the Institute for Mathematics and its Applications at the University of Minnesota ran a special program on "Semiconductors." The four weeks were broken into three major topic areas:

1. Semiconductor technology computer-aided design and process modeling during the first week (July 15–19, 1991).
2. Semiconductor device modeling during the second and third weeks (July 22–August 2, 1991).
3. Circuit analysis during the fourth week (August 5–9, 1991).

This organization was natural since process modeling provides the geometry and impurity doping characteristics that are prerequisites for device modeling; device modeling, in turn, provides static current and transient charge characteristics needed to specify the so-called compact models employed by circuit simulators. The goal of this program was to bring together scientists and mathematicians to discuss open problems, algorithms to solve such, and to form bridges between the diverse disciplines involved.

The program was championed by *Farouk Odeh* of the IBM T. J. Watson Research Center. Sadly, Dr. Odeh met an untimely death. We have dedicated the proceedings volumes to him.

In this volume, we have combined the papers from the device modeling portion (weeks 2 and 3) of the program.

In 1991, semiconductor device modeling for practical engineering problems was largely based on the so-called drift-diffusion equations, a Poisson equation for the electrostatic potential coupled with advection-diffusion transport equations for the electrons and holes (in silicon, for example). Another popular model equation is the Boltzmann transport equation (BTE) of which the drift-diffusion equations are an approximation. For sufficiently small structures or III-V (like GaAs) devices, some of the assumptions of the drift-diffusion model are incorrect. Alternate derivatives of the BTE, such as energy-balance (or energy-transport) and hydrodynamic models, are of considerable interest. In fact, Dr. Odeh made a number of influential contributions to the hydrodynamic model and algorithms for it. The papers in this volume describe a variety of models and effectual techniques for dealing with them.

W. M. Coughran, Jr.
Murray Hill, New Jersey

Julian Cole
Troy, New York

Peter Lloyd
Allentown, Pennsylvania

Jacob White
Cambridge, Massachusetts

CONTENTS

SEMICONDUCTORS, PART II

Device Modeling

SEMICONDUCTORS, PART I

Process Modeling

Circuit Simulation

Dedication

Farouk Odeh (1933 - 1992)

Please refer to *Semiconductors, Part I*, IMA Volume # 58 for the complete dedication to Farouk Odeh.

LIST OF PARTICIPANTS

Aarden, J.	University of Nijmegen
Baccarani, Giorgo	University of Bologna
Bennett, Herbert	NIST
Biswas, Rana	Iowa State University
Blakey, Peter	Motorola Corporation
Borucki, Leonard	Motorola Corporation
Buergler, Josef	ETH Zurich
Casey, Michael	University of Pittsburgh
Cercignani, Carlo	Politecnico di Milano
Cole, Dan	IBM GPD
Cole, Julian	Rensselaer Polytechnic Institute
Coughran, Jr., William	AT&T Bell Labs
Cox, Paul	Texas Instruments
Degond, Pierre	Ecole Polytechnique
Gaal, Steven	University of Minnesota
Gardner, Carl	Duke University
Gartland, Chuck	Kent State University
Gerber, Dean	IBM
Giles, Martin	University of Michigan
Glodjo, Arman	University of Manitoba
Gnudi, Antonio	Universita Degli Studi Di Bologna
Grubin, Harold	Scientific Research Associates
Hagan, Patrick	Los Alamos National Lab
Hamaguchi, Satoshi	IBM
Henderson, Mike	IBM
Jerome, Joseph W.	Northwestern University
Johnson, Michael	IBM
Kalachev, Leonid	Moscow State University
Kerkhoven, Thomas	University of Illinois, Urbana
King, John	University of Nottingham
Kundert, Ken	Cadence Design Systems
Langer, Erasmus	Technical U. Vienna
Law, Mark	University of Florida
Leimkuhler, Ben	University of Kansas
Liniger, W.	IBM
Liu, H.C.	National Research Council, Ottawa
Liu, Sally	AT&T Bell Labs
Liu, Xu-Dong	UCLA
Lloyd, Peter	AT&T Bell Labs
Lojek, Robert	Motorola
Lumsdaine, Andrew	MIT
Makohon, Richard	University of Portland

Meinerzhagen, Berndt	Technischen Hochschule Aachen
Melville, Robert	AT&T Bell Labs
O'Malley, Robert E.	Rensselaer Polytechnic Institute
Odeh, Farouk	IBM
Palusinski, O.	University of Arizona
Perline, Ron	Drexel University
Petzold, Linda R.	University of Minnesota
Pidatella, Rosa Maria	Citta' Universitaria, Italy
Pillage, Larry	University of Texas
Please, Colin	Southhampton University
Poupaud, Frederic	University of Nice
Reyna, Luis	IBM
Richardson, Walter	University of Texas at San Antonio
Ringhofer, Christian	Arizona State University
Rose, Donald J.	Duke University
Rudan, Massimo	University of Bologna
Ruehli, Albert	IBM
Schmeiser, Christian	TU-Wien-Austria
Seidman, Tom	U.of Maryland-Baltimore County
Sever, Michael	Hebrew University
Singhal, K.	AT&T Bell Labs
So, Wasin	IMA
Souissi, Kamel	IBM
Strojwas, Andre	Carnegie Mellon University
Suto, Ken	Tohoku University
Szmolyan, Peter	TU-Wien-Austria
Tang, Henry	IBM
Thomann, Enrique	Oregon State University
Venturino, Ezio	University of Iowa
Vlach, Jeri	University of Waterloo
Ward, Michael	Stanford University
White, Jacob	MIT
Wrzosek, Darek	University of Warsaw
Young, Richard A.	University of Portland

ON THE CHILD-LANGMUIR LAW FOR SEMICONDUCTORS

N. BEN ABDALLAH* AND P. DEGOND*

1. Introduction. The design of many high technology components in solid-state electronics, in vacuum diode technology or in high power hyperfrequency amplification requires an accurate description of charged-particle transport. Among all the possible models, the Vlasov or the Boltzmann equations, coupled with the Poisson or Maxwell equations for the fields, provide the most accurate description of the physics of charged - particle transport. The numerical simulation of these models is an important tool for the designers.

The modelling of the injection of particles from a metallic cathode into the vacuum is particularly difficult, because of the existence of a space-charge boundary layer close to the cathode. A similar boundary layer lies at the junction between the highly doped source region of a semiconductor device, and its lowly doped active region, when the junction is direct biased. A realistic simulation relies on a correct description of the injection of carriers, because it determines the current which flows through the device. Up to now, in Monte-Carlo or particle simulations of semiconductors, much time is spent in the computation of the source regions which are physically uninteresting, just to provide an accurate description of the injection of particles. The same results could probably be obtained at a lower cost by modelling the injection process by adequate boundary conditions. The purpose of this paper is to present some ideas in this direction which are inspired from similar studies in vacuum diode simulations.

The modelling of vacuum diodes which operate under very large biases can be done by means of a perturbation analysis. A small parameter ε naturally appears as the ratio of the thermal energy of the particles that are injected at the cathode, over the external applied potential. An adequate scaling of the stationary Vlasov-Poisson equations provides a singular perturbation problem associated with this small parameter ε. Formally letting ε tend to zero yields a reduced problem which merely amounts to suppose that the particles are emitted from the cathode with a vanishing velocity. This reduced problem was investigated for the first time by Langmuir and Compton [1], who showed that in this context, the current intensity which flows through the diode cannot exceed a limiting value, often referred to as the "Child–Langmuir" current. They also give an explicit formula for the reduced problem in the one-dimensional cartesian case and provide us with various approximation formulas in the cylindrically or spherically symmetric cases.

The mathematical analysis of this problem first started with a study of the boundary-value problem for stationary Vlasov-Poisson equations in the one-dimensional cartesian case [2]. The perturbation problem and its convergence towards the reduced problem of [1] was analyzed in [3], in the same one-dimensional cartesian

*Centre de Mathematique et Leurs Applications, ENS-Cachan, 61, avenue du President Wilson, 94235 CACHAN Cedex.

geometry. Then, a numerical algorithm for the practical computation of the reduced solution was proposed in [4].

The passage to higher dimensions and more complicated models was initiated in [5,6], in which the well-posedness of the boundary value problems for the stationary Vlasov-Poisson, Vlasov-Maxwell and Vlasov-Poisson-Boltzmann equations are proved in any dimension. This analysis explicitly applies to semiconductor models. Then, by combining the ideas of [5,6], and of [3], the perturbation problem for the stationary cylindrically or spherically symmetric Vlasov-Poisson equation was investigated in [7].

In this paper, we present the first application of this perturbation analysis to the semiconductor Boltzmann equation. We will consider a simplified one dimensional device which consists of two highly doped N^+ regions on each side of a lowly doped N^- region. Such an $N^+ - N^- - N^+$ device closely resembles a vacuum diode, where the metallic cathode and anode are replaced by the N^+ zones, and the vacuum region, by the N^- zone. Assuming for the moment that the collisions of the carriers with the crystal lattice defects are negligible, we can use the same system of stationary Vlasov-Poisson equations in both cases and investigate the perturbation problem. Strictly speaking, the reduced solution will only be relevant, either for large direct biases, or low lattice temperatures (in other words, small thermal injection velocity). The reduced problem has been solved in this collisionless case by Shur and Eastman [8]. In particular, they investigate the effect of a non-vanishing doping density in the N^- region, on the current voltage characteristics.

However, the collisionless approximation is not valid in the realistic situations. To investigate the effect of collisions, Shur and Eastman [9] proposed a different model, based on a simplified one-dimensional hydrodynamic model consisting of two equations of momentum and energy balance. In this paper, we show that the first approach can be carried on: we use an adequate scaling of the Vlasov-Poisson-Boltzmann equation of semiconductors which yields a perturbation problem. The reduced problem can be explicitly written, and the proof of its well-posedness is in progress. It reduces to the Langmuir and Compton [1] or Shur and Eastman [8] solution when the collision frequency vanishes. It also exhibits the same features; namely the current intensity cannot exceed a limiting value which now depends on the collision frequency.

The original paper of Shur and Eastman [8,9] motivated a lot of studies in semiconductor physics to determine whether the interesting properties of their model device could be experimentally observed and used for practical purposes. It turned out that the answer was not clear, essentially because a lot of complex and still poorly understood physical phenomena are involved in the operation of a real device. However, the model seems to correctly account for the injection phenomenon. In large scale 2D or 3D particle simulations of vacuum devices, the Child-Langmuir model is implemented as a boundary condition which guarantees that the emission phenomenon is correctly described [10,11]. In the same spirit, a new boundary condition has been presented in [4] for which a detailed mathematical analysis is available. These boundary conditions could be used and adapted to Monte-Carlo or deterministic particle simulations of semiconductor devices.

The outline of the paper is as follows: in section 2, we will present the vacuum diode problem, which is simpler and for which a quite complete mathematical theory is available. In section 3, we will investigate the semiconductor case.

2. The vacuum diode problem. We start with the Vlasov-Poisson equation for the plane diode [2,3]. Let $F(X,V)$, $\phi(X)$ and $N(X)$ respectively denote the distribution function, the electric potential and the electron density for $X \in [0, L]$, $V \in \mathbf{R}$, where L is the length of the diode. The Vlasov-Poisson system reads:

$$V\frac{\partial F}{\partial X} + \frac{e}{m}\frac{d\phi}{dX}\frac{\partial F}{\partial V} = 0, \quad x \in [0, L], \quad V \in \mathbf{R}, \tag{2.1}$$

$$\frac{d^2\phi}{dX^2} = \frac{e}{\varepsilon_0}N(X), \quad X \in [0, L], \tag{2.2}$$

$$N(X) = \int_{-\infty}^{\infty} F(X,V)\,dV, \quad X \in [0, L] \tag{2.3}$$

where $-e$ and m are the charge and the mass of the electrons, ε_0 is the vacuum permittivity. The emission at the cathode $X = 0$ is described by a given emission profile $G(V) \geq 0$, $V \in [0, \infty)$. We suppose that no particle is emitted at the anode $X = L$. We take the cathode potential as the reference potential, and prescribe the anode potential ϕ_L. The corresponding boundary conditions are thus:

$$F(0, V) = G(V), \quad V > 0, \tag{2.4}$$

$$F(L, V) = 0, \quad V < 0, \tag{2.5}$$

$$\phi(0) = 0, \quad \phi(L) = \phi_L > 0. \tag{2.6}$$

The existence and the uniqueness theory for the problem (2.1)–(2.6) is developed in [2]. In particular, we recall that the existence of solutions is proved under some mild assumptions on the decay of G when $V \to +\infty$; and that the uniqueness is guaranteed if G is decreasing from $[0, \infty)$ onto $[0, \infty)$.

From the positivity of F and from (2.2), it follows that the potential ϕ is a strictly convex function. Thus, ϕ admits a unique minimum ϕ_a at a point $X = a$, and, given the boundary condition (2.6), we have $\phi_a \leq 0$. Thus, two cases can occur: either $\phi_a = 0$, which implies $a = 0$ and $\phi \geq 0$ on $[0, 1]$, or $\phi_a < 0$, which implies $a > 0$ and $\phi < 0$ on $(0, a]$. The phase portrait of the characteristics (cf. figure 1) of the Vlasov equation (2.1) shows that in the first case, all the particles which are emitted from the cathode actually cross the device and reach the anode. In this case, the total current density J:

$$J = -e \int_{-\infty}^{\infty} VF(X,V)\,dV \tag{2.7}$$

is equal to the injected current:

$$J = J_G,$$

$$J_G = -e \int_{0}^{\infty} VG(V)\,dV, \tag{2.8}$$

On the other hand, in the second case, the potential barrier ϕ_a reflects some fraction of the injected particles towards the cathode, and the current J is such that

$$|J| < |J_G|.$$

More precisely, the only particles which contribute to the current are those which are emitted from the cathode with a larger kinetic energy than the barrier potential energy; and we have the formula:

$$J = -e \int_{\sqrt{\frac{2e|\phi_a|}{m}}}^{\infty} V G(V)\,dV.$$

Now, the first case arises when the flux of injected particles is small, and the second case, when this flux is large. Indeed, in [2], it is proven that, given a normalized emission profile $G_0(V)$ such that

$$\int_0^{\infty} G_0(V)\,dV = 1,$$

there exists a real number $\lambda_0 > 0$, such that the problem (2.1)–(2.6) with $G(V) = \lambda G_0(V)$ has the following behaviour:

    (i) If $\lambda \leq \lambda_0$, we are in the first case: $\phi \geq 0$; $J = J_G$

    (ii) If $\lambda > \lambda_0$, we are in the second case: $a > 0$; $\phi < 0$ on $(0, a]$; $|J| < |J_G|$.

The second case is that of a space-charge limited current, and the interval $(0, a]$ is the space-charge region.

Up to now, we have depicted a fairly general situation. However, in many cases, the diode operates in a regime where the electrons are extracted from the cathode with a very low thermal energy (typically $1eV$), compared with the applied bias (typically several Mega Volts), but where, on the other hand, the density of emitted particles is large, so that the total injected current is finite. We define the thermal emission velocity V_G by:

$$(2.9) \qquad V_G = \left(\int_0^{\infty} V^2 G(V)\,dV \Big/ \int_0^{\infty} G(V)\,dV \right)^{1/2}.$$

The thermal emission energy is small compared with the applied bias if:

$$\frac{mV_G^2}{2} \ll e\phi_L$$

or

$$(2.10) \qquad V_G \ll V_L, \quad V_L = \sqrt{\frac{2e\phi_L}{m}}.$$

So, we introduce a "small" parameter ε by:

$$(2.11) \qquad \varepsilon = \frac{V_G}{V_L} \ll 1.$$

To express that the emission density is large, and that the emission current is finite, it is more convenient to introduce a scaling.

We shall use L, V_L and ϕ_L as characteristic length, velocity and potential scales. We introduce auxiliary units of density $\overline{N}$, current density $\overline{J}$, and distribution function $\overline{F}$, according to

$$(2.12) \qquad \overline{N} = \frac{\varepsilon_0 \phi_L}{e L^2}, \quad \overline{J} = e \overline{N} V_L, \quad \overline{F} = \frac{\overline{N}}{V_L},$$

and use the following change of variables and unknowns:

$$(2.13) \qquad \begin{cases} X = Lx, & V = V_L v, \quad \phi = \phi_L \varphi, \\ N = \overline{N} n, & J = -\overline{J} j, \quad F = \overline{F} f. \end{cases}$$

Furthermore, we introduce a dimensionless profile $g(v)$, and express $G(V)$ according to

$$(2.14) \qquad G(V) = \frac{\overline{F}}{\varepsilon^2} g \left(\frac{V}{V_G} \right).$$

The expression (2.14) simply means that V_G is the characteristic velocity associated with G, and the factor ε^2 insures that the injected current J_G remains independent of ε, in units of $\overline{J}$. The expression (2.14) can be rewritten:

$$(2.15) \qquad \begin{aligned} G(V) &= \overline{F} g^\varepsilon \left(\frac{V}{V_L} \right), \\ g^\varepsilon(v) &= \frac{1}{\varepsilon^2} g \left(\frac{v}{\varepsilon} \right). \end{aligned}$$

Now the scaling (2.13) together with (2.15) transforms the problem (2.1)–(2.6), into the following perturbation problem for the Vlasov-Poisson equation:

$$(2.16) \qquad v \frac{\partial f^\varepsilon}{\partial x} + \frac{1}{2} \frac{d\varphi^\varepsilon}{dx} \frac{\partial f^\varepsilon}{\partial v} = 0, \quad x \in [0,1], \quad v \in \mathbf{R},$$

$$(2.17) \qquad \frac{d^2 \varphi^\varepsilon}{dx^2} = n^\varepsilon(x), \quad x \in [0,1],$$

$$(2.18) \qquad n^\varepsilon(x) = \int_{-\infty}^{\infty} f^\varepsilon(x,v)\,dv, \quad x \in [0,1],$$

$$(2.19) \qquad f^\varepsilon(0,v) = g^\varepsilon(v) = \frac{1}{\varepsilon^2} g \left(\frac{v}{\varepsilon} \right), \quad v > 0,$$

$$(2.20) \qquad f^\varepsilon(1,v) = 0, \quad v < 0,$$

$$(2.21) \qquad \varphi^\varepsilon(0) = 0, \quad \varphi^\varepsilon(1) = 1.$$

The dimensionless current density j^ε is now given by

$$(2.22) \qquad j^\varepsilon = \int\limits_{-\infty}^{\infty} v f^\varepsilon(x,v)\,dv$$

while the dimensionless emission density n_g^ε, emission current j_g^ε, and thermal velocity v_g^ε, are given by

$$(2.23) \qquad n_g^\varepsilon = \int\limits_{0}^{\infty} g^\varepsilon(v)\,dv = \varepsilon^{-1} n_g, \quad n_g = \int\limits_{0}^{\infty} g(v)\,dv,$$

$$(2.24) \qquad j_g^\varepsilon = \int\limits_{0}^{\infty} v g^\varepsilon(v)\,dv = j_g, \quad j_g = \int\limits_{0}^{\infty} v g(v)\,dv,$$

$$(2.25) \qquad v_g^\varepsilon = (n_g^\varepsilon)^{-1/2} \left(\int\limits_{0}^{\infty} v^2 g^\varepsilon(v)\,dv \right)^{1/2}$$

$$= \varepsilon n_g^{-1/2} \left(\int\limits_{0}^{\infty} v^2 g(v)\,dv \right)^{1/2}$$

We notice that the density of injected particles is large $(0(\varepsilon^{-1}))$, the inflow current is finite and the thermal injection velocity is small $(0(\varepsilon))$. This is consistent with the assumptions of our asymptotics, as previously exposed.

The perturbation problem (2.16)–(2.21) will be called the "Child-Langmuir perturbation problem", and the reduced solution (f,φ) is the limit of $(f^\varepsilon,\varphi^\varepsilon)$ when ε tends to zero. We are now interested in characterizing the reduced solution. We will define the dimensionless Child-Langmuir current j_{CL} by:

$$(2.26) \qquad j_{CL} = \frac{4}{9}.$$

Then, we have the following:

LEMMA 2.1. *[3]: Let j be a real number, $j \in [0,\infty)$. The problem:*

$$(2.27) \qquad \begin{cases} \frac{d^2\varphi}{dx^2} = \frac{j}{\sqrt{\varphi}}, & x \in [0,1], \\ \varphi(0) = 0, & \varphi(1) = 1 \end{cases}$$

has a solution if and only if j satisfies:

$$(2.28) \qquad 0 \le j \le j_{CL} = \frac{4}{9}.$$

Moreover, the solution is unique and satisfies $\varphi \ge 0$ on $[0,1]$. Additionally, if $j = j_{CL}$, we have $d\varphi/dx(0) = 0$.

We can establish the convergence theorem, and characterize the reduced solution:

THEOREM 2.2. *[3]: Let $(f^\varepsilon, \varphi^\varepsilon)$ be the solution of the perturbation problem (2.16)–(2.21). Then, we have $f^\varepsilon \to f$ and $\varphi^\varepsilon \to \varphi$ as $\varepsilon \to 0$, where f and φ are given by:*

$$(2.29) \qquad f(x,v) = \frac{j}{\sqrt{\varphi(x)}}\delta(v - \sqrt{\varphi(x)})$$

$$(2.30) \qquad j = \min(j_g, j_{CL})$$

and φ is the unique solution of problem (2.27). Moreover, f is a distributional solution of the Vlasov equation (2.16), and the converges hold in the weak star topology of bounded measures for f^ε, and in the $C^1([0,1])$ topology for φ^ε.

We just give a few indications on this theorem, and refer to [3] for its complete proof. It is clear from (2.19) that the support of the boundary datum g concentrates on the single point $\{v = 0\}$ as ε goes to zero. Thus, it is natural to think that the support of the reduced solution f concentrates on the characteristics issued form $v = 0$ at the point $x = 0$. The equation of this characteristics is given by

$$v^2 - \varphi(x) = \text{ constant,}$$

where we have $v = 0$ and $\varphi(x) = 0$ at $x = 0$. Thus, the constant is zero, and its equation is finally

$$(2.31) \qquad v(x) = \pm\sqrt{\varphi(x)}.$$

This imposes $\varphi \geq 0$. Only the $+$ sign in (2.31) has to be considered, because we are interested in particles which start from the cathode and reach the anode. Thus, the reduced solution can be written

$$(2.32) \qquad f(x,v) = n(x)\delta(v - \sqrt{\varphi(x)})$$

where $n(x)$ is an unknown density for the time being. Now, the total current j associated with (2.32) is independent of x, as can be seen by integrating (2.16) with respect to v. Thus, we have

$$(2.33) \qquad j = n(x)\sqrt{\varphi(x)} = \text{ constant },$$

and the expression (2.29) follows. Now, Poisson's equation (2.17), (2.21) together with (2.33) gives rise to the problem (2.27), and lemma 2.1 insures that $j \leq j_{CL}$. For obvious reasons, j is also smaller than the injection current j_g, so that we have

$$(2.34) \qquad j \leq \min(j_g, j_{CL}).$$

That the inequality (2.34) is actually an equality follows from a careful analysis of the convergence problem that cannot be detailed here.

The relation (2.30) expresses that, when the emission current j_g is larger than the Child-Langmuir current j_{CL}, the current j which flows through the diode is limited to the value j_{CL}, whatever j_g or the emission profile g can be. To understand why it is so, in [3], it is proved that, when $j_g \geq j_{CL}$ and ε is small enough, the potential φ^ε exhibits a strictly negative minimum value $\varphi_a^\varepsilon < 0$ at a point $a^\varepsilon > 0$. In other words, a potential barrier establishes which reflects some fraction of the particles towards the cathode (cf. figure 1, case 2). In [3], the following asymptotics is proved:

$$(2.35) \qquad a^\varepsilon = 0(\varepsilon^{3/2}), \quad |\varphi_a^\varepsilon| = 0(\varepsilon^2).$$

It shows that the space-charge region is actually a boundary layer which is shrunk to zero when ε tends to zero and that the potential barrier vanishes at the same time. This explains why the reduced potential φ is positive, as stated in Lemma 2.1. However, the reduced solution keeps a reminiscence of the effect of the potential barrier and of the particles reflection in the current limitation. Indeed, in the limiting process, the potential barrier height φ_a^ε "automatically" adjusts so as to allow a flow of particles, just equal to j_{CL}.

To conclude this section, we express the Child-Langmuir current in terms of the physical variables:

$$(2.36) \qquad \begin{aligned} J_{CL} &= j_{CL}\overline{J} \\ &= \frac{4}{g}\varepsilon_0\sqrt{\frac{2e}{m}}\frac{\phi_L^{3/2}}{L^2}. \end{aligned}$$

This formula can be found in [1], and is often referred to as the 3/2-power law for the current-voltage characteristics of the vacuum diode.

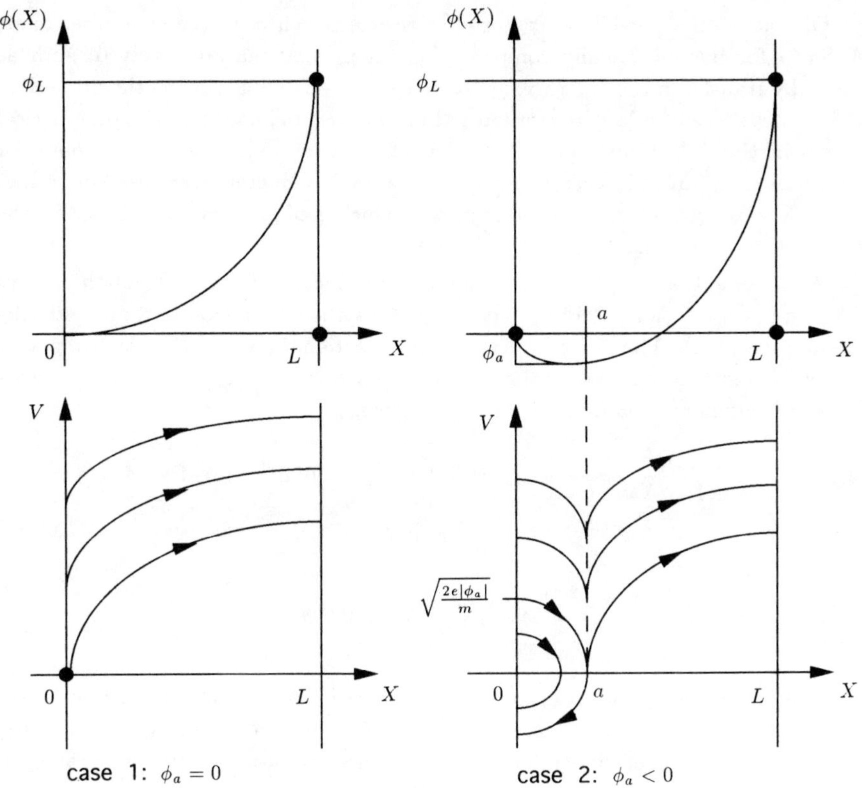

Figure 1: Potential and Phase Portrait of the characteristics
for the stationary 1-D Vlasov-Poisson problem.

3. The semiconductor case. [12]. We now consider a one dimensional unipolar semiconductor structure, which consists of two highly doped N^+ regions on each side of a lowly doped N^- region. In such a structure, the N^+ regions approximately behave like metallic contacts: the electric potential is nearly constant all along the N^+ zones; the injection of the carriers is entirely determined by the potential barrier, and the space-charge layer which surrounds the $N^+ - N^-$ junction on the source side essentially extends into the N^- zone; the injection of carriers at the drain contact is negligible. Therefore, in such a structure, the N^- region can be modelled just like the vacuum diode, by assuming that the injection of carriers at the $N^+ - N^-$ junctions can be described by a given emission profile $G(V)$ on the source side, and can be neglected on the drain side.

Of course, this model is very crude. Even in such a simple structure, made of *GaAs* for instance, many complex physical phenomena are involved, such as intervally transfer in the N^- region close to the $N^- - N^+$ junction on the drain side, and the presence of a long relaxation tail of particles belonging to the peripherical valleys in the N^+ drain region [13]. Therefore, our model does not account for the detailed features of electron transport in a semiconductor structure, but rather, focuses the analysis on the injection process, which is of primary importance for the global behaviour of many devices.

We assume that the $N^+ - N^-$ junctions are located at $X = 0$ (for the source side) and $X = L$ (for the drain side), and that the N^- region is represented by the interval $[0, L]$. The electron distribution function $F(X, V)$, the electric potential $\phi(X)$ and the electron density $N(X)$ satisfy the system of stationary Vlasov-Poisson-Boltzmann equations of semiconductors:

$$(3.1) \qquad V \frac{\partial F}{\partial X} + \frac{e}{m} \frac{d\phi}{dX} \frac{\partial F}{\partial V} = Q(F), \quad X \in [0, L], \quad V \in \mathbb{R},$$

$$(3.2) \qquad \frac{d^2 \phi}{dX^2} = \frac{e}{\varepsilon_0} N(X), \quad X \in [0, L],$$

$$(3.3) \qquad N(X) = \int_{-\infty}^{\infty} F(X, V) dV, \quad X \in [0, L].$$

We denote by $Q(F)$, the collision operator which models the interaction of the electrons with the crystal defects. The reader will find in [14, 15, 16] a fairly complete description of these interactions. In this paper, we will restrict our analysis to the relaxation time model

$$(3.4) \qquad Q(F) = -\frac{1}{\mathcal{J}}(F(X, V) - N(X) M_T(V))$$

where $\mathcal{J} > 0$ is the relaxation time, $M_T(V)$ is the normalized Maxwellian distribution associated with the lattice temperature T

$$(3.5) \qquad M_T(V) = \frac{1}{\sqrt{\frac{2\pi k_B T}{m}}} \exp\left(-\frac{mV^2}{2k_B T}\right),$$

and k_B is the Boltzmann constant. In the Poisson equation (3.2), we neglected the doping density, which is a fairly good assumption when the ratio N^+/N^- of the doping densities of the highly and the lowly doped regions is large enough [17]. Nevertheless, the incorporation of the doping density in the present analysis does not present major difficulties and is left to the reader. ε_0 now denotes the medium permittivity.

The system (3.1)-(3.3) is supplemented with the following boundary conditions:

$$(3.6) \qquad F(0, V) = G(V), \quad V > 0,$$

$$(3.7) \qquad F(L, V) = 0, \quad V < 0,$$

$$(3.8) \qquad \phi(0) = 0, \quad \phi(L) = \phi_L > 0.$$

These boundary conditions have been justified in the introduction of this section. More precisely, ϕ_L is the applied bias, and since the potential is nearly constant in the N^+ regions, it applies entirely at the boundary of the N^- region. Now $G(V)$ is the injection profile. Since the N^+ region on the source side is close to a state of thermal equilibrium with the crystal lattice, it is natural to assume that

$$(3.9) \qquad G(V) = N^+ M_T(V), \quad V > 0$$

where N^+ is the doping density of the N^+ region. On the other side, the injection is negligible, which implies (3.7).

The existence of solutions of the system (3.1)-(3.3), (3.6)-(3.8) is mathematically proven in [6]. Its numerical solution has been achieved by iterative methods in [17, 18], by particle methods in [19, 20], and by Monte-Carlo methods (see [14] and references therein).

In this paper, we are concerned with the Child-Langmuir asymptotics, as presented in section 2. Here, according to (3.5), (3.9), the thermal emission velocity clearly coincides with the lattice thermal velocity

$$(3.10) \qquad V_G = \sqrt{\frac{k_B T}{m}}.$$

Thus, the Child-Langmuir asymptotics is valid if (see (2.10)).

$$(3.11) \qquad k_B T \ll 2e\phi_L,$$

i.e. for large applied biases or for low temperatures. According to (2.11), we let

$$(3.12) \qquad \varepsilon = \sqrt{\frac{k_B T}{2e\phi_L}} \ll 1.$$

For instance, if $T = 77$ Kelvin and $\phi_L = 1$ Volt, we find $\varepsilon \simeq 0.06$.

As in the vacuum diode problem, we will use the scaling (2.12)-(2.13) to find the singular perturbation problem associated with the Child-Langmuir asymptotics. Given the expressions (3.9), (3.10), we may express the incoming distribution $G(V)$ according to (2.14), with

$$(3.13) \qquad g(v) = j_g \exp\left(-\frac{v^2}{2}\right),$$

provided that the ratio $N^+/\overline{N}$ of the density in the N^+ region over the density unit $\overline{N}$ satisfies

$$(3.14) \qquad \frac{N^+}{\overline{N}} = \frac{\sqrt{2\pi}\, j_g}{\varepsilon}.$$

In other words, $\overline{N}/N^+$ must be an $0(\varepsilon)$ quantity. In a $GaAs$ structure, with $N^+ = 10^{24} m^{-3}$, $\phi_L = 1$ Volt, $L = 0.4\mu m$, we have $\overline{N}/N^+ \simeq 0.004$, which meets the preceding requirements.

Finally, we introduce a dimensionless relaxation time τ, by

$$(3.15) \qquad \tau = \frac{V_L \mathcal{J}}{L}$$

so that the collision operator (3.4) can be written

$$(3.16) \qquad Q(\overline{F}f) = -\frac{V_L \overline{F}}{L} \frac{1}{\tau} \left[f - n \frac{1}{\varepsilon} M_0 \left(\frac{v}{\varepsilon} \right) \right]$$

where $M_0(v)$ is the dimensionless Maxwellian distribution

$$(3.17) \qquad M_0(v) = \frac{1}{\sqrt{2\pi}} \exp\left(-\frac{v^2}{2} \right).$$

Finally, with (3.13), (3.16) and (3.17), the perturbation problem associated with the Child-Langmuir asymptotics is written in the semiconductor case:

$$(3.18) \qquad v \frac{\partial f^\varepsilon}{\partial x} + \frac{1}{2} \frac{d\varphi^\varepsilon}{dx} \frac{\partial f^\varepsilon}{\partial v} = -\frac{1}{\tau} \left(f^\varepsilon - n^\varepsilon \frac{1}{\varepsilon} M_0 \left(\frac{v}{\varepsilon} \right) \right),$$
$$x \in [0, 1], v \in \mathbf{R},$$

$$(3.19) \qquad \frac{d^2 \varphi^\varepsilon}{dx^2} = n^\varepsilon(x), \quad x \in [0, 1],$$

$$(3.20) \qquad n^\varepsilon(x) = \int_{-\infty}^{\infty} f^\varepsilon(x, v) dv, \quad x \in [0, 1],$$

$$(3.21) \qquad f^\varepsilon(0, v) = g^\varepsilon(v) = \frac{\sqrt{2\pi} j_g}{\varepsilon^2} M_0 \left(\frac{v}{\varepsilon} \right), \quad v > 0,$$

$$(3.22) \qquad f^\varepsilon(1, v) = 0, \quad v < 0,$$

$$(3.23) \qquad \varphi^\varepsilon(0) = 0, \quad \varphi^\varepsilon(1) = 1.$$

We will now report on some partial results and conjectures concerning the asymptotic limit $\varepsilon \to 0$ of the perturbation problem (3.18)-(3.23). We are first concerned with the determination of the reduced solution $(f, \varphi) = \lim(f^\varepsilon, \varphi^\varepsilon)$, $\varepsilon \to 0$. The formal limit of equations (3.18)-(3.20), (3.22), (3.23) gives rise to the system:

$$(3.24) \qquad v \frac{\partial f}{\partial x} + \frac{1}{2} \frac{d\varphi}{dx} \frac{\partial f}{\partial v} = -\frac{1}{\tau}(f - n\delta(v)),$$

$$(3.25) \qquad \frac{d^2 \varphi}{dx^2} = n(x), \quad x \in [0, 1],$$

$$(3.26) \qquad n(x) = \int_{-\infty}^{\infty} f(x, v) dv, \quad x \in [0, 1],$$

$$(3.27) \qquad f(1, v) = 0, \quad v < 0,$$

$$(3.28) \qquad \varphi(0) = 0, \quad \varphi(1) = 1,$$

where $\delta(v)$ is the delta-function; the formal limit of (3.21) can be expressed by:

$$(3.29) \qquad \text{Support } \{ f(0, v), v \geq 0 \} = \{ v = 0 \}.$$

In the collisionless case ($\tau = \infty$; section 2), the condition (3.29) forces the solution to be a positive measure supported by the characteristics issued form the point $(x, v) = (0, 0)$. In the present case ($\tau < +\infty$), it is natural to think that the solution will exhibit the same features. This means that the particles fly from the cathode to the anode, following this characteristics. However, some particles will undergo a collision in their way to the anode. Given the shape of the collision operator at the right hand side of (3.24), the collision sends the velocity of the particles back to zero. Thus, after the collision, the particle follows another characteristics, namely the one issued from the point $(y, 0)$, where y is the location of the collision (cf. figure 2). This new characteristic is given by the equation:

$$v^2 - \varphi(x) = \text{constant} = -\varphi(y)$$

that is

$$(3.30) \qquad v = \sqrt{\varphi(x) - \varphi(y)}$$

(for obvious reasons, only the $+$ sign has to be retained). Of course, once the particle is on the secondary characteristics (3.30), it can suffer a second collision which sends it to a third characteristic and so on.

Therefore, in addition to a positive measure supported by the principle characteristics (i.e. issued from $(0,0)$), the solution $f(x,v)$ also contains a superposition of the contributions of each of the secondary characteristics (3.30), for $y \in (0, 1]$. Therefore, we write it

$$(3.31) \qquad f(x,v) = n_1(x)\delta(v - \sqrt{\varphi(x)}) + \int_0^x \bar{n}_2(x,y)\delta(v - \sqrt{\varphi(x) - \varphi(y)})dy,$$

where $n_1(x)$ is the density of particles carried by the principal characteristics and $\bar{n}_2(x,y)dy$ is the density of particles carried by the bundle of characteristics issued between the points $(y, 0)$ and $(y + dy, 0)$. Mathematically speaking, the integral on the right hand side of (3.31) can be viewed as a change of variables.

We introduce

$$(3.32) \qquad j_1(x) = n_1(x)\sqrt{\varphi(x)},$$

$$(3.33) \qquad \bar{j}_2(x,y) = \bar{n}_2(x,y)\sqrt{\varphi(x) - \varphi(y)},$$

$$(3.34) \qquad n_2(x) = \int_0^x \bar{n}_2(x,y)dy$$

$$(3.35) \qquad j_2(x) = \int_0^x \bar{j}_2(x,y)dy = \int_0^x \bar{n}_2(x,y)\sqrt{\varphi(x) - \varphi(y)}dy$$

j_1 is the current flowing along the principal characteristics, n_2 and j_2 are the density and current carried by all the secondary characteristics. We have

$$(3.36) \qquad n_1(x) + n_2(x) = n(x)$$

$$(3.37) \qquad j_1(x) + j_2(x) = j = \text{constant} .$$

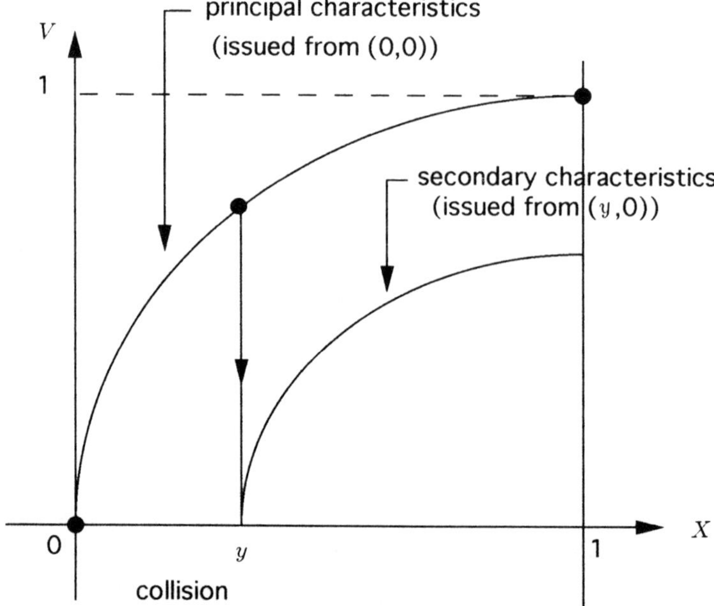

Figure 2: Schematics of the reduced solution in the semiconductor case ($\tau < \infty$).

Now, inserting the expression (3.31) into the equation (3.24), we find the following equations:

$$(3.38) \qquad \frac{dj_1}{dx} + \frac{n_1(x)}{\tau} = 0, \quad x \in [0,1],$$

$$(3.39) \qquad \frac{\partial}{\partial x}\bar{j}_2(x,y) + \frac{\bar{n}_2(x,y)}{\tau} = 0, \quad 0 < y < x < 1,$$

$$(3.40) \qquad \bar{j}_2(y,y) = \frac{1}{\tau}n(y).$$

The equation (3.38) means that the current decreases as one moves along the principal characteristics, because of the collisions. The same is true for each one of the secondary characteristics (equation (3.39)). Finally, equation (3.40) specifies that the current carried by one of the secondary characteristics at its starting point y, is made of the contribution of all the particles which have collided at this point, either from the principal or from all the secondary characteristics (cf. (3.36)). If we assume that $\bar{j}_2(x,y)$ is a smooth function, we deduce from (3.35) that

$$(3.41) \qquad \lim_{x \to 0} j_2(x) = 0$$

and thus from (3.37), that

$$(3.42) \qquad j_1(0) = j$$

The total current j flowing through the device will be assumed arbitrary for the moment. Thus, (3.42) provides the Cauchy condition for the differential equation (3.38).

Then, with (3.32) and (3.33), we can write the solutions of the equations (3.38)-(3.40), (3.42) according to:

$$(3.43) \qquad j_1(x) = j \exp\left(-\frac{1}{\tau} \int_0^x \frac{dz}{\sqrt{\varphi(z)}}\right)$$

$$(3.44) \qquad \bar{j}_2(x,y) = \frac{n(y)}{\tau} \exp\left(-\frac{1}{\tau} \int_y^x \frac{dz}{\sqrt{\varphi(z) - \varphi(x)}}\right).$$

For a given φ, the equation (3.43) gives a closed expression of j_1 and, with (3.32), the expression of n_1 follows:

$$(3.45) \qquad n_1(x) = \frac{j}{\sqrt{\varphi(x)}} \exp\left(-\frac{1}{\tau} \int_0^x \frac{dz}{\sqrt{\varphi(z)}}\right).$$

The situation is different for $\bar{j}_2$ since $\bar{j}_2$ depends on n which in turn depends on $\bar{j}_2$ via equations (3.34) and (3.33). From now on, we let:

$$(3.46) \qquad g_\varphi(x,y) = \exp\left(-\frac{1}{\tau} \int_y^x \frac{dz}{\sqrt{\varphi(z) - \varphi(y)}}\right).$$

Using (3.34), (3.33) and (3.44), we obtain

$$(3.47) \qquad n_2(x) - \int_0^x \frac{g_\varphi(x,y)}{\tau\sqrt{\varphi(x) - \varphi(y)}} n_2(y)dy = \int_0^x \frac{g_\varphi(x,y)}{\tau\sqrt{\varphi(x) - \varphi(y)}} n_1(y)dy.$$

Since n_1 has been previously determined by (3.45), equation (3.47) is an integral equations for n_2. In [12], we prove that, under suitable assumptions on φ, the equation (3.47) admits a unique solution for any value of j.

We now investigate the coupling with the Poisson equation, which gives rise to the following problem:

$$(3.48) \qquad \frac{d^2\varphi}{dx^2} = n(x), \quad x \in [0,1],$$

$$(3.49) \qquad \varphi(0) = 0, \quad \varphi(1) = 1,$$

$$(3.50) \qquad n(x) = n_1(x) + n_2(x)$$

where n_1 is given (3.45) and n_2 is the solution of (3.47), and where j in (3.45) is an arbitrary constant. The following conjecture is inspired from the existence of the Child-Langmuir current in the collisionless case (Lemma 2.1), and is yet partially proven.

LEMMA 3.1. *(Conjecture, cf. [12]): There exists a function $\tau \in (0, \infty) \rightarrow j_{CL}(\tau) \in (0, \infty)$, such that the problem (3.48)-(3.5), (3.45), (3.47) has a solution if and only if j satisfies:*

$$0 \leq j \leq j_{CL}(\tau). \tag{3.51}$$

Moreover, the solution is unique and satisfies $\varphi \geq 0$ on $[0, 1]$. Additionally, if $j = j_{CL}(\tau)$, we have $d\varphi/dx(0) = 0$.

Of course, $j_{CL}(\tau)$ is the Child-Langmuir current in the collisional case, and depends on the relaxation time τ. It can be determined as the solution of a particular problem which will not be detailed here (cf. [12]).

The second part of the conjecture concerns the convergence of the perturbed solution $(f^\varepsilon, \varphi^\varepsilon)$ to the reduced solution. We state

THEOREM 3.2. *(Conjecture, cf. [12]): Let $(f^\varepsilon, \varphi^\varepsilon)$ be the solution of the problem (3.18)-(3.23). Then, we have $f^\varepsilon \rightarrow f$; $\varphi^\varepsilon \rightarrow \varphi$ as $\varepsilon \rightarrow 0$, where f is given by (3.31),*

$$j = \min(j_g, j_{CL}(\tau)) \tag{3.52}$$

and n_1, n_2, φ are given by Lemma 3.1. Moreover, f is a distributional solution of (3.24) and the convergences hold in the weak star topology of bounded measures for f^ε, and in the $C^0([0, 1])$ topology for φ^ε.

In terms of the physical variables, the Child-Langmuir current is given by

$$J_{CL}(\tau) = j_{CL}(\tau) \varepsilon_0 \sqrt{\frac{2e}{m}} \frac{\phi_L^{3/2}}{L^2} \tag{3.53}$$

This shows that, in the regime of validity of the Child-Langmuir asymptotics, the current – voltage characteristics is – up to a multiplicative factor – that of a vacuum diode.

4. Conclusion. The last remark of the preceding section shows that our model is not realistic for practical devices, and the reason is that many important physical phenomena have been neglected in this model. However, it shows that, as far the injection of particles in the active region is concerned, the behaviour of a semiconductor device is close to that of a vacuum diode. Thus, the numerical algorithms for the modelling of the injection phenomenon which are used in large scale vacuum diode codes could be efficient in semiconductor device simulations. One of these algorithms [4] has been mathematically analyzed. Its relevance for semiconductor applications will be investigated in a forthcoming paper.

REFERENCES

[1] I. LANGMUIR AND K.T. COMPTON, *Electrical Discharges in Gases: Part II, Fundamental Phenomena in Electrical Discharges*, Rev. Mod. Phys. 3 (1931), pp. 191–257.

[2] C. GREENGARD AND P.A. RAVIART, *A Boundary Value Problem for the Stationary Vlasov-Poisson Equations: the plane diode*, Comm. Pure Appl. Math. 43 (1990), pp. 473–507.

[3] P. DEGOND AND P.A. RAVIART, *An Asymptotic Analysis of the One-Dimensional Vlasov-Poisson System: the Child-Langmuir Law*, Report no. 197, CMAP, Ecole Polytechnique, 1989, to appear in Asymptotic Analysis.

[4] P. DEGOND AND P.A. RAVIART, *On a Penalization of the Child-Langmuir emission condition for the one dimensional Vlasov-Poisson equation*, preprint, to appear in Asymptotic Analysis.

[5] F. POUPAUD, *Solutions stationnaires des Equations de Vlasov-Poisson*, C.R. Acad. Sci. Paris 311 (1990), pp. 307–312.

[6] F. POUPAUD, *Boundary Value Problems for the Stationary Vlasov-Maxwell systems*, preprint, to appear in "Forum Mathematicum".

[7] P. DEGOND, S. JAFFARD, F. POUPAUD AND P.A. RAVIART, *The Child-Langmuir asymptotics of the Vlasov-Poisson Equation for Cylindrically or Spherically Symmetric Diodes*, manuscript.

[8] M.S. SHUR AND L.F. EASTMAN, *Ballistic Transport in Semiconductor at Low Temperatures for Low-Power High-Speed Logic*, IEEE Trans. Electron Dev. ED-26 (1979), pp. 1677–1683.

[9] M.S. SHUR AND L.F. EASTMAN, *Near Ballistic Election Transport in GaAs Devices at 77° K*, Solid-State Election 24, pp. 11–18.

[10] W. HERRMANSFELDT, *electron Trajectory Program*, SLAC Technical Report 166, sept. 1973.

[11] T. WEISTERMAN, *A particle-in-Cell Method as a tool for Diode Simulations*, Nuclear Instruments and Methods in Physics Research A263 (1988), pp. 271–279.

[12] N. BEN ABDALLAH AND P. DEGOND, *The Child-Langmuir asymptotics of the Boltzmann Equation of Semiconductors*, in preparation.

[13] H.U. BARANGER AND J.W. WILKINS, *Ballistic Electrons in a Submicron Structure, the Distribution Function and Two-Valley Effects*, Physica B134 (1985), pp. 470–474.

[14] L. REGGIANI (editor), *Hot-Electron Transport in Semiconductors*, Springer, Berlin, 1985.

[15] B. NICLOT, P. DEGOND AND F. POUPAUD, *Deterministic Particle Simulations of the Boltzmann Transport Equation of Semiconductors*, J. Comput. Phys. 78 (1988), pp. 313–349.

[16] P. DEGOND AND F.J. MUSTIELES, *A Deterministic Particle Method for the Kinetic Model of Semiconductors: the Homogeneous Field Model*, to appear in Solid-State Electron.

[17] H.U. BARANGER, *Ballistic Electrons in a Submicron Semiconducting Structure: A Boltzmann Equation Approach*, Ph.D Thesis, Cornell University, January 1986, p. 40.

[18] H.U. BARANGER AND J.W. WILKINS, *Ballistic Structure in the Electron Distribution Function of Small Semiconducting Structures: General Features and Specific Trends*, Phys. Rev. B36 (1987), pp. 1487–1502.

[19] P. DEGOND AND F. GUYOT-DELAURENS, *Particle Simulations of the Semiconductor Boltzmann Equation for One-Dimensional Inhomogeneous Structure*, J. Comput. Phys. 90 (1990), pp. 65–97.

[20] F. DELAURENS AND F.J. MUSTIELES, *A New Deterministic Particle Method for Solving Kinetic Transport Equations: The Semiconductor Boltzmann equation Case*, Preprint.

A CRITICAL REVIEW OF THE FUNDAMENTAL SEMICONDUCTOR EQUATIONS

G. BACCARANI*, F. ODEH**, A. GNUDI*, AND D. VENTURA*

Abstract. Physical models for numerical device simulation are reviewed in an engineering perspective, and their derivation from more general physical principles is examined. More specifically, the limitations of the basic semiconductor equations within the drift-diffusion model is discussed and its validity range is assessed with emphasis on silicon devices. An approximate solution to the Boltzmann Transport Equation (BTE) is obtained by using the Hilbert expansion method near equilibrium conditions; however, in contrast to previous treatments, the drift and diffusion term are linked together in order to overcome the usual limitation of small external fields. The solution is carried out up to the third term of the expansion, thus leading to a correction of the usual drift-diffusion current density. This correction yields a nonlinear velocity-field relationship and accounts for velocity overshoot.

1. Introduction. The drift-diffusion model of current transport in semiconductors is to date well established [1]. Numerical solvers based on the above model have successfully been used to predict new device effects and to help the designing of advanced processes and devices [2-3]. Yet, its mathematical justification under general non-homogeneous conditions is still unclear, since its derivation from the Boltzmann Transport Equation (BTE) usually requires the assumption of a small electric field and small carrier gradients [4]. Even in the simplest electron device, i.e. the $p - n$ junction, such a condition is almost never fulfilled: fields as large as 10^5 V/cm, well exceeding $k_B T/q^\lambda$ (where λ is the carrier mean-free path) are typically found even under equilibrium conditions. It is therefore important to assess more carefully the validity of the model from a mathematical standpoint.

The main aim of this paper is to identify the limits of the drift-diffusion model in the context of device simulation. We use the Hilbert expansion method [4] to derive the model but, in contrast to previous approaches, we link the drift and diffusion terms together in order to properly scale their combination near equilibrium conditions. In doing that, the validity of the model is assessed and found to hold so long as the gradient of the quasi-Fermi potential is small compared with $(k_B T/q^\lambda)$. In the second part of the paper, we consider the equations governing the higher-order terms, and solve them up to third order. Hence, we find the correction to the current density.

The paper is organized as follows: section 2 defines the mathematical problem and discusses the limitations of the BTE in the context of electron transport in semiconductors. The scaling of the BTE is performed in section 3. Section 4 is devoted to the evaluation of the first-order solution of the BTE, and to a discussion of its limitations. The second- and third-order solutions to the BTE are derived in section 5, where the correction to the current density is identified. Finally, conclusions are drawn in section 6.

---

*Dipartimento di Elettronica, Informatica e Sistemistica, Università di Bologna, viale Risorgimento 2, 40136 Bologna, Italy.
**IBM Research Division, T.J. Watson Research Center, Yorktown Heights, NY 10598, USA.

2. Semiconductor equations. The starting point of a first-principles transport theory is the quantum mechanical Liouville-Von Neumann equation for the density matrix. Under a number of simplifying assumptions, the latter reduces to the well-known Boltzmann Transport Equation (BTE) reported below

$$(2.1) \qquad \frac{df}{dt} = \frac{\partial f}{\partial t} + \mathbf{u} \cdot \operatorname{grad} f \mp \frac{q}{h} \mathbf{E} \cdot \operatorname{grad} f = Sf$$

where $f(\mathbf{r}, \mathbf{k}, t) d^3 r\, d^3 k$ represents the number of electrons/holes in the elementary volume $d^3 r\, d^3 k$ of the phase space at the time t, and $\mathbf{u}$ is the group velocity. In (2.1) the total derivative df/dt is evaluated along the classical trajectory $\mathbf{r} = \mathbf{r}(t)$, $\mathbf{k} = \mathbf{k}(t)$, i.e. in the absence of collisions. The $\mp$ sign refers to the electron and hole cases, respectively. The collision term on the RHS of (2.1) represents the rate of change of f due to collisions; its expression is the following

$$(2.2) \qquad Sf = \int [S(\mathbf{r}, \mathbf{k}', \mathbf{k}) f(\mathbf{r}, \mathbf{k}', t) - S(\mathbf{r}, \mathbf{k}, \mathbf{k}') f(\mathbf{r}, \mathbf{k}, t)] d^3 k',$$

provided a non degenerate electron gas is assumed. In (2.2) $S(\mathbf{r}, \mathbf{k}, \mathbf{k}') d^3 k'$ represents the scattering probability per unit time from the state $\mathbf{k}$ at $\mathbf{r}$ to any other state $\mathbf{k}'$ contained in $d^3 k'$. In what follows we shall assume that, due to the symmetry properties of the S function, the operator S is even, i.e.

$$(2.3) \qquad S[f(-\mathbf{k})] = (Sf)(-\mathbf{k})$$

Actually, the above assumption is only correct for isotropic scattering mechanisms, such as acoustic- and optical-phonon interactions. If impurity scattering plays a role, we shall assume an equivalent isotropic impurity scattering which provides the same momentum relaxation time as the real anisotropic scattering mechanism.

In (2.1), the electric field $\mathbf{E} = -\operatorname{grad} \varphi$ is governed by Poisson's equation

$$(2.4) \qquad \operatorname{div}(\epsilon_s \operatorname{grad} \varphi) = q(p - n + N_D - N_A)$$

where φ is the electric potential, ϵ_s the semiconductor dielectric constant, n and p the electron and hole concentrations, N_D and N_A the donor and acceptor impurity densities, respectively. It should be mentioned that, within a device, $\mathbf{E}$ may be large and fast varying both in space and time, depending upon the nature of the device and its boundary conditions.

The problem of device simulation is thus defined by a couple of BTEs (2.1) for both electrons and holes and by the coupling equation (2.4). For the sake of simplicity, we shall consider only the BTE for electrons and neglect generation-recombination effects, which play a negligible role for most micron-size devices. From the definition of the distribution function f, the electron concentration and the current density are given by

$$(2.5a) \qquad n(\mathbf{r}, t) = \int f(\mathbf{r}, \mathbf{k}, t) d^3 k$$

$$(2.5b) \qquad \mathbf{J}_n(\mathbf{r}, t) = -q \int \mathbf{u} f(\mathbf{r}, \mathbf{k}, t) d^3 k.$$

Equation (2.1) basically represents a continuity equation in the phase space which relies upon the single-particle approximation. The basic assumptions underlying equations (2.1) and (2.2) can be summarized as follows (see e.g. [5]):

i) The single-particle approximation holds, i.e. the interaction among carriers is assumed to be weak.

ii) The band theory and the effective-mass theorem apply to the semiconductor.

iii) The electric field is slowly varying over a scale length comparable with the physical dimensions of the wave packet describing the motion of the particle; therefore a classical description of electron dynamics is possible via the Ehrenfest theorem.

iv) Only point collisions are considered in time and in space. Therefore, no energy is gained by the particle from the electric field during collisions.

v) Electron-electron interaction is neglected in the scattering integral (2.2).

vi) The scattering probability is independent of the external field.

vii) The electron/hole gas is not degenerate.

viii) The effect of the magnetic field is neglected.

In essence, equation (2.1) relies upon a semi-classical picture of carrier transport in semiconductors. Electron motion results from a sequence of drifts in the electric field, followed by scattering events. The drifting time, the type of scattering process, and the final state are random quantities which can be expressed in terms of transition rates due to the various processes; however, the free flight is deterministic and depends on the spatial distribution of the electric field.

The BTE is an integro-differential equation which, unfortunately, allows for a closed-form solution only in extremely special cases. The most successful technique to attack the problem is based on a Monte Carlo approach [6-7], whereby hot-electron effects could be successfully interpreted in most semiconductor materials. Other techniques based on a variational approach [8] or iterative procedures [9-11] could help in solving certain specific problems, but turned out to be of a scarse general applicability.

If equation (2.1) is integrated over the first Brillouin zone, we find a continuity equation in the physical space, namely

$$\text{(2.6)} \qquad \frac{\partial n}{\partial t} - \frac{1}{q} \operatorname{div} \mathbf{J}_n = 0$$

where equations (2.5) have been accounted for. In order that equation (2.6) can be self-consistently solved with Poisson's equation (2.4), an expression for the current density $\mathbf{J}_n$ must be found against φ, n and their derivatives. In principle, this can be achieved by looking for approximate solutions to equation (2.1).

3. Scaling of the BTE. The approximate, or asymptotic, analysis of a differential equation is usually accomplished by first introducing a scaling process, the aim of which is that of assessing the relative importance of the various terms and of identifying the operating conditions under which some terms can be considered of a higher order with respect to others. In doing so, a set of normalizing constants, possibly related tot he physical processes being described by the equation, must be identified. Here a problem arises, for equations (2.1) and (2.4) do not uniquely identify a characteristic time or length: the latters are rather determined by the specific nature of the device under consideration and its boundary conditions. The scaling which follows pertains to the regime of "near equilibrium" in a sense which is given later.

We assume as a characteristic length the device active length L and as a characteristic time the transit time of the carriers travelling at the velocity $v_d = \mu(k_B T/qL)$; hence, the transit time turns out to be $\tau_t = L/v_d$. Next, we normalize the group velocity $\mathbf{u}$ by the thermal velocity $v_{th} = (k_B T/m^*)^{1/2}$ and the crystal momentum $\mathbf{k}$ by $k_{th} = (m^*/\hbar)v_{th}$. Finally, the scattering rates are normalized by τ_{po}^{-1}, where

$$(3.1) \qquad \tau_{po} = \frac{\langle \tau_{co}(\mathbf{u}), u^2 f_0 \rangle}{\langle 1, u^2 f_0 \rangle}$$

and $\tau_{co}(\mathbf{u})$ is the inverse scattering rate for an undoped semiconductor material. In (3.1) brackets denote an averaging, i.e. a scalar product of L_2 functions over the velocity space. Also, f_0 is the equilibrium distribution function. We now define the following normalized physical quantities:

$$t^* = t/\tau_t; \quad \mathbf{r}^* = \mathbf{r}/L; \quad \mathbf{u}^* = \mathbf{u}/v_{th}; \quad \mathbf{k}^* = \mathbf{k}/k_{th}; \quad S^* = \tau_{po} S$$

and, by substituting the normalized variables into (2.1), we find

$$(3.2) \qquad \epsilon^2 \frac{\partial f^*}{\partial t^*} + \epsilon(\mathbf{u}^* \cdot \mathrm{grad}_* f^* - \mathbf{E}^* \cdot \mathrm{grad}_* f^*) = S^* f^*$$

where f^* is the new distribution function defined in the $(\mathbf{r}^*, \mathbf{k}^*)$ space, $\mathbf{E}^*$ is the normalized field

$$(3.3) \qquad \mathbf{E}^* = \frac{qL\mathbf{E}}{k_B T}$$

and

$$(3.4) \qquad \epsilon = \sqrt{\frac{\tau_{po}}{\tau_t}} = \frac{\lambda_o}{L}$$

where $\lambda_o = v_{th}\tau_{po}$ is the carrier mean-free path for an undoped semiconductor. We shall attempt an approximate solution of equation (3.2) under the assumptions that: i) $\epsilon \ll 1$ and ii) the device is weakly perturbed with respect to equilibrium. This means that the applied voltage is not too large, and we defer for now the estimate of an acceptable value for the voltage itself. We want to stress, however, that we do not impose any constraint on the local values of the force acting upon the carriers, apart from those which ensure the applicability of the Ehrenfest theorem; nor do we assume a slowly-varying impurity concentration in (2.4) and within the scattering operator S^*. From a mathematical standpoint, this implies that the two terms in parenthesis on the l.h.s. of equation (3.2) may be individually large. These terms, however, cancel in equilibrium, and it is reasonable to assume that under assumption of a small perturbation their algebraic sum remains of the same order as that of f^*.

4. First-order solution to the BTE. In this paragraph we attempt an approximate solution to the BTE and, in order to simplify the analysis, we assume spherical energy bands. For the sake of convenience, we omit henceforth the superscript * to denote normalized variables. It is now worth noting that $\lambda_o \simeq 10\,nm$ in silicon and, assuming $L \simeq 1\,\mu m$, then $\epsilon \simeq 10^{-2}$. The form of equation (3.2) then suggests an expansion of the f function in a power series of the form [4]

$$(4.1) \qquad f = \sum_{j=0}^{\infty} \epsilon^j f^{(j)}$$

and, by substituting (4.1) into (3.2) and equating coefficients of the same power, we find the following set of equations

$$(4.2a) \qquad S f^{(0)} = 0$$

$$(4.2b) \qquad S f^{(1)} = \mathbf{u} \cdot \operatorname{grad} f^{(0)} - \mathbf{E} \cdot \operatorname{grad} f^{(0)}$$

$$(4.2c) \qquad S f^{(2)} = \frac{\partial f^{(0)}}{\partial t} + \mathbf{u} \cdot \operatorname{grad} f^{(1)} - \mathbf{E} \cdot \operatorname{grad} f^{(1)}$$

$$\ldots\ldots\ldots\ldots\ldots\ldots$$

Substituting the power series expansion (4.1) into equation (2.5a) gives

$$(4.3a) \qquad n(\mathbf{r}, t) = \sum_{j=0}^{\infty} n^{(j)}(\mathbf{r}, t)$$

where

$$(4.3b) \qquad n^{(j)}(\mathbf{r}, t) \equiv \epsilon^j \int f^{(j)}(\mathbf{r}, \mathbf{k}, t) d^3 k$$

The solution of equation (4.2a) is

$$(4.4) \qquad f^{(0)}(\mathbf{r}, \mathbf{k}, t) = \exp(-\varepsilon(\mathbf{k})) g(\mathbf{r}, t)$$

where the normalized energy ε is related to the crystal momentum by the band function

$$(4.5) \qquad \gamma(\varepsilon) = \varepsilon(1 + \alpha\varepsilon) = \frac{1}{2}\mathbf{k} \cdot \mathbf{k}$$

and $g = g(\mathbf{r}, t)$ is an arbitrary function of position and time. The arbitrariness of the g function comes from assumption iv) above that only point collisions may occur in time and in space.

From (4.3b), the zero-order solution $f^{(0)}$ may be expressed as

$$(4.6) \qquad f^{(0)}(\mathbf{r}, \mathbf{k}, t) = N_c^{-1} \exp(-\varepsilon(\mathbf{k})) n^{(0)}(\mathbf{r}, t)$$

where

$$(4.7) \qquad N_c = (2\pi)^{3/2} \left(1 + \frac{15}{4}\alpha\right)$$

The term $(15/4)\alpha$ in equation (4.7) represents the correction due to the non-parabolicity of the band. Because of the small value of α, the above correction is typically of the order of 5%. Under the condition that the r.h.s. of equations (4.2) belong to the range of the operator S, $f^{(j)}$ may be formally represented as

$$(4.8) \qquad f^{(j)} = S^{-1}\left(\frac{\partial f^{(j-2)}}{\partial t} + \mathbf{u}\cdot\operatorname{grad} f^{(j-1)} - \mathbf{E}\cdot\operatorname{grad} f^{(j-1)}\right)$$

with $j = 1, 2, 3\ldots$, where we use the notation $f^{(-1)} \equiv 0$, and where the functions $f^{(j)}$ are uniquely determined by the normalizing conditions

$$(4.9) \qquad \langle 1, f^{(j)}\rangle = 0 \quad \text{for } j \geq 1$$

Using (4.9) one obtains

$$(4.10) \qquad n(\mathbf{r}, t) = n^{(0)}(\mathbf{r}, t)$$

We note that, since the r.h.s. of (4.2) are alternatively even and odd functions in $\mathbf{k}$, the parity properties of the S operator ensure us that $f^{(j)}$s are alternatively even and odd as well.

If we consider now equation (4.2b), then using (4.6) and (4.10) leads to the equation

$$(4.11) \qquad S f^{(1)} = N_c^{-1} \exp(-\varepsilon(\mathbf{k}))\mathbf{u}\cdot(\operatorname{grad} n + n\mathbf{E})$$

whose solution is

$$(4.12) \qquad f^{(1)}(\mathbf{r}, \mathbf{k}, t) = -N_c^{-1}\tau_c(\mathbf{r}, \mathbf{k})\exp(-\varepsilon(\mathbf{k}))\mathbf{u}\cdot(\operatorname{grad} n + n\mathbf{E})$$

where $\tau_c(\mathbf{r}, \mathbf{k})$ is the inverse scattering rate normalized to τ_{po}. The $\mathbf{r}$ dependence of τ_c is due to the position-dependent impurity concentration. From the definition of the current density (2.5b) we find

$$(4.13) \qquad \mathbf{J}_n^{(1)}(\mathbf{r}, t) = -\int \varepsilon\mathbf{u}f^{(1)}(\mathbf{r}, \mathbf{k}, t)d^3k = \varepsilon D_n(\mathbf{r})(\operatorname{grad} n + n\mathbf{E})$$

where the normalized diffusivity is

$$(4.14a) \qquad D_n(\mathbf{r}) = \tau_p(\mathbf{r})(1 - 5\alpha)$$

and $\tau_p(\mathbf{r})$ is a suitable weighted average of the collision time given by

$$(4.14b) \qquad \tau_p(\mathbf{r}) = \frac{\langle\tau_c(\mathbf{r}, \mathbf{k})u^2\rangle}{\langle u^2\rangle}$$

and may be identified with the momentum relaxation time. The correction factor 5α accounts for the non-parabolicity of the band. The latter, which is usually neglected in elementary theory, turns out to be of the order of 6%. Using (4.13) we can express $f^{(1)}$ as a function of $\mathbf{J}_n^{(1)}$, namely

$$(4.15) \qquad f^{(1)}(\mathbf{r}, \mathbf{k}, t) = -N_c^{-1}\tau_c(\mathbf{r}, \mathbf{k})\exp(-\varepsilon(\mathbf{k}))\mathbf{u} \cdot \frac{1}{\epsilon D_n(\mathbf{r})}\mathbf{J}_n^{(1)}$$

In order to conclude that $f^{(0)} + \epsilon f^{(1)}$ represents a meaningful 1st order solution to equation (3.2), we must compare the relative magnitudes of $f^{(0)}$ and $\epsilon f^{(1)}$. Such a comparison is made problematic by the nature of the two functions at hand, which turn out to be even and odd, respectively. By observing that the mean quadratic value of each component of the group velocity is 1, it is reasonable to demand that

$$(4.16) \qquad \left\langle \left| \epsilon \frac{f^{(1)}\mathbf{u}}{f^{(0)}} \right| \right\rangle \ll 1$$

which, from (4.12), reads

$$(4.17) \qquad \left| \epsilon\tau_p(\mathbf{r})\frac{1}{n}(\operatorname{grad} n + n\mathbf{E}) \right| \ll 1$$

In order to better clarify the meaning of (4.17), we define a normalized quasi-Fermi potential φ_n in the standard fashion

$$(4.18) \qquad n = n_i \exp(\varphi - \varphi_n)$$

Inequality (4.17) then becomes

$$(4.19) \qquad |\epsilon\tau_p(\mathbf{r})(\operatorname{grad}\varphi_n)| \ll 1$$

which dictates that the change in the normalized quasi-Fermi potential φ_n over a carrier mean-free path $\lambda = \tau_p(\mathbf{r})\lambda_o$ is $o(1)$. Equivalently, (4.19) implies that the carrier velocity is much smaller than thermal velocity, i.e., $v \ll v_{th}$. Expression (4.19), on the other hand, does not imply that the electric potential variation over a carrier mean-free path is $o(1)$. Such a condition is often violated within real devices even in equilibrium. If φ_n is a smooth function of position, as is the case in forward-biased $p - n$ junctions, the limiting condition on the applied voltage V_a turns out to be

$$(4.20) \qquad V_a \ll \frac{1}{\epsilon}$$

which is not especially restrictive. On the other hand, (4.19) is likely to be violated when one considers diffusion near an absorbing boundary, such as carrier transport within the base of a bipolar transistor. In this case, the distribution function tends to become fairly skewed and highly distorted with respect to equilibrium, which is reflected by the increase of the average carrier velocity. Under such circumstances, equation (4.13) is expected to break down, and higher-order corrections to the distribution, which we derive below, will become more important.

5. Second-order solution to the BTE. Having determined $f^{(0)}$ and $f^{(1)}$, we can now address equation (4.2c). The solvability condition for (4.2c), namely

$$\langle 1, S f^{(2)} \rangle = 0 \tag{5.1}$$

implies that

$$\frac{\partial n}{\partial t} - \frac{1}{\epsilon} \operatorname{div} \mathbf{J}_n^{(1)} = 0 \tag{5.2}$$

which is equivalent to (2.5) so long as $\mathbf{J}_n^{(1)} \simeq \mathbf{J}_n$.

For the solution of equation (4.2c), we assume a one-dimensional geometry. From equations (4.6), (4.10) and (4.15) we find:

$$
\begin{aligned}
S f^{(2)} = N_c^{-1} \exp(-\varepsilon) \Bigg\{ &\left[\frac{\partial n}{\partial t} - \frac{1}{\epsilon} \frac{\tau_c(x,k)}{D_n(x)} u_x^2 \frac{\partial J_{nx}^{(1)}}{\partial x} \right] + \\
&\frac{1}{\epsilon} \frac{\tau_c(x,k)}{D_n(x)} \left[\gamma'^{-1} - \left(1 - \frac{1}{\tau_c} \frac{d\tau_c}{d\varepsilon} + \gamma'^{-1} \gamma'' \right) u_x^2 \right] E_x J_{nx}^{(1)} \Bigg\}
\end{aligned}
\tag{5.3}
$$

Equation (5.3) has been obtained under the assumption that the ratio τ_c/τ_p, and therefore D_n/τ_p, is actually independent of x. Such a condition would be exact if τ_c could be expressed as the product of a function of position times a function of energy. Under such circumstances, the averaging process (4.14b) leading to τ_p would not alter the x-dependence of τ_c. Strictly speaking, such a condition is not fulfilled when impurity scattering plays an important role; on the other hand, it is reasonable to assume a weak x-dependence of the above ratio, so that equation (5.3) is expected to be near-exact.

The r.h.s. of equation (5.3) is an even function of $\mathbf{k}$ and, since S is an even operator, $f^{(2)}$ must be an even function of $\mathbf{k}$. Thus, we may expand $f^{(2)}$ in spherical harmonics as follows

$$f^{(2)}(x, \mathbf{k}, t) = f_0^{(2)}(x, k, t) + f_2^{(2)}(x, k, t) \frac{1}{2}(3\cos^2\theta - 1) \tag{5.4}$$

where some obvious symmetry on $f^{(2)}$ is implied. Defining

$$A \equiv N_c^{-1} \exp(-\varepsilon) \left(\frac{\partial n}{\partial t} + \frac{1}{\epsilon} \frac{\tau_c(x,k)}{D_n(x)} \gamma'^{-1} E_x J_{nx}^{(1)} \right), \tag{5.5a}$$

$$B \equiv N_c^{-1} \exp(-\varepsilon) \frac{1}{\epsilon} \frac{\tau_c(x,k)}{D_n(x)}$$

$$\left[\frac{\partial J_{nx}^{(1)}}{\partial x} + \left(1 - \frac{1}{\tau_c} \frac{d\tau_c}{d\varepsilon} + \gamma'^{-1} \gamma'' \right) E_x J_{nx}^{(1)} \right] u^2 \tag{5.5b}$$

where u is the velocity magnitude, we write equation (5.3) in the form

$$S f^{(2)} = A - B \cos^2\theta = \left(A - \frac{1}{3}B \right) - \frac{2}{3}B \left(\frac{1}{2}(3\cos^2\theta - 1) \right) \tag{5.6}$$

and balance the coefficients of the spherical harmonics of order 0 and 2, to obtain

$$(5.7a) \qquad (Sf^{(2)})_0 = \left(A - \frac{1}{3}B\right)$$

$$(5.7b) \qquad (Sf^{(2)})_2 = \frac{2}{3}B$$

Using the energy-conserving properties of the scattering mechanisms, the l.h.s. of (5.7) may now be explicitly evaluated. For (5.7a), only inelastic scattering (i.e., optical phonon) does actually contribute and we find

$$
\begin{aligned}
(5.8) \qquad c_{op} \Big\{ & N_{op}^+ g(\varepsilon + \varepsilon_{op}) f_0^{(2)}(\varepsilon + \varepsilon_{op}) - [N_{op}^+ g(\varepsilon - \varepsilon_{op}) + N_{op} g(\varepsilon + \varepsilon_{op})] f_0^{(2)}(\varepsilon) + \\
& N_{op} g(\varepsilon - \varepsilon_{op}) f_0^{(2)}(\varepsilon - \varepsilon_{op}) \Big\} = \left(A - \frac{1}{3}B\right)
\end{aligned}
$$

where we have defined $f_0^{(2)}(\varepsilon) \equiv g(\varepsilon) \equiv 0$ for $\varepsilon < 0$. In (5.8) ε_{op} is the normalized optical-phonon energy; $N_{op} = [\exp(\varepsilon_{op}) - 1]^{-1}$ is the phonon occupation number; $N_{op}^+ = N_{op} + 1$; c_{op} is the optical-phonon scattering probability per unit time, and $g(\varepsilon)$ is the density-of-states function given by

$$(5.9) \qquad g(\varepsilon) = \frac{2}{\sqrt{\pi}} \gamma^{1/2} \gamma'$$

Since equation (5.8) is a difference equation with variable coefficients, an exact solution would generally require a numerical approach. In what follows, we shall assume that $g(\varepsilon + \varepsilon_{op})$ and $g(\varepsilon - \varepsilon_{op})$ may be approximated by $g(\varepsilon)$ at the l.h.s. of (5.8). Such an approximation is expected to be reasonable for large values of the energy ε, due to the weak ε dependence of the g function. Furthermore, it does not alter the solution of the associated homogeneous equation, which turns out to be proportional to $f^{(0)}$. Under the above simplifying assumption, equation (5.8) becomes

$$
\begin{aligned}
(5.10) \qquad & f_0^{(2)}(\varepsilon + \varepsilon_{op}) - (1 + \delta) f_0^{(2)}(\varepsilon) + \delta f_0^{(0)}(\varepsilon - \varepsilon_{op}) = \\
& [c_{op} N_{op}^+ g(\varepsilon)]^{-1} \left(A - \frac{1}{3}B\right) \equiv F(\varepsilon)
\end{aligned}
$$

In (5.10) δ has been defined as follows

$$(5.11) \qquad \delta = (N_{op}/N_{op}^+) = \exp(-\varepsilon_{op})$$

Using the expressions for A, B given by (5.5), the r.h.s. of (5.10) turns out to be

$$
\begin{aligned}
(5.12) \qquad F(\varepsilon) = \left(\frac{2}{\sqrt{\pi}} c_{op} N_{op}^+ N_c \gamma^{1/2} \gamma'\right)^{-1} \Bigg\{ & \left[\frac{\partial n}{\partial t} - \frac{2\varepsilon}{3\varepsilon \gamma'^2} \frac{\tau_c(x,\varepsilon)}{D_n(x)} \frac{\partial J_{nx}^{(1)}}{\partial x}\right] + \\
& \frac{1}{\epsilon} \frac{\tau_c(x,\varepsilon)}{D_n(x)} \left[\gamma'^{-1} - \frac{2}{3}\left(1 - \frac{1}{\tau_c} \frac{d\tau_c}{d\varepsilon} + \gamma'^{-1} \gamma''\right) \varepsilon \gamma'^{-2}\right] E_x J_{nx}^{(1)} \Bigg\} \exp(-\varepsilon)
\end{aligned}
$$

In principle, equation (5.10) can now be solved by means of the z-transform technique. However, to get a transparent form of the solution $f_0^{(2)}$, we make some simplification of the rather complicated function $F(\varepsilon)$. First we assume parabolic bands and obtain

$$(5.13) \quad F(\varepsilon) = \left(\frac{2}{\sqrt{\pi}}c_{op}N_{op}^+N_c\right)^{-1}\left\{\left[\frac{\partial n}{\partial t} - \frac{2\varepsilon}{3\epsilon}\frac{\tau_c(x,\varepsilon)}{\tau_p(x)}\frac{\partial J_{nx}^{(1)}}{\partial x}\right] + \frac{1}{\epsilon}\frac{\tau_c(x,\varepsilon)}{\tau_p(x)}\left[1 - \frac{2}{3}\left(1 - \frac{1}{\tau_c}\frac{d\tau_c}{d\varepsilon}\right)\varepsilon\right]E_x J_{nx}^{(1)}\right\}\varepsilon^{-1/2}\exp(-\varepsilon)$$

Next, we assume steady-state conditions, so that the first term in brackets equals zero. Finally, some energy dependence for $\tau_c(\varepsilon)$ must be considered. In analogy with previous assumptions, we take

$$(5.14a) \quad \tau_c(x,\varepsilon) = \tau_0(x)\varepsilon^{-\beta}$$

with

$$(5.14b) \quad \beta \simeq \frac{1}{2}$$

The resulting expression for $F(\varepsilon)$ is

$$(5.15) \quad F(\varepsilon) = \frac{\pi/4}{\epsilon c_{op}N_{op}^+N_c}(\epsilon^{-1} - 1)\exp(-\varepsilon)E_x J_{nx}^{(1)}$$

The singularity in the above expression is due to (5.14) and the simplifying assumption that $g(\varepsilon + \varepsilon_{op}) \simeq g(\varepsilon)$, which was used to get a difference equation with constant coefficients from (5.8). Therefore, we cannot expect the r.h.s. of (5.15) to be correct at low energy. On the other hand, we are looking for a solution of (5.10) only for $\varepsilon \geq \varepsilon_{op}$ and, for large energies, ε^{-1} can be neglected with respect to 1, so that, if we drop the term ε^{-1} we still expect our solution to be asymptotically correct. Hence, the simplified final form of the difference equation (5.10) becomes

$$(5.16) \quad \begin{aligned} f_0^{(2)}(\varepsilon + \varepsilon_{op}) - (1 + \delta)f_0^{(2)}(\varepsilon) + \delta f_0^{(2)}(\varepsilon - \varepsilon_{op}) = \\ -\frac{\pi/4}{\epsilon c_{op}N_{op}^+N_c}\exp(-\varepsilon)E_x J_{nx}^{(1)} \end{aligned}$$

As shown in Appendix A, the solution of (5.16) which also satisfies the normalizing condition (4.9), is

$$(5.17) \quad f_0^{(2)} = \frac{\pi/4}{\epsilon N_c c_{op}\varepsilon_{op}}\left(\varepsilon - \frac{3}{2}\right)\exp(-\varepsilon)E_x J_{nx}^{(1)}$$

Now $f_2^{(2)}$ may be trivially obtained from (5.7b), and turns out to be

$$(5.18) \quad \begin{aligned} f_2^{(2)} = \frac{2}{3}\tau_c(x,\varepsilon)B = \\ \frac{\sqrt{\pi}}{\epsilon N_c}\left(\varepsilon^{1/2} + \frac{1}{2}\varepsilon^{-1/2}\right)\exp(-\varepsilon)E_x J_{nx}^{(1)} \end{aligned}$$

Then, from (5.4), the expression of the $f^{(2)}$ function becomes

$$f^{(2)}(x, k, \theta) = \frac{\pi/4}{\epsilon N_c c_{op}\varepsilon_{op}} \left(\varepsilon - \frac{3}{2}\right) \exp(-\varepsilon) E_x J_{nx}^{(1)} +$$

(5.19)

$$\frac{\sqrt{\pi}}{\epsilon N_c} \left(\varepsilon^{1/2} + \frac{1}{2}\varepsilon^{-1/2}\right) \exp(-\varepsilon) E_x J_{nx}^{(1)} \left(\frac{1}{2}(3\cos^2\theta - 1)\right)$$

Equation (5.19) holds in the limit of $\varepsilon \gg \varepsilon_{op}$, and breaks down for $\varepsilon \simeq \varepsilon_{op}$. Nevertheless, the form of equation (5.19) allows us to conclude that the average energy in excess of $3/2$ turns out to be proportional to the local product $E_x J_{nx}^{(1)}$, as may be seen by evaluating its statistical average

$$n\langle\varepsilon\rangle = \frac{1}{N_c} \int \varepsilon \exp(-\varepsilon) d^3k + \frac{(\pi/4)\epsilon E_x J_{nx}^{(1)}}{N_c c_{op}\varepsilon_{op}} \int \left(\varepsilon^2 - \frac{3}{2}\varepsilon\right) \exp(-\varepsilon) d^3k$$

and therefore

(5.20)
$$n(\langle\varepsilon\rangle - 3/2) = \frac{3\pi}{8} \frac{\epsilon E_x J_{nx}^{(1)}}{c_{op}\varepsilon_{op}} = \tau_w \epsilon E_x J_{nx}^{(1)}$$

where the energy relaxation time τ_w has been defined as

(5.21)
$$\tau_w = \frac{3\pi}{8} (c_{op}\varepsilon_{op})^{-1}$$

Apart from the numerical coefficient $(3\pi/8)$, equation (5.21) lends itself to an easy interpretation: since $N_{op}^+ = N_{op} + 1$, the following identity holds:

(5.22)
$$c_{op}\varepsilon_{op} = N_{op}^+ c_{op}\varepsilon_{op} - N_{op} c_{op}\varepsilon_{op}$$

showing that $(c_{op}\varepsilon_{op})$ may be interpreted as the difference between the emission and absorption scattering probabilities per unit time, multiplied by the optical-phonon energy. Therefore, $n\langle\varepsilon\rangle(c_{op}\varepsilon_{op})$ represents the rate at which energy relaxes to equilibrium. From (5.21) it appears that, to second order in the distribution function, the energy relaxation time is a constant, independent of the carrier energy. Monte Carlo simulations of current transport in semiconductors show that τ_w is indeed a very weak function of energy. We also notice that equation (5.20) is a simple form of energy balance under the assumption of a negligible divergence of the energy flow $\mathbf{S}_n$. Since $\mathrm{div}\,\mathbf{J}_n = 0$, the above assumption is justified for small values of the temperature gradient, which is again consistent with the assumption of a small perturbation with respect to equilibrium.

We now consider the equation for $f^{(3)}$, which is

(5.23)
$$Sf^{(3)} = u_x \frac{\partial f^{(2)}}{\partial x} - E_x \frac{\partial f^{(2)}}{\partial k_x}$$

Since $f^{(3)}$ is an odd function in $\mathbf{k}$, the l.h.s. is simply given by

(5.24)
$$Sf^{(3)} = -\frac{f^{(3)}}{\tau_c(x, k)}$$

and the solution of equation (5.23) is

$$(5.25) \qquad f^{(3)} = -\tau_c(x,k)\left(u_x\frac{\partial f^{(2)}}{\partial x} - E_x\frac{\partial f^{(2)}}{\partial k_x}\right)$$

From (5.25), the second-order correction to the current density can now be evaluated

$$(5.26) \qquad J_{nx}^{(3)} = \int \epsilon^3 f^{(3)} u_x d^3k$$

and the corrected expression of the current density reads

$$(5.27) \qquad J_{nx} \simeq J_{nx}^{(1)}\left\{1 + \epsilon^2\left[\eta_n\frac{\partial E_x}{\partial x} - \zeta_n E_x^2\right]\right\}$$

where

$$(5.28a) \qquad \eta_n = \frac{\epsilon}{E_x J_{nx}^{(1)}}\int -\tau_c(x,k)f^{(2)}(x,k,\theta)u_x^2 d^3k$$

$$(5.28b) \qquad \zeta_n = \frac{\epsilon}{E_x J_{nx}^{(1)}}\int -\tau_c(x,k)\frac{\partial f^{(2)}}{\partial k_x}u_x d^3k$$

From (5.27) we notice that the correction in brackets is second order in ϵ, and that it is composed of two terms: the first one, proportional to $\partial E_x/\partial x$, was first suggested by Thornber [12] based on heuristic considerations, and it was shown to account for non-static transport effects such as velocity overshoot. The second one, proportional to E_{nx}^2, is a second-order correction to the low-field mobility.

It is worth mentioning that, although equation (5.19) is only approximate, the linear dependence of $f^{(2)}$ upon the product $E_x J_{nx}^{(1)}$ is exact, because the averaging process implied by equation (5.8) will inherit the functional dependence of the r.h.s. on $E_x J_{nx}^{(1)}$. A final consideration is that the present procedure can in principle be extended to the multi-dimensional case, leading to a tensor relationship which generalizes equation (5.27).

6. Summary and conclusions. In this work we have reexamined the derivation of the drift-diffusion model from the BTE. Such a derivation is obtained by means of the Hilbert expansion near equilibrium conditions. Compared with previous approaches, we do not require that the electric field and the carrier gradients are individually small, since such a condition is never fulfilled in real devices. Rather, we assume that the algebraic sum of the drift and diffusion terms within the BTE is $O(\epsilon)$ or, alternatively, that the carrier average velocity is small compared with the thermal velocity. We then solve the set of resulting equations up to the third order in ϵ, and determine under general conditions a correction to the usual expression of current density.

Such a correction is composed of two terms: one of them is proportional to the square of the electric field, and accounts for mobility degradation near $k_B T/q\lambda$. The second one is proportional to $\partial E_x/\partial x$ and accounts for non-static effects such as velocity over- and undershoot occurring in response to a sharp change of the electric field. Finally, a byproduct of the present treatment is an expression for the energy relaxation time, which turns out to be independent of the carrier average energy.

Appendix A. Consider the sequence of sample $n\varepsilon_{op}$, with $n = 0, 1, 2, \ldots$, and take the unilateral z-transform of equation (5.16). We find

$$\text{(A.1)} \qquad \hat{f}_0^{(2)}(z)[z - (1+\delta) + \delta z^{-1}] - z f_0^{(2)}(0) = \widehat{F}(z)$$

where

$$\hat{f}_0^{(2)}(z) = \sum_{n=0}^{\infty} f_0^{(2)}(n\varepsilon_q) z^{-n}$$

and

$$\widehat{F}(z) = \sum_{n=0}^{\infty} F(n\varepsilon_q) z^{-n}$$

are the unilateral z-transforms of the unknown function $f_0^{(2)}(\varepsilon)$ and the r.h.s. $F(\varepsilon)$, respectively. From (A.1) we get

$$\text{(A.2)} \qquad \hat{f}_0^{(2)}(z) = \frac{z\widehat{F}(z)}{z^2 - (1+\delta)z + \delta} + \frac{z^2 f_0^{(2)}(0)}{z^2 - (1+\delta)z + \delta}$$

The roots of the denominator in (A.2) are $z_1 = \delta$, $z_2 = 1$, so that equation (A.2) takes the form

$$\text{(A.3)} \qquad \hat{f}_0^{(2)}(z) = \frac{z\widehat{F}(z)}{(z-\delta)(z-1)} + f_0^{(2)}(0)\left\{1 + \frac{(1+\delta)z - \delta}{(z-\delta)(z-1)}\right\}$$

Also, from the expression (5.16) of $F(\varepsilon)$, namely

$$F(\varepsilon) = -F_o \exp(-\varepsilon)$$

we find that

$$\text{(A.4)} \qquad \widehat{F}(z) = -F_o \frac{z}{z + \exp(-\varepsilon_q)}$$

Standard procedures then yield the anti-transform of (A.3) in the form

$$\text{(A.5)}$$
$$f_0^{(2)}(\varepsilon) = \frac{F_o}{(1-\delta)^2}\left\{(2-\delta)\frac{\varepsilon}{\varepsilon_{op}}\exp(-\varepsilon) - \frac{\varepsilon - \varepsilon_{op}}{\varepsilon_{op}}\exp(-\varepsilon) - u_h(\varepsilon - \varepsilon_{op})\right\} +$$
$$f_0^{(2)}(0)\left\{\delta_d(0) + \frac{1}{(1-\delta)}u_h(\varepsilon - \varepsilon_{op}) - \frac{\delta}{(1-\delta)}\exp(-\varepsilon)\right\}$$

where $u_h(\varepsilon)$ is the Heaviside unit function, and $\delta_d(\varepsilon)$ is the discrete delta function. By imposing on $f_0^{(2)}(\varepsilon)$ the property of vanishing asymptotically, we finally obtain equation (5.17).

REFERENCES

[1] W.V. VAN ROOSBROECK, *Theory of flow of electrons and holes in germanium and other semiconductors*, Bell Syst. Tech. J., vol. 29 (1950), pp. 560–607.

[2] S. SELBERHERR, *Analysis and simulation of semiconductor devices*, Springer Verlag, Vienna, 1984.

[3] W.L. ENGL, *Process and device modeling*, North Holland, 1986.

[4] P.A. MARKOWICH, C.A. RINGHOFER, C. SCHMEISER, *Semiconductor equations*, Springer Verlag, Wien, 1990.

[5] G. BACCARANI, *Physics of submicron devices*, in Large Scale Integrated Circuit Technology: State of the art and prospects, Eds.: L. Esaki and G. Soncini, Nijhoff, The Hague (1982), pp. 647–669.

[6] W. FAWCETT, A.D. BOARDMAN, S. SWAIN, *Monte Carlo determination of electron transport properties in Gallium Arsenide*, J. Phys. Chem. Solids, vol. 31 (1970), pp. 1963–1990.

[7] C. CANALI, C. JACOBONI, F. NAVA, G. OTTAVIANI, A. ALBERIGI QUARANTA, *Electron drift velocity in silicon*, Phys. Rev. B, vol. 12 (1975), pp. 2265–2284.

[8] M. KOHLER, *Behandlung von Nichtgleichgewichtsvorgängen mit Hilfe eines Extremal Prinzips*, Zeitschrift für Physik, vol. 124 (1948), pp. 772–789.

[9] P.A. WOLFF, *Theory of multiplication in silicon and germanium*, Phys. Rev. B, vol. 95 (1954), pp. 1415–1420.

[10] G.A. BARAFF, *Distribution function and ionization rates for hot electrons in semiconductors*, Phys. Rev., vol. 128 (1962), pp. 2507–2517.

[11] D.L. RODE, *Electron mobility in direct-gap polar semiconductors*, Phys. Rev. B, vol. 2 (1970), pp. 1012–1024.

[12] K.K. THORNBER, *Current equations for velocity overshoot*, IEEE Electron Device Lett., vol. EDl-3 (1982), pp. 69–71.

PHYSICS FOR DEVICE SIMULATIONS AND ITS VERIFICATION BY MEASUREMENTS

HERBERT S. BENNETT AND JEREMIAH R. LOWNEY*

Abstract. The motivations for using computers to simulate the electrical characteristics of transistors are discussed. Our work and that of others in the area of device physics and modeling are described. We compare conventional device physics with an alternative approach to device physics that is more directly traceable to quantum-mechanical concepts. We then apply this new approach to quasi-neutral regions, space-charge regions, and regions with high levels of carrier injection. Examples of applying quantum-mechanically-based device physics to energy band diagrams for bipolar transistors are given. The limits for using theoretical results from uniform media in numerical simulations of devices with large concentration gradients are discussed. Calculations of the effective intrinsic carrier concentrations for gallium arsenide and silicon are also given along with published data. In addition, calculations of the mobilities for GaAs that are based in part on quantum-mechanical phase shifts are compared with published data. We then conclude with a discussion of the requirements for verifying and calibrating device simulators for the submicrometer domain.

Key words. gallium arsenide, heavy doping effects, donor-ion-carrier interactions, carrier-carrier interactions, Fermi energy, screening radii, effective intrinsic carrier concentrations, scattering mechanisms, phonons, plasmons, and carrier mobilities

I. Motivation and Introduction

Even though the major emphasis of this paper is to provide the basis for increased understanding of why devices physically behave as they do, we give here some general statements as to why device simulations are beneficial to the electronics industry.

In the past, experimental iterations for designing devices were economically acceptable. Device simulations to make incremental changes in a given process technology were done after product introduction. This approach to engineering is no longer competitive. The costs (process changes, multiple mask sets, circuit design changes, time, and personnel) of experimental iterations are very high. Also, numerical simulations reduce considerably the time for communication among those who develop processes and design new devices and circuits (technology) and those who produce new products (systems). This communication time for information to diffuse among diverse specialties is often overlooked. However, it might be essential in determining which of several competing companies is first to market a new product.

Device simulations predict quantitatively the advantages and disadvantages of competing technologies and thereby suggest the specific technology for greater investment of resources. They also may suggest ways to improve incrementally the performance of a device based upon a given technology which may have been selected by numerical simulations.

*Semiconductor Electronics Division, National Institute of Standards and Technology, Gaithersburg, MD 20899

One goal of simulating the operating characteristics of devices is to relate their electrical performance to the properties of the materials from which they are made. Key to accomplishing this goal is the device physics used in models for carrier transport. These models contain parameters that describe carrier mobilities and lifetimes and the effects of electric fields and high concentrations of dopant ions and carriers on such quantities. Conventional procedures for determining model parameters are based on interpretations of electrical measurements, rely very much upon empirical relations, and give acceptable results for transistors with dimensions greater than about a micrometer. Such empirical procedures may not give reliable results for smaller transistors and they most likely will not be adequate for future devices that have features sizes less than about 0.2 micrometers.

When dimensions are less than about a micrometer, one also needs device physics, based on first principles, to understand problems that arise in making reliable devices and to develop strategies to overcome design limits. These problems include isolation associated with very densely packed devices and transient and steady-state phenomena associated with radiation and electrically induced failure mechanisms. Addressing these challenges requires careful evaluation of: 1) the physics that is used in numerical simulations of devices, 2) the dependence of complex numerical algorithms on physical models, 3) the interpretation of electrical and optical measurements to give the many parameters that affect device performance, and 4) the verification methods used for both physical models and numerical simulations. These parameters include band structure changes, band edge discontinuities, anisotropic mobilities, recombination lifetimes, and interface state energies and densities.

Developing computationally efficient computer programs that simulate the operation of solid-state devices is another goal of workers in this area. Achieving this goal requires compromises between the sophistication of solid-state physics and the pragmatic demands of electrical engineering. There are three classes of numerical simulations for devices: equivalent circuits, compact device models, and detailed device models. In this paper, we describe only numerical simulations based on detailed device models. Detailed models are based on doping profiles and numerical solutions to the coupled nonlinear semiconductor device equations with appropriate boundary conditions [1]. Such models allow for the explicit description of the device physics associated with doping profiles and carrier lifetimes, concentrations, and mobilities. The semiconductor equations are solved self-consistently by either finite element or finite difference procedures and include Poisson's equation for conservation of charge, continuity equations for holes and electrons, and several constitutive equations.

Many device simulators for bipolar and field effect transistors require physical models and associated input parameters that describe how carrier transport varies with carrier concentrations, ionized dopant densities, and temperature. Confidence in using the predictions from device simulators for manufacturing is achieved by comparing predictions with measurements on devices. The motivation for our performing these first-principles calculations for device models is to gain improved physical insights about the many scattering mechanisms that affect carrier transport at 300 K. By so doing, we may then contribute to reducing the number of unknown physical parameters in numerical simulations that predict the electrical performance of devices such as bipolar transistors and solar cells. Device manufacturers may then have greater confidence when they use simulators to design products.

We compare in Section II conventional and improved device physics and discuss the key input parameters for numerical simulations. In Section III, we apply this improved device physics to quasi-neutral regions, space-charge regions, and regions with high levels of carrier injection. We also discuss the limits for using theoretical results from uniform material in numerical simulations of devices with large concentration gradients. Section IV contains calculations of the effective intrinsic carrier concentrations for gallium arsenide and silicon and mobilities for GaAs. Section V has examples of NIST work that applies the above quantum-mechanically-based device physics to energy band diagrams and to silicon and GaAs bipolar transistors.

The results of this paper are useful not only for device modeling but also for interpreting optical measurements such as photoluminescence and absorption. The doping density and carrier dependences, predicted by the multiscattering formalism and many-body theory, of the changes in the conduction band edge, valence band edge, and effective intrinsic carrier concentration are discussed. The conduction and valence band edge changes due to carriers interacting with the dopant ions are calculated by Klauder's third and fifth levels of approximation [2], while the changes due to the carriers interacting with other carriers are calculated according to Abram's formalism [3].

II. Materials Properties and Device Physics

for Numerical Simulations

Table I contains a list of symbols used in this paper and their definitions.
TABLE I: List of symbols.

LATIN

a	lattice constant
a_0	Bohr radius (0.529×10^{-8} cm)
a_0^*	effective Bohr radius
a_r	reduced screening radius
c	carrier concentration
D	defect or trap density
$\overline{D}$	variational determinant
E	carrier energy with respect to its band extremum
$\mathbf{E}$	vector electric field
E_b	energy of bound state associated with defects or traps
E_F	Fermi energy
E_G	bandgap
E^∞	hydrogenic, unscreened bound state energy
f_0	Fermi function
F_n	nth order Fermi-Dirac integral
G_t	Gibbs free energy of trap level

$\hbar$	Planck's constant
k	carrier wavenumber
k_B	Boltzmann Constant
L	scattering operator
m_c^*	conduction band effective mass
m_v^*	valence band effective mass
m_x^*	conduction or valence band effective mass
m_0	free electron mass
$M(E_F)$	function of the Fermi energy based on Eq. (3)
N	net doping density
N_A	acceptor density
N_D	donor density
N_I	ionized dopant density
N_{ph}	phonon occupation number
N_{SRH}	SRH reference density (fitting parameter)
N_X	total dopant density
n	electron density
n_i	intrinsic carrier concentration
n_{ie}	effective intrinsic carrier concentration
$\overline{n}$	equilibrium electron density
p	hole density
$\overline{p}$	equilibrium hole density
R_{XX}	average distance between dopants atoms
r_c	critical screening radius
$r_{\mu n}$	electron mobility ratio
$r_{\mu p}$	hole mobility ratio
r_s	screening radius
T	Temperature
$v_n(maj)$	majority electron speed in n-type material
$v_n(min)$	minority electron speed in p-type material
$v_p(maj)$	majority hole speed in p-type material
$v_p(min)$	minority hole speed in n-type material
W_B	base width
W_C	collector width
W_E	emitter width
x	reduced carrier energy
$\mathbf{x}$	position vector
z_l	reduced phonon energy
z_p	reduced plasmon energy

GREEK

α	fitting parameter
δ_l	phase shift for the lth partial wave
ΔE_c	effective change in conduction band edge
ΔE_v	effective change in valence band edge
$\Delta E_G(N_I, c)$	change in bandgap
ΔE_G^e	effective bandgap narrowing (fitting parameter) from electrical measurements
ϵ_0	static dielectric constant
ϵ_∞	optical dielectric constant
η	reduced Fermi energy
$\mu_n(maj)$	majority electron mobility in n-type material
$\mu_n(min)$	minority electron mobility in p-type material

$\mu_p(maj)$	majority hole mobility in p-type material
$\mu_p(min)$	minority hole mobility in n-type material
Φ	variational function
τ_n	electron lifetime
τ_p	hole lifetime
τ_{xD}	theoretical SRH lifetime
τ'_{xSRH}	empirical SRH lifetime
τ_{x0}	prefactor for theoretical SRH lifetime
τ'_{x0}	prefactor (fitting parameter) for empirical SRH lifetime

Several of the materials properties that are needed for numerical simulations based on detailed device models are given in Table II.

Table II. Material Properties Required for Numerical Simulations of Devices.

Band Structure

Bandgap, band edge changes, and separations between conduction band minima for GaAs

Density of States

distorted (nonparabolic) density of states for each band

Equilibrium Carrier Densities

$$n_{ie}^2(N_D) = np \text{ and } n_{ie}^2(N_A) = np$$

Transport Parameters for Majority and Minority Carriers

$$\mu_n(\mathbf{E}, N_D) \text{ and } \mu_n(\mathbf{E}, N_A)$$
$$v_n(\mathbf{E}, N_D) \text{ and } v_n(\mathbf{E}, N_A)$$
$$\mu_p(\mathbf{E}, N_D) \text{ and } \mu_p(\mathbf{E}, N_A)$$
$$v_p(\mathbf{E}, N_D) \text{ and } v_p(\mathbf{E}, N_A)$$

Recombination Properties

Auger recombination lifetimes for holes and electrons

Shockley-Read-Hall (defect) lifetimes for holes and electrons

Generation Properties

Impact ionization rates and tunnel rates

Dimensional Factors

$$W_E, W_B, \text{ and } W_C$$

Dopant Density Profiles

$$N_D(\mathbf{x}) \text{ and } N_A(\mathbf{x})$$

Defect Density Profiles

$D(\mathbf{x})$ and the extent to which $D(\mathbf{x})$ depends on N_D and N_A

Contact and Interface Properties

Interface trap density, Fermi energy, and work function at Schottky-barrier interfaces

Interface trap densities at ohmic contacts and other interfaces

As devices become smaller and operate more closely to their design limits, larger fractions of their active regions will have high concentrations of either dopants, carriers, or both. A major effort in developing the appropriate device physics is to describe high concentration effects in a manner that is tractable with today's computers. The methods used by researchers to address this problem may be classified according to two approaches.

II.A. Modeling Approaches. The first approach, which we call conventional (or empirical) device physics (CDP), is based primarily on electrical measurements. Many of the input parameters to describe high concentration effects are determined by interpretations of electrical measurements on the devices being modeled or on similar devices. Ambiguous results may occur when extracting model parameters from electrical measurements on devices. Such extractions for model parameters usually are based on a lower level device model and hence become dependent on the lower level device model itself. Empirical procedures give acceptable predictions for the effects of small variations in processing and in device geometry and dimensions. They are specific to a given device, and often are a fast way to parameterize a given device. However, models built in this way may not lead to a fundamental understanding of the physical mechanisms responsible for the performance of the device and may suggest wrong strategies to increase device performance.

Even though the above empirical parameters may predict measured current-voltage characteristics of devices to acceptable accuracy, they are not necessarily the physically correct ones. Using them to simulate effects is questionable. Hence, the alternatives given below are recommended when numerically simulating the electrical behavior of VLSI transistors that have regions with high concentrations of either carriers or dopant ions.

The second approach, which we call first-principles or improved device physics (IDP), is based on the interpretation of nonelectrical measurements. It also relies more on the results from theoretical calculations. When possible, it is preferable to obtain input parameters by first-principles calculations that are verified directly by alternative measurements such as the use of optical absorption or photoluminescence to determine bandgap narrowing. Although absorption and photoluminescence measurements do not require lower level device models for their interpretation, they do require sophisticated analyses to give the perturbed densities of states. First-principles procedures give acceptable results for larger changes in fabrication processes and in device geometry and dimensions than do conventional procedures [4-6]. They also have a much higher probability to lead to a fundamental understanding of device behavior.

First-principles procedures relate dopant and carrier concentrations directly to band structure changes and use scattering theory to calculate separate values for minority and majority carrier mobilities. The calculated values for these two mobilities may differ by factors of 4 at the same doping density. First-principles procedures are more likely to suggest correct strategies for increasing performance, to suggest novel devices, and to yield reliable and accurate models that are applicable to several different processing technologies (e.g., $1.0 \ \mu\text{m} \geq W_B \geq 0.1 \ \mu\text{m}$). However, first-principles approaches require consistent sets of well-characterized, research quality samples; require a substantially more interdisciplinary effort and the necessary financial support of such an infrastructure; and often require more computer time.

II.B. Carrier Transport and Densities. There are two main ways to describe carrier transport in devices with submicrometer sized features. The first way is to describe carrier transport in terms of an effective intrinsic carrier concentration, n_{ie}. This is an acceptable way when only the carrier concentrations are needed. Most device models, see for example those in references [7] and [8], use this first way and relate carrier transport indirectly to the band structure through n_{ie}. This may not always be the best procedure for the advanced devices being made now.

The second way is to describe carrier transport directly in terms of the individual bands. This is the preferred way when values for the internal electric fields are required in addition to the carrier concentrations. The internal fields determine carrier transport, and in most cases differ from fields derived from applied voltages. A few device models, such as those in references [9–12], use this second way. These models formulate the transport of carriers directly in terms of the effective band edge changes and thereby relate transport directly to the band structure.

As stated in the last paragraph, most drift-diffusion models for devices formulate the transport of carriers in terms of n_{ie}. Usually in such models, the law of mass action for the equilibrium carrier concentration is modified for heavily doped semiconductors by the expression,

$$(1) \qquad n_{ie}^2 = \overline{n} \cdot \overline{p} = n_i^2 \exp\left(\Delta E_G^e / k_B T\right),$$

where $\overline{n}$ and $\overline{p}$ are the equilibrium electron and hole densities in the regime for which the law of mass action is not a valid approximation, i.e., when the carrier density exceeds 10^{17} cm^{-3} in n-type GaAs and 10^{18} cm^{-3} in p-type GaAs and silicon. The fitting parameter, ΔE_G^e, is called an effective bandgap narrowing and is determined by the interpretation of electrical measurements on semiconductor devices. The calculation of n_{ie} from Eq. (1) may lead to self-consistent, but physically convoluted, descriptions of devices that contain high concentrations of either dopants or carriers. Note that the effect of degeneracy is implicitly included in the effective bandgap narrowing.

Using a compact model and assuming Boltzmann statistics and transparent bases, Slotboom and deGraaff [13] interpreted their electrical measurements and obtained n_{ie} in terms of an empirical expression. They also assumed that the minority mobility equals the majority mobility at the same doping density. Their resulting expressions for n_{ie} and ΔE_G^e as functions of the doping density are an internally consistent description of the processing technology on which their transistors were based, namely, doping densities less than 2.5×10^{19} cm^{-3} and emitter-base junction depths greater than 1 μm. Subsequent researchers have used Slotboom and de Graaff's empirical expression for doping densities beyond the range for which it was derived, namely, much greater than 2.5×10^{19} cm^{-3}.

In the past, the concept of an effective intrinsic carrier concentration that depends on the doping density served device modeling well. However, as devices became more advanced, this concept had its limitations and required great care in its use. The concept of an intrinsic carrier concentration n_i for nondegenerate, lightly doped semiconductors is based on the law of mass action for ideal, noninteracting gases [14,15]. This law is modified for heavily doped and highly interacting semiconductors by utilizing expression (1). The "effective bandgap" change ΔE_G^e depends on the doping density and on the particular compact model [13,16,17] used to interpret electrical measurements on electronic devices. However, for state-of-the-art devices, both the doping density and the carrier densities alter the band structures. Thus, the effective bandgap is a function of both, not just the dopant density as most device models assume.

Many of the empirical relations [18,19] for the dependence on doping density of ΔE_G^e, n_{ie}, $\mu_p(min)$, $\mu_n(min)$, τ_p, and τ_n are derived indirectly from the interpretation of electrical measurements on electronic devices. These interpretations themselves depend in part on the same quantities. Such electrical results have a high probability to be convoluted with the quantities being determined and usually should not be applied to cases outside the range of values for which the empirical relations have been verified experimentally.

Swanson, del Alamo, and their coworkers [4] and del Alamo and Wagner [20] were among the first to attempt to remove the convoluted nature of parameters based entirely on the interpretations of electrical measurements. Key to accomplishing this was their use of optical measurements to determine mobility. This work has brought about much better agreement between photoluminescence data and electrical data as well as theory for ΔE_G^e, but the values obtained electrically are still somewhat larger than either values obtained from photoluminescence or from theory above 5×10^{19} cm^{-3} in silicon.

Conventional models also contain the assumption that hole (electron) mobility in n-type material equals the hole (electron) mobility in p-type material. Such assumptions are not generally valid. For example, holes scatter differently from donors than from acceptors because repulsive and attractive potentials of the same strength do not scatter equally. Hence, above densities of about 10^{17} cm^{-3} in GaAs and Si, the majority and minority mobilities for the same carrier type are not equal. The better numerical simulations then should contain four mobilities (two for holes and two for electrons) and should have separate input parameters for each material type.

II.C. Carrier Lifetimes at Low Injection Levels. The Shockley-Read-Hall (SRH) lifetime [21,22], τ'_{xSRH}, for conventional device physics is given by empirical relations such as,

$$(2) \qquad \tau'_{xSRH} = \tau'_{x0}/\left(1 + (N(z)/N_{SRH})\right),$$

where $N_{SRH} = 5 \times 10^{16}$ cm^{-3} and where x denotes n or p. The authors of reference [18] remark that empirical relations such as Eq. (2) give the best results for devices which have deep emitter-base junctions located more than 3 μm from the surface.

An alternative way to compute the SRH lifetime is to return to the original work of Shockley, Read, and Hall. They give expressions for the lifetimes of holes in n-type material that contains one-carrier trap centers such as donor centers with a deep lying energy level relative to the bottom of the conduction band. As an example, for low level injection, the SRH lifetime is given by the expression [23],

$$\tau_{pD}(theory) = \tau_{p0} \left[1 + \exp\left((G_t - E_F)/kT\right)\right]$$

$$(3) \qquad + \tau_{n0} \left[n_{ie}^2/\overline{n}^2 \left(1 + \exp\left(E_F - G_t)/kT\right)\right)\right].$$

The lifetime of holes injected into heavily doped n-type material is τ_{p0}, and the lifetime of electrons injected into heavily doped p-type material is τ_{n0}. Equation (3) is valid for n-type material, $\bar{p}/\bar{n} \ll 1$, and for the neutral region where $n' \simeq p'$, if the traps do not alter their densities appreciably relative to n'. The excess carrier densities are denoted by prime. The Gibbs free energy may depend upon the doping density N_D because of bandgap narrowing and includes the degeneracy effects of the states associated with the traps. Given the Gibbs free energy and the quantities τ_{p0} and τ_{n0}, $\tau_{pD}(theory) = \tau_0 \times M(E_F)$.

The quantities ΔE_G, ΔE_c, ΔE_v, n_{ie}, $\mu_p(min)$, $\mu_n(min)$, and τ_x for silicon have been calculated as functions of the doping density by self-consistent methods that are directly derivable from quantum mechanics and that are independent of empirical fits to electrical measurements [24]. These quantum-mechanical calculations for these quantities should be considered as a unit. That is, subsets of these seven quantities determined by quantum-mechanical methods should not be combined in device models with quantities determined by interpreting electrical measurements. In many cases, values for these quantities differ substantially from previous empirical values derived by interpreting measurements from electronic devices. For some doping profiles and junction depths, the errors in the empirical values compensate one another and give fortuitous agreement with measured electrical behavior. Also, the effective bandgap values given by the interpretation of measurements on electronic devices are fitting parameters and do not correspond to the band structure of semiconductors and to the bandgap values obtained by the interpretation of absorption and photoluminescence measurements.

II.D. Comparison of CDP and IDP. Table III summarizes the major differences between the CDP and IDP approaches for the physical quantities that are needed in numerical simulations. In CDP, the effective intrinsic carrier concentrations, n_{ie}, mobilities, μ, and lifetimes, τ, are obtained from the interpretation of electrical measurements. The mobility of holes in n-type material is assumed equal to the mobility of holes in p-type material at the same dopant density. A similar equality is assumed for electrons. And the minority carrier lifetime due to Shockley-Read-Hall processes is also assumed to be a three-parameter (τ_0', α, and N_{SRH}) fit to electrical data, whereas in IDP, the equilibrium electron and hole concentrations are calculated from first principles (quantum mechanics) in terms of a self-consistent distorted band structure (bandgap renormalization). The latter is verified by comparison with measurements of optical absorption coefficients for heavily doped materials. The minority mobilities are calculated from scattering theory and are found for silicon to be greater than majority mobilities. The SRH lifetime is given by first-principles calculations as a function of the Fermi energy. The major physical quantities presented in Table III for IDP and CDP should be considered as a single unit and not separately. That is, subsets of these quantities from the CDP column must not be combined with quantities from the IDP column in numerical simulations and vice versa.

Table III: Comparison of Conventional and Improved
Device Physics.

Physical Quantity	Approaches	
	Conventional Device Physics (CDP)	Improved Device Physics (IDP)

n_{ie} or effective bandgap narrowing Δ_G^e	electrical measurements	first-principles calculations independent measurements
minority carrier mobility	$\mu_{min} = \mu_{maj}$	$\mu_{min} \neq \mu_{maj}$
SRH lifetime	$\frac{\tau_0'}{1+(N/N_{SRH})^\alpha}$	$\tau_0 M(E_F)$

III. Physically More Correct Device Physics

III.A. Quasi-Neutral Region. In the quasi-neutral region of devices, the majority carrier density is nearly equal to the net doping density, and the minority carrier density is much smaller than either the net doping or majority carrier densities. For this case, both dopant-ion-carrier interactions and carrier-carrier interactions must be considered since they are generally competitive.

The approach here is to seek independent determinations for the quantities of ΔE_G, ΔE_c, ΔE_v, n_{ie}, $\mu_p(min)$, $\mu_n(min)$, τ_p, and τ_n. That is, whenever possible, the effects of high dopant and carrier concentrations on ΔE_G, ΔE_c, ΔE_v, n_{ie}, $\mu_p(min)$, $\mu_n(min)$, τ_p, and τ_n are calculated from first principles and not by extraction from the performance of semiconductor devices. The results for ΔE_G, ΔE_c, ΔE_v, and n_{ie} in n-type and p-type material are given in references [25] and [26] for silicon and GaAs, respectively.

Bandgap narrowing, the term used to emcompass the quantities ΔE_G, ΔE_c, ΔE_v, and n_{ie}, is an important topic in device modeling because it affects the density of minority carriers and the magnitudes of the internal fields. It arises from 1) the interaction of dopant ions with the bands, and 2) the interactions among the carriers themselves. The first effect is treated most accurately according to the multiscattering theory of Klauder [2]. The second effect requires many-body theory as given by Abram et al. [3]. In uncompensated neutral material, the carrier density is equal to the ionized dopant density, and the narrowing may then be determined as a function of dopant density only. If this restriction is not satisfied, however, the solutions become a function of both dopant and carrier densities. In many regions of devices and under various operating conditions, quasi-neutrality will be violated and bandgap narrowing must be computed as a function of both dopant and carrier density.

When the same kind of analyses based on quantum mechanics is applied to minority mobilities [27,28], the results are that the mobility ratio of minority and majority electrons or holes for silicon and GaAs, $r_{\mu n} = \left(\mu_n(min)/\mu_n(maj)\right)$ and $r_{\mu p} = \left(\mu_p(min)/\mu_p(maj)\right)$, respectively, at the same doping density are not unity as usually assumed in most device models.

Among the many physical quantities which enter device models, the minority-carrier recombination lifetimes have considerable uncertainty. These uncertainties are due in part to imprecise knowledge of how the defect densities depend on doping densities and of how fabrication methods, device geometries, and two- and three-dimensional effects alter the lifetimes.

III.B. Space-Charge Region. Appreciable bandgap narrowing can occur in space-charge regions in devices because the unscreened dopant ions interact very strongly with conduction and valence bands. The carrier-carrier interactions become negligible while the dopant-carrier interactions are enhanced by the reduced free-carrier screening. Room temperature measurements of capacitance versus voltage of heavily doped p-n junction silicon diodes showed a large decrease of the built-in gradient voltage with increasing substrate doping density [29]. These heavily doped linearly graded junctions had very narrow space-charge regions so that the densities of both donors and acceptors were nearly equal in the space-charge region. A plot of inverse capacitance cubed versus reverse bias voltage yielded an excellent straight line for all the diodes tested so that the intercept voltage was well defined. The theoretical values for the intercept voltage, called the gradient voltage, may be obtained from the work of Chawla and Gummel [30] and are less than the full built-in voltage because of the voltage drops across the Debye-tail regions. A comparison between the theoretical and measured values of the gradient voltage showed that the measured values were considerably smaller than the theoretical ones. Not only were they smaller, but they decreased with increasing substrate doping density, in contrast to customary theoretical predictions which tend to increase with the doping density.

The reduction in gradient voltage is caused by bandgap narrowing [31]. A simple theory based on a rigid band shift yielded an equality between the reduction of gradient voltage and that of the energy gaps. A logarithmic plot of the effective bandgap narrowing versus substrate doping density is shown in Fig. 1. The data are nearly linear with a slope of approximately 0.5. However, it is important to include the impurity gradient a in a complete parameterization of the data so that one cannot assume that for all graded junctions with the same substrate doping the bandgap narrowing would be the same. The complete specification of the samples is given in Table IV. The data displayed in Fig. 1 are therefore specific to the diodes tested. The magnitude of these data is much larger than has been previously observed for bandgap narrowing in neutral material [32]. This result may be expected because of the lack of screening of the ions in the space-charge region in contrast to the free-carrier screening which occurs in neutral material. A schematic of the effect of bandgap narrowing in the space-charge region is shown in Fig. 2. Note that the potential drop across the Debye-tail regions is greatly enhanced and that the potential drop across the fully depleted region, the gradient voltage, is concomitantly reduced.

Table IV: Results of C versus V Measurements on Heavily Doped Silicon Diodes. See also Figure 1.

Sample	N_b (cm^{-3})	a(cm^{-4})	$V_g^e(V)$	$V_g^t(V)$	$\Delta V_g(V)$
1	$1.4\times10^{19}(n)$	2.4×10^{23}	0.32	0.816	0.50
2	$6.0\times10^{18}(p)$	8.6×10^{22}	0.46	0.781	0.32
3	$4.5\times10^{18}(n)$	1.1×10^{23}	0.46	0.789	0.33
4	$3.2\times10^{18}(p)$	9.9×10^{22}	0.54	0.786	0.25
5	$7.2\times10^{17}(p)$	3.7×10^{22}	0.65	0.752	0.10
6	$7.2\times10^{17}(p)$	2.8×10^{22}	0.65	0.742	0.09
7	$2.9\times10^{17}(p)$	1.3×10^{22}	0.65	0.716	0.07

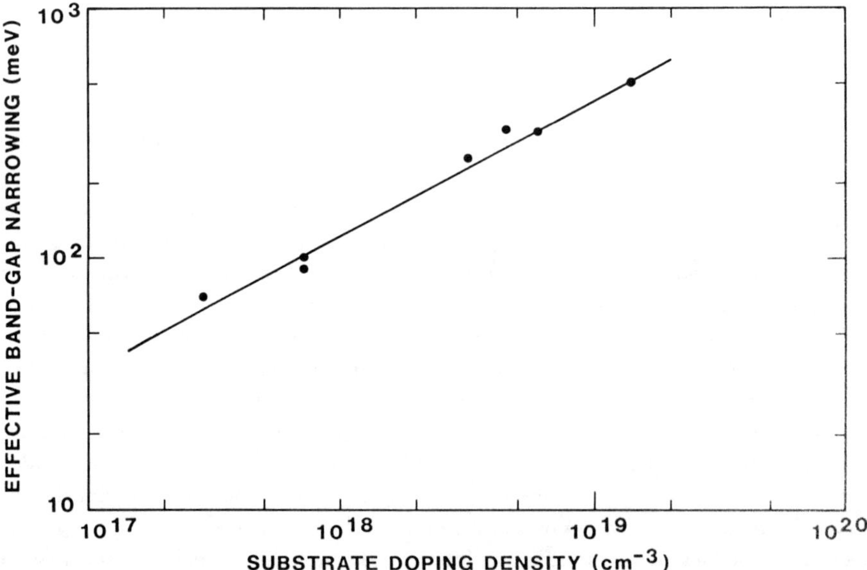

Fig. 1. Results of measurements of inverse capacitance cubed versus voltage on linearly graded silicon diodes. The effective bandgap narrowing is equal to the difference between the theoretical and observed intercept voltage for diodes with the given substrate doping densities. (from Reference [31])

The amount of bandgap narrowing can vary throughout the space-charge region because of variations in both dopant and carrier density, and a full two-dimensional calculation that treats the directions parallel and perpendicular to the junction differently would be needed to solve the problem exactly. However, it is important that device engineers be aware of this phenomenon because it can affect carrier generation, tunneling, and deep-level measurements involving space-charge regions as well as the measured built-in voltage.

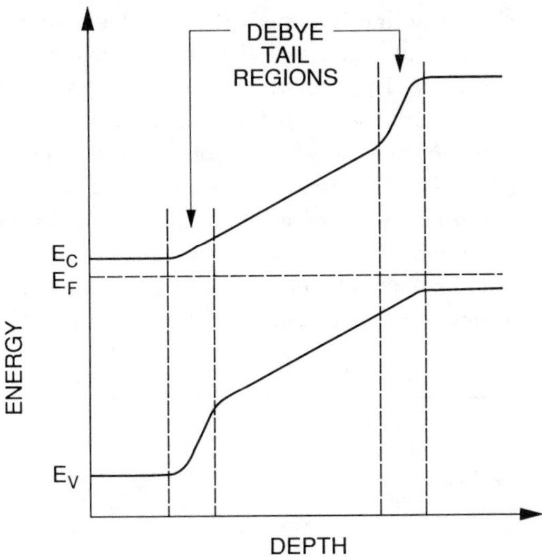

Fig. 2. A schematic of an n^+p^+ junction showing the bandgap narrowing associated with carrier depletion in the space-charge region. The two Debye-tail regions occur between the dashed lines. (from Reference [31])

Recent work on a wide range of silicon and GaAs diodes [33] has shown similar shifts in the built-in voltage. There have also been measurements on heavily doped silicon MOS capacitors [34] that have shown comparable shifts in the voltage corresponding to the onset of inversion. The measured shifts agree very well with those in Table IV for comparable doping densities. However, such direct comparisons are difficult because of the difference between MOS capacitors and pn junctions physically.

III.C. High Injection. High levels of injection occur frequently in semiconductor devices. Examples range from collector regions in bipolar transistors to photoconductive switches [35]. The resulting electron-hole plasmas can reach densities that have important effects on the valence and conduction bands of the semiconductor. Similar situations can occur in bulk material under intense optical excitation as occurs in photoluminescence [36].

For the first set of calculations, it is assumed that the dopant density is sufficiently small that the effects due to the interactions of the carriers with dopant ions are negligible, a condition that requires the dopant density to be well below 10^{17} cm^{-3}. The range of electron-hole plasma density considered is from 5×10^{16} to 1×10^{19} cm^{-3}. These calculations were performed at room temperature, which requires an extension of the many-body theory to cases for which full degeneracy does not apply [3,37].

Figure 3 shows the energy shifts of the conduction and valence band edges due to an electron-hole plasma in lightly doped silicon. The conduction band edge is lowered while the valence band edge rises. The sum of the absolute values of these shifts gives the narrowing of the gap, which is also shown. Figure 4 shows the shifts as a function of wavenumber k in the bands for a plasma density of 1×10^{18} cm^{-3}. The arrows indicate the wavenumbers in the valence and conduction bands corresponding to $4\ k_B T$. Note that the narrowing increases for increasing k values corresponding to an energy of several $k_B T$ into the bands, and therefore, the average bandgap narrowing measured experimentally is larger than at the edge. The cusp-like feature at the maximum energy shift for each band may be less acute due to lifetime-broadening of the states at room temperature [38].

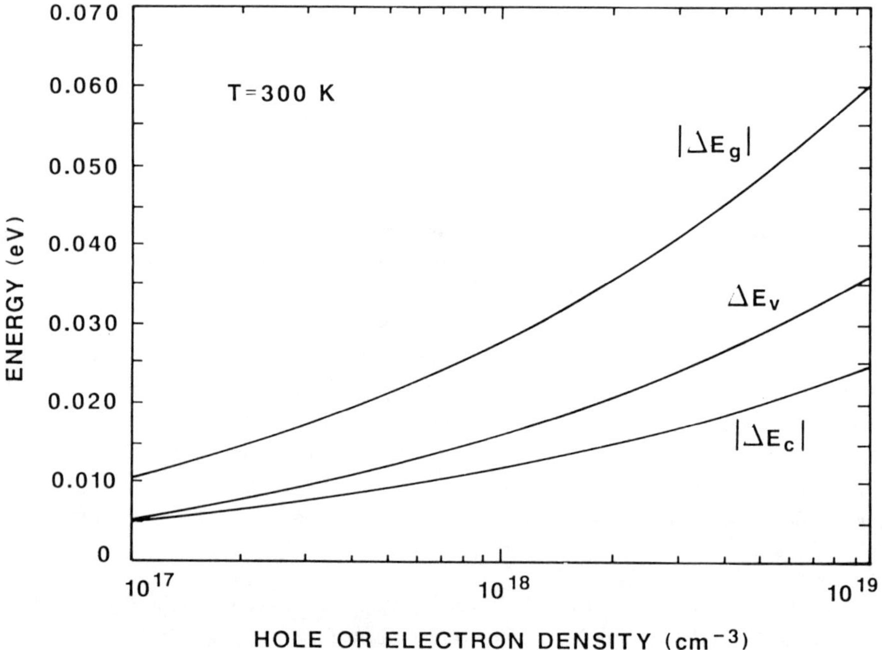

Fig. 3. Band-edge shifts for the conduction band, $-\Delta E_c$, valence band, ΔE_v, and their absolute sum, $-\Delta E_g$, at room temperature as a function of electron-hole plasma density in silicon. The conduction band edge shifts downward in energy, the valence band upward, and the bandgap is narrowed by the sum of these two energy shifts. (from Reference [37])

Effective rigid-band shifts can be found by integrating the Fermi-Dirac distribution function over the bands and computing the change in Fermi energy associated with the modification of the density of states due to the many-body effects. The results for the bandgap narrowing determined by either the band edge shifts or the effective shifts are given in Table V for both silicon and GaAs.

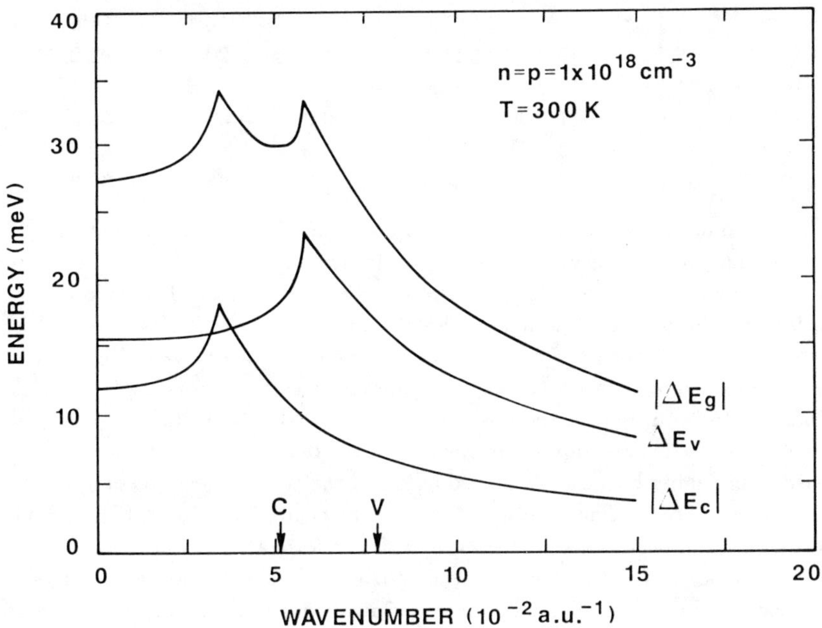

Fig. 4. Energy shifts as a function of wavenumber for the conduction band, $-\Delta E_c$, valence band, ΔE_v, and their absolute sum, $-\Delta E_g$, at room temperature for an electron-hole plasma density of 1×10^{18} cm^{-3} in silicon. (from Reference [37])

Table V: Values for the Plasma-Induced Band-Edge Shifts, $\Delta E_{c,v,g}$, and Effective Shifts, $\Delta E^e_{c,v,g}$, of the Conduction and Valence Bands and Bandgaps, Respectively, for Si and GaAs at 300 K.

$n=p$ (cm^{-3})	$-\Delta E_c$ (meV)	$-\Delta E^e_c$ (meV)	ΔE_v (meV)	ΔE^e_v (meV)	$-\Delta E_g$ (meV)	$-\Delta E^e_g$ (meV)
Silicon						
5×10^{16}	3.8	4.0	4.1	4.7	7.9	8.7
1×10^{17}	5.1	5.6	5.6	6.5	10.7	12.1
2×10^{17}	6.8	7.7	7.7	8.8	14.5	16.5
5×10^{17}	9.7	11.3	11.3	13.0	21.0	24.3
1×10^{18}	12.5	14.8	15.0	16.9	27.5	31.7
2×10^{18}	15.8	18.9	19.7	21.6	35.5	40.5
5×10^{18}	21.3	25.0	27.4	29.1	48.7	54.1
1×10^{19}	26.3	30.0	34.5	35.9	60.8	65.9
Gallium Arsenide						
5×10^{16}	4.0	4.1	4.1	4.5	8.1	8.6
1×10^{17}	5.4	5.6	5.6	6.0	11.0	11.6

2×10^{17}	7.1	7.6	7.6	8.0	14.7	15.6
5×10^{17}	9.7	10.9	11.1	11.3	20.8	22.2
1×10^{18}	11.7	14.0	14.8	14.7	26.5	28.7
2×10^{18}	13.6	17.5	19.4	19.0	33.0	36.5
5×10^{18}	18.9	23.5	27.3	26.2	46.2	49.7
1×10^{19}	28.6	29.2	34.9	33.1	63.6	62.3

In the second set of calculations, we explore the situation in heavily doped material in the presence of an electron-hole plasma. Both the effects of carrier-carrier and dopant-carrier interactions have been included [2,39]. The solid curves in Fig. 5 show the density of states for an acceptor density of 1.5×10^{18} cm^{-3} in uninjected silicon [39], with a screening radius $r_s = 63$ a$_0$, while those in Fig. 6 show the situation after injection of an excess hole plus electron density of 8.5×10^{18} cm^{-3}, $r_s = 25.5$ a$_0$. In both figures, the dashed curves show the unperturbed, parabolic density of states. The bandgap narrowing due to the dopant-carrier interaction has decreased appreciably because of the reduction of the free-carrier screening radius by the injected carriers. The lowering of the conduction band edge in Fig. 6 could not be determined because it was so small. Therefore, this modification of the density of states by injected carriers needs to be included in the device physics for numerical simulations. This example is for silicon, but similar effects occur in other semiconductors such as GaAs. The modifications needed to extend these calculations to other semiconductors is straightforward and only involves changing the appropriate material parameters. The importance of the modifications of the density of states due to high-level injection have been discussed recently by de Lyon et al. [40] with regard to GaAs heterojunction bipolar transistors. Therefore, device models must be modified to include both the direct effect of high injection on the band structure through carrier-carrier interactions and the indirect effect on the dopant-carrier interaction through the free-carrier screening radius.

The effects of high-injection levels on the conduction and valence bands of silicon and GaAs have been determined according to theory based on first principles. The results show important narrowing of the energy gap by the injected electron-hole plasma as well as a reduction in the dopant-carrier interaction because of a reduction in the free-carrier screening radius. Interestingly, these two effects tend to compensate each other somewhat in heavily doped and heavily injected material. Since the operating characteristics of a bipolar device are very sensitive to the band gap, device models need to include these effects in order to model a device correctly throughout its operating regime. It is also necessary to include these effects in the operation of lasers [41] and in the interpretation of photoluminescence data [36,42].

The minority-carrier recombination lifetimes and majority- and minority-carrier or depletion mobilities are is affected as well by injection level. The values appropriate for space-charge regions, regions with low-level injection, and those with high-level injection, are very different in general. A discussion of the dependence of lifetime on carrier density is given in reference [43] in terms of the Shockley-Read-Hall theory for recombination through traps. No specific reference exists for the effects on mobility. Therefore device models must recompute the lifetime for the

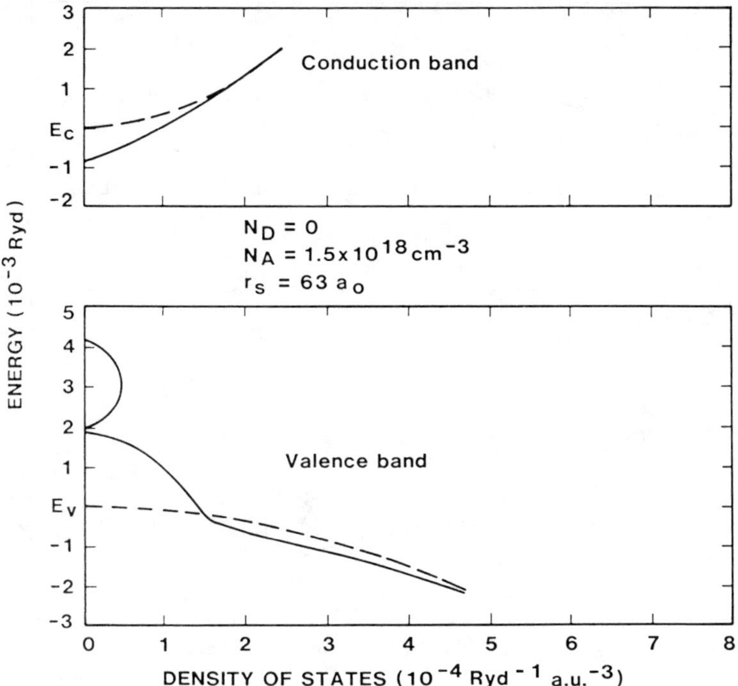

Fig. 5. Density of states of the conduction and valence bands of a neutral substrate with an acceptor density of 1.5×10^{18} cm^{-3} at room temperature in silicon. The dashed curves are for the unperturbed bands. (from Reference [37])

various regions in a device for each operating point.

III.D. Limitations of Theory Based on Uniform Media. Industry projections show that the feature size of semiconductor devices will shrink over the next ten years to dimensions on the order of or smaller than 0.25 μm. This will be accomplished in part by changing from shallow planar structures to those that penetrate more deeply into the substrate, for example, trench isolation of devices. As a result of this further microminiaturization, the spatial gradients of dopant atoms will increase, and the behavior of devices will be more sensitive to quantum-mechanical effects.

Numerical simulations for the electrical behavior of such devices are presently based on uniform theories. However, limitations exist on the validity of such theories when applied to devices with large gradients in dopant and carrier densities. We have investigated the limitations that result from applying the results from the theory of uniform-bulk material (silicon and gallium arsenide) to devices with steep density gradients [44]. Two criteria exist for using the results from the theory of uniform-bulk material on a point-by-point basis along a dopant or carrier density profile. The first states that the logarithmic derivative of the carrier density with

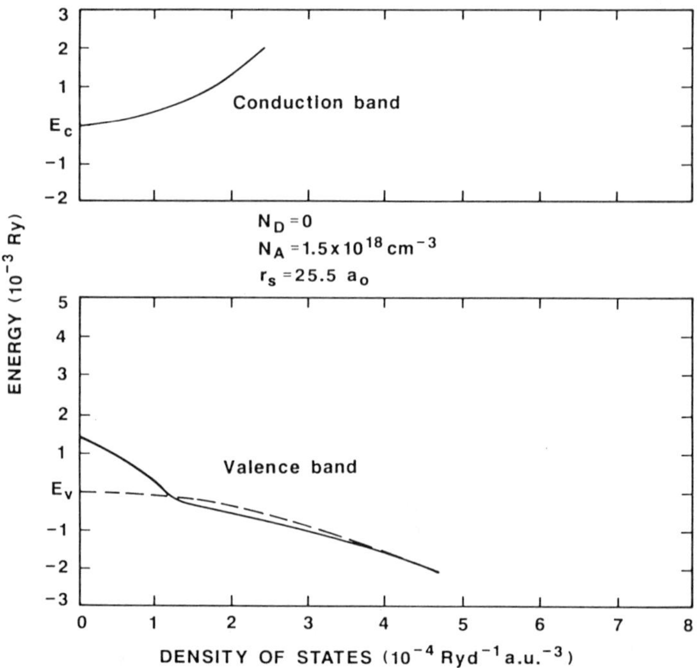

Fig. 6. Density of states of the conduction and valence bands of a high-injection layer with an acceptor density of 1.5×10^{18} cm^{-3} and an excess electron plus hole density of 8.5×10^{18} cm^{-3} at room temperature in silicon. The dashed curve is for the unperturbed band. (from Reference [37])

respect to position must be smaller than 5% of the inverse of the free-carrier screening radius. The other is that the logarithmic derivative of the dopant density with respect to position must be less than 5% of the inverse of the effective Bohr radius. These criteria are based on our previous theoretical calculations that require both the screening radius and effective Bohr radius to be nearly constant within a volume containing the grid used in the computations. Therefore, there is an upper bound to the variation of dopant and carrier density with position for which uniform theory is valid. Other quantities, such as mobilities and lifetimes, that also enter numerical simulations will have similar upper bounds.

A further complication due to a dopant gradient is the built-in electric field that accompanies it. This field leads to a potential barrier for electrons and holes such that their wavefunctions slowly decay into the energy gap beyond the point where the kinetic energy of the carriers equals zero. This effect constitutes a further narrowing of the energy gap, and preliminary results show that this narrowing effect can become a significant fraction of the bandgap narrowing in uniform material for devices with emitter regions on the order of 0.1 μm in silicon at low temperatures. The above validity ratios also show that devices on this order cannot be described by uniform theory. In n-type gallium arsenide, the situation is even worse because the electron effective mass is only one-fifth of that in silicon. For the n-type region in gallium arsenide devices, the approximation of quasi-uniformity breaks down at dimensions on the order of 0.5 μm. A full two-dimensional quantum-mechanical treatment then becomes necessary to distinguish between the directions parallel and perpendicular to the doping gradient. Such calculations require state-of-the-art computers and numerical simulations.

IV. n_{ie} AND MOBILITIES FROM QUANTUM MECHANICS

IV.A. Effective Instrinsic Carrier Concentrations. Figure 7 from reference [45] gives the values of n_{ie}/n_i for n-type and p-type GaAs as functions of dopant density. The effective intrinsic carrier concentration for a given dopant density is n_{ie} and the intrinsic carrier concentration is n_i. The donor densities therein range from 10^{15} cm^{-3} to 10^{19} cm^{-3} and the acceptor densities range from 10^{15} cm^{-3} to 10^{20} cm^{-3}. Figure 8 from reference [45] gives the values of n_{ie}/n_i for p-type silicon as a function of acceptor densities from 10^{13} cm^{-3} to 10^{20} cm^{-3}.

The details for the fifth- and third-level approximations on which the data in Figures 7 and 8 are based are given in reference [26] and in the references contained therein. The perturbed densities of states for electrons and holes are calculated from self-energy expressions given in reference [26]. These calculations incorporate the Thomas-Fermi expression for the screening radius, charge neutrality condition, and full Fermi-Dirac statistics to the compute Fermi energy and the screening radius for given values of ionized impurity concentration and temperature. For a given temperature (300 K) and dopant density, the screening radius is calculated with unperturbed densities of states and full ionization for the dopants. The results given here are for uncompensated material.

The n_{ie}/n_i ratios must approach 1 in the limit of very low dopant densities. At the intermediate dopant densities, a small amount of deionization ("freeze out") occurs due to the existence of bound states associated with the dopants. One feature of the intermediate range of dopant densities is that the bound-state energy varies from its value for the isolated bound-state energy to zero. A more physically rigorous method than the one outlined below requires many iterations of the Klauder's fifth level of approximation [2] to determine the binding energies self-consistently with the perturbed densities of states, screening radius, and the Fermi energy. Even for today's supercomputers, this is not practical.

Therefore, we use a simpler but physically valid approach for the lower dopant densities. The critical screening is defined to be that screening radius r_c at which the bound-state energy goes to zero. The results of the numerical calculations of the Schroedinger wave equation with isolated screened Coulomb potentials [46,47] is that $r_c = 0.84(E^\infty \epsilon)^{-1}$, where $E^\infty = 13.6 m_x^*/\epsilon^2$ is the hydrogenic, unscreened bound state energy in eV according to the point ion approximation. The bound state energy for a given screening radius that is greater than or equal to r_c is, according to reference [46],

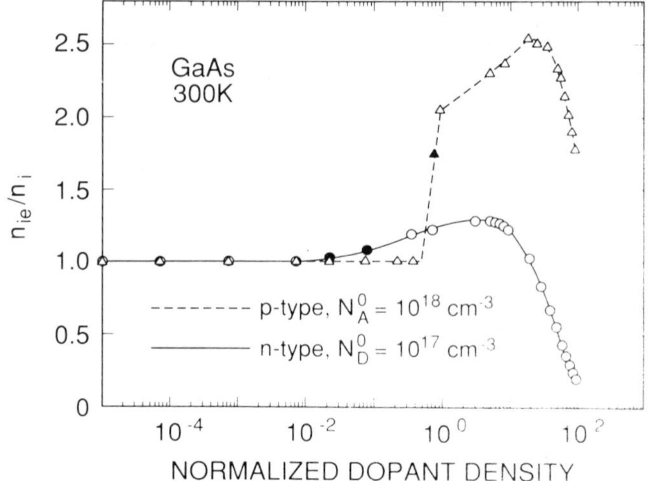

Fig. 7. n_{ie}/n_i ratios for n- and p-type GaAs as functions of normalized dopant density. The donor and acceptor densities are given by the normalized dopant density times $N_D^0 = 10^{17}$ cm^{-3} and $N_A^0 = 10^{18}$ cm^{-3}, respectively. The solid circles and triangles denote the interpolation points for n-type and p-type GaAs, respectively.

$$(4) \qquad E_b = 1.08 E^\infty [1 - (r_c/r_s)^2].$$

The relation (4) for the self-consistent energy of the bound state in terms of the screening radius is valid only when the dopant atoms are far enough apart so that they do not interact with one another, and the concept of a bound state localized on one atom is meaningful. This condition requires that the average distance between the dopant atoms, $R_{XX} = N_X^{-1/3}$, is greater than the effective Bohr radius $a_0^* = (\epsilon/m_x^*)a_0$. We use here the condition that

$$(5) \qquad R_{XX} \geq 10 \times a_0^*$$

so that the wave functions midway between two dopant atoms are decreased by two orders of magnitude from their values at their respective centers.

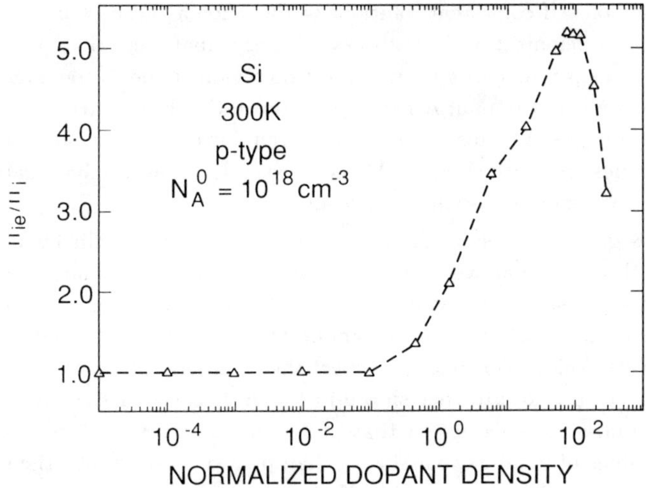

Fig. 8. n_{ie}/n_i ratio for p-type silicon as a function of normalized dopant density. The acceptor density is given by the normalized dopant density times $N_A^0 = 10^{18}$ cm^{-3}.

Our method for the lower densities is to solve the Thomas-Fermi expression for the screening radius in terms of the Fermi energy self-consistently with the charge neutrality condition. We use the unperturbed densities of states to calculate the electron and hole concentrations at thermal equilibrium. The GaAs model parameters for the calculations are the values at 300 K of the unperturbed effective masses $m_c^* = 0.067m_0$ and $m_v^* = 0.572m_0$, intrinsic bandgap energy $E_G = 1.424$ eV, and static dielectric constant $\epsilon_0 = 12.9$. For these values, the hydrogenic, unscreened bound state energy in the point ion approximation is 0.0055 eV below the conduction band for n-type and 0.048 eV above the valence band for p-type. These two values are in good agreement with measured values for many dopants [48]. The critical screening radius is 162 a_0 for n-type and 18.9 a_0 for p-type.

Figure 7 gives the combined results of the calculations for n_{ie}/n_i based on Klauder's third or fifth level approximation for the higher densities [27] and based on self-consistent solutions to the Thomas-Fermi relation for the screening radius and charge neutrality for the lower densities. The decade for which n_{ie}/n_i begins to differ from 1 in n-type GaAs is 10^{15} cm^{-3}, while for p-type, it is 10^{17} cm^{-3}. The solid circles and triangles denote the interpolation points that smooth the transition between the method based on Klauder's approximations for self-energies and the method based on self-consistent solutions to the Thomas-Fermi relation and charge neutrality. The deionization in both cases is very small, less than 1%. The rapid change in n_{ie} for p-type GaAs in the transition region is due to the large amount of bandgap narrowing at dopant densities just above the disappearance of the bound state. For n-type GaAs, this change is greatly reduced because the corresponding dopant densities are much smaller. At these lower densities, the bandgap narrowing and bound-state energies are also much smaller.

Figure 8 gives the results for p-type silicon that we obtain by using similar methods to those given above for GaAs. The point-ion approximation on which our calculations are based is very good for boron-doped silicon. Central cell and multi-valley corrections for n-type silicon complicate the results somewhat, especially in the low-density region. But one can expect the results for phosphorus- and arsenic-doped silicon to be close to that shown in Fig. 8, because of the closeness of their isolated ground-state energies to that of boron. The reason that p-type silicon has larger values of band-gap narrowing than p-type GaAs is that the shifts of the conduction band of GaAs are less than that of silicon because of the smaller electron effective mass for GaAs.

IV.B. Mobilities for GaAs. It is important to have accurate values for the majority and minority mobilities of electrons and holes in GaAs to model GaAs devices. There has been a tendency in the past to assume that they are equal, which can lead to incorrect interpretation of device data for such quantities as band-gap narrowing or recombination lifetime. As has been shown previously [27,28], the minority mobility may differ from the majority mobility by factors of 4 or more. This difference was shown in Ref. [27] for silicon by using a partial-wave, phase-shift analysis instead of the usual Born approximation, which is insensitive to the sign of the charge. The Born approximation is especially poor for ionized impurity scattering because of the relatively strong scattering of long-wavelength carriers, which have low energies and therefore violate the validity condition for the Born approximation. Differences also occur between minority and majority mobilities because of carrier-carrier scattering, which is a second-order scattering mechanism for majority carriers, but a primary one for minority carriers.

In reference [28], we have included in our calculations of the mobilities for GaAs, all the important scattering mechanisms: polar optic phonon, nonpolar optic phonon (holes only), plasmon scattering, acoustic phonon, piezoelectric, ionized impurity, and carrier-carrier. The Boltzmann transport equation was solved by the variational procedure outlined in Walukiewicz *et al.* [28]. This method avoids the use of the relaxation-time approximation that is invalid for mechanisms that involve energy transfers comparable to or greater than $k_B T$. Matthiesson's rule is also not used since it is not valid for GaAs, as shown in Ref. [49]. The scattering rates are summed prior to the variational solution. The result is a highly accurate calculation of the majority and minority mobilities, with the only limitations having to do with the wavefunction nature of the carriers that act as scattering centers.

We calculated in reference [28] both the scattering of carriers by ionized dopants and the scattering of minority carriers by majority carriers in terms of quantum mechanical phase shifts. We computed the phase shifts from the asymptotic behavior of the partial wave solutions to the Schrödinger equation with a screened Coulomb potential [50]. Plasmon scattering also was included in the low-field minority mobility. The effect of plasmons is large for the minority electron mobility, but negligible for the minority hole mobility because of the very small electron mass.

The results of these calculations agree well with the existing measurements and show some interesting structure at high doping levels because of the complexities of the carrier-carrier and plasmon scattering mechanisms. Some recent experiments support these findings, and further measurements in these interesting regions could verify these results and help determine the best values for these subtle quantum-mechanical effects. The equations that govern each mechanism are discussed in the following paragraphs.

We use the variational method outlined in Ref. 49 to solve the transport equation. This method is based on a maximum entropy principle and allows the mobility to be represented by the ratio of determinants. The elements in the matrices are transport-related integrals with the variational functions Φ represented by a power series in energy with powers between 0 and 3. We extend the upper power from 2 to 3 to obtain greater accuracy than in Ref. 49, which used only powers 0 to 2. The quantities used for the materials constants are given in Table VI. Note that h_{14} is the only nonzero piezoelectric tensor component and C_t is the transverse elastic constant. These values are our best estimates based on Ref. 49, Sze [48], and Reggiani [51].

Table VI. List of Materials Parameters [49], [48], and [51]

Name	Symbol	Value
electron effective mass	m_e^*	0.067
heavy-hole effective mass	m_h^*	0.572
low-frequency dielectric constant	ϵ_0	13.1
high-frequency dielectric constant	ϵ_∞	10.9
optical phonon energy	$\hbar\omega$	36 meV
acoustic deformation potential	$E_1(\text{elec.})$	7 eV
acoustic deformation potential	$E_1(\text{hole})$	3.5 eV
longitudinal elastic constant	C_l	14.03×10^{11} dyn/cm^2
piezoelectric coefficient	$h_{14}^2(3/C_l + 4/C_t)$	2.39×10^{-2}
optical deformation potential	d_0	29.9 eV

mass density	ρ_0	5.31 g/cm^3
lattice constant	a	5.65×10^{-8}cm

The mobility, μ, is given in cm^2/V·sec by the equation:

$$(6) \qquad \mu = 1.17 \times 10^{15} \frac{1}{m^* F_{1/2}(\eta)} \frac{\overline{D}_{3/2,3/2}}{\overline{D}},$$

where $F_n(\eta)$ is the n-th order Fermi-Dirac integral with reduced Fermi energy η = $E_F/k_B T$. The terms $\overline{D}_{3/2,3/2}$ and $\overline{D}$ refer to the determinants that contain the variational integrals. Note that this and subsequent equations differ somewhat from those in references [28] and [49] because we write here the prefactors so that the equations are applicable to both polar and non-polar materials. The equations in references [28] and [49] are applicable only for polar materials for which $\epsilon_0 \neq \epsilon_\infty$. All of our equations were derived from first principles according to the theory of Howarth and Sondheimer [52] and Ehrenreich [53]. The variational integrals contain the scattering operator L(C) given in its most general form by:

$$(7) \quad L(C) = -L_{pop}(C) - L_{nop}(C) - L_{pl}(C) + C(x)x^{3/2}\left(\frac{1}{\tau_{ac}} + \frac{1}{\tau_{pel}} + \frac{1}{\tau_{imp}} + \frac{1}{\tau_{cc}}\right),$$

where $C(x)$ is a variational function, which in this work is given by x^s, where s is an integer s between 0 and 3. The reduced carrier energy is $x = E/k_B T$, where the carrier energy, E, is with respect to its band extremum.

The remaining quantities in Eq. (7) are defined and discussed in the following paragraphs.

IV.B.1 Screened Optic Phonon (Polar) Scattering. The operator for screened optic phonon (polar) scattering, L_{pop}, is given by the sum of two terms, L_{pop}^a and L_{pop}^e, that correspond respectively to the absorption and emission of a photon:

$$(8) \qquad L_{pop}(C) = L_{pop}^a(C) + L_{pop}^e(C),$$

where for electrons,

$$L_{pop}^a(C) = A(f_+/f)(N_{ph} + 1)\times$$
$$(9) \qquad \{C_+[(R_+ + a_r)S_+ - a_r R_+ T_+ - 4U_+]/4 - xC(S_+ - a_r T_+)/2\},$$

$$L_{pop}^e(C) = Ah(x - z_l)(N_{ph} f_-/f)\times$$
$$(10) \qquad \{C_-[(R_- + a_r)S_- - a_r R_- T_- - 4U_-]/4 - xC(S_- - a_r T_-)/2\}.$$

and

$$(11) \qquad A = 5.20\times10^{13} z_l (m^*T)^{1/2} (\frac{1}{\epsilon_\infty} - \frac{1}{\epsilon_0}).$$

The several quantities in Eqs. (8) to (10) are,
$f_+ = f_0(x + z_l), f_- = f_0(x - z_l), f = f_0(x), C_+ = C(x + z_l), C_- = C(x - z_l),$
$C = C(x), R_+ = 2x + a_r + z_l, R_- = 2x + a_r - z_l, U_+ = [x(x+z_l)]^{1/2}, U_- = [x(x-z_l)]^{1/2},$
$T_+ = 4U_+/(R_+^2 - 4U_+^2), T_- = 4U_-/(R_-^2 - 4U_-^2), S_+ = ln[(R_+ + 2U_+)/(R_+ - 2U_+)],$
and $S_- = ln[(R_- + 2U_-)/(R_- - 2U_-)]$, where $z_l = \hbar\omega/k_BT$ is the reduced phonon
energy, a_r is the reduced screening energy, $a_r = \hbar^2/(2m^* r_s^2 k_B T)$, and r_s is the
Thomas-Fermi screening length generalized to all carrier densities by a direct computation of the appropriate Fermi-Dirac integral as in Eq. (9) of Ref. [49]. The
function $h(x - z_l)$ is the unit step function; $f_0(x) = [\exp(x - \eta) + 1]^{-1}$ is the Fermi
function; and $N_{ph} = [\exp(z_l) - 1]^{-1}$ is the phonon occupation number.

The operator $L_{pop}(C)$ is generalized to deal with holes by dividing the entire
right hand sides of Eqs. (9) and (10) by 2.3 to account for the p-like symmetry of
the hole wavefunctions (a factor of 2) and the effect of the light hole band (a factor
of 1.15) [54]. The p-like symmetry of the hole wavefunctions leads to a marked
decrease in the overlap function associated with the scattering matrix element. The
effect of the light holes is small because their fractional contribution is small.

IV.B.2 Nonpolar optic phonon scattering. The operator for the nonpolar
optic phonon scattering is given by the sum of two terms as in Eq. (8).

$$(12) \qquad L_{nop}(C) = L_{nop}^a(C) + L_{nop}^e(C),$$

where
(13)
$$L_{nop}^a = B(f_+/f)(N_{ph}+1)\{C_+[(q_2^2-q_1^2)(1+z_l/2x)/2-(q_2^4-q_1^4)/8k^2]-C(q_2^2-q_1^2)/2\},$$

(14)
$$L_{nop}^e = Bh(x-z_l)(N_{ph}f_-/f)\{C_-(p_2^2-p_1^2)(1-z_l/2x)/2-(p_2^4-p_1^4)/8k^2]-C(p_2^2-p_1^2)/2\},$$

where $q_1^2 = 2m^*m_0[(x+z_l)^{1/2}-x^{1/2}]^2k_BT/\hbar^2, q_2^2 = 2m^*m_0[(x+z_l)^{1/2}+x^{1/2}]^2k_BT/\hbar^2,$
$p_1^2 = 2m^*m_0[(x - z_l)^{1/2} - x^{1/2}]^2k_BT/\hbar^2, p_2^2 = 2m^*m_0[(x - z_l)^{1/2} + x^{1/2}]^2k_BT/\hbar^2,$
$k^2 = 2m^*m_0E/\hbar^2$, and

$$(15) \qquad B = \frac{7.8 \times 10^{13} \hbar^2 d_0^2 (m^*T)^{1/2}}{4\pi a^2 \rho_0 z_l k_B^2 T^2 e^2}.$$

This term, $L_{nop}(C)$, which only applies to holes because of their p-like symmetry, is divided by 1.2 to take into account the effect of the light holes [54]. This
term does not apply to electrons because their s-like symmetry does not satisfy the
selection rules.

IV.B.3. Plasmon scattering. The operator for plasmon scattering of electrons is given below. There is a cutoff wavenumber q_c for plasmons above which the collective mode cannot exist [55-57] because the electron gas cannot support a wave with a wavelength much shorter than the free-carrier screening radius, which defines the characteristic response length for the electron gas. This cutoff wavenumber is on the order of the inverse screening length. We vary q_c over its expected range from $0.25/r_s$ to $1.0/r_s$ in the following calculations. The couplings of the plasmon to the polar optic phonon modes have been neglected because there is no simple theory for the hybrid modes and because such coupling should not have much effect on the overall numerical result for the mobility. If one were to deal with these hybrid modes, it would be necessary to associate one hybrid mode with a phonon-like mode and another hybrid mode with a plasmon-like mode in order to carry out the calculations. For these reasons, we decided to deal with the pure uncoupled modes throughout. We define four squared wavenumbers:

$q_1^2 = 2m^*m_0[(x+z_p)^{1/2} - x^{1/2}]^2 k_B T/\hbar^2, q_2^2 = 2m^*m_0[(x+z_p)^{1/2} + x^{1/2}]^2 k_B T/\hbar^2,$
$p_1^2 = 2m^*m_0[(x-z_p)^{1/2} - x^{1/2}]^2 k_B T/\hbar^2,$ and $p_2^2 = 2m^*m_0[(x-z_p)^{1/2} + x^{1/2}]^2 k_B T/\hbar^2,$

where $z_p = \hbar\omega_p/k_B T$ and the plasma frequency $\omega_p = [4\pi N_I e^2/(m^*\epsilon_\infty)]^{1/2}$. We assume full ionization with no compensation. If all four wavenumbers are less than q_c then $L_{pl}(C)$ is given by the sum of the following two terms as in Eq. (8).

$$(16) \qquad L_{pl}(C) = L_{pl}^a(C) + L_{pl}^e(C),$$

where for electrons,

$$(17) \qquad L_{pl}^a(C) = P(f_+/f)(N_{ph}+1)\{C_+[R_+ S_+ - 4U_+]/4 - xCS_+/2\},$$

$$(18) \qquad L_{pl}^e(C) = Ph(x - z_p)(Nf_-/f)\{(C_-[R_- S_- - 4U_-]/4 - xCS_-/2\},$$

where $R_+ = 2x + z_p, R_- = 2x - z_p, U_+ = [x(x+z_p)]^{1/2}, U_- = [x(x-z_p)]^{1/2},$
$S_+ = ln[(R_+ + 2U_+)/(R_+ - 2U_+)], S_- = ln[(R_- + 2U_-)/(R_- - 2U_-)]$ and

$$(19) \qquad P = 5.20 \times 10^{13} z_p (m^*T)^{1/2}/\epsilon_\infty.$$

However, if $q_2 < q_c$, then $S_+ = ln(q_c^2/q_1^2)$ and $4U_+ = \hbar^2 q_c^2/2m^*m_0 - \{2x + z_p - 2[(x+z_p)x]^{1/2}\}k_B T$. If $q_1 < q_c$, then $S_+ = 0$ and $U_+ = 0$. Similarly, if $p_2 < q_c$, then $S_- = ln(q_c^2/p_1^2)$ and $4U_- = \hbar^2 q_c^2/2m^*m_0 - \{2x - z_p - 2[(x-z_p)x]^{1/2}\}k_B T$. If $p_1 < q_c$, then $S_- = 0$ and $U_- = 0$. The dependence of the cutoff wavenumber and plasma frequency on doping density leads to interesting phenomena for plasmon scattering such as a relative minimum in the minority mobility as a function of dopant density. The plasmon scattering is generalized to holes by dividing L_{pl} by 2.3, the same as for polar optic phonons.

IV.B.4. Relaxation times. The remaining scattering mechanisms are treated in terms of a relaxation time τ because the change in energy associated with a scattering event is less than $k_B T$.

IV.B.4.1. Acoustic phonon. Acoustic phonon scattering is given by the following expression:

$$(20) \qquad \tau_{ac}^{-1} = 4.167 \times 10^{19} E_1^2 (m^* T)^{3/2} x^{1/2} / C_l,$$

which is generalized to holes by dividing the right-hand side (r.h.s.) of Eq. (20) by 1.2 to account for the contribution of light holes [54].

IV.B.4.2. Piezoelectric (acoustic-mode) scattering. Piezoelectric (acoustic-mode) scattering is given by the following expression:

$$(21) \qquad \tau_{pel}^{-1} = 9.45 \times 10^{11} h_{14}^2 \left(\frac{3}{C_l} + \frac{4}{C_t} \right) (T m^* / x)^{1/2},$$

which is generalized to holes by dividing the r.h.s by 2.3 as for optic phonons. There is no reference for this value. We have used it because the wavelength (q) dependence of the squared piezoelectric scattering matrix element differs from that for polar optic phonons by only a factor of q in the denominator; i.e., the former varies as q^{-1} and the latter varies as q^{-2}. This somewhat weaker inverse q dependence could lead to a larger value than 2.3, but since this is already a relatively large value for the effect of the overlap factor, we felt it was probably sufficient. In any case the effect of piezoelectric scattering on the mobility is very small, and is sometimes neglected [54].

IV.B.4.3. Ionized impurity scattering. Ionized impurity scattering is treated both in terms of the Brooks-Herring theory [58] and in terms of a phase-shift calculation [27]. Brooks-Herring (BH) theory is based on the Born approximation, which is valid at high carrier energies relative to the interaction potential. The BH theory is not valid for the carriers near the band extrema, which are strongly scattered by the screened Coulomb potential used to model the dopant ions. Therefore, our phase shift calculation, based on a maximum of fifty partial waves, is a much more accurate method since it is valid for all carrier energies.

The relaxation time from the BH theory is:

$$(22) \quad \tau_{imp,BH}^{-1} = \frac{2.415 N_I}{\epsilon_0^2} (m^*)^{-1/2} (xT)^{-3/2} [ln(1 + 4x/a_r) - 4x/a_r(1 + 4x/a_r)^{-1}].$$

The corresponding relaxation time from the phase shift method is:

$$(23) \qquad \tau_{imp}^{-1} = \frac{4\pi \hbar N_I}{k m^* m_0} \Sigma_{l=1}^{\inf} l \sin^2(\delta_{l-1} - \delta_l),$$

where the lth phase-shift, δ_l, is obtained from the asymptotic form of the lth partial wave solution to the Schrödinger equation with a screened Coulomb potential [50].

The upper bound in the phase-shift sum is determined by the stopping criterion that the next term in the series is less than 1 percent of the leading term. The upper limit of 50 satisfied this desired numerical accuracy for the cases presented. The number of needed partial waves increased from about 10 to 40 as the doping density decreased over the calculated ranges. The Schrödinger wave equation was solved by the Adams-Bashforth method [59] for an initial value problem with the endpoint determined by the potential's having decayed to a negligible value. The inverse tau values were obtained for 100 values of wavenumber $k = [2m^*m_0 E/\hbar^2]^{1/2}$ over the range required by the mobility-defining integrals ($10k_BT$ above and below the Fermi energy or $10 \; k_BT$ into the band). A parabolic spline fit was used for interpolation.

It is assumed in the model that scattering occurs by isolated centers, which is not the case if the screening radius becomes much greater than half the separation distance between the ions. This condition was met sufficiently well in these calculations. The ionized impurity scattering for holes is calculated in the same manner as that for electrons without dividing by the overlap factor or the light-hole factor as was done for the phonon scattering operators for holes. We have made this approximation based on the relative strength of low-angle scattering for this mechanism and the fact that the overlap factor due to the p-wave nature of holes goes to unity for small angles. We have neglected the effect of light holes since their contribution to the conductivity associated with this mechanism is very small [54].

IV.B.4.4. Carrier-carrier scattering.

IV.B.4.4. Carrier-carrier scattering. We now discuss the scattering mechanisms that deal with carrier-carrier interactions, which are different for majority and minority carriers [60]. One would think that majority carrier-majority carrier scattering would not have any effect because momentum is conserved by the system of carriers. However, there is a second-order effect, which is very significant numerically and results from the redistribution of energy and momentum among the carriers. At room temperature this effect reduces the mobility associated with ionized impurity scattering the most [61]. We only consider the correction to this mechanism. Were it not for majority carrier-majority carrier scattering, high-energy carriers would not experience much scattering due to ionized impurities, but they do because of their interactions with low energy carriers that have been scattered by these impurities. Thus, there is an overall decrease of the mobility due to this majority carrier-majority carrier scattering.

We used the variational method of Appel [62] to compute a correction factor for ionized-impurity scattering (Brooks-Herring theory). The ratio of the mobility with and without majority carrier-majority carrier scattering, which we call the carrier-carrier scattering factor (CCSF), is given by Fig. 1 in Ref. [62] as a function of the parameter δ, the ratio of the thermal electron wavelength to the screening radius, $[h/(2m^*m_0 k_B T)^{1/2}]/r_s$. This expression was derived for nondegenerate statistics in Ref. [62]. We generalized it for all statistics by replacing the average thermal energy,

$3k_BT/2$, with the Fermi energy when the Fermi energy is greater than the average thermal energy. This is an upper limit for the effect when degeneracy applies, because some of the carriers become unable to take part in the scattering process as a result of the Pauli exclusion principle. In the limit of extreme degeneracy (low temperature) this whole effect therefore goes away.

The theory of Luong and Shaw [63] uses a multiscattering quantum-mechanical approach to calculate CCSF. Their result of 0.632 is independent of temperature and carrier density and is valid only for the limit of nondegenerate statistics. We compare the result of Appel with the result of Luong and Shaw, which we believe to be a lower limit for the mobility. The independence of Luong and Shaw's CCSF [63] on screening radius may result from approximations in their theory that are valid only in the limit of screening radii large compared with the average carrier wavelength. We use the same results for CCSF from references [62] and [63] for both majority electrons and majority holes, as we do for the ionized impurity scattering itself.

For minority carriers the situation is somewhat different because momentum is not conserved by the scattered system. The scattering of minority carriers by majority carriers is treated the same as ionized impurity scattering except that the reduced mass is used for the scattered carrier, with the scattering carrier having infinite mass in the center of mass system. Both the Brooks-Herring and phase-shift approaches were used to calculate the relaxation times, which follow Eqs. (22) and (23). The Pauli exclusion principle again affects the minority carrier-majority carrier scattering because majority carriers that are well below the Fermi energy have few available final states into which they may scatter. In degenerate semiconductors at low temperature, this effect greatly reduces this scattering mechanism as has been seen in two-dimensional systems [64]. The actual calculation would require a many-body theory for the scattering of an assembly of minority carriers off majority ones. Tractable formalisms for the coupled system with degeneracy included do not exist. Instead we treat the scattering particles as classical particles and approximate the effect by removing those states with energies below the Fermi energy. Thus final states were considered available with occupancy factors below 0.5 and filled with occupancy factors above 0.5. This is a somewhat crude approximation, but it serves the purpose of showing the size of the phenomenon. Again both electrons and holes are treated in the simple band approximation.

We estimate the error associated with this approximation for the mobility associated with carrier-carrier scattering to be between 20 and 30 percent, depending on the degree of degeneracy and the relative importance of the distribution of carriers that lies between nearly full degeneracy and nearly nondegenerate statistics. This would translate into a ten percent error in the total mobility. Therefore, we consider the final results to be only a good estimate that should stimulate further measurements and calculations.

IV.B.5. Mobility results.

The method for obtaining the carrier mobility for a given doping density follows a set procedure. First the Fermi energy and free carrier screening radius are found by solving the charge-neutrality condition by iteration and then computing the integral obtained from Thomas-Fermi theory for the screening radius [49]. The Fermi integrals $F_j(x)$ have been evaluated in terms of the series solutions given by Van Halen and Pulfrey [65], which are accurate to better than 1 part in 10^5. Better values than appear in Ref. [65] were used in order to obtain this accuracy. As defined in Ref. [65] with the range of x values in parentheses: for j=$-1/2$, a_7= 0.00029111, (0-5/2) and 0.00024904, (5/2-5); a_8 = -0.00015907, (0-5/2) and $-$.00001300, (5/2-5); and a_9 = 0.00001832, (0-5/2) and 0.0000002985 (5/2-5). For j=1/2, a_7=0.0001335, (0-2) and -0.00003581, (2-4) [66]. Full ionization was assumed throughout. The phase shifts were then calculated and the relaxation times were fit with B-splines as a function of wavenumber. Finally, the mobility was calculated according to the equations derived in the previous section. Our theoretical calculations for the four mobilities are expected to be upper limits to the measured mobilities because there may be additional scattering mechanisms due to defects that were not included in the calculations.

The majority electron mobility in n-type GaAs is shown in Fig. 9. The experimental data from the literature [67] are Hall mobilities and can differ from the calculated drift mobility by the Hall factor, which is close to but not exactly one. Lee and Look [68] have treated the Hall factor in GaAs; but because of the difficulties in calculating this quantity accurately and consistently with our theory, we have not included any correction due to this effect, which we expect to be at most on the order of 10%. We show four curves for the theoretical mobility in order to show the size of the various effects we are including. The uppermost curve corresponds to the Born expression for ionized impurity scattering with no inclusion of the effect of electron-electron scattering (CCSF=1.). Generally this curve lies above the data. The second uppermost curve shows the effect of using the phase-shift analysis to obtain the ionized impurity scattering, and the result is a lowering of the mobililty by at most ten percent.

The next curve down includes the effect of electron-electron scattering as obtained from the theory of Appel [62] generalized approximately to full statistics. The mobility is lowered, but not enough to fit the data. The lowest curve is obtained with the expression of Luong and Shaw [63] and agrees best with the data, but is suspect because of both its lack of dependence on screening radius and, as is true as well for Appel's work, the neglect of Pauli exclusion for the case of degeneracy. However, there is another mechanism that permits the electrons to exchange energy and momentum, namely, the plasmon interaction. This is an effect that will be significant just as the single-particle interactions diminish [55]. Although no theory exists to account for this effect, we propose that it may account for the large electron-electron interaction observed in the data. We do not agree with other explanations for the relatively low electron mobilities at large doping concentrations. Walukiewicz *et al.* [49] have invoked partial compensation, which is now thought to be a small effect in these materials. Meyer and Bartoli [67] have neglected the effect of screening on the optic phonons, which results in their lower theoretical mobilities at large doping. We believe that this screening must be included, and that the lowering may be related to electron-electron interactions, both single-particle and

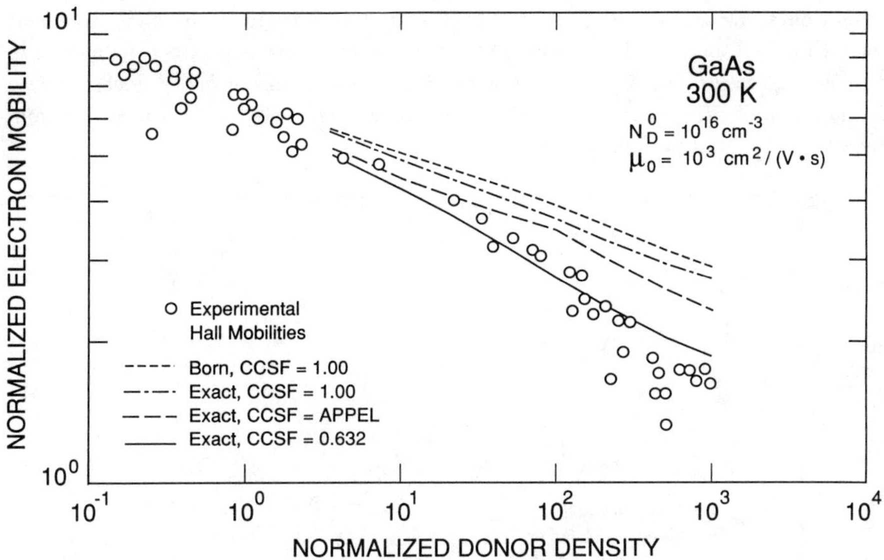

Fig. 9. Majority electron mobility as a function of donor density. The open circles are the experimental data given in Fig. 6 of Ref. [67]. The curve with short dashes gives the calculated mobility when the Born approximation is used with the carrier-carrier scattering factor (CCSF) equal to 1. The other three curves give the mobilities when the exact phase shifts are used with values of CCSF equal to 1 (short dash-long dash), 0.632 (solid), and the generalization of the theory of Appel [62] (long dash). The mobility and donor densities have been normalized to μ_0 and N_D^0, respectively.

collective (plasmon induced) ones. Of course, the effect of compensation and the Hall factor must be considered as well.

The minority electron mobility in p-type GaAs is shown in Fig. 10. The experimental data in Fig. 10 are drift mobilities from the literature measured by time-of-flight or high-frequency cutoff methods [69-72]. The uppermost curve is for the Born approximation for both ionized impurity and electron-hole scattering. The next curve down is for the phase-shift method for both of these mechanisms. The difference is small because the increase of the mobility for ionized impurity scattering is nearly compensated by the decrease of the mobility for electron-hole scattering. The next curve shows the effect of the electrons scattering off hole plasmons. The lower limit of 0.25 has been used for the plasmon cutoff factor (PCF), which equals $q_c^2 r_s^2$. Good agreement with the data occurs for the next curve with PCF = 1.0, the upper limit, except possibly above 1×10^{19} cm^{-3}. Our theory predicts a rise in the mobility due to the increase in the plasmon frequency in this doping range. An even greater rise is shown in the last curve that deducts the hole density below the Fermi energy from the electron-hole scattering process. The very recent data of Furuta *et al.* [71] and Lovejoy *et al.* [72] are in qualitative agreement with this predicted rise in mobility. We propose that more experiments be done in this doping range to determine if the mobility rises. If the mobility does not rise, it may be indicative of the onset of scattering by other defects that can result from processing at high doping densities.

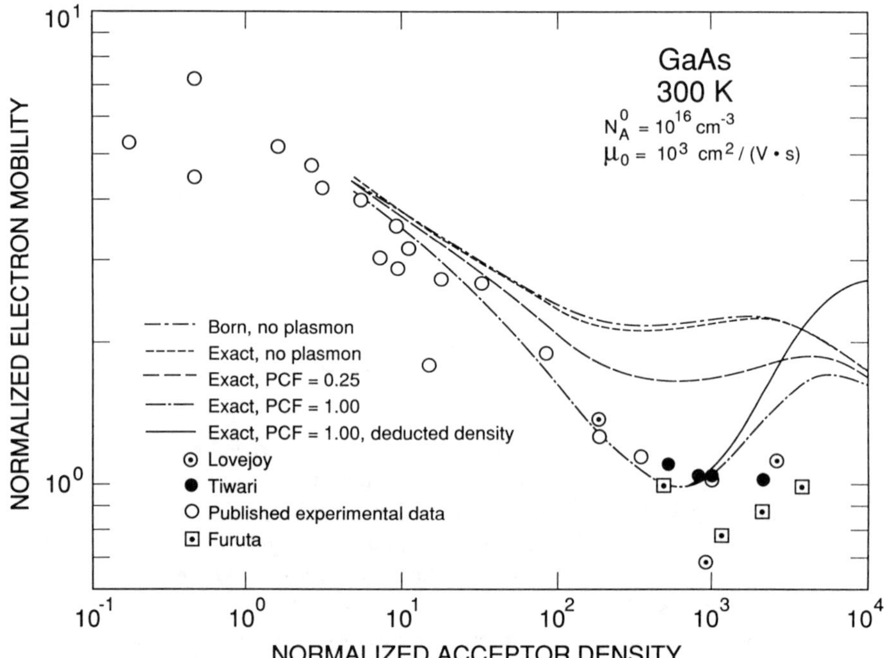

Fig. 10. Minority electron mobility as a function of acceptor density. The open circles are the experimental data given in Fig. 2 of Ref. [69]. The solid circles, circled dots, and squared dots are the more recent experimental data from Refs. [70-72], respectively. The curve with short and long dashes gives the calculated mobility when the Born approximation is used with no plasmon scattering. The curve with short dashes gives the calculated mobility when the exact phase shifts are used with no plasmon scattering. The remaining three curves give the mobilities when the exact phase shifts are used with plasmon scattering included. The plasmon cut-off factors of 0.25 (long dash) and 1.0 (dash-dot) are used. The solid curve gives the mobility that results when the density of majority carriers below the Fermi energy is deducted from the total density available for carrier-carrier scattering. The mobility and acceptor densities have been normalized to μ_0 and N_A^0, respectively.

Our results for the majority hole mobility are given in reference [28]. Overall the agreement between theory and experiment is similar to that indicated by Fig. 9. The results for the minority hole mobility are also presented in reference [28]. Because of the difficulty in measuring minority hole mobilities, one should expect larger error bars than for the other cases. Therefore, the overall agreement between theory and experiment is reasonably good.

Figure 11 gives the ratios of the minority-carrier mobility to the majority- carrier mobility for electrons and holes. The values of $\mu_n(p-type)/\mu_n(n-type)$ are given by the solid curve, and the values of $\mu_p(n-type)/\mu_p(p-type)$ are given by the dashed curve. These ratios are calculated from the curves with the label "exact, PCF=1, deducted density" for the minority mobilities and from the curves with the label "exact, CCSF=APPEL" for the majority mobilities. Unlike the mobility ratios for silicon [27], the mobility ratios for GaAs are less than 1. They have values between 0.38 and 0.86 for electrons and between 0.27 and 0.61 for holes over the dopant density range considered in this paper. Since GaAs has more complicated scattering processes than occur in silicon and has strong asymmetry between its electron and hole effective masses, the results are very different from those for silicon.

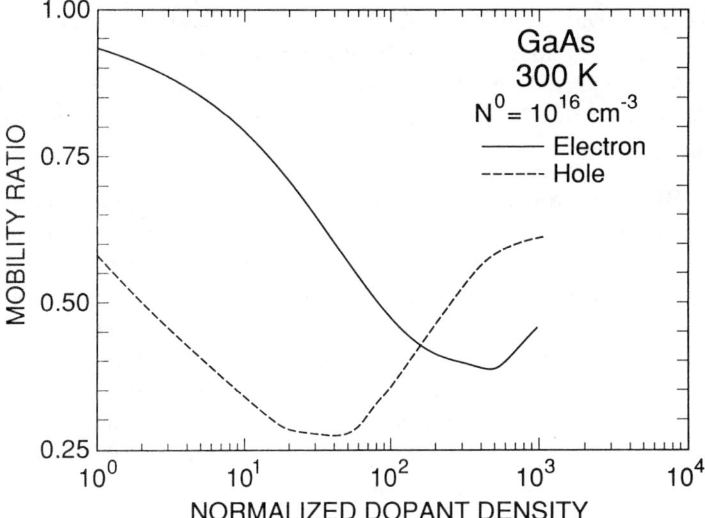

Fig. 11 Electron and hole mobility ratios as functions of the dopant density. The values of $\mu_n(p-type)/\mu_n(n-type)$ are given by the solid curve and the values of $\mu_p(n-type)/\mu_p(p-type)$ are given by the dashed curve. The dopant densities have been normalized to N^0 equal to 1×10^{16} cm^{-3}.

The above approach for calculating scattering mechanisms also has implications for the scattering cross sections used in Monte-Carlo calculations. These cross sections are the basis for computing the carrier-velocity versus electric-field relations that are used in device simulators, which are usually based on the drift-diffusion equations. Our expressions can be easily modified to obtain the appropriate scattering rates, with the formulas involving the phase shifts converted to yield total cross sections. We expect that application of our results to these calculations should yield more accurate values for the velocity-field relations.

V. APPLICATIONS

V.A. Heterojunction bipolar transistors – Energy band diagrams. The foregoing band changes have significant implications for understanding device behavior from energy band diagrams. Consider, for example, a typical heterojunction bipolar transistor that has an n-type $Ga_{0.7}Al_{0.3}As$ emitter, $N_D = 6 \times 10^{17}$ cm^{-3}, $W_E = 0.25$ μm,; a p-type GaAs base, $N_A = 10^{19}$ cm^{-3}, $W_B = 0.1$ μm; and an n-type GaAs collector. The collector contains two epitaxial layers. The first layer has $N_D = 5 \times 10^{16}$ cm^{-3}, $W_{C1} = 0.5$ μm, and the second layer has $N_D = 4 \times 10^{18}$ cm^{-3}, $W_{C2} = 0.6$ μm. Figure 12 shows the energy band diagram for this heterostructure. The dashed lines are the results based on the unperturbed band structure. It is this set of lines that has been generally used to design devices or to explain device operation. The solid lines are the results based on distorted bands that include the effects of high concentrations of carriers and dopants. As can be seen, bandgap narrowing affects the emitter, base, and collector regions. Both Figs. 7 and 12 should be considered together. They show that the emitter and collector have partial cancellation between the competing effects of degeneracy and bandgap narrowing. However, the main effect in the base is the increase in n_{ie} due to bandgap narrowing.

V.B. Bipolar transistors. Emitters in homojunction bipolar transistors are heavily doped to reduce the reverse injection of minority carriers from the base to the emitter and to increase the efficiency of the emitter. However, the bandgap narrowing and bulk recombination in the emitter that result from high doping concentrations place a practical limit on the emitter doping, beyond which performance does not improve with increased emitter doping. This practical limit on the emitter doping then places an upper limit for the base doping if acceptable current gains are to be maintained. However, to improve high-frequency performance, bases can be doped more heavily than suggested by the upper limit associated with high gains.

As shown above, whenever carrier densities or ionized dopant densities exceed 10^{17} cm^{-3} in GaAs or 10^{18} cm^{-3} in silicon at 300 K, the carrier-dopant-ion and carrier-carrier interactions lead to appreciable changes in the band structure and density of states for the carriers. These changes must be included in the interpretation of optical (absorption and luminescence) measurements and in the device physics used to simulate transistors numerically, particularly those with heavily doped, submicron structures. They also affect greatly the parasitic processes such as latch-up in complementary metal-oxide-semiconductor (CMOS) field-effect transistors (FETs).

V.B.1. Silicon bipolar transistors. We used the improved device physics (IDP) developed at NIST to compare the predictions of conventional device physics (CDP) and IDP in numerical simulations of silicon bipolar transistors with emitter

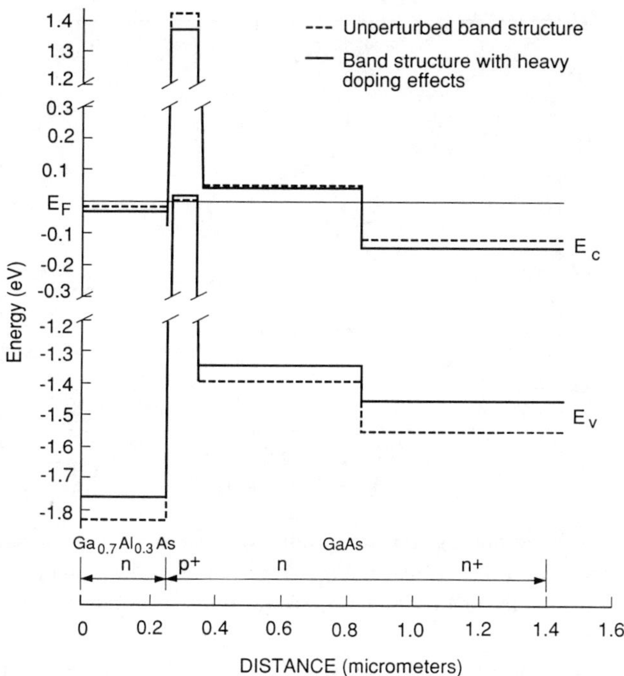

Fig. 12. Energy band diagram for a typical high-frequency, heterojunction bipolar transistor.

and base widths that vary from 10 μm to 0.16 μm [5,6]. For large 10-μm devices, CDP and IDP both agree with measured I-V characteristics. However, for small 0.16-μm devices, only the IDP gives agreement with measurements. The latter was the result of an NIST-Tektronix collaboration that verified experimentally the IDP concepts for devices with junction depths as small as 0.16 μm and with doping densities in excess of 2×10^{20} cm^{-3}.

When the IDP is incorporated into device analysis codes such as SEDAN [73,74] and then used to compute the electrical performance of npn transistors, the predicted values agree with the measured values of the current-voltage characteristics and dc common emitter gains for devices with emitter-base junction depths between 10 μm and 0.16 μm. The net doping profile for device C, a diffused npn transistor, is shown in Fig. 13. The emitter is As-doped, the base is B-doped, and the collector is P-doped. The profiles for the emitter and base are the results from SIMS data. The collector profile is the result from the interpretation of spreading resistance measurements.

Researchers [75] report that, when available, profiles for heavily doped, shallow structures obtained by methods such as SIMS should be used in detailed device models. Profiles obtained by methods that determine the concentration of carriers from the interpretation of such measurements as spreading resistance may give erroneous junction depths.

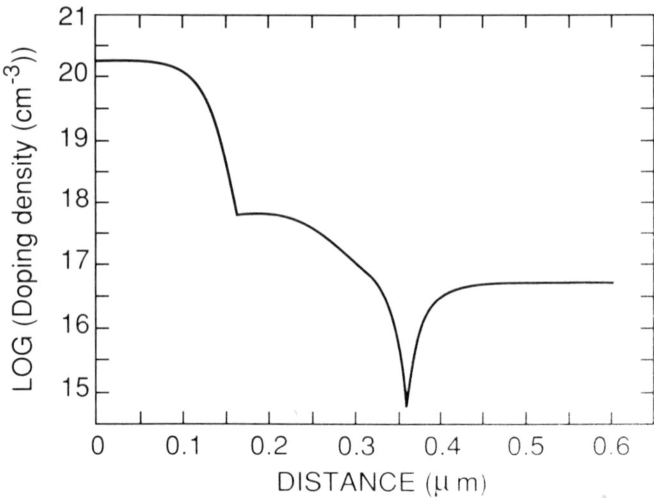

Fig. 13. Net doping density profile for device C. The boron density near the surface is 3.5×10^{18} cm^{-3} and the emitter sheet resistance is 45 Ω/square. (from Reference [6])

Figure 14 shows that device physics that follows from quantum mechanics (i.e., that does not depend on empirical fits to electrical measurements and that has been verified by independent optical measurements) predicts the correct current-voltage (I-V) characteristics and the dc common emitter gains of submicron bipolar transistors.

V.C. GaAs bipolar transistors. As for the above case of silicon, using the best available physical models is essential for predictive numerical simulations of GaAs bipolar transistors. Recently, the above theoretical calculations of n_{ie} and mobilities for GaAs at 300K in Section IV have been implemented in a two-dimensional, drift-diffusion simulator [76]. In order to compare predicted and measured DC common emitter gains, several npn GaAs homojunction bipolar transistors with different but heavily doped bases and emitters were fabricated by molecular beam epitaxy. The predicted gains of 8, 25, and 46 for these transistors agreed very well with their measured gains of 9, 22, and 42 at high current, respectively. Without using the new theoretical data for n_{ie} but setting n_{ie} equal every where to the intrinsic carrier concentration, n_i, the predicted gains became 4, 14, and 27, respectively. Sensitivity analyses on mobilities, lifetimes, and n_{ie} showed that physically correct n_{ie} and mobilitiy values are very important for predictive simulations that contain

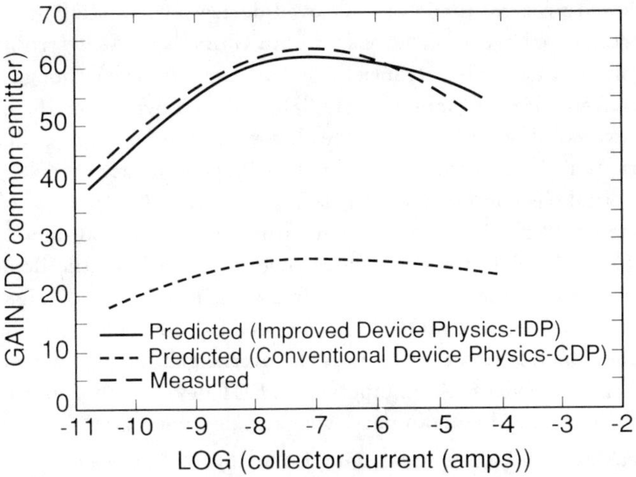

Fig. 14. DC common emitter gains as functions of the collector current. The long-dashed, short-dashed, and solid curves give the measured values, the values predicted by conventional device physics, and the values predicted by the improved device physics, respectively. (from Reference [6])

no variational parameters. The other key quantities such as dopant density profiles and minority carrier mobilities and lifetimes were all verified experimentally by measurement methods that do not depend on the use of lower level device models.

VI. CONCLUSIONS

VI.1. Device physics. In this paper, we have shown the significance of using the best possible device physics when numerically simulating devices. Bipolar models based on drift-diffusion equations are now capable of predicting the characteristics of silicon devices over a much broader range of device dimensions and doping profiles. For example, use of the first-principles approach gives correct predictions for npn transistors with base widths from 0.2 μm to 10 μm. We also have shown that bipolar models should include separate and distinct values of parameters for the n-type and p-type regions as functions of carrier and dopant-ion concentrations. That is, detailed device models should contain: minority and majority hole mobilities, minority and majority electron mobilities, and either the effective intrinsic carrier concentrations for n-type and p-type material or, preferably, the effective conduction and valence band edge changes for both n-type and p-type material.

We reviewed progress made on improved device physics (IDP) for GaAs. The results reported here have applications for both GaAs/AlGaAs heterojunction bipolar transistors and field-effect transistors. For example, even though the channel regions of GaAs MESFETs may be only lightly doped, whenever the electron concentrations exceed about 10^{17} cm^{-3}, the changes in the band edges due to carrier-carrier interactions in the channel should be included in device models. These shifts alter the internal electric fields in the devices. Also, the authors of reference [77] have examined the effect of bandgap narrowing on modulation doped field effects transistors (MODFETs). Bandgap narrowing occurs in both the depleted, doped region and the conducting channel. They find that the effect of bandgap narrowing is most severe for p-channel MODFETs in which it leads to a parasitic conducting channel in the doped region that reduces the overall gain of the intrinsic transistor. Hence, with silicon devices, GaAs bipolar and FET device models must be modified to include the effects of bandgap narrowing and degeneracy.

The measured and theoretical data for many of the electrical and material properties of GaAs devices are not adequate for reliable engineering without detailed verification by separate, nonelectrical measurements. The input quantities for detailed-device models of GaAs transistors contain many more unknown parameters than those for silicon due to the nearby L and X conduction band minima and the presence of optical phonons. Incorporating adequate physical concepts in GaAs device models requires new measurement techniques (perhaps based on ultrafast spectroscopy) for mobilities and lifetimes as functions of electric fields, dopant density, carrier energy, and carrier density in processed GaAs. Specially designed test structures will be needed to resolve many of the uncertainties concerning the physical concepts to be incorporated into GaAs device models.

VI.2. Intercomparisons and verification. We conclude this paper by briefly discussing some issues concerned with the benchmarking and calibration/verification of device simulators. Similar issues also pertain to process simulations. Benchmarking, calibrations, and intercomparisons ("round robins") would improve the quality of process and device simulators and provide improved predictive simulators. However, efforts are now fragmented and somewhat like a cottage-industry. Current trends in CAD make it very important for the CAD community to reach a consensus among its many competing options and to chart a more stable course for its future.

Benchmarking and calibration/verification are essential for marketplace competition. Benchmarking is the comparison of outputs from device simulators that purport to calculate the same quantities when all of the codes are given "identical" inputs. Verification/calibration involves building special devices or structures, measuring their electrical or optical properties, and verifying the predictions of device simlulators by comparison with measurements that are interpreted with the best physics available.

Given that a single institution, is not likely to have all the resources to address the issues and to provide predictive, high-quality process and device simulators, collaborative efforts among key players are essential. Such collaborative efforts should strive to accomplish the following:

1. Design specialized structures and devices that will be used to test the physical models in computer codes for simulating semiconductor processing and the electrical performance of devices.

2. Verify the predictions of these process and device simulations by performing measurements on specialized structures and devices.

For example, a collaboration could occur among institutions with capabilities in drift-diffusion simulators, Monte-Carlo simulators, experimental device physics, device fabrication, and theory. Specialized structures would be designed and fabricated to provide independently verified, input parameters for these simulators. Devices would be designed to test the physical models in these simulators and to test the performance of the simulators for accuracy. This testing would be done by intercomparisons among participating institutions for which the same device would be measured and tested as several sites. Results would be guidelines on which mechanisms dominate in a given process or device class; critical parameters to measure for verifying numerical simulations of processes and devices; and high confidence levels in predictions from numerical simulations.

Acknowledgments. We thank M. Lundstrom of Purdue University, West Lafayette, Indiana, and M. Tomizawa and T. Ishibashi of Nippon Telegraph and Telephone Corporation, Atsugi, Japan, for helpful discussions and for providing preprints of their work in this area.

REFERENCES

1. J. L. BLUE AND C. L. WILSON, IEEE Trans. Electron Devices **ED–30**, 1056 (1983).
2. J. R. Klauder, Ann. Phys. **14**, 43 (1961).
3. R. A. Abram, G. N. Childs, and P. A. Saunderson, J. Phys. **C17**, 6105 (1984).
4. J. del Alamo, S. Swirhun, and R. M. Swanson, Proceedings of the IEDM, 290 (1985).
5. H. S. Bennett, IEEE Trans. Electron Devices **ED–30**, 920 (1983).
6. H. S. Bennett and D. E. Fuoss, IEEE Trans. Electron Devices **ED–32**, 2069 (1985).
7. M. Kurata and J. Yoshida, IEEE Trans. Electron Devices **ED–31**, 467 (1984).
8. P. M. Asbeck, D. L. Miller, R. Asatourian, and C. G. Kirkpatrick, IEEE Electron Device Letters **EDL–3**, 403 (1982).
9. M. S. Adler, Solid-State Electronics **26**, 387 (1983).
10. S. P. Gaur, P. A. Habitz, Y. J. Park, R. K. Cook, Y.-S. Huang, and L. F. Wagner, IBM Journal of Research and Development **29**, 242 (1985).
11. M. S. Lundstom, R. J. Schwartz, and J. L. Gray, Solid-State Electronics **24**, 195 (1981).
12. A. H. Marshak, Solid-State Electronics **31**, 1551 (1988).
13. W. Slotboom and H. C. deGraaff, Solid-State Electronics **19**, 586 (1976).
14. H. B. Callen, *Thermodynamics*, New York: Wiley & Sons, 1960, p. 207.
15. F. Reif, *Fundamentals of Statistical and Thermal Physics*, New York: McGraw-Hill, 1965, p. 324.
16. D. D. Tang, IEEE Trans. Electron Devices **ED–27**, 563 (1980).
17. A. W. Weider, Proceedings of the IEDM, 460 (1978).
18. M. S. Adler and G. E. Possin, IEEE Trans. Electron Devices **ED–28**, 1053 (1981) and references therein.
19. G. E. Possin, M. S. Adler, and B. J. Baliga, IEEE Trans. Electron Devices **ED–31**, 3 (1984).
20. J. Wagner and J. A. del Alamo, J. Appl. Phys. **63**, 425 (1988).
21. G. E. Possin, M. S. Adker, and B. J. Baliga, Proceedings of the ASTM Symposium on Lifetime Factors in Silicon, ASTM STP 712, 192 (1980).

22. D. J. Roulston, N. D. Arora, and S. G. Chamberlain, IEEE Trans. Electron Devices **ED–29**, 284 (1982).

23. W. Shockley and W. T. Read, Jr., Phys. Rev. **87**, 835 (1952).

24. H. S. Bennett, Solid-State Electronics **28**, 193 (1985).

25. H. S. Bennett, J. Appl. Phys. **59**, 2837 (1986).

26. H. S. Bennett and J. R. Lowney, J. Appl. Phys. **62**, 521 (1987).

27. H. S. Bennett, Solid-State Electronics **26**, 1157 (1983).

28. J. R. Lowney and H. S. Bennett, J. Appl. Physics **69**, 7102 (1991).

29. J. R. Lowney and W. R. Thurber, Electron. Lett. **20**, 142 (1984).

30. B. R. Chawla and H. K. Gummel, IEEE Trans. Electron Devices **ED–18**, 178 (1971).

31. J. R. Lowney, Solid-State Electronics **28**, 187 (1985).

32. H. S. Bennett, J. Appl. Phys. **55**, 3582 (1984).

33. S. C. Jain, R. P. Mertens, P. Van Mieghem, M. G. Mauk, M. Ghannam, G. Borghs, and R. Van Overstraeten, Proceedings of the IEEE 1988 Bipolar Circuits and Technology Meeting, 1988, J. Jopke, Ed., p. 195.

34. H. C. Chen, S. S. Li, and K. W. Teng, Solid-State Electronics **32**, 339 (1989).

35. A. Neugroschel, J. S. Wang, and F. A. Lindholm, IEEE Electron Device Letters, **EDL–6**, 253 (1985).

36. M. Capizzi, S. Modesti, A. Frova, J. L. Staehli, M. Guzzi, and R. A. Logan, Phys. Rev. B **29**, 2028 (1984).

37. J. R. Lowney, Proceedings of the IEEE 1988 Bipolar Circuits and Technology Meeting, 1988, J. Jopke, Ed., p. 188.

38. P. T. Landsberg and D. J. Robbins, Solid-State Electronics **28**, 137 (1985).

39. J. R. Lowney, J. Appl. Phys. **59**, 2048 (1986).

40. T. J. de Lyon, H. C. Casey, Jr., and A. J. SpringThorpe, J. Appl. Phys. **65**, 2530 (1989).

41. A. Yariv, *Quantum Electronics*, New York: Wiley & Sons, 1967, p. 282.

42. M. C. Wu, Y. K. Su, K. Y. Cheng, and C. Y. Chang, Solid-State Electronics **31**, 251 (1988).

43. J. R. Lowney, R. D. Larrabee, and W. R. Thurber, IEEE Proceedings of the Custom Integrated Circuits Conference, May 1983, p. 152.

44. J. R. Lowney and H. S. Bennett, J. Appl. Phys. **65**, 4823 (1989).

45. H. S. Bennett and J. R. Lowney, Solid-State Electronics **33**, 675 (1990).

46. J. R. Lowney, A. H. Kahn, J. L. Blue, and C. L. Wilson, J. Appl. Phys. **52**, 4075 (1981).

47. J. R. Lowney, J. Appl. Phys. **64**, 4544 (1988).

48. S. M. Sze, *Physics of Semiconductor Devices*, New York: Wiley & Sons, 1981, 2nd edition, p.21 and p. 850.

49. W. Walukiewicz, L. Lagowski, L. Jastrzebski, M. Lichtensteiger, and H. Gatos, *J. Appl. Phys.* **50**, 899 (1979).

50. J. R. Lowney and H. S. Bennett, *J. Appl. Phys.* **53**, 433 (1982).

51. L. Reggiani, *Hot-Electron Transport in Semiconductors* (Springer-Verlag, New York, 1985), p.7ff.

52. D. J. Howarth and E. H. Sondheimer, *Proc. Roy. Soc. London* A **219**, 53 (1953).

53. H. Ehrenreich, *Phys. Rev.* **120**, 1951 (1960).

54. J. D. Wiley in *Semiconductors and Semimetals*, Ed. by Willardson and Beer (Academic Press, New York, 1974) Vol. 10, p. 91.

55. P. Lugli and D. K. Ferry, *Appl. Phys. Lett.* **46**, 594 (1985).

56. R. Katoh, M. Kurata, and J. Yoshida, *IEEE Trans. Electron Devices* ED-36, 846, (1989).

57. M. E. Kim, A. Das, and S. D. Senturia, *Phys. Rev.* B **18**, 6890 (1978).

58. H. Brooks and C. Herring, *Phys. Rev.* **83**, 879 (1951).

59. L. F. Shampine and H. A. Watts, *DEPAC-Design of a User Oriented Package of ODE Solvers*, Sandia National Laboratories Technical Report, SAND-79-2374, 1979.

60. W. Walukiewicz, J. Lagowski, L. Jastrzebski, and H. C. Gatos, *J. Appl. Phys.* **50**, 5040 (1979).

61. D. Chattopadhyay, *J. Appl. Phys.* **53**, 3330 (1982).

62. J. Appel, *Phys. Rev.* **125**, 1815 (1962).

63. M. Luong and A. W. Shaw, *Phys. Rev.* B **4**, 2436 (1971).

64. R. A. Hopfel, J. Shah, P. A. Wolff, and A. C. Gossard, *Phys. Rev.* B **37**, 6941 (1988).

65. P. Van Halen and D. L. Pulfrey, *J. Appl. Phys.* **59**, 2264 (1986).

66. D. L. Pulfrey, private communication.

67. J. R. Meyer and F. J. Bartoli, *Phys. Rev.* B **36**, 5989 (1987).

68. H. J. Lee and D. C. Look, *J. Appl. Phys.* **54**, 4446 (1983).

69. H. Ito and T. Ishibashi, *J. Appl. Phys.* **65**, 5197 (1989).

70. S. Tiwari and S. L. Wright, *Appl. Phys. Lett.* 56, 563 (1990).

71. T. Furuta and M. Tomizawa, *Appl. Phys. Lett.* 56, 824 (1990).

72. M. L. Lovejoy, B. M. Keyes, M. E. Klausmeier-Brown, M. R. Melloch, R. K. Ahrenkiel, and M. S. Lundstrom, "Time-of-Flight Measurements of Zero-Field Electron Diffusion in P^+-GaAs," *Extended Abstracts for the 22nd International Conference of Solid State Devices and Materials*, Sendai, Japan, pages 613 -616 (1990).

73. SEDAN Semiconductor Device Analysis, Stanford University, Stanford, California, January 1980 version.

74. These identifications do not imply recommendation or endorsement by the National Institute of Standards and Technology.

75. J. Albers, P. Roitman, and C. L. Wilson, IEEE Trans. Electron Devices **ED−30**, 1453 (1983).

76. M. Tomizawa, T. Ishibashi, H. S. Bennett, and J. R. Lowney, Extended Abstracts of the 1991 VLSI Process and Device Modeling Workshop, Oiso, Japan, May 1991 and submitted for publication.

77. D. R. Myers, J. A. Lott, J. R. Lowney, J. F. Klem, and C. P. Tigges, Proceedings of the 1990 International Electron Devices Meeting 90CH2865-4, 759 (1990).

AN INDUSTRIAL PERSPECTIVE ON SEMICONDUCTOR TECHNOLOGY MODELING

PETER A. BLAKEY* AND THOMAS E. ZIRKLE*

1. Introduction.

The central activities of semiconductor technology simulation are semiconductor process simulation and semiconductor device simulation. Process simulation is used to predict the physical structure that results from a specified sequence of processing steps. Device simulation is used to predict the electrical characteristics of a specified physical structure. Semiconductor technology simulation is often referred to as technology CAD or simply TCAD. It uses results and techniques from several traditional academic disciplines. These include applied mathematics, physics, chemistry, chemical engineering, electrical engineering, material science, and computer science.

A central problem of TCAD is communication across activities and between disciplines. The purpose of this paper is to present an industrial perspective on semiconductor technology modeling in general, and device simulation in particular, in a way that makes the concerns and priorities of industry understandable to mathematically oriented researchers. The first part of the paper provides a general technical and economic perspective on the field of TCAD. The second part focuses in more detail on technical issues associated with device simulation.

2. Technical perspectives on TCAD.

2.1 TCAD as a component of the IC design process.

The position of TCAD within the spectrum of integrated circuit design activities is depicted in figure 1. The activities that are not part of TCAD include system and subsystem design, circuit design, layout, verification of adherence to physical design rules, and verification of timing. These activities are referred to as "higher level" CAD. The information processed in higher level CAD often involves one-zero (on-off switch) level abstractions of electrical behavior, and polygon representations of physical structures. Computers are used to manage the vast quantities of individually simple information elements. In contrast, computers are used in TCAD in response to the nonlinearity and complexity of the equations required by the models.

Higher level CAD is primarily associated with the design of products to be fabricated in relatively mature production technologies. TCAD is usually associated with the development of new technologies or with process control for production technologies. Higher level CAD and TCAD overlap in the area of circuit simulation. This is a final, high- level activity of TCAD, and an initial low-level activity of higher level CAD.

---

*Modeling and Simulation Group, Motorola Advanced Technology Center, 2200 W. Broadway Road, Mesa, AZ 85202, USA.

The implementation of higher level CAD and TCAD tools often centers on issues and algorithms centered in computer science. TCAD also involves physics, chemistry, and numerical techniques. Higher level CAD is presently larger and more mature than TCAD efforts.

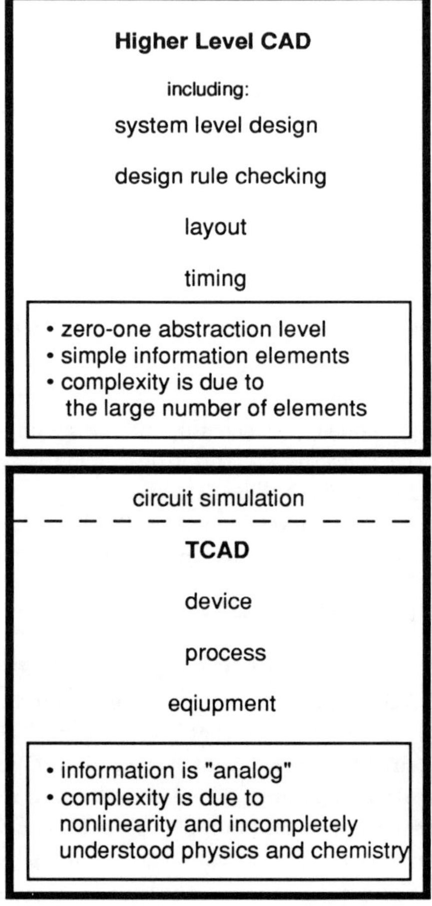

Figure 1. CAD divisions in VLSI design activities.

2.2 Conceptual models of TCAD activities.

Three useful models of TCAD are: a feed-forward information flow model, a tool development activity model, and an integration level model. Each model captures different aspects of the field.

2.2.1 The feed-forward information flow model.
Figure 2 shows a feed-forward information flow model of TCAD activities. The central activities are process and device simulation. Process simulation is sometimes divided into steps with "wafer-internal" consequences (e.g. oxidation, diffusion, and ion implantation), and steps with "wafer-external" consequences (e.g. lithography and etch). This division has, historically, influenced the functionality incorporated into individual process simulators. The output of process simulation is one key source of the structural information required as input by device simulators (other options include experimental measurements and idealized "made-up" structures). The output of device simulation is a possible source of input to parameter extraction routines used for developing device models for use in circuit simulators. This link between device simulation and circuit oriented device models is often neglected.

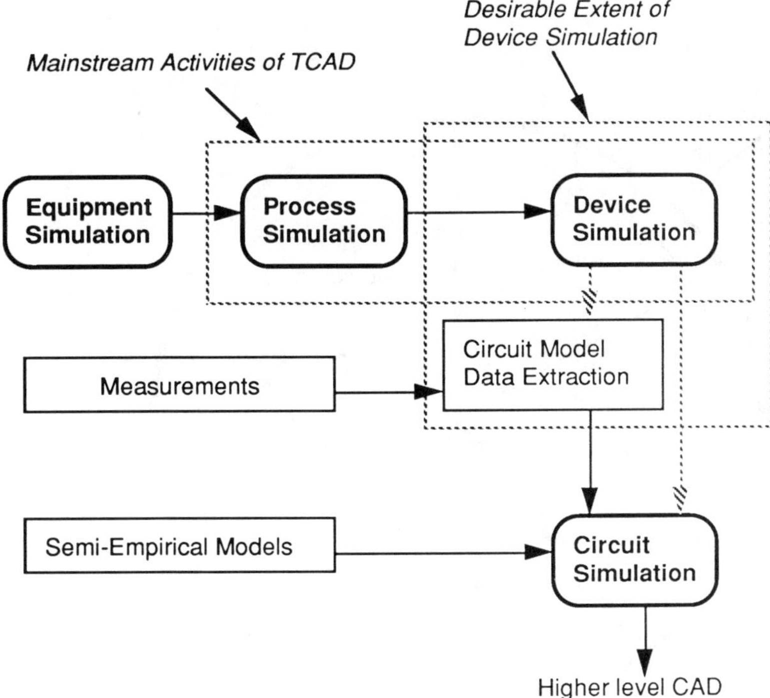

Figure 2. Information feed forward flow model.

A new aspect of process simulation is becoming important. This is concerned with mass transport to the wafer, and is commonly referred to as equipment simulation. It may focus on the "feature scale", predicting the impact of mass transport, chemical reactions and physical processes on the development of topographical features. It may also focus on the "wafer scale", yielding information about conditions as a function of the wafer position in a reactor and of position on the wafer surface. In this case, the information generated can be passed along to the "feature scale" simulators.

2.2.2 The tool development activity "pie" model. Another view of TCAD is obtained by focusing on the activities involved in implementing a TCAD simulator. Figure 3 shows these activities in the form of a "pie" whose slices align approximately with traditional academic disciplines. The natural rotation is clockwise, starting from the engineering issues that motivate the application. The physics (and perhaps chemistry) pertinent to the application must be assessed and expressed mathematically, very often as a set of coupled, nonlinear, partial differential equations. Data collection, analysis, and visualization capabilities must be considered. These activities form the design of functionality. Numerical techniques must be selected to solve the mathematical equations, and software engineering issues associated with the design and coding of the simulators must be considered. These two activities form the design of the implementation.

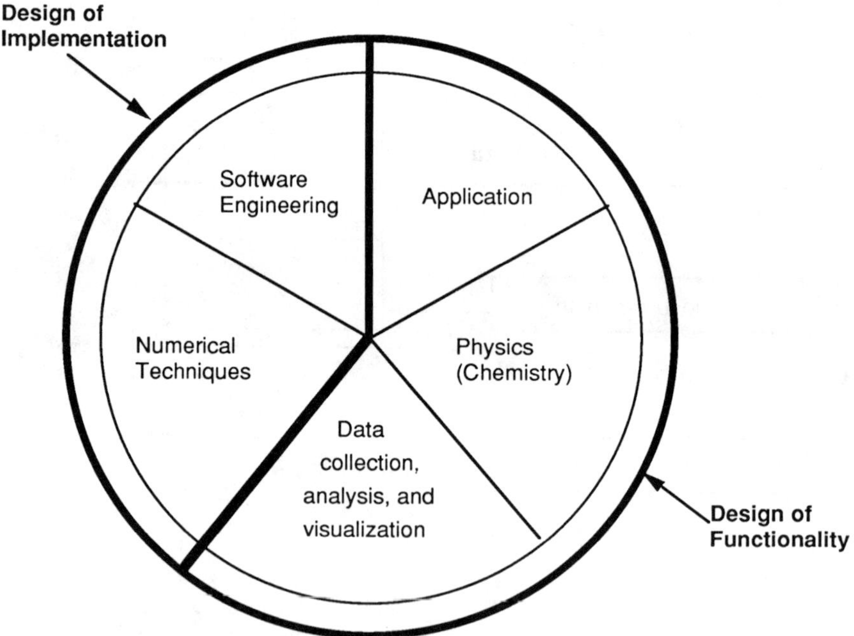

Figure 3. Tool development pie model.

The design and implementation of a well thought out simulator can have a very satisfying artistic element. However, simulator implementation seldom proceeds in a strictly linear fashion. There is iteration between the design steps as tradeoffs between generality, efficiency (accuracy), and ease of implementation are assessed. Once the simulator is written, additional physics, improved data collection, analysis, and visualization capabilities, and different numerical techniques are then often grafted on to it. The overall utility of the resulting simulator is greatly influenced by the weakest link in the chain of activities involved.

The pie model is obviously useful to those concerned with the implementation of TCAD simulators. It is also a useful representation for specialized researchers who wish to place their work in context.

2.2.3 The integration level model. Another view of TCAD focuses on the extent to which the use of multiple tools in combination is automated. A representation of this view is shown in figure 4. The innermost circle corresponds to past practice: much of the required input was in the language of the tool implementation domain, not of the application domain; each tool used its own data format, and a text oriented user interface; data transfer between tools was the responsibility of the user; and multiple run scheduling (e.g. for design and optimization) was the responsibility of the user. The middle ring corresponds to the mainstream of present day practice: interfaces between most major tools are widely available; user interfaces are becoming more graphics oriented and standardized; and there is some automatic scheduling of multiple runs for optimization and statistical design.

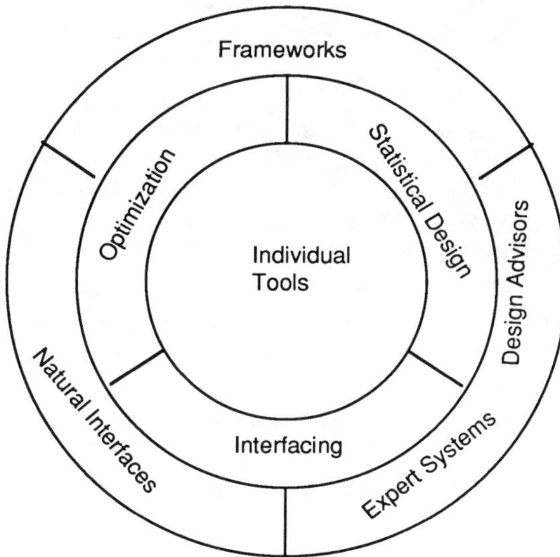

Figure 4. Tool integration model.

The outer ring corresponds to current and anticipated future developments. One clear trend is the development of frameworks. Tasks such as gridding and viewing data are common to all simulators. Duplication of effort is avoided if these capabilities are implemented just once and then used as required by a range of simulators. This promotes standardization and higher quality implementations of specialized services.

Another trend is the use of natural language interfaces. The goal is to permit users to employ the terminology of the problem domain (laboratory or equipment terms) when providing input to the simulators, in order to make the tools as accessible as possible. A natural language interface will normally be the top level of a hierarchy in which lower levels provide the user with increasing control and access to detail. Such an organization permits users to access the tools in a manner commensurate with their expertize.

A final trend is toward the the use of expert systems. The field of technology development is so broad and multi-disciplinary that no individual can have a grasp all the details. However, it is possible to capture a broad range of specialized knowledge in an expert system. This can be used to critique the design efforts of individuals or groups, automatically flagging areas of concern and suggesting possible improvements. One example would be a system that could assess reliability and degradation associated with hot carrier effects. The specialized body of knowledge associated with this issue is reasonably well developed, but most engineers in industry are not familiar with it.

3. Economic perspectives on TCAD.

TCAD activities in industry are impacted by conflicting economic considerations. The potential economic benefits of TCAD are huge, but achieving them is difficult. This section provides a perspective on these issues.

3.1 Potential benefits of TCAD.

The attraction of TCAD is that it can accelerate the development of semiconductor technologies. The basic way to develop a new technology is to perform numerous experiments to determine a sequence of processing steps and conditions that lead to satisfactory results. A previous technology normally provides a starting point for such experiments. The problems with this approach are that it is slow since each experimental cycle takes several weeks and that it is expensive, since major capital resources and considerable manpower are required to perform the experiments. (A state of the art wafer processing facility presently costs between $100 million and $1 billion, and a single development laboratory may employ several hundred engineers and technicians.) The promise of TCAD is that it can reduce the number of experiments required to develop a new technology, thereby saving both time and money. Even moderate reductions in development times can save tens of millions of dollars.

3.2 Difficulties in achieving the benefits.

It is not easy to achieve all the potential benefits of TCAD. The reasons for this include: imperfections in the tools (in both accuracy and ease of use); inappropriate use of the tools; and difficulties in measuring the benefits of simulation.

3.2.1 Tool limitations. Modern technologies typically involve more than one hundred processing steps. A present limitation is that the physics and chemistry of some processing steps is not yet well understood. Even in cases where the physics or chemistry is reasonably well understood, the model parameters are often not known for all conditions of interest. When simulating an entire process the errors and uncertainties in the simulation compound at each step. It is consequently not yet possible to obtain accurate predictions of final structures by simulating a state-of-the-art process from start to finish.

Existing tools are also relatively difficult to use, especially for process and device engineers who have many other job responsibilities. User interfaces are often crude, vary greatly from tool to tool, and require the user to have some knowledge of simulator implementation techniques as well as knowledge of the problem domain. These difficulties encourage potential users to find other means to answer their questions.

3.2.2 Inexperienced tool usage. The industrial use of TCAD tools is often rather unsophisticated. Much effort focuses on adjusting input parameters in order to match experimental data. This exercise can provide a "calibration" for subsequent "what-if" studies, and the required departures from default parameter values provide insight into how well the underlying physics is understood. However the naive adjustment of an inappropriate subset of parameters can, and often does, lead to dangerously misleading results.

A correct view of the tools can lead to useful application of simulation avoiding common pitfalls. One of the most important uses of simulation is obtaining insight. An electrical measurement can provide an exact result for a terminal current at a given bias, but gives no insight into the details of the internal distributions of charges, electric potential and electric field. Understanding the internal behavior is often the key to modifying a structure in a way which will lead to desired changes in external characteristics. Insight is also the basis for developing analytic theories and engineering "rules of thumb". Engineers become more experienced and effective as they develop insight and intuition. This process can be greatly accelerated by the appropriate use of TCAD tools. If users reject a particular tool solely because results do not match experimental practice, the potential benefit of reduced design cycle time is not achieved.

Even when the absolute numbers predicted by TCAD simulators do not match experiment very well, calculated trends may provide a useful guide in experimental or statistical design. TCAD tools can also be used to identify the approximate location of an optimum in a design space, after which experiments are used for the final optimization. There are occasions when simulation is the only source of numbers required by other parts of the design process: e.g. when preliminary circuit design is to be performed concurrently with process development; and when quantities useful in the design process are relatively simple to calculate, but difficult to measure. In these cases simulation data is used until (or unless) better data becomes available.

3.2.3 Inappropriate concentrations of tool use. The effective use of TCAD requires thought, care, and judgment. As yet too few people have the necessary combination of skills, training, and inclination. Tool use is often the responsibility of a small group of simulation specialists, but such people may be inadequately coupled to applications. The most effective class of TCAD user is normally the small minority of applications oriented engineers who have learned to use the tools well. An appropriate mission of TCAD specialists is to enlarge and support this type of user community.

3.2.4 Difficulties in measuring the benefits. The benefits obtained from using TCAD tools are seldom fully appreciated. When improved insight is obtained as a result of using TCAD or when a new idea suggested by simulation is verified experimentally, the role played by simulation is frequently forgotten or under-appreciated. Simulation often demonstrates why an apparently promising idea will not work, but there is little incentive to publicize failed ideas and so the contribution of simulation in avoiding wasted resources (manpower, time, and materials) is rarely acknowledged. Even with the effective use of TCAD, the pressure on experimental facilities is never reduced because the freed capacity is simply used for additional development efforts.

3.2.5 Miscellaneous factors. Simulation efforts suffer from pervasive misconceptions and lofty expectations. Novice users often hope that simulation will be a "silver bullet". They sometimes become overly disillusioned and negative when their initial expectations are disappointed. Senior managers seldom have personal experience of TCAD. They are easily frustrated by difficulties involved in assembling a TCAD group of adequate size and struggle when trying to define the mission and structure the activities of TCAD groups.

3.3 The situation facing industrial TCAD groups.

The activities appropriate for a TCAD group in industry fall into three categories: user service and support (the assessment, acquisition, and installation of simulators, and training and supporting engineers who use the tools); advanced use of the tools; and developing new physical models and improved solution techniques. User service and support usually absorbs most of the available effort, even of people hired because of their potential contributions to the other missions.

TCAD groups throughout the IC industry have too few people to cover all potential TCAD activities. Management knows that there are problems, but may not understand them fully. Typical responses include: repeatedly requiring justification for the continued existence of TCAD activities; reorganizing and rationalizing TCAD activities across an organization; and attempting to make TCAD groups self-supporting by commercializing the tools developed in house. None of these responses has proved effective. On the contrary, they can promote an environment that makes it harder to assemble a critical mass of competent, motivated professionals.

It is becoming clear that resource limitations preclude, for all but the largest organizations, the servicing in-house of the entire range of TCAD activities. Very few industrial groups are able or are willing to indulge in the ab initio development of major TCAD tools, an activity that requires several man years of effort with no guarantee of an eventual payoff. A trend is developing towards contracting out as many services as possible to specialized external organizations. This is especially apparent in the case of user service and support. Most codes are initially developed at U.S. and European universities. They need to be ported to different platforms, provided with better user interfaces, and enhanced in other ways. It is much more cost-effective for this work to be done once, by a specialist vendor, rather than repeatedly by every TCAD group in industry. A specialist vendor can also exploit specialization and economies of scale to do a better job of training users, fixing bugs, and other forms of user support. Reduced responsibility for "generic" support activities can free up industrial TCAD groups for company-specific applications-oriented activities.

4. A framework for assessing device simulators.

A useful framework for assessing the functionality of particular device simulators may be obtained starting from the activity "pie" model of section 2.2. Additional hierarchical decomposition of the slices of the pie leads to a checklist of implementation and functionality options. Such decompositions will be outlined for the case of device simulation for three slices: physics, numerical techniques, and software engineering. Such decompositions are not unique. The ones presented here are actually quite shallow in the sense that only one or a few layers of decomposition are performed. Even so they have proved useful for bringing structure to an otherwise complicated topic.

4.1 Physics.

Semiconductor physics provides the underlying models for device simulation. When "physics" is broadly interpreted as those factors that influence the form of the underlying equations, then spatial dimensionality and time dependency become additional, if trivial, sections of a top level decomposition. Semiconductor physics can be divided into bulk physics and surface/interface physics. Each category can be subdivided into charge transport and generation and recombination processes. The following decomposition is then obtained.

A. Semiconductor Physics

   I. Bulk Effects

      a. Charge Transport

      Quantum transport models

      Quasi-free particle approximation models

      Scattering process level models

        Monte Carlo models

      Boltzmann Transport Equation Moment Models

        Energy and Momentum conserving models

        Energy conserving and "Temperature" models

      Drift-Diffusion Model

      b. Generation and Recombination Processes

Direct recombination

Shockley-Read-Hall recombination

Auger recombination

Impurity ionization/carrier freeze-out

Thermal generation

Impact ionization

Band to band tunneling

Optical generation

II. Surface and Interface Effects

Surface scattering

Thermionic emission

Field emission

Thermionic-field emission

Quantum mechanical transmission and reflection at interfaces

Interface state trapping/detrapping

Surface state trapping/detrapping

Surface recombination and generation

B. Spatial Dimensionality

I. One Dimension

II. Two Dimensions

III. Three Dimensions

C. Time Dependence

I. DC

II. Small-signal (frequency domain perturbation analysis)

III. Large-Signal Time-Domain Transient

4.2 Numerical Techniques.

One possible top level decomposition of numerical techniques associated with device simulation is into: nonlinear solution strategy; linear subproblem solution strategy; discretization techniques; and mesh generation. Some subcategories are given below. No attempt has been made to subdivide the discretization category beyond finite difference and finite element, since it is not obvious which subcategorizations would be most useful.

A. Nonlinear Solution Technique

I. Degree of Coupling

Full Coupled

Partially Coupled

Decoupled (Gummel)

II. Damping Strategy

B. Linear Subproblem Solution Technique

I. Direct

Full Gauss Elimination

Sparse Matrix Techniques

II. Indirect

Preconditioner

Accelerator

Multi-grid strategy

C. Discretization Techniques

I. Finite Difference

- - - - - -

II. Finite Element

- - - - - -

D. Mesh/Grid Construction

I. Grid type

Tensor product

Terminated line

Triangulated

- - - - - -

II. Grid specification

User controlled

Automatic

III. Adaptive meshing techniques

Error estimation strategy

Node insertion/deletion strategy

4.3 Software Engineering.

One possible top-level decomposition of software engineering issues is into five areas: logical and physical design; implementation language; operating system; adherence to standards; and hardware environment. A second level of decomposition is outlined below.

A. Logical and Physical Design

Data Structures

Algorithms

Modules

B. Implementation Language

   Procedural (e.g. FORTRAN, C)

   Pure Object-Oriented (e.g. Smalltalk, Eiffel)

   Hybrid (e.g. C++)

C. Operating System

   Multi-platform (e.g. Unix)

   Proprietary (e.g. TSO, VMS)

D. Adherence to Standards

   GUI (e.g. OSF/MOTIF, Open Look)

   Graphics Primitives (e.g. X-Windows, GKS, PHIGS)

   Networking

E. Hardware Environment

   Scalar

   Vector

   Multiprocessor

   Massively parallel

5. The mathematical problems of TCAD.

Mathematical issues are a subset of TCAD problems. From an industrial point of view the most pressing problems are centered not so much in mathematics as in the areas of process physics and chemistry, software engineering, and tool integration. As mentioned earlier, uncertainties in physical models and in the values of parameters make it virtually impossible to obtain accurate results from "open loop" simulation of a complete state-of-the-art process. The industrial use of process simulation tools currently focuses, not surprisingly, on local "closed loop" investigations of individual process steps. This actually mirrors experimental development activities, much of which is performed as a set of decoupled subtasks. Progress in developing improved models and accurate parameters is painfully difficult and slow. More rapid progress is taking place in some other non- mathematical areas of TCAD research and development. Framework development is a particularly active area, and modern software engineering techniques are being used to make individual tools and modules more robust, reliable, extensible, and maintainable.

The more challenging and potentially most interesting mathematical problems of TCAD are in the areas of process and equipment simulation. Process simulation involves moving boundaries, the coupled nonlinear diffusion of multiple species, viscous flow, and stress. Equipment simulation adds factors such as fluid dynamics, mass transport, and chemical kinetics. Relatively little mathematical analysis has been performed of the equations that arise in process and equipment modeling. This area could be very active for the next few years. Device simulation, in contrast, has already been rather well picked over. This is especially true for the mathematics associated with the drift-diffusion approximation. There is still some scope for establishing the underlying properties of the systems of equations that result from using energy conserving models of charge transport.

Areas of applied mathematics that impact simulator implementation are always of interest. Improved iterative techniques for solving systems of nonlinear equations are needed. Improved reliability of convergence, preferably guaranteed for an arbitrary initial guess, is a very high priority and increased efficiency is also valuable. Most nonlinear problems are solved as a sequence of linear subproblems, so improved techniques for solving systems of linear systems are of interest. Increased numerical efficiency and reduced memory storage are useful and reliable convergence can be an issue for some of the systems that occur (e.g. in small-signal frequency domain perturbation analysis). Iterative methods that employ preconditioning and acceleration are well established among implementors of TCAD tools. There may be scope for greater exploitation of multigrid methods. Another important area is gridding and meshing: non-specialist users should not be required to specify and manipulate grids. Algorithms and data structures for adaptive gridding are consequently of great interest as are algorithms for related tasks such as the constrained Delauney triangulation of nonconvex regions with internal boundaries. There is also interest in improved discretization schemes that provide more attractive tradeoffs between accuracy, efficiency, and generality.

Much high quality mathematical research is never exploited in applications because it is demonstrated only for a simple, highly idealized, "toy" problem. Numerical techniques are more likely to be exploited when they are demonstrated for a nontrivial class of problems, and within the framework provided by a popular TCAD tool. This also facilitates benchmarking and direct comparisons between different techniques. It may be uninteresting for a researcher to spend days or weeks becoming familiar with an existing code, but the vastly increased probability of the underlying research being exploited means that funding agencies should strongly encourage this.

TCAD requires participants to straddle traditional academic boundaries. The communication barriers between mathematicians, engineers, physical scientists, and computer scientists are high. Specialists must learn at least some of the language of the disciplines they interact with. From an industrial point of view the problem with TCAD is not a shortage of specialists in the subfields, but rather a shortage of generalists to perform the syntheses involved in implementing and using TCAD tools.

6. Conclusions.

The field of semiconductor technology modeling has been reviewed in an attempt to provide a context for the mathematical research that impacts TCAD. Some economic and managerial issues were discussed because these influence the selection, prioritization, and funding of research activities. The overview provided here may also be of use to, and facilitate communication among, the many other constituencies involved in TCAD research, development, and applications.

7. Acknowledgments.

The opinions expressed in this paper are those of the authors, but many of the insights were obtained with the assistance of colleagues at Motorola. It is therefore a pleasure to acknowledge useful technical discussions with Dr. Len Borucki, Mr. Ik-Sung Lim and Dr. Kuntal Joardar of the Advanced Technology Center, and with Dr. Marius Orlowski, Dr. Ravi Subrahmanyan and Mr. Matthew Noel of the Advanced Products Research and Development Laboratory. The authors are also grateful to Dr. Ed Hall and Dr. Charlie Meyer for strong managerial support and interesting discussions on a broad range of general issues.

COMBINED DEVICE-CIRCUIT SIMULATION FOR ADVANCED SEMICONDUCTOR DEVICES

J. F. BÜRGLER*, H. DETTMER*, C. RICCOBENE*
W. M. COUGHRAN, JR.[†], AND W. FICHTNER*

Abstract. To develop and optimize semiconductor devices it is common practice to advocate numerical device modeling by using simulation programs. Typically the simulation of a unit is split into levels, starting with process simulation (to study the effect of wafer-processing steps) and device simulation (to study the electrical behavior), to circuit simulation (to study the electrical features of an ensemble of devices), leading possibly to the logic simulation of a whole unit. Unfortunately, this approach makes it very hard to model the interaction of the device and the circuit in the transient case: therefore, in order to optimize the transient behavior of a particular device it is necessary to include the effects of both the drive circuit (i.e., control part), as well as the load circuit. During the last year, we have developed a software environment for combined device-circuit simulation studies dedicated to semiconductor power devices, magnetic field sensors, and BiCMOS structures augmented with simple circuits. In our work we discuss some critical points in detail. These include the grid generation and adaptation, the scaling of the unknowns (i.e., how to deal with high voltages), the numerical procedures used to solve the transient problem (with special attention to supercomputer architectures), the assembly of the device (element assembly versus edge assembly) and circuit equations as well as the physical models used. We show that it is essential to properly chose the numerical techniques to obtain accurate and reliable results. Studies of the turn-off of power devices, the switching of BiCMOS structures, and of magnetic field sensors indicate the success of the current approach.

1. Introduction. Very large scale integration (VLSI) has evolved at an enormous rate, progressing from hundreds of components on an integrated circuit (IC) in the 1960s to more than one million devices today. This staggering advance in complexity can be attributed to two major factors: the innovations in microelectronics technology, and the progress in the development and availability of computer-aided design (CAD) tools.

In the application of CAD tools, simulation plays a very important role. For all levels of the CAD hierarchy — architecture to logic, circuit and layout - simulation has become a universally accepted procedure in analysis or synthesis. At the physical level, the use of software tools in the development of new processes and novel device structures present a particularly worthwhile alternative to the experimental trial-and-error approach. These tools are commonly divided into process, device and circuit simulation packages, depending on their area of applicability. The results obtained from computer calculations provide the user with considerable insight in to the physics of processing, device behavior and circuit performance.

For many microelectronic applications, the behavior of a critical component is not simply determined by the performance of the underlying semiconductor device, but by the interactions of this device with its environment. A typical example for this phenomenon might be the switching behavior of a semiconductor power device (e.g., a thyristor). During turn-on and turn-off, the current-voltage relation of this thyristor is completely governed by the external circuitry, in particular the performance of the

---

* Swiss Fed. Inst. of Technology, 8092 Zurich, Switzerland
† AT&T Bell Laboratories, Murray Hill, NJ 07974, USA

free-wheeling diode and the parasitic inductances of the wiring and the load. Another important example concerns the design and optimization of I/O pad structures for VLSI circuits. In this case, the influence of the external elements such as the bonding wire and the inductance of the package are of major concern.

The use of software tools for the simulation of combined device/circuit structures is by no means a new trend. Some years ago, the MEDUSA [1] program from RWTH Aachen was announced that allows the numerical modeling of one-dimensional bipolar devices. These devices can be embedded into external circuits made up of lumped elements. More recently, the well-known two-dimensional device simulator PISCES [2] has been extended by several groups in both academia [3,4] and industry [5].

In this paper, we give a detailed description of our approach to combined circuit/device simulation. The paper is organized as follows: the principles of device simulation are illuminated in section 2. Section 3 discusses circuit simulation in detail and section 4 combines both into one common frame. Some numerical aspects are discussed in section 5 and the results in section 6 show the power of the approach chosen. Finally, section 7 draws some conclusions.

2. Device simulation: Equations and discretization. We are considering semiconductor devices in a range of operation where the drift-diffusion model [6] is appropriate. This is true for devices operated at room temperature with characteristic dimensions on the order of microns. Typically this applies to bipolar devices with base lengths in the range mentioned, as well as to, i.e., MOSFETs with channel lengths in the same range. In particular the model is applicable to power devices.

The drift-diffusion model can easily be derived as the lowest order moments, in fact the first order moments, of the Boltzmann transport equation for the distribution function for the electrons and holes respectively [7]. We find for the current continuity equations for both electrons and holes

$$(1) \qquad -\nabla \cdot \mathbf{J}_n + q \left(R + \frac{\partial n}{\partial t} \right) = 0,$$

$$(2) \qquad \nabla \cdot \mathbf{J}_p + q \left(R + \frac{\partial p}{\partial t} \right) = 0.$$

Here we introduced the electron and hole current densities $\mathbf{J}_n$ and $\mathbf{J}_p$, as well as the electron and hole densities n and p, respectively; q is the absolute value of the electron charge and R the generation-recombination rate.

Using either thermodynamic considerations [8] or the approximate solution to the Boltzmann transport equation [9], the current relation can be obtained by appropriately integrating over $\mathbf{k}$-space the respective group velocity times the charge to yield in the case of a weak external magnetic field

$$(3) \qquad \mathbf{J}_n + \mu_n^\star \mathbf{J}_n \times \mathbf{B} = q \left(\mu_n\, n\, \mathbf{E}_n + D_n\, \nabla n \right),$$

$$(4) \qquad \mathbf{J}_p + \mu_p^\star \mathbf{J}_p \times \mathbf{B} = q \left(\mu_p\, p\, \mathbf{E}_p - D_p\, \nabla p \right).$$

Here we introduced the effective electric field $\mathbf{E}_n$ ($\mathbf{E}_p$) to account for degeneracy effects, the mobility and diffusivity μ_n (μ_p) and D_n (D_p) for electrons and holes, respectively. Diffusivity and mobility are coupled via the Einstein relation $D_n =$

$kT\,\mu_n/q$ $(D_p = kT\,\mu_p/q)$. $\mu_n^\star$ $(\mu_p^\star)$ denotes the Hall mobility for electrons which can be written as $\mu_n^\star = \mu_n\,r_n$ $(\mu_p^\star = \mu_p\,r_p)$ where r_n (r_p) represents the Hall scattering factor which is 1.1 (0.7) for electrons (holes). The effective electric fields for electrons and holes are given by

$$(5) \qquad \mathbf{E}_n \;=\; \mathbf{E} - \frac{kT}{q}\nabla\left(\ln n_{ie,n}\right),$$

$$(6) \qquad \mathbf{E}_p \;=\; \mathbf{E} + \frac{kT}{q}\nabla\left(\ln n_{ie,p}\right),$$

with the corresponding effective intrinsic densities $n_{ie,n}$ and $n_{ie,p}$.

The electric field in the presence of a charge distribution is determined by the Poisson equation

$$(7) \qquad -\nabla\cdot\left(\epsilon\,\nabla\psi\right) \;=\; -q\left(n - p - N\right),$$

where ϵ denotes the dielectric constant, ψ is the electrostatic potential determining the electric field $\mathbf{E} = -\nabla\psi$, and N is the net doping concentration. With the appropriate boundary conditions the system (1)–(4) and (7) can be solved uniquely. The boundary conditions for homogeneous Neumann boundaries are

$$(8) \qquad \mathbf{E}\cdot\mathbf{n} \;=\; \sigma\,R\,(\mathbf{B}\times\mathbf{E})\cdot\mathbf{n} + \frac{\sigma_n^2\,R_n}{\sigma}\,(\mathbf{B}\times\nabla\xi_n)\cdot\mathbf{n} - \frac{\sigma_p^2\,R_p}{\sigma}\,(\mathbf{B}\times\nabla\xi_p)\cdot\mathbf{n},$$

$$(9) \qquad \mathbf{J}_n\cdot\mathbf{n} \;=\; 0,$$

$$(10) \qquad \mathbf{J}_p\cdot\mathbf{n} \;=\; 0.$$

Here, $\mathbf{n}$ denotes the outward unit normal on the the boundary of the simulation domain. For Dirichlet boundary parts

$$(11) \qquad \psi \;=\; \psi_{appl} + \frac{kT}{q}\,\sinh^{-1}\left(\frac{N}{2\,n_{ie}}\right),$$

$$(12) \qquad n \;=\; \sqrt{\frac{N^2}{4} + n_{ie}^2} + \frac{N}{2},$$

$$(13) \qquad p \;=\; \sqrt{\frac{N^2}{4} + n_{ie}^2} - \frac{N}{2},$$

where $\sigma = \sigma_n + \sigma_p$ with $\sigma_n = q\,\mu_n\,n$ and $\sigma_p = q\,\mu_p\,p$ denotes the ambipolar electric conductivity, $R = \left(\sigma_n^2\,R_n + \sigma_p^2\,R_p\right)/\sigma^2$ is the ambipolar Hall coefficient with $\sigma_n\,R_n = -\mu_n\,r_n$ and $\sigma_p\,R_p = \mu_p\,r_p$ (see [10]). Furthermore $\xi_n = \psi - \phi_n$ and $\xi_p = -(\psi - \phi_p)$ with the quasi-Fermi level ϕ_n and ϕ_p for electrons and holes given by

$$(14) \qquad n \;=\; n_{ie,n}\exp\frac{q\left(\psi - \phi_n\right)}{kT},$$

$$(15) \qquad p \;=\; n_{ie,p}\exp\frac{-q\left(\psi - \phi_p\right)}{kT}.$$

Through this relation the equations (1)–(4) and (7) comprise a highly nonlinear system of coupled partial differential equations whose solution, in general can be found only numerically. Because of the ever shrinking dimensions of typical semiconductor devices used in integrated circuits like MOSFETs, DRAMs, etc., three-dimensional (3-d)

TABLE 1
Scaling factors for basic units.

description	scaling factor		
	symbol	numerical value	unit
Length	l_i	3.358×10^{-3}	cm
Mass	$qU_T n_i l_i^5$	2.619×10^{-20}	g
Time	l_i^2	11.275×10^{-6}	s
Current	$qn_i l_i$	7.969×10^{-12}	A

effects become more and more important [11]. However, a wide variety of devices can still be considered as essentially 2-d; examples are sensor and high power devices with typical dimensions from several to a few hundred microns. In the following we concentrate on the 2-d case; however, most of the results can be applied in a straight forward manner to the 3-d case. Special attention must be given to grid generation [12,13] and the numerical solution of linear systems of equations [14,15].

The physical parameters like the mobilities, the generation/recombination rates, and the effective intrinsic densities are taken from standard physical models as described in [16].

It is common practice and highly recommended from a numerical point of view to use scaled variables instead of the ones used in (1)–(7). We apply the well known scaling procedure after de Mari [17,18], where time is scaled like the length squared, in order to get a dimensionless diffusivity, the voltages are scaled by the thermal voltage $U_T = kT/q$, the charge densities are scaled by the intrinsic charge density qn_i, and the length is scaled by the Debye length l_i. Table 1 represents the scaling factors for the basic units from which the derived quantities can be easily computed.

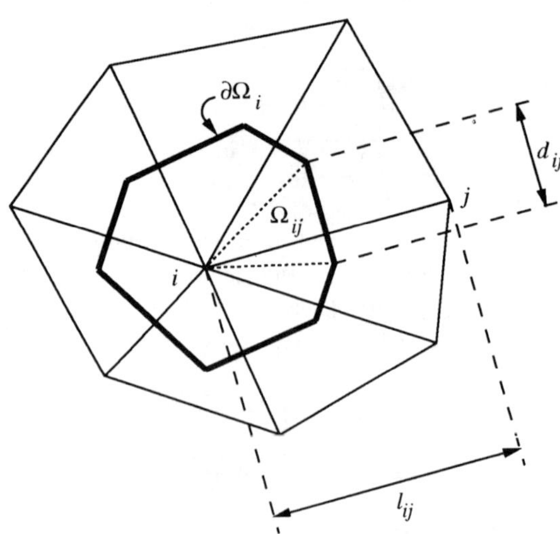

FIG. 1. *Box around node i.*

Several discretization schemas have been proposed to solve the system of equations

(1,2,7) numerically [19,20]. For the range of operation we are primarily interested in (high-voltage, high-current regimes) we decided to apply the well known box method [21,22,23] with current densities along the edges according to 1-d Scharfetter and Gummel scheme [24].

In applying the box method the simulation domain is first triangulated, i.e., split into a disjoint union of triangles. Our grid generator [25] creates only nonobtuse triangles, i.e., triangles with angles smaller or equal than $\pi/2$. This avoids the obtuse angle problem well known in literature [2]: unphysical spikes in the solution can be avoided. The dual grid, also called Voronoi grid [26], represents the boxes around each node in the triangulation (Fig. 1).

To show the complexity of the grid we use a BiCMOS structure with the doping distribution in Fig. 2. On the left of the trench an npn bipolar transistor and p-channel MOSFET (far left) are shown. On the right, next to the trench, there is a pnp bipolar transistor and an n-channel MOSFET (far right). The grid used for the computation is refined according to the doping concentration (Fig. 3). A blow-up of the left MOSFET is shown in Fig. 4.

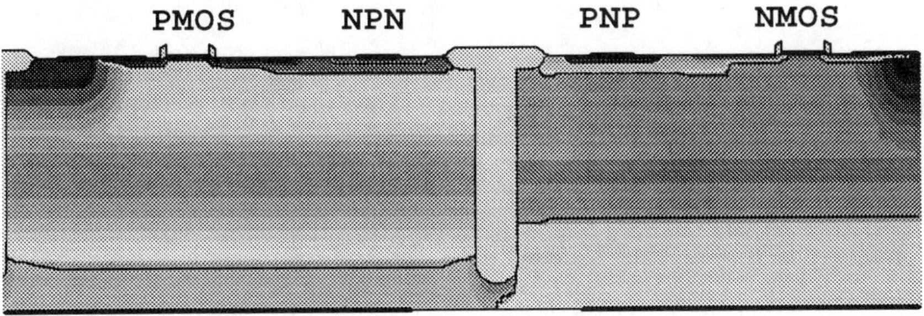

FIG. 2. *Structure and doping distribution for BiCMOS.*

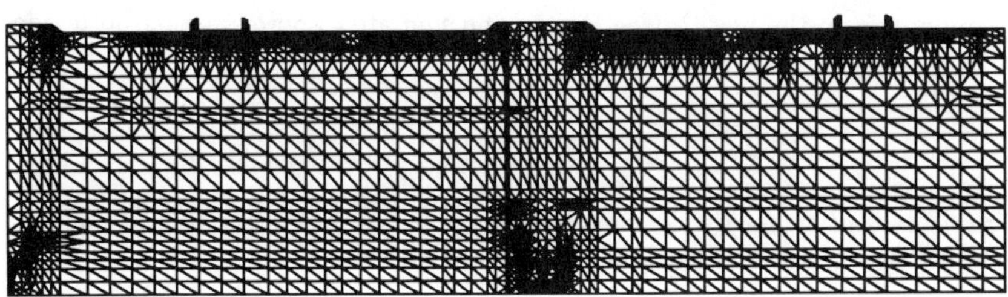

FIG. 3. *Grid for BiCMOS.*

This example shows that it is absolutely necessary to have a grid with locally varying point density. Here, and also in power device simulation (see section 6 for a MCT example), the ratio of the largest to the smallest edge in a reasonable triangulation can easily be on the order of 10^4–10^5. This, together with the large number of grid points (up to 10^4), has a significant influence on the condition number of the Jacobian matrices [27].

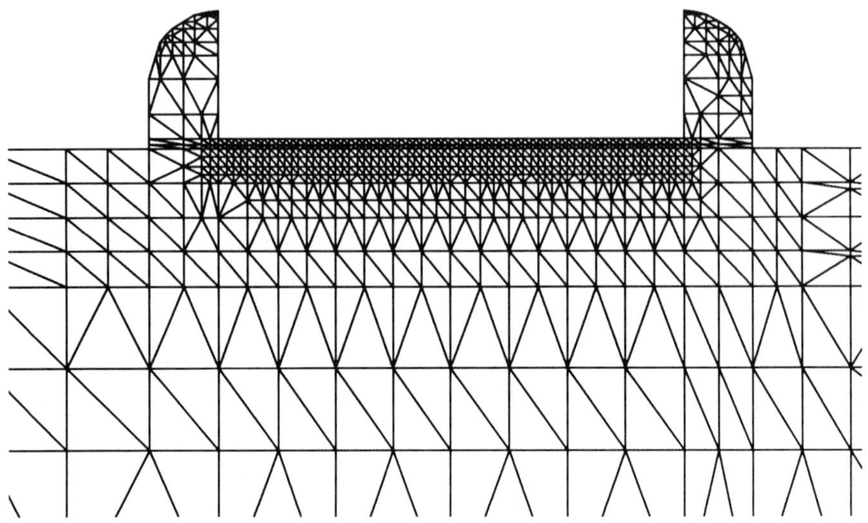

FIG. 4. *Blow-up of grid for left* MOSFET *of* BiCMOS.

The box method can be applied to any conservation law in differential or integral form such as (1), (2), and (7) by applying the Gauss theorem to convert the integral of a divergence of a vector field over some volume to a surface integral. In particular for the electron continuity equation (3) the following equations are equivalent

$$
(16) \qquad -\nabla \cdot \mathbf{J}_n + q\left(R + \frac{\partial n}{\partial t}\right) \;=\; 0 \quad \text{in } \Omega \subset \mathbb{R}^2,
$$

$$
(17) \qquad -\int_\Omega \nabla \cdot \mathbf{J}_n \, d\Omega + q \int_\Omega \left(R + \frac{\partial n}{\partial t}\right) d\Omega \;=\; 0, \quad \text{and}
$$

$$
(18) \qquad -\int_{\partial\Omega} \mathbf{J}_n \cdot d\mathbf{n} + q \int_\Omega \left(R + \frac{\partial n}{\partial t}\right) d\Omega \;=\; 0.
$$

We apply this to the box Ω_i (see Fig. 1) and find after approximating both sides of (18)

$$
(19) \qquad -\sum_\tau \sum_{j \neq i} J^\tau_{n,ij} \, d^\tau_{ij} + q\,\mu(\Omega^\tau_i)\left(R^\tau_i + \frac{dn_i}{dt}\right) \;=\; 0.
$$

Here we used $\mu(\Omega^\tau_i)$ for the the area of the intersection of the box Ω_i and the triangle τ and R^τ_i for the corresponding source term at node i.

The flux $J^\tau_{n,ij}$ along the edge from node i to node j in triangle τ is obtained under the usual assumptions of Scharfetter and Gummel [24] as

$$
(20) \qquad J^\tau_{n,ij} \;=\; q\frac{\mu^\tau_{n,ij}}{l_{ij}}\left[n_j\, B(u_{ji}) - n_i\, B(u_{ij})\right],
$$

where $B(x) = x/(\exp x - 1)$ denotes the Bernoulli function. Moderate modifications have to be made if $\mathbf{B} \neq 0$ [28]. Unfortunately, the discretization scheme proposed therein does not conserve the current; a property strongly desired by engineers. However, our computations have shown that this does not significantly influence the results for weak fields (up to 1 Tesla in Si): the current is still conserved with an error smaller than a fraction of a percent with respect to the terminal currents.

The discretization of the Poisson equation (7) is similar: the component of the field along an edge is approximated by $E_{ij} = u_{ij}/l_{ij}$, where $u_{ij} = u_i - u_j$.

After applying the discretization procedure outlined above to the equations (3), (4), and (7), we end up with a system of first order ordinary differential equations of the form

$$(21) \qquad \mathbf{F}_d(t) = \frac{d}{dt}\mathbf{q}_d(t) + \mathbf{f}_d(t) \;\;=\;\; 0,$$

where the individual components of $\mathbf{F}_d$ correspond to the electron and hole continuity equations and the Poisson equation for each node in the semiconductor device; $\mathbf{f}_d$ represents the steady-state part, and $\mathbf{q}_d$ denotes the densities.

3. Circuit simulation: Equations. So far we have considered the semiconductor device with simple boundary conditions. The next step is to briefly review the basics in circuit simulation in order to tackle the combined device-circuit problem.

When dealing with circuits, it is very helpful to introduce the notion of a graph; this is a set of nodes and edges where the edges are connected to the nodes. If the edges are directed we have a directed graph. Each circuit can be represented by a directed graph (the direction of an edge defines the direction of the current flow) (see [29, page 76ff]). Corresponding to a directed graph, there is an incidence matrix, i.e., a unique description of the graph, equivalent to the graph itself. Figure 5 shows a simple graph with four nodes and six directed edges. Without loss of generality we assume in the sequel that node zero is grounded. Therefore, because the voltage is known, it is no longer necessary to consider this node.

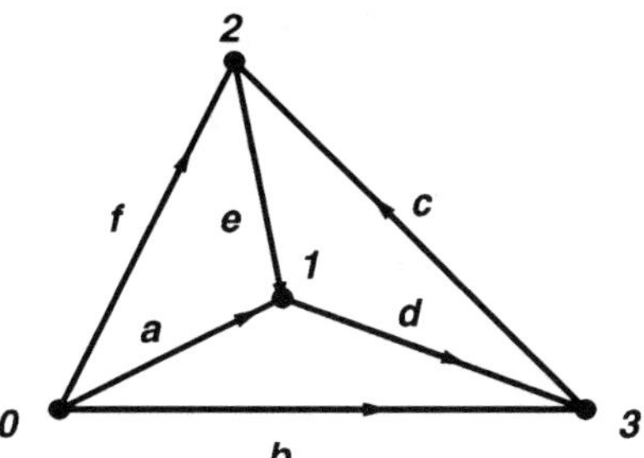

FIG. 5. *Directed graph of simple circuit.*

The incidence matrix consists of as many rows as there are nodes (excluding node zero) and of as many columns as there are edges in the graph. If edge i points from node j to node k we put one in column i, row j and minus one in column i, row k.

For the graph in Fig. 5 we obtain the following incidence matrix

$$
(22) \qquad A \;=\; \begin{matrix} 1 \\ 2 \\ 3 \end{matrix}
\begin{pmatrix}
 & a & b & c & d & e & f \\
-1 & 0 & 0 & 1 & -1 & 0 \\
0 & 0 & -1 & 0 & 1 & -1 \\
0 & -1 & 1 & -1 & 0 & 0
\end{pmatrix}.
$$

The physical laws that govern the behavior of a circuit are the Kirchhoff current law (KCL), a direct consequence of charge conservation $\nabla \cdot \mathbf{J} = 0$, the Kirchoff voltage law (KVL), a direct consequence of the Maxwell equation $\nabla \times \mathbf{E} = -\partial \mathbf{B}/\partial t$, and the constitutive equations for the circuit elements, which describe their behavior, i.e., voltage-current relations. It is easy to verify that these equations can be written as

$$
(23) \qquad
\begin{pmatrix}
1 & 0 & -A^T \\
Y_b & Z_b & 0 \\
0 & A & 0
\end{pmatrix}
\begin{pmatrix}
\mathbf{V}_b \\ \mathbf{I}_b \\ \mathbf{V}_n
\end{pmatrix}
=
\begin{pmatrix}
0 \\ \mathbf{W}_b \\ 0
\end{pmatrix},
$$

which is known as the sparse tableau formulation in literature [29,30]. The unknowns in this formulation are the branch voltages, $\mathbf{V}_b$, the node voltages, $\mathbf{V}_n$, the branch currents, $\mathbf{I}_b$, and the term arising from i.e., independent current sources, $\mathbf{W}_b$. In general, (23) comprises a differential-algebraic system of equations with index not necessarily one [31]. We can easily reduce the number of unknowns by substituting the branch voltages by the nodal voltages with

$$
(24) \qquad \mathbf{V}_b = A\mathbf{V}_n
$$

to get

$$
(25) \qquad
\begin{pmatrix}
Y_b A^T & Z_b \\
0 & A
\end{pmatrix}
\begin{pmatrix}
\mathbf{V}_n \\ \mathbf{I}_b
\end{pmatrix}
=
\begin{pmatrix}
\mathbf{W}_b \\ 0
\end{pmatrix}.
$$

If we can further eliminate the branch currents, i.e., if there are only voltage controlled circuit elements, then we end up with the well-known nodal formulation

$$
(26) \qquad A Y_b A^T \mathbf{V}_n \;=\; -A \mathbf{J}_b.
$$

The circuit elements which we can handle in the nodal formulation are for example; resistors, capacitances, independent current sources, voltage controlled current sources, etc.

To overcome this limited number of circuit elements, modified nodal formulation was introduced: here we distinguish between good elements, i.e admissible elements in nodal formulation, and bad elements, whose current can not be expressed in terms of the nodal voltages. For those elements, the branch currents are kept in the formulation as additional unknowns.

By elimination or conversion of inadmissible elements into equivalent circuits of admissible elements, nodal formulation can still be advocated. This technique is used in CAzM [32] and described in [30] in detail for independent current sources and inductors: in the first case the current and one nodal voltage are eliminated, in the second case an equivalent circuit consisting of two voltage-controlled current sources and one capacitor is constructed.

4. Combining device and circuit simulation. So far we have dealt with the device and the circuit separately. This section combines both parts by the appropriate handling of the interface between the two parts.

Typically, a device is connected to an external circuit only via its contacts. In order to by able to model physically relevant contacts we introduce the possibility to have nonideal contacts with a distributed resistance and capacitance. The contact node, c, is therefore connected to a particular node on the device, l, via R_l and C_l, respectively (see Fig. 6). Depending on the size of R_l (or C_l) the interface nodes l are no longer on an equipotential line.

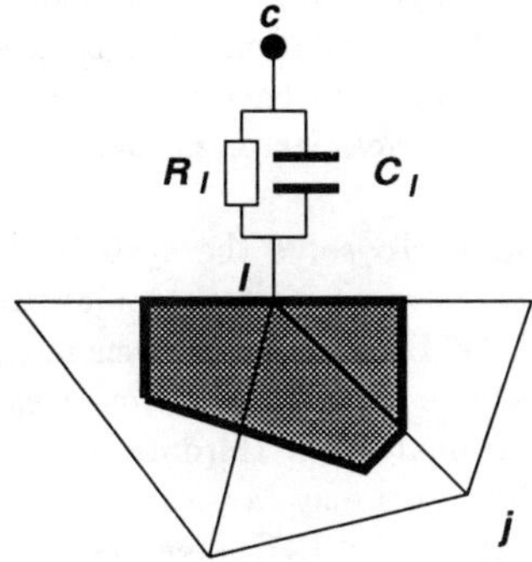

FIG. 6. *Interface node l connected to contact c.*

The equations for the combined device-circuit problem consist of the Poisson and the electron and hole current continuity equations of the internal and homogeneous Neumann nodes of the device, the current continuity equations for the interface nodes, i.e., nodes on contacts connected to external circuit elements, and the circuit equations

$$(27) \qquad \mathbf{F}_d\left(\mathbf{z}_d, \mathbf{z}_i\right) \;=\; \frac{d}{dt}\mathbf{q}_d\left(t\right) + \mathbf{f}_d\left(\mathbf{z}_d, \mathbf{z}_i\right) = 0$$

$$(28) \qquad \mathbf{F}_i\left(\mathbf{z}_d, \mathbf{z}_i, \mathbf{z}_c\right) \;=\; \frac{d}{dt}\mathbf{q}_i\left(t\right) + \mathbf{f}_i\left(\mathbf{z}_d, \mathbf{z}_i, \mathbf{z}_c\right) = 0$$

$$(29) \qquad \mathbf{F}_c\left(\mathbf{z}_i, \mathbf{z}_i\right) \;=\; \frac{d}{dt}\mathbf{q}_c\left(t\right) + \mathbf{f}_c\left(\mathbf{z}_i d, \mathbf{z}_c\right) = 0$$

Here, the subscripts d, i, and c refer to the device, the interface, and the circuit, respectively. The Jacobian matrix, neglecting transient contributions, has the form

$$(30) \qquad \begin{pmatrix} \dfrac{\partial \mathbf{f}_d}{\partial \mathbf{z}_d} & \dfrac{\partial \mathbf{f}_d}{\partial \mathbf{z}_i} & 0 \\[2ex] \dfrac{\partial \mathbf{f}_i}{\partial \mathbf{z}_d} & \dfrac{\partial \mathbf{f}_i}{\partial \mathbf{z}_i} & \dfrac{\partial \mathbf{f}_i}{\partial \mathbf{z}_c} \\[2ex] 0 & \dfrac{\partial \mathbf{f}_c}{\partial \mathbf{z}_i} & \dfrac{\partial \mathbf{f}_c}{\partial \mathbf{z}_c} \end{pmatrix}$$

The Jacobian matrix emerging from the linearization procedure is structurally, but usually not numerically, symmetric and eventually contains zero diagonal elements in the circuit diagonal block, $\partial \mathbf{F}_c / \partial \mathbf{z}_c$. Gaussian elimination without pivoting would therefore not work unless an approximate Newton method with diagonal damping [33], i.e., two parameter damping is used.

Currently, we use a block elimination strategy, where the critical circuit diagonal block is solved by a Gaussian elimination with partial pivoting; in addition a number of back-solves with the device diagonal block, $\partial \mathbf{F}_c / \partial \mathbf{z}_c$, have to be made. This is reasonable as the number of unknowns in the circuit is typically less than 100 and the number of back solves is usually less than 10. Furthermore, when advocating a direct solver, as is typically done in 2-d simulations, the back solves do not significantly increase the run time as the forward and backward substitution is at least one order of magnitude faster than the numerical factorization.

5. Numerical techniques. To solve the system of stiff ordinary differential equations (27)–(29), we apply a composite trapezoidal-backward differentiation formula second-order method (TR-BDF2) [34]. This scheme is a second order, L-stable, one step method, which is easy to restart. The time step is controlled by the local truncation error (LTE) proportional to the third derivative of $\mathbf{q}$ and is approximated by divided differences (see [35]). This may cause severe errors in the LTE near steady state solutions. As an alternative, the BDF step could be recomputed using a TR step and then Milne's device [36] can be employed as a LTE estimator [34].

In steady-state and for each TR or BDF step a nonlinear system of equations has to be solved. Here, they are solved by means of a damped Newton strategy either applying parameter selection after Bank and Rose [33] or Deuflhard and others [37].

In each iteration of the damped Newton method a system of linear equations arises. They are solved either by sparse Gaussian elimination or by an iterative method. In the first case two different methods can be applied; a standard sparse matrix package such as the one in PLTMG [38] or the supernode code [39] especially designed to perform well on Cray supercomputers. Whereas the first method applies a minimum degree ordering the second method reorders in supernodes which can then be processed in parallel. The iterative solution methods are based on the package for iterative linear solvers (PILS) [40] which allows to choose between various preconditioners and iterative solvers. A comparison between the cpu-time used for different solvers is depicted in Table 2. This corresponds to a steady-state, coupled solution of a MOSFET with $V_{gs} = 5.0$V and $V_{ds} = 0.1$V. The initial guess is the solution to slightly changed $V_{ds} = 0.05$V. The solution variables are the electrostatic potential and the electron and hole quasi-Fermi levels, respectively. The Jacobian matrices are so badly conditioned that only numerical dropping with a tolerance of $\varepsilon = 0.01$ combined with CGSTAB [41] converged. Thanks to the good preconditioner, typically only some ten iterations have been necessary. Note that the total time includes the assembly, as well as the solution time of the linearized system for a total of four Newton iterations; however, the assembly time is the same for all methods.

This is one of the cases we found direct methods superior to iterative methods. However, in most situations iterative linear solvers are more efficient. Also the number of unknowns is quite small and we are certainly below the crossover point of the efficiency of direct and iterative methods.

Total Cray Y-MP uniprocessor cpu-time for one nonlinear solve.

	total cpu-time in s		
nodes	smp	super	PILS
1173	5.991	4.137	9.833
2204	13.982	7.570	17.695
4008	26.445	14.723	43.163
8041	74.129	33.214	126.950

6. Illustrative examples. We first consider a bipolar device sensitive to external magnetic field. Devices like this are used for earth magnetic field measurements, the reading of magnetic tapes and disks or the recognition of magnetic ink patterns of bank notes and credit cards. Other applications include contactless switching, linear and angular displacement detection, potential-free current detection, and integrated watt-meters. The device considered here is a vertical bipolar transistor.

Fig. 7 depicts the doping concentration in a cross section of the symmetrically arranged bipolar transistors. Because of the Lorentz force, the charged carriers are deflected on their path from emitter (E) to the collectors (C1, C2) or vice versa. This causes different collector currents. The difference is a measure of the strength of the applied magnetic field.

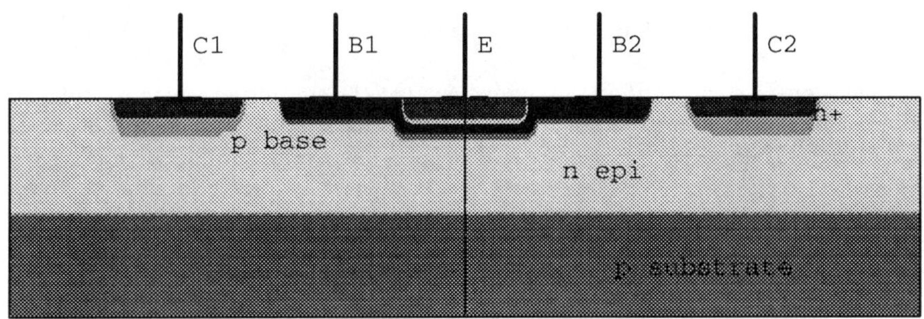

FIG. 7. *Structure and doping concentration of magneto transistor.*

The grid for the left half of the device used for the numerical simulation is depicted in Fig. 8. The total grid consists of 3479 vertices and 6718 triangles. A constant current of 10μA was assumed at the otherwise floating base contacts (B1, B2). The distributed resistance at these contacts was assumed negligible. Therefore, two independent current sources constituted the external circuit. The emitter, substrate, and collector voltages applied were 0, 0, and 5V, respectively.

The simulation program was run on a Cray Y-MP or on a Convex C200 both with 2 processors. The solution time for one particular working point — fixed magnetic field — varied from 8s to 14s on a Cray Y-MP and from 120s to 260s on a Convex C200.

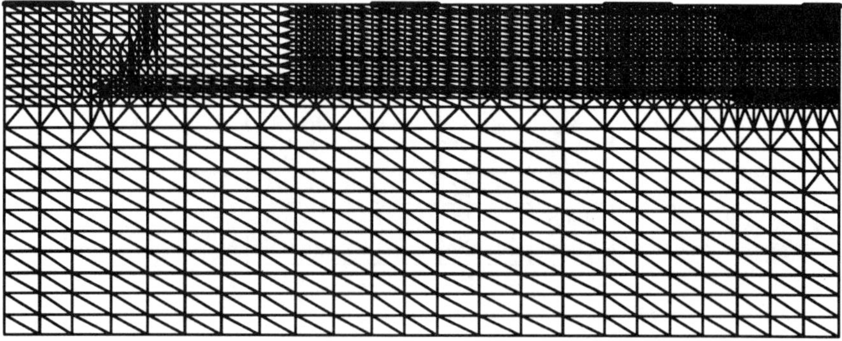

FIG. 8. *Left half of grid for vertical magneto transistor.*

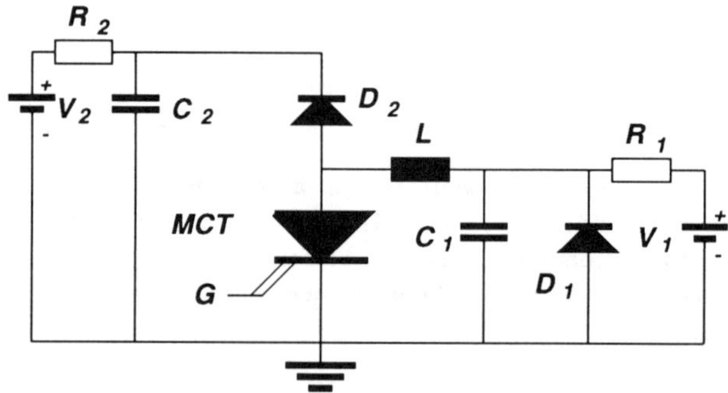

FIG. 9. MCT *attached with test circuit.*

Simulations like this verified the conjecture that the emitter efficiency modulation is responsible for the sensitivity of the vertical Hall device [10]. This is in contrast to the horizontal Hall device, where carrier deflection due to the Lorentz force plays the major rule.

The second example demonstrates turn-off simulations of an MOS controlled thyristor (MCT). An MCT is a four layer structure, i.e., two diodes in series, with integrated MOS transistors to switch the thyristor on or off. Power devices embrace a wide spectrum of different applications, ranging from conventional analog circuits that are operated at non-standard voltage levels, to smart power circuits, and to circuits composed of parallel arrangements of similar devices, such as the MCT. This device is hoped to be the future replacement of the gate turn-off thyristor (GTO) which has serious short-comings such as the need for strong protection and gate drive circuits and the restrictions in frequency. Typically several thousand identical cells are integrated on a wafer. Figure 10 depicts the doping concentration of the top 10μm of the MCT; the simulated structure extends to the full thickness of the wafer, i.e., to 500μm.

The vastly different feature sizes in an MCT establish a tremendous challenge to the numerical techniques and algorithms used. Not only do we have ratios of largest

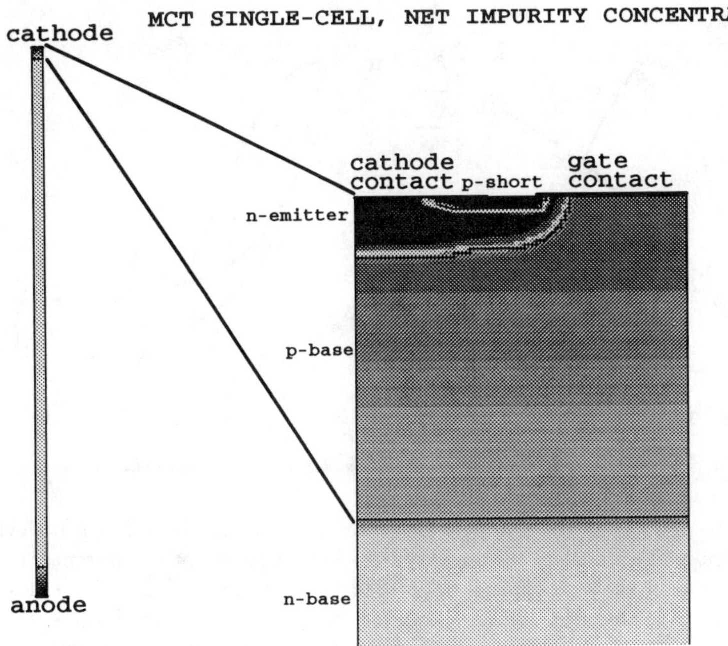

FIG. 10. *Doping concentration in top region of* MCT.

to smallest edge lengths up to five order of magnitudes, but also a large number
of grid points (up to 10^4), and in addition, drastically changing physical variables.
Remember that, on top of the MCT, MOSFETs with typical gate lengths of a few
microns are used for controlling the device. Also, in the off-state, the blocking voltage
is on the order of several thousand volts, whereas in the on-state the current density
is very high (hundreds of A/cm^2). All these facts contribute to the condition number
of the problem and might degrade the convergence behavior.

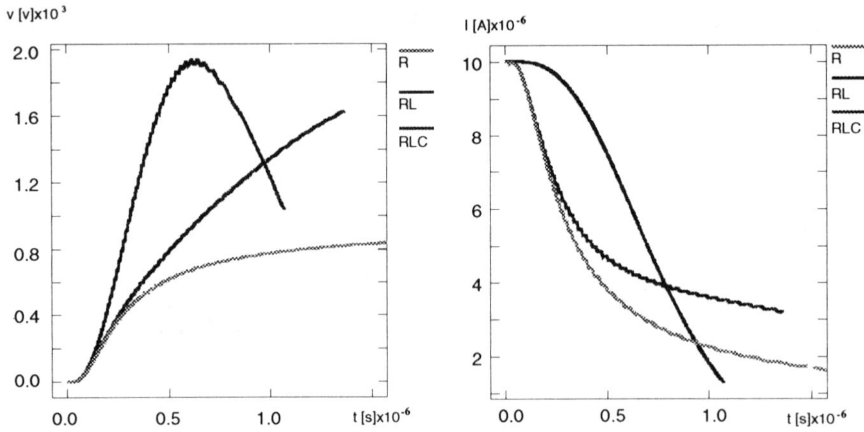

FIG. 11. *Voltage (left) and current (right) under different load conditions.*

In order to investigate the transient behavior, such as the turn-off of this device, under different load conditions, it necessary to include the appropriate external circuit elements. Figure 9 shows a typical setup used for testing (see [42]) of the device. Here, the gate drive circuit, G, is not shown for the sake of convenience; however, the load circuit consists of an inductance, L, and a resistor R_1. The voltage sources are $V_1 \approx 30$V and $V_2 \approx 1000$V, respectively.

Figure 11 depicts the voltage (left) and the corresponding current (right) at the anode under different load conditions. The figure clearly indicates that the load has a tremendous impact on the output characteristic of the device.

As a third example we show the transient simulation results of a single phase half bridge inverter in CMOS technology. This device converts a DC input voltage into a symmetrical AC output voltage of desired magnitude and frequency. For a fixed DC input voltage a variable output voltage is obtained by varying the gain of the inverter normally accomplished by pulse-width-modulation control in the inverter. The output voltage waveforms of ideal inverters should be sinusoidal. However, the waveforms of practical inverters contain certain harmonics. For low and medium power applications quasi-square-wave voltages may be acceptable. Inverters are used in induction heating variable AC motor driver standby power supplies.

The principle of the single half bridge inverter can be explained by Fig. 12. If only power MOSFET M2 is turned on by means of V2, the voltage across the load is $V_{cc}/2$. If on the other side M1 is turned on, the voltage across the load is $-V_{cc}/2$. The voltage is switched between these values by a logic circuit - not shown in the figure — which establishes the input voltages $V1$ and $V2$, respectively. To avoid short circuit conditions, this logic circuit has to be designed such that $M2$ and $M1$ are not both in on-state at the same time. For an inductive load, i.e., an AC drive, the load current can not change immediately with the output voltage. If $M2$ is turned off, the load current would continue to flow trough the diode integrated in $M1$ and the lower half of the DC source. Similarly, if $M1$ is turned off, the current flows through the diode

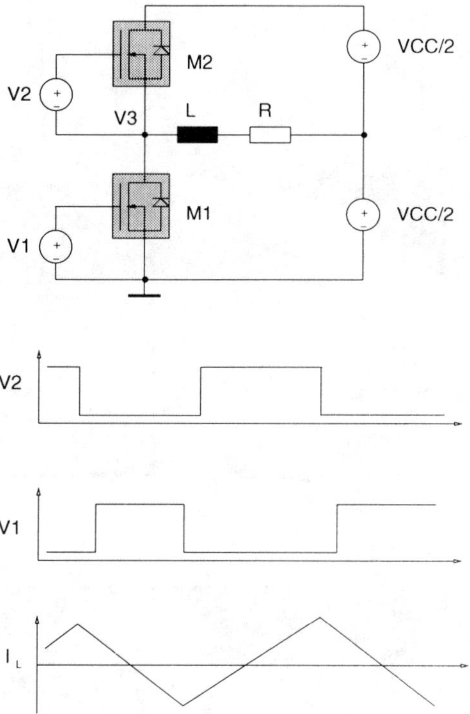

FIG. 12. *Half bridge inverter schematics.*

integrated in $M2$ and the upper half of the DC source.

Figure 13 shows the doping concentration in the two power MOSFETs M1 and M2. A blow-up of the source region of the right device is depicted in Fig. 14. The grid is very fine in order to resolve the small structures. Fig. 15 depicts the potential: at the beginning of the transient simulation when M1 is turned on; at $t = 5\mu s$ when M2 is turned on; and at $t = 10\mu s$ when again M1 is turned on. The current in the inductive load for a full cycle is shown in Fig. 16. The computed current slightly deviates from the ideal given in Fig. 12. The simulation was performed on a Cray Y-MP using approximately 11 hours of CPU-time for the 3400 combined TR-BDF2 time steps.

7. Conclusion. There are many situations where standard device simulation is no longer sufficient for device characterization. Our approach is to include external circuit elements to study the coupling between device and circuit. This is especially important for heavily loaded circuit elements, where the feedback of the load is no longer negligible. Typical examples are power devices. Also smaller problems such as magneto sensors can be simulated properly only if the device can be augmented by a number of simple circuit elements such as independent current sources and distributed contact resistances. The device simulation program, *gensim*, used here solves the combined device-circuit problem consistently for steady-state quasi-stationary and transient conditions.

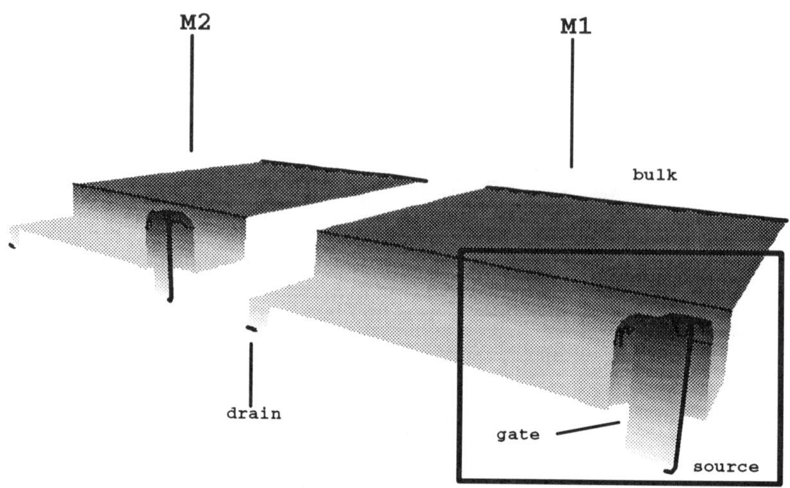

FIG. 13. *Doping concentration.*

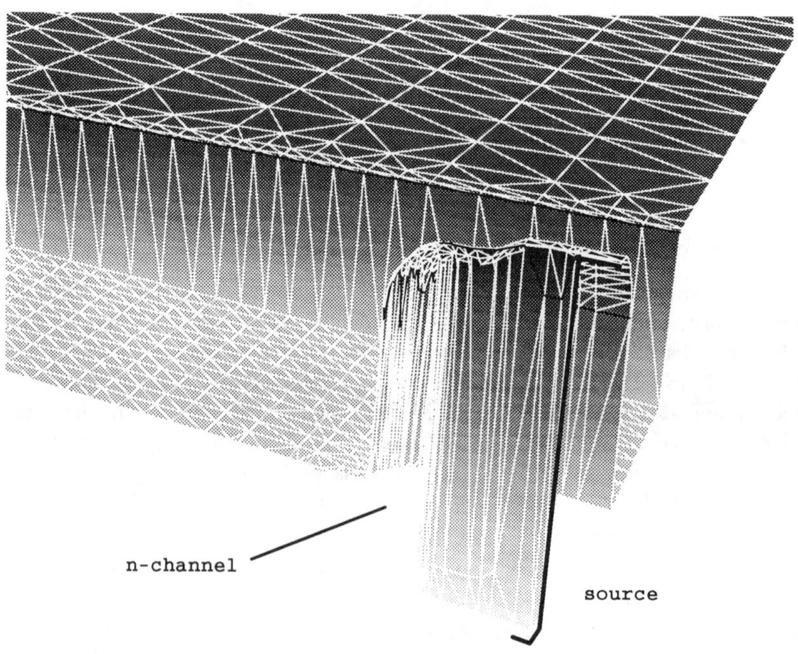

FIG. 14. *Blow-up of right source region.*

Acknowledgments. The present work would not have been possible without the help of several people and organizations. The computer centers of ETH Zurich and ETH Lausanne have generously allowed us access to the Cray Y-MP and Cray-2 machines, respectively.

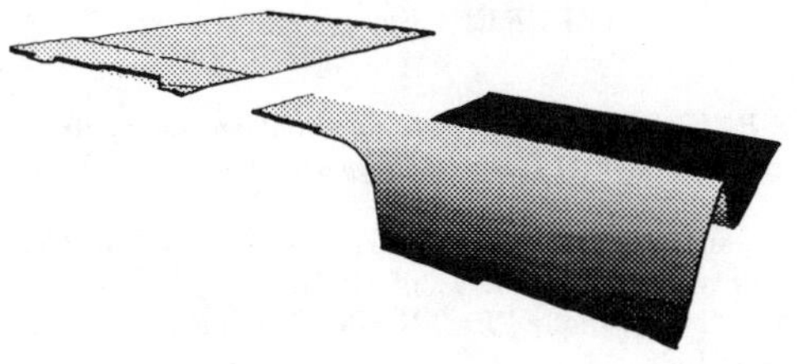

$t = 0\ \mu\mathrm{s}$

$t = 5\ \mu\mathrm{s}$

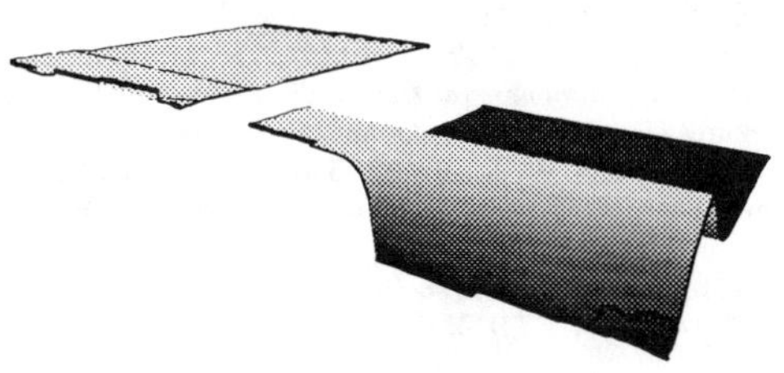

$t = 10\ \mu\mathrm{s}$

FIG. 15. *Electrostatic potential during switching.*

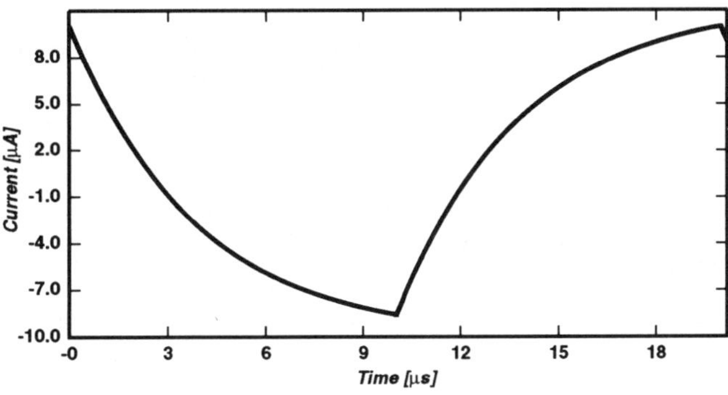

FIG. 16. *Transient inductor current.*

REFERENCES

[1] W. L. Engl, R. Laur, and H. K. Dirks, "MEDUSA - a simulator for modular circuits," *IEEE Trans. on Computer Aided Design of Integrated Circuits and Systems*, vol. CAD-1, pp. 85–93, 1982.

[2] M. R. Pinto, C. S. Rafferty, and R. W. Dutton, *PISCES II: Poisson and Continuity Equation Solver*. Stanford University, Stanford, CA 94305, 1984.

[3] J. G. Rollins and J. Choma, "Mixed-mode PISCES-SPICE coupled circuit and device solver," *IEEE TCAD*, vol. 7, Aug 1988.

[4] K. Johansson, "Incorporation of an external circuit in the semiconductor simulation program PISCES," Master's thesis, Asea Brown Boveri, Corporate Research Center, 5400 Baden, Switzerland, 1989.

[5] "TMA product announcement: PISCES II-B." TMA's Newsletter on Developments in Semiconductor Technology Simulation, vol. 3, no. 4, July-August 1991.

[6] G. Baccarani, M. Rudan, R. Guerrieri, and P. Ciampolini, "Physical models for numerical device simulation," in *Proc. of the Comett-Euroform*, (DEIS-University of Bologna, Bologna, Italy), Mar. 1991.

[7] K. Bløtekjær, "Transport equations for electrons in two-valley semiconductors," *IEEE Trans. on Electron Devices*, vol. ED-17, no. 1, pp. 38–47, 1970.

[8] G. Wachutka, "Rigorous thermodynamic treatment of heat generation and conduction in semiconductor device modeling," *IEEE Transactions on CAD*, vol. 9, no. 11, pp. 1141–49, 1990.

[9] O. Madelung, *Introduction to Solid State Theory*. New York: Springer-Verlag, 1978.

[10] C. Riccobene, G. Wachutka, and J. F. B. H. Baltes, "2d numerical modeling of dual collector magnetotransistors: Evidence for emitter efficiency modulation," in *Proceedings of the Fifth Conference of EUROSENSORS*, 1991 (submitted Jan 25).

[11] P. Conti, G. Heiser, and W. Fichtner, "Three-dimensional transient simulation of complex silicon devices," *Jap. J. of Appl. Phys. Lett.*, december 1990. An abridged version appeared in *Extended Abstracts of the 1990 Int. Conf. on Solid State Devices and Materials*, pages 143 -146, Sendai, Japan, 1990.

[12] P. Conti, N. Hitschfeld, and W. Fichtner, "Ω – an octree-based mixed element grid allocator for adaptive 3d device simulation," *IEEE Trans. on CAD/ICAS*, september 1991. in press.

[13] P. Conti, *Grid Generation for Three-dimensional Device Simulation*. PhD thesis, ETH Zürich, 1991. published by Hartung-Gorre Verlag, Konstanz, Germany.

[14] G. Heiser, C. Pommerell, J. Weis, and W. Fichtner, "Three-dimensional numerical semiconductor device simulation: Algorithms, architectures, results," *IEEE Trans. on CAD/ICAS*, Sept. 1991. in press.

[15] G. Heiser, *Design and Implementation of a Three Dimensional General Purpose Semiconductor Device Simulator*. PhD thesis, ETH-Zürich, 1991. publ. by Hartung Gorre Verlag, Konstanz, Germany.

[16] J. F. Bürgler, "GENSIM: GENeral SIMulation Program for Two-Dimensional Semiconductor Devices," Tech. Rep. 12, Integrated Systems Laboratory ETHZ, Switzerland, 1991.

[17] A. D. Mari, "An Accurate numerical steady-state one-dimensional solution of the p-n Junction," *Solid-State Electronics*, vol. 11, pp. 33–58, 1968.

[18] A. D. Mari, "An Accurate numerical one-dimensional solution of the p-n Junction under arbitrary transient conditions," *Solid-State Electronics*, vol. 11, pp. 1021–1053, 1968.

[19] J. F. Bürgler, R. E. Bank, W. Fichtner, and R. K. Smith, "A New Discretization Scheme for the Semiconductor Current Continuity Equations," *IEEE Trans. CAD*, vol. 8, no. 5, pp. 479–89, 1989.

[20] J. F. Bürgler, *Discretization and Grid Adaptation in Semiconductor Device Modeling*. PhD thesis, Swiss Federal Institute of Technology, Zürich, 1990.

[21] R. S. Varga, *Matrix Iterative Analysis*. Englewood Cliffs: Prentice-Hall, 1962.

[22] E. M. Buturla, P. E. Cottrell, B. M. Grossman, and K. A. Salsburg, "Finite-Element Analysis of Semiconductor Devices: The FIELDAY Program," *IBM J. Res. Develop.*, vol. 25, no. 4, pp. 218–231, 1981.

[23] R. E. Bank, D. J. Rose, and W. Fichtner, "Numerical Methods for Semiconductor Device Simulation," *IEEE Trans. Electr. Dev.*, vol. ED-30, pp. 1031–1041, 1983.

[24] D. L. Scharfetter and H. K. Gummel, "Large-Signal Analysis of a Silicon Read Diode Oscillator," *IEEE Transactions on Electron Devices*, vol. ED-16, pp. 64–77, 1969.

[25] S. Mueller, K. Kells, and W. Fichtner, "Automatic Rectangle-based Adaptive Mesh Genera-

tion without Obtuse Angles," *special issue of IEEE Trans. CAD, Proceedings NUPAD III*, vol. submitted, 1990.

[26] G. Voronoi, "Nouvelle application des paramètres continus a la théorie des formes quadratiques," *J. reine angew. Mathematik*, vol. 134, pp. 198–287, 1908.

[27] U. Ascher, P. A. Markowich, C. Schmeiser, H. Steinrück, and R. Weiss, "Conditioning of the Steady State Semiconductor Device Problem," *SIAM J. Appl. Math.*, vol. 49, no. 1, pp. 165–185, 1989.

[28] A. Nathan, *Carrier Transport in Magnetotransistors*. PhD thesis, University of Alberta, Edmonton, Alberta, 1988.

[29] J. Vlach and K. Shinghal, *Computer Methods for Circuit Analysis and Design*. New York: Van Nostrand Reinhold Company, 1983.

[30] J. K. White and A. Sangiovanni-Vincentelli, *Relaxation Techniques for the Simulation of VLSI Circuits*. Boston: Kluwer Academic Publishers, 1987.

[31] E. Griepentrog, "The index of differential-algebraic equations and its significance for the circuit simulation," in *Mathematical Modelling and Simulation of Electrical Circuits and Semiconductor Devices* (R. E. Bank, R. Bulirsch, and K. Merten, eds.), pp. 11–25, Birkhäuser Verlag, 1990.

[32] W. M. Coughran, Jr., E. Grosse, and D. J. Rose, "CAzM: A Circuit Analyzer with Macromodeling," *IEEE Transactions on Electron Devices*, vol. ED-30, pp. 1207–1213, 1983.

[33] R. E. Bank and D. J. Rose, "Global Approximate Newton Methods," *Numer. Math.*, vol. 37, pp. 279–295, 1981.

[34] R. E. Bank, W. M. Coughran, Jr., W. Fichtner, E. H. Grosse, D. J. Rose, and R. K. Smith, "Transient Simulation of Silicon Devices and Circuits," *IEEE Trans. CAD*, vol. CAD-4, pp. 436–451, 1985.

[35] G. Dahlquist and A. Björck, *Numerical Methods*. Series in automatic computation, Englewood Cliffs, New Jersey: Prentice-Hall, 1974.

[36] J. D. Lambert, *Computational Methods in Ordinary Differential Equations*. London, New York: John Wiley & Sons, 1973.

[37] P. Deuflhard, "Global inexact newton methods for very large scale nonlinear problems," Tech. Rep. SC 90-2, Konrad-Zuse-Zentrum für Informationstechnik, Berlin, Feb. 1990.

[38] R. E. Bank, *PLTMG: A Software Package for Solving Elliptic Partial Differential Equations* Users' Guide 6.0. Society for Industrial and Applied Mathematics, 1990.

[39] G. Ng and B. W. Peyton, "A supernodal cholesky factorization algorithm for shared-memory multiprocessors," tech. rep., Oak Ridge National Laboratory, Apr. 1991.

[40] C. Pommerell and W. Fichtner, "PILS: An iterative linear solver package for ill-conditioned systems," in *Supercomputing '91*, (Albuquerque, NM), ACM-IEEE, Nov. 1991.

[41] H. A. van der Vorst, "Bi-CGSTAB: A fast and smoothly converging variant of BI-CG for the solution of nonsymmetric linear systems," *SIAM J. Sci. Stat. Comput.*, vol. 13, 1992. In press.

[42] F. Bauer, E. Halder, K. Hofmann, H. Haddon, P. Roggwiller, T. Stockmeier, J. Bürgler, W. Fichtner, S. Müller, M. Westermann, J.-M. Moret, and R. Vuilleumier, "Design aspects of MOS-controlled thyristor elements: Technology, simulation and experimental results," *IEEE Trans. on Electron Devices*, vol. 38, pp. 1605–1611, July 1991.

METHODS OF THE KINETIC THEORY OF GASES RELEVANT TO THE KINETIC MODELS FOR SEMICONDUCTORS

CARLO CERCIGNANI*

Abstract. A survey of the mathematical techniques for the linear Boltzmann equation relevant to the kinetic approach to the theory of semiconductor devices is presented. Particular emphasis is given to the use of variational techniques and the computation of the Green's function in a constant electric field.

1. Introduction. When the transport of charges in a semiconductor is considered on a sufficiently large time scale, then the motion of the carries is decidedly influenced by the short range interactions with crystal lattice, which can be described, in a classical picture of the electron gas, by particle collisions. This situation, which occurs in high-density integrated circuits explains why there has been an increasing interest in understanding the mathematics of electron gas in submicron structures [1]. The basic tool, in this situation, is given by the Boltzmann equation [2], which may exclude the short range interactions between carriers, which only play a role when the particle density is very large, but can incorporate the Pauli exclusion principle, if necessary. Here we shall not take into account the fact that we deal with quasi particles, rather than particles and use the velocity variable ξ rather than the wave vector k. The main changes when passing to quasiparticles are the use of the effective mass m_* in place of the mass in free space m and the periodicity in k, related to the existence of the Brillouin zones. Also we shall assume that we deal with just one species; the extension to a mixture of carriers is trivial, if cumbersome.

Since the Boltzmann equation for classical gases has been used for several years in the study of flight in the upper atmosphere and similar problems occur in the transport of neutrons and radiation, we can try to borrow some of the methods and results [3, 4].

One first qualitative result arises from the influence of the rarefaction parameter ℓ/d (called Knudsen number by the aereodynamicists), *i. e.* the ratio between the mean free path ℓ and a typical macroscopic length d (the length scale over which the state of the gas changes significantly even in the diffusion approximation). A study of how the situations change with this parameter leads to the identification of the following five regimes:

(1) $\ell/d \ll 1$ (say $\ell/d \leq 10^{-2}$). This is the *diffusion-drift regime* with standard boundary conditions at the contacts.

(2) $10^{-2} \leq \ell/d \leq 10^{-1}$. This is the diffusion regime corrected by some effects near the boundaries, due to the presence of a layer of the thickness of a few mean free paths, where the diffusion approximation is not accurate. This is

---

*Dipartimento di Matematica, Politecnico di Milano, Piazza Leonardo da Vinci 32, 20133 Milano, Italy.

called the slip regime in rarefied gas dynamics, because the gas slips over the boundary rather than having zero velocity there. Here it might be called the *jump regime* unless one finds or has already found a better name; in fact the density of electrons appears to have a discontinuity at the boundary that would be negligible in the usual diffusion-drift approximation.

(3) $10^{-1} \le \ell/d \le 10$. This is the typical regime where the Boltzmann equation is of paramount importance. Since this regime is intermediate between the diffusion-drift (without or with jump) and the next two regimes, where the collisions are a perturbation or negligible at all, it might be called *transition regime* (this is the name it has been given by aerodynamicists). Other possible names could be *kinetic theory regime* or *Boltzmann regime*.

(4) $10 \le \ell/d \le 100$. In this regime the collisions with the lattice become rare and can be treated as a perturbation of the next regime. It might be called *nearly ballistic regime* (the corresponding name in rarefied gas dynamics is *nearly free-molecular regime*).

(5) $\ell/d \ge 100$. In this regime the collisions with the lattice are completely negligible. This already has a name in semiconductor theory: the *ballistic regime* (the corresponding name in rarefied gas dynamics is *free-molecular regime*).

It is to remarked that even in regime 1 there are regions where the diffusion approximation is invalid: these are narrow layers near the boundaries or interfaces. For very small values of ℓ/d, however, these regions are extremely tiny and what happens there has practically no influence on what occurs elsewhere.

2. Basic equations. The Boltzmann equation for the electron gas in a semiconductor may be written as follows [1]:

$$(2.1) \quad \frac{\partial f}{\partial t} + \xi \cdot \frac{\partial f}{\partial x} + \frac{q}{m} E \cdot \frac{\partial f}{\partial \xi} = \int_{\mathbf{R}^3} [s(\xi', \xi; x) f'(1 - \eta f) - s(\xi, \xi'; x) f(1 - \eta f')] d\xi'$$

where $f = f(x, \xi, t)$ is the distribution function, a function of position x, velocity ξ and time t, normalized in such a way as to give the number density when integrated over $\mathbf{R}^3$ with respect to ξ. Here $s(\xi', \xi; x)$ is the probability density for a particle with velocity ξ' to be scattered in the volume element $d\xi$ about ξ, q is the electric charge (negative for electrons), m the mass of the carriers and $\eta = h^3/2m^3$ for particles with spin $1/2$ ($h = 2\pi\hbar$ is the Planck constant). The electric field E is the sum of the external field (applied or produced by ions) and the field produced by carriers. In Eq. (2.1) f' means $f(x, \xi', t)$.

We remark that $\eta = 1.92 \, 10^{-10} m^6 \, sec^{-3} = 192 \, cm^6 sec^{-3}$. Since the order of magnitude of f is $n(m/2\pi k_B T)^{3/2}$, where n is the number density, T the temperature and k_B the Boltzmann constant, $f \ll 1/\eta$ means:

$$n \ll h^{-3}(2m^3)(2\pi k_B T/m)^{3/2} = 2(2\pi m k_B T/h^2)^{3/2}$$

$$\cong 2.5 \, 10^{25} m^{-3} = 2.5 \, 10^{19} \, cm^{-3}.$$

The principle of detailed balance yields:

$$(2.2) \quad s(\xi', \xi; x) f'_e(1 - \eta f_e) = s(\xi, \xi'; x) f_e(1 - \eta f'_e)$$

where f_e is the Fermi-Dirac distribution at the temperature of the lattice.

Equation (2.1) is nonlinear, but has a nonlinearity different from the nonlinearity occurring in the Boltzmann equation for gases; here the nonlinearity can make life easier because it ensures that f is bounded ($f \leq 1/\eta$). In many applications (lightly doped semiconductors) $f \ll 1/\eta$ and the nonlinear term can be neglected

$$(2.3) \qquad \frac{\partial f}{\partial t} + \xi \cdot \frac{\partial f}{\partial x} + \frac{q}{m} E \cdot \frac{\partial f}{\partial \xi} = \int_{\mathbf{R}^3} [s(\xi',\xi;x)f' - s(\xi,\xi';x)f]d\xi'$$

When using this approximation it is useful to remark that Eq. (2.2) is equivalent to

$$(2.4) \qquad s(\xi',\xi;x)M' = s(\xi,\xi';x)M$$

where

$$(2.5) \qquad M(\xi) = (m/2\pi k_B T)^{3/2} \exp(-m\xi^2/2k_B T)$$

is the Maxwellian distribution at the temperature of the lattice. Equation (2.3) is linear and many properties of linear transport equations can be applied. It is, of course, easy to prove theorems of existence and uniqueness for both initial and boundary value problems.

3. The variational principle. One property which gives important information by means of simple approximate solutions is the trivial variational principle for the steady case in a given electric field. This can be obtained by an argument analogous to that used in the kinetic theory of gases [3-6]. Let us rewrite Eq. (2.3) for the steady case in the form

$$(3.1) \qquad Df = Lf + S$$

where we have added a source term S and let

$$(3.2) \qquad Df = \xi \cdot \frac{\partial f}{\partial x} + \frac{q}{m} E \cdot \frac{\partial f}{\partial \xi}$$

$$(3.3) \qquad Lf = \int_{\mathbf{R}^3} [s(\xi',\xi;x)f' - s(\xi,\xi';x)f]d\xi'$$

Let us introduce the scalar product in $L^2(\mathbf{R}^3 \times \mathcal{D})$:

$$(3.4) \quad (f,g) = (2\pi k_B T/m)^{3/2} \int_{\mathbf{R}^3} \int_{\mathcal{D}} f(x,\xi)g(x,\xi)[M(\xi)]^{-1} \exp[qV(x)/k_B T]d\xi dx$$

where V is the electric potential, such that $E = -\partial V/\partial x$ and $\mathcal{D}$ the domain occupied by the carriers. It is then clear that

$$(3.5) \qquad (f, Lg) = (Lf, g)$$

where Eq. (2.4) has been used, and

$$(3.6) \qquad (f, PDg) = (PDf, g) + (f, Pg)_+ - (f, Pg)_-$$

where P is the parity operator in velocity space $(Pf(\xi) = f(-\xi))$ and

$$(3.7)$$
$$(f, g)_\pm$$
$$= (2\pi k_B T/m)^{3/2} \int_{\mathbf{R}^3_\pm} \int_{\partial \mathcal{D}} f(x, \xi) g(x, \xi) \exp[m\xi^2/2k_B T + qV(x)/k_B T] |\xi \cdot n| d\xi d\sigma$$

is a scalar product for the restriction of the traces along the boundary $\partial \mathcal{D}$ of $\mathcal{D}$ to the half spaces $\mathbf{R}^3_\pm$, defined by $\pm \xi \cdot n > 0$ (here $n = n(x)$ is the normal at $x \in \partial \mathcal{D}$ pointing into $\mathcal{D}$). Since P commutes with L for typical short-range interactions, Eqs. (3.2) and (3.3) imply:

$$(3.8) \qquad (f, P(Dg - Lg)) = (P(Df - Lf), g) + (f, Pg)_+ - (f, Pg)_-$$

This can give rise to variational principle in the following way. Define for any function $\tilde{f}$ in the domain of $(D - L)$

$$(3.9) \qquad J_0(\tilde{f}) = (\tilde{f}, P(D\tilde{f} - L\tilde{f} - 2S))$$

the first variation of J_0 is

$$(3.10)$$
$$\delta J_0 = (\delta\tilde{f}, P(D\tilde{f} - L\tilde{f} - 2S)) + (\tilde{f}, P(D\delta\tilde{f} - L\delta\tilde{f}))$$
$$= (\delta\tilde{f}, P(D\tilde{f} - L\tilde{f} - 2S)) + (P(D\tilde{f} - L\tilde{f}), \delta\tilde{f}) + [(\tilde{f}, P\delta\tilde{f})_+ - (\tilde{f}, P\delta\tilde{f})_-]$$

Thus if $\tilde{f} = f$, a solution of Eq. (3.1),

$$(3.11) \qquad \delta J_0 = (f, P\delta\tilde{f})_+ - (f, P\delta\tilde{f})_-$$

From this equation we see that if $\tilde{f} = f + \delta\tilde{f}$, where $\delta\tilde{f}$ vanishes on the boundary, then $\delta J_0 = 0$ and we have a variational principle. In this form, however, the variational principle is not useful. The boundary conditions are usually in the form

$$(3.12) \qquad f^+ = Af^- + f_0$$

where f_0 is a given function, f^+ and f^- the restrictions of the traces on the boundary to $\pm \xi \cdot n > 0$, while A is a linear operator, which thanks to a reciprocity law at the wall, satisfies

$$(3.13) \qquad (f, PAg)_- = (PAf, g)_-$$

Then

$$(3.14) \quad \begin{aligned} (f, P\delta\tilde{f})_+ &= (Af, P\delta\tilde{f})_+ + (f_0, P\delta\tilde{f})_+ = (PAf, \delta\tilde{f})_- + (f_0, P\delta\tilde{f})_+ \\ &= (f, PA\delta\tilde{f})_- + (Pf_0, \delta\tilde{f})_- \end{aligned}$$

Equation (3.11) then becomes:

$$(3.15) \qquad \delta J_0 = (Pf_0, \delta\tilde{f})_- - (f, P(\delta\tilde{f} - A\delta\tilde{f}))_-$$

If we now let

$$(3.16) \qquad J_1(\tilde{f}) = (\tilde{f}, P(\tilde{f} - A\tilde{f} - 2f_0))_-$$

we have

$$(3.17) \qquad \delta J_1 = (\delta\tilde{f}, P(\tilde{f} - A\tilde{f} - 2f_0))_- + (\tilde{f}, P(\delta\tilde{f} - A\delta\tilde{f}))_-$$

and if $\tilde{f} = f$, a solution of Eq. (3.1) with the boundary conditions (3.12), we have:

$$(3.18) \qquad \delta J_1 = -(Pf_0, \delta\tilde{f})_- + (f, P(\delta\tilde{f} - A\delta\tilde{f}))_-$$

and if we let $J = J_0 + J_1$, we have that

$$(3.19) \qquad\qquad\qquad \delta J = 0$$

characterizes the solutions of Eq. (3.1) with the boundary conditions (3.12). Generally speaking $J(\tilde{f})$ will not have a maximum nor a minimum at $f = \tilde{f}$. Yet, the variational principle can be exploited to obtain accurate estimates of $J(f)$, starting from rather inaccurate trial functions. Let us check now the meaning of $J(f)$ in the case when the source term S is zero. We have

$$(3.20) \qquad J(f) = J_0(f) + J_1(f) = -(f, Pf_0)_-$$

Thus the meaning of $J(f)$ depends on the boundary source f_0. Let us take a simple example, a one-dimensional problem in which at $x = 0$ and $x = L$ there is emission and (absorption) of carriers with a Maxwellian distribution and different potential (say 0 and $\bar{V}$). In this case the boundary degenerates into a pair of points and

$$(3.21) \quad J(f) = c_0 \int_{\mathbf{R}^3_-} f^-(0, \xi)|\xi \cdot n|d\xi - c_L \exp(q\bar{V}/k_B T) \int_{\mathbf{R}^3_-} f^-(L, \xi)|\xi \cdot n|d\xi$$

where $n = i$ (the unit vector of the x-axis) at $x = 0$ and $-i$ at $x = L$, while c_0 and c_L are the constants of proportionality of f^+ to $M(\xi)$ at $x = 0$ and $x = L$ respectively.

Let us remark that the particle flow j, proportional to the electric current, is the same at both $x = 0$ and $x = L$ because of the steadiness of the current. Then if j is taken to be positive when the net flow is from $x = 0$ to $x = L$:

$$(3.22)$$

$$j = \int_{\mathbf{R}^3} f(0,\xi)\xi \cdot n\, d\xi = \int_{\mathbf{R}^3_+} f^+(0,\xi)|\xi \cdot n|d\xi - \int_{\mathbf{R}^3_-} f^-(0,\xi)|\xi \cdot n|d\xi$$

$$= c_0\sqrt{\frac{k_B T}{2\pi m}} - \int_{\mathbf{R}^3_-} f^-(0,\xi)|\xi \cdot n|d\xi = -c_L\sqrt{\frac{k_B T}{2\pi m}} + \int_{\mathbf{R}^3_-} f^-(L,\xi)|\xi \cdot n|d\xi$$

and

$$(3.23) \qquad J(f) = -[c_0 + c_L \exp(q\bar{V}/k_B T)]j + [c_0^2 - c_L^2 \exp(q\bar{V}/k_B T)]\sqrt{\frac{k_B T}{2\pi m}}$$

Thus an accurate estimate of $J(f)$ provides an accurate estimate of the net current j.

Let us consider now a simpler problem, which has, however, a great importance, since it provides the boundary conditions for the diffusion-drift approximation, when we are in the regime, that has been called the *jump regime* in Sect. 1. We shall assume that the device is large enough for the diffusion-drift approximation to apply a few mean free paths away from the boundary, but we want to calculate the correct density at the boundary. To this end we consider a half-space problem. The semiconductor occupies a half-space ($x > 0$) bounded by a metal. In this case the boundary is just a point, but we must transform the problem in order to avoid trivialities or divergences. Also we adopt a relaxation time approximation:

$$(3.24) \qquad Lf = [n(x)M(\xi) - f(x,\xi)]/\tau$$

where τ is the relaxation time. The boundary condition is, as above $f = c_0 M(\xi)$.

We can use the variational principle in a naïve way in order to determine the best function in a given class of trial functions. We can also link, however, the value of the number of density $n(x)$ at $x = 0$, $n(0)$, to $J(f)$ and thus obtain a much more accurate estimate of the jump $n(0) - c_0$. To this end we note that we can write a generic trial function of the form

$$(3.25) \qquad \tilde{f} = [\bar{n}(x) + c_0 \exp(qEx/k_B T)]M(\xi) + \left(\frac{dn}{dx} - \frac{qE}{k_B T}n\right)N(\xi) + \tilde{g}(\xi,x)$$

where $\tilde{g}$ is a new arbitrary function, while $\bar{n}(x)$ differs from $n(x)$ by a constant and is uniquely determined by the constancy of the electric current

$$(3.26) \qquad \frac{dn}{dx} - \frac{qE}{k_B T}n = c$$

(c being given) and the boundary condition $\bar{n}(0) = 0$ and

$$(3.27) \quad N(\xi) = (m/2\pi k_B T)^{3/2}\frac{m}{qE}\exp[-m(\xi_2^2 + \xi_3^2)/2k_B T$$

$$- m\xi_1/qE\tau)]\int_{-\infty\,\mathrm{sgn}(qE)}^{\xi_1} \exp(-m\xi_1'^2/2k_B T + m\xi_1'/qE\tau)\xi_1'\, d\xi_1'$$

(remark that $\int N(\xi)d\xi = 0$).

Since the first two terms in Eq. (3.25) give an exact solution of Eq. (3.1), in terms of g the problem becomes

$$Dg = Lg$$

with the boundary condition

(3.29) $$g^+ = -cN(\xi)$$

which can be solved by considering another functional

(3.30) $$J(\tilde{g}) = (\tilde{g}, P(D\tilde{g} - L\tilde{g})) + (\tilde{g}, P(\tilde{g} + 2cN))_-$$

When $\tilde{g}$ becomes the exact solution, $J(g) = c(g, PN)_- = -c^2(N, PN)_+ - c(g, PN)_0$, where

(3.31) $$(g, PN)_0 = \int_{\mathbf{R}^3} g(\xi)N(-\xi)/M(\xi)\xi_1 \, d\xi$$

Let us use for any x the notation

(3.32) $$(g, PN)_x = \int_{\mathbf{R}^3} \exp(-qEx/k_BT)g(\xi)N(-\xi)/M(\xi)\xi_1 \, d\xi$$

If we multiply Eq. (3.28) by PN and integrate with respect to ξ, we find

(3.33) $$\frac{d}{dx}(g, PN)_x = 0$$

where we have used the relation

(3.34) $$\frac{q}{m}E\frac{\partial N}{\partial \xi_1} = LN - M\xi_1$$

and the fact that $\int_{\mathbf{R}^3} g(\xi)\xi_1 \, d\xi$ vanishes at infinity and hence throughout because of mass conservation.

Hence $(g, PN)_x$ is constant with respect to x and can be evaluated at any position; in particular at infinity. Then we have:

(3.35) $$(g, PN)_0 = (g, PN)_{as} = [n(0) - c_0](M, PN)_0$$

and

(3.36) $$J(g) = -c^2(N, PN)_+ - c[n(0) - c_0](M, PN)_0$$

a relation that links $J(g)$ to the jump $n(0) - c_0$.

Thus we see that we are able to link the value $J(f)$ of the functional $J(\tilde{f})$ to simple quantities having a relevant physical meaning; this is very important, because an error ϵ in f becomes an error ϵ^2 in the evaluation of $J(f)$.

4. Variational calculations. Let us see how the variational method works in practice. Let us take a trial function which is simple enough and yet sufficiently general to describe the diffusion regime and the ballistic regime. To simplify (and without any loss of generality), we assume $q\bar{V} < 0$. We then take

$$
\begin{aligned}
\tilde{f} = (m/2\pi k_B T)^{3/2} \exp(&-m\xi^2/2k_B T - qV(x)/k_B T)[a_1 \\
&+ a_2\Theta(\xi)\Theta(m\xi^2/2k_B T + qV(x)/k_B T) \\
&+ a_3\Theta(-\xi)\Theta(-m\xi^2/2k_B T - q(V(x) - \bar{V})/k_B T) \\
&+ a_4\exp(qV(x)/k_B T) + a_5 N(\xi)]
\end{aligned}
\tag{4.1}
$$

where Θ denotes Heaviside's step function and a_i ($i = 1, 2, 3, 4, 5$) are five constants to be determined by the variational principle. To this end one must compute

$$
J(\tilde{f}) = (\tilde{f}, P(D\tilde{f} - L\tilde{f})) + (\tilde{f}, P(\tilde{f} - A\tilde{f} - 2f_0))_-
\tag{4.2}
$$

when $\tilde{f}$ is given by Eq. (4.1). We obtain a second degree expression in the a_i's; if we take the derivatives of this expression with respect to the a_i's themselves and put them equal to zero, we arrive at a system of linear equations, which gives the best values of the constants. If we put these back in the functional, we obtain the approximation to J and hence to j.

Since this is too complicated for a lecture, let us work out the simpler half-space problem that was outlined in the previous section.

We assume that the electric field is constant ($V = -Ex$) and take the following trial function:

$$
\tilde{f} = n(x)M(\xi) + \left(\frac{dn}{dx} - \frac{qE}{k_B T}n\right) N(\xi)
\tag{4.3}
$$

which expresses the fact that the current is constant. We remark that the trial function in Eq. (4.3) is an exact solution of the Boltzmann equation, but not of the problem, because it will not satisfy the boundary conditions; if we keep c (the constant in Eq. (3.26)), fixed, the only freedom we have is another constant in $n = n(x)$. It is this constant, say $n(0)$, that can be varied; the variational principle gives the recipe for fixing it.

As we pointed out in the previous section, it is better to work with the function $\tilde{g}$, defined by Eq. (3.25). If we now take as a trial function:

$$
\tilde{g} = a\exp(qEx/k_B T)M(\xi)
\tag{4.4}
$$

(this is the same as above, with $n(0) = a$), then a is the constant to be varied and we have

$$
J(\tilde{g}) = a^2(M, M)_- + 2ca(M, PN)_-
\tag{4.5}
$$

If we impose $dJ/da = 0$, we obtain

$$
a = -c(M, PN)_-/(M, M)_-
\tag{4.6}
$$

and $J(g) = -c^2(M, PN)_-^2/(M, M)_-$. Then Eq. (3.36) gives

$$
n(0) - c_0 = c[(M, PN)_-^2/(M, M)_- - (N, PN)_+]/(M, PN)_0
\tag{4.7}
$$

and we obtain the best determination of the density jump.

5. Structure of the one-dimensional solution in a constant electric field. Let us consider the one-dimensional Boltzmann equation in the relaxation time approximation

$$(5.1) \qquad \xi\frac{\partial f}{\partial x} + \frac{q}{m}E\frac{\partial f}{\partial \xi} = [n(x)M(\xi) - f]/\tau$$

where now ξ is the velocity component along the x-axis, $n(x)$ and τ (as previously) the number density and the (constant) relaxation time, while $M(\xi)$ the one-dimensional Maxwellian

$$(5.2) \qquad M(\xi) = (\beta/\pi)^{1/2}\exp(-\beta\xi^2) \qquad [\beta = m/(2k_BT)]$$

In analogy with the gas dynamic case [3, 4, 7] and following the lines of a joint paper with C. Toepffer [8], let us look for solutions with separated variables

$$(5.3) \qquad f = g_\lambda(\xi)\exp(qE\lambda x/m)$$

and let us normalize g_λ to unity

$$(5.4) \qquad \int g_\lambda(\xi)d\xi = 1$$

Then $g = g_\lambda$ will satisfy

$$(5.5) \qquad \lambda\xi g + \frac{\partial g}{\partial \xi} = \nu(M(\xi) - g)$$

where

$$(5.6) \qquad \nu = m/\tau q E$$

Equation (5.5) can be solved to yield

$$(5.7) \qquad \begin{aligned} g &= A\exp(-\nu\xi - \lambda\xi^2/2) \\ &\quad + (\beta/\pi)^{1/2}\exp(-\nu\xi - \lambda\xi^2/2)\nu\int_0^\xi \exp(\nu\xi' + \lambda\xi'^2/2 - \beta\xi'^2)d\xi' \end{aligned}$$

where the constant A and the limits of integration should be chosen in order to ensure that g is bounded and satisfies Eq. (5.4). We remark that λ and g might be complex, since the solutions we are looking for have a physical significance only as a basis for an expansion. For any $\mathfrak{Re}\,\lambda < 2\beta$, the integral in Eq. (5.7) converges when the lower or upper bound are taken to be $\pm\infty$. The first term requires, however, $\mathfrak{Re}\,\lambda > 0$ (unless $A = 0$). Let us check when we have an acceptable solution at $\pm\infty$. For $2\beta > \mathfrak{Re}\,\lambda > 0$ both terms go to zero; the extremes of the interval also belong to the spectrum because the exponentials oscillate but do not diverge there. Let us now impose the normalization condition:

(5.8)

$$A \int_{-\infty}^{+\infty} \exp(-\nu\xi - \lambda\xi^2/2)d\xi$$

$$+ (\beta/\pi)^{1/2}\nu \int_{-\infty}^{+\infty} d\xi \exp(-\nu\xi - \lambda\xi^2/2) \int_0^\xi \exp(\nu\xi' + \lambda\xi'^2/2 - \beta\xi'^2)d\xi' = 1$$

We obtain

$$(5.9) \quad A = (\lambda/2\pi)^{1/2} \exp(\nu^2/4\lambda^2) \left[1 + \beta^{1/2}\nu^2 \int_0^1 (1-t)([\beta - \lambda/2)t^2 \right.$$

$$\left. + \lambda/2]^{-3/2} \exp\{(\nu^2(1-t)^2/4[(\beta - \lambda/2)t^2 + \lambda/2]\}dt \right]$$

The case $\lambda = 0$ is singular and must be dealt with separately; the solution in this case was discussed by Trugman and Taylor [9].

Thus the spectrum fills the strip $0 \leq \mathfrak{Re}\ \lambda \leq 2\beta$ and we need an area integral to represent the general solution. We remark that $\lambda = 2\beta$ corresponds to the barometric distribution.

6. The Green's function. Let us consider again the one-dimensional Boltzmann equation in the relaxation time approximation

$$(6.1) \qquad \xi\frac{\partial f}{\partial x} + \frac{q}{m}E\frac{\partial f}{\partial \xi} = [n(x)M(\xi) - f]/\tau$$

where ξ is the velocity component along the x-axis, $n(x)$ and τ the number density

$$(6.2) \qquad n(x) = \int f d\xi$$

and the (constant) relaxation time, while $M(\xi)$ is the one-dimensional Maxwellian:

$$(6.3) \qquad M(\xi) = (\beta/\pi)^{1/2} \exp(-\beta\xi^2)$$

The Green's function solves

$$(6.4) \qquad \xi\frac{\partial G}{\partial x} + \frac{q}{m}E\frac{\partial G}{\partial \xi} = [n_G(s;\xi')M(\xi) - G]/\tau + \delta(s)\delta(\xi - \xi')$$

where

$$(6.5) \qquad n_G(s;\xi') = \int G(\xi, s; \xi')d\xi$$

Following again Cercignani and Toepffer [8], let us consider the Fourier transform $\tilde{G}(\xi, q; \xi')$ of G such that

$$(6.6) \qquad G(\xi, q; \xi') = (2\pi)^{-1} \int \exp(iqs)\tilde{G}(\xi, q; \xi')dq$$

Equation (6.4) yields:

$$(6.7) \qquad iq\xi\tilde{G} + \alpha\frac{\partial\tilde{G}}{\partial\xi} = [\tilde{n}_G(q;\xi')M(\xi) - \tilde{G}]/\tau + \delta(\xi - \xi')$$

where

$$(6.8) \qquad \alpha = \frac{q}{m}E$$

Henceforth we shall assume $\alpha > 0$ without any loss of generality. Let now $H(\xi, q)$ be the solution of

$$(6.9) \qquad iq\xi H + \alpha\frac{\partial H}{\partial\xi} + H/\tau = 0$$

such that $H(0,q) = 1$ for $\xi = 0$. Explicitly

$$(6.10) \qquad H(\xi, q) = \exp[-\xi/(\alpha\tau) - iq\xi^2/(2\alpha)]$$

Then Eq. (6.7) delivers:

$$
\begin{aligned}
\tilde{G}(\xi, q; \xi') &= \alpha^{-1}H(\xi, q)\int_{-\infty}^{\xi}[M(\xi'')\tilde{n}_G(q;\xi')/\tau + \delta(\xi - \xi')]/H(\xi'', q)d\xi'' \\
(6.11) \qquad &= (2\pi\tau)^{-1}\tilde{n}_G(q,\xi')H(\xi, q)[\beta/(\beta - iq/\alpha)]^{1/2}(\mathrm{erf}\{[\beta/2 - iq/(2\alpha)]^{1/2}\xi \\
&\quad - [2(\beta - iq/\alpha)]^{-1/2}/(\alpha\tau)\} \pm 1)\exp\{[2(\beta - iq/\alpha)]^{-1}(\alpha\tau)^{-2}\} \\
&\quad + \alpha^{-1}\exp[-(p - p')/(\alpha\tau) - iq(p^2 - p'^2)]\Theta(\xi - \xi')
\end{aligned}
$$

where ± 1 applies according to whether the square root are drawn on the first or second sheet (this does not change $\tilde{G}$) end erf is the error function.

The consistency condition (6.5) (or rather, its Fourier transform) delivers:

$$
\begin{aligned}
\tilde{n}_G(q;\xi') &= [\alpha H(\xi', q)]^{-1}\int_{\xi'}^{\infty}H(\xi, q)d\xi\left\{1\right.\\
&\quad \left. -(\alpha\tau)^{-1}\int_{-\infty}^{\infty}H(\xi, q)\int_{-\infty}^{\xi}M(\xi'')/H(\xi'', q)d\xi'']\right\}^{-1} \\
(6.12) \qquad &= \alpha^{-1}\exp[\xi/(\alpha\tau)iq\xi^2/(2\alpha)][\pi\alpha/2iq]^{1/2}\left(1 - \mathrm{erf}\left\{[iq/(2\alpha)]^{1/2}\xi'\right.\right.\\
&\quad \left.\left. -[1/(2iq\alpha)]^{1/2}\right\}\right)\exp[(2iq\alpha\tau^2)^{-1}]\left\{1 - (\alpha\tau)^{-1}[\pi\beta\alpha/[2iq(\beta\right.\\
&\quad \left. - iq/\alpha)]\right]^{1/2}\left(1 - \mathrm{erf}\{\beta^{1/2}[2(iq\alpha)(\beta\right.\\
&\quad \left.\left. -iq/\alpha)]^{-1/2}\tau^{-1}\}\right)\exp\left\{\beta[2(iq\alpha)(\beta - iq/\alpha)]^{-1}\tau^{-2}\right\}\right\}
\end{aligned}
$$

Thus the Fourier transform is completely determined, although it must be admitted that its expression is rather cumbersome.

It may be useful to examine the asymptotic behaviour of the Green's function for short distances. This means that the parameter τ is large (in terms of s) and we can solve by treating the integral term as a perturbation. Then

$$(6.13) \qquad G(\xi, q; \xi') = G^{(0)}(\xi, s; \xi') + G^{(1)}(\xi, s; \xi') + \cdots$$

where:

$$(6.14) \qquad \begin{aligned} \xi \frac{\partial G^{(0)}}{\partial x} + \alpha \frac{\partial G^{(0)}}{\partial \xi} &= -G^{(0)}/\tau + \delta(s)\delta(p - p') \\ \xi \frac{\partial G^{(1)}}{\partial x} + \alpha \frac{\partial G^{(1)}}{\partial \xi} &= [n_G^{(0)}(s; \xi')M(\xi) - G^{(1)}]/\tau \end{aligned}$$

$$\cdots = \cdots$$

The solution at zeroth order is

$$(6.15) \quad G^{(0)}(\xi, s; \xi') = \alpha^{-1} \exp[-(\xi - \xi')/(\alpha\tau)]\Theta(p - p')\delta(s - (m/2\alpha)(\xi^2 - \xi'^2))$$

From this we compute

$$
\begin{aligned}
(6.16) \\
n_G^{(0)}(s; p') =& [\beta/\pi]^{1/2} \int \alpha^{-1} \exp[-(\xi - \xi')/(\alpha\tau) \\
& - \beta\xi^2/(2m)]\Theta(\xi - \xi')\delta(s - (m/2\alpha)(\xi^2 - \xi'^2))dp \\
=& [\beta/\pi]^{1/2} \left\{ [2/(\xi'^2 + 2\alpha s)^{1/2}] \exp[(\xi' - [\xi'^2 + 2\alpha s]^{1/2})/(\alpha\tau) - \beta m\xi'^2 \right. \\
& - \alpha s]\Theta(\xi' + 2\alpha s)[1 - \Theta(-s)\Theta(\xi')] + [2/(\xi'^2 + 2\alpha s)^{1/2}] \exp[(\xi' \\
& \left. + [\xi'^2 + 2\alpha s]^{1/2})/(\alpha\tau) - \beta m\xi'^2 - \alpha s]\Theta(\xi' + 2\alpha s)\Theta(-s)\Theta(-\xi') \right\}
\end{aligned}
$$

Inserting this in equation for $G^{(1)}$ we can compute the first order correction.

7. Concluding remarks. We have surveyed some of the mathematical techniques for the linear Boltzmann equation relevant to the kinetic approach to the theory of semiconductor devices. Particular emphasis has been given to the use of variational techniques and the computation of the Green's function in a constant electric field, but the general philosophy of the paper is that one can gain a lot from an analytical study of the Boltzmann equation, even if the final goal is to develop a numerical code.

REFERENCES

[1] H. L. GRUBIN, K. HESS, G. J. IAFRATE AND D. K. FERRY (EDS), *Physics of Semiconductor Devices*, Plenum, New York, 1984.

[2] S. M. SZE, *Physics of semiconductor devices*, 2nd edition, Wiley, New York, 1981.

[3] C. CERCIGNANI, *The Boltzmann Equation and its Application*, Springer, New York, 1988.

[4] C. CERCIGNANI, *Mathematical Methods in Kinetic Theory*, 2nd revised edition, Plenum Press, New York, 1990.

[5] C. CERCIGNANI, *A Variational Principle for Boundary Value Problems in Kinetic Theory*, J. Stat. Phys., 1 (1969), pp. 297–311.

[6] C. CERCIGNANI AND C. D. PAGANI, *A Variational Approach to Boundary Value Problems in Kinetic Theory*, Phys. Fluids, 9 (1966), pp. 1167–1173.

[7] C. CERCIGNANI, *Elementary Solutions of the Linearized Gas Dynamics Boltzmann Equation and their Application to the Slip Flow Problem*, Ann. Phys., 20 (1962), pp. 219–233.

[8] C. CERCIGNANI AND C. TOEPFFER, *in preparation*.

[9] S. A. TRUGMAN AND A. J. TAYLOR, *Analytic solution of the Boltzmann equation with applications to electron transport in inhomogeneous semiconductors*, Phys. Rev. B, 33 (1986), pp. 5575–5584.

SHOCK WAVES IN THE HYDRODYNAMIC MODEL FOR SEMICONDUCTOR DEVICES

CARL L. GARDNER[*]

Abstract. Numerical simulations of a family of steady-state electron shock waves (parametrized by the amount of heat conduction) in a one micron Si semiconductor device at 77 K are presented, using a steady-state upwind method. The electron shock wave has a finite width which scales linearly with the amount of heat conduction.

Comparisons of the hydrodynamic simulations with a Monte Carlo simulation of Laux using the DAMOCLES program are also presented. The hydrodynamic prediction of an electron shock wave [1] in Si at 77 K is confirmed by the DAMOCLES simulation of the Boltzmann equation. Good agreement between the two different methods for simulating the electron shock wave can be obtained by adjusting the amount of heat conduction in the hydrodynamic model.

1. Introduction. The hydrodynamic model treats the propagation of electrons in a semiconductor as the flow of a charged, heat conducting gas in an electric field. As such, the hydrodynamic model should perhaps be called *electrogasdynamics* (with heat conduction).

The electron gas has a soundspeed c [2] given by

$$(1) \qquad c = \sqrt{T/m}$$

where T is the electron temperature in energy units and m is the effective electron mass. The electron flow may be either subsonic or supersonic. In the case of a transition from supersonic flow to subsonic flow, an electron shock wave will in general develop in the hydrodynamic model.

The shock wave actually enhances electron flow in the channel of an $n^+ - n - n^+$ diode, allowing for higher electron densities, velocities, and currents [1]. The shock profile is no more dramatic than the effects of junctions in the diode.

In Si at $T_0 = 300$ K, $c = \sqrt{\alpha}\ 1.3 \times 10^7$ cm/s where $T = \alpha T_0$, while at $T_0 = 77$ K, $c = \sqrt{\alpha}\ 6.6 \times 10^6$ cm/s. The electron saturation velocity $v_s \approx 10^7$ cm/s at 300 K, and $v_s \approx 1.25 \times 10^7$ cm/s at 77 K. Thus it is difficult (though not impossible [1]) to produce an electron shock wave in Si at room temperature, but easy at liquid nitrogen temperature [1].

I will present numerical simulations of a family of steady-state electron shock waves (parametrized by the amount of heat conduction) in a one micron semiconductor device at 77 K, using a steady-state upwind method. The electron shock wave has a finite width which scales linearly with the amount of heat conduction. I will present numerical evidence that the shock width goes to zero as the amount of heat conduction in the model goes to zero.

I will also present comparisons of the hydrodynamic simulations with a Monte Carlo simulation of Laux using the DAMOCLES [3] program. The DAMOCLES simulation of the Boltzmann equation confirms the hydrodynamic prediction of an

---

[*] Research supported in part by the National Science Foundation under grant DMS-8905872. Department of Computer Science, Duke University, Durham, NC 27706.

electron shock wave [1] in Si at 77 K. Good agreement between the two different methods for simulating the electron shock wave can be obtained by adjusting the amount of heat conduction in the hydrodynamic model.

2. The hydrodynamic model. The hydrodynamic equations are

$$(2) \qquad \frac{\partial n}{\partial t} + \nabla \cdot (n\mathbf{v}) = \left(\frac{\partial n}{\partial t}\right)_c$$

$$(3) \qquad \frac{\partial \mathbf{p}}{\partial t} + \mathbf{v}\nabla \cdot \mathbf{p} + \mathbf{p} \cdot \nabla\mathbf{v} = -en\mathbf{E} - \nabla(nT) + \left(\frac{\partial \mathbf{p}}{\partial t}\right)_c$$

$$(4) \qquad \frac{\partial W}{\partial t} + \nabla \cdot (\mathbf{v}W) = -en\mathbf{v} \cdot \mathbf{E} - \nabla \cdot (\mathbf{v}nT) - \nabla \cdot \mathbf{q} + \left(\frac{\partial W}{\partial t}\right)_c$$

$$(5) \qquad \nabla \cdot (\epsilon\nabla\phi) = -e(N_D - N_A - n), \quad \mathbf{E} = -\nabla\phi$$

where n is the electron density, $\mathbf{v}$ is the velocity, $\mathbf{p} = mn\mathbf{v}$ is the momentum density, e (> 0) is the electronic charge, $\mathbf{E}$ is the electric field, T is the temperature in energy units, $W = \frac{3}{2}nT + \frac{1}{2}mnv^2$ is the energy density, $\mathbf{q}$ is the heat flow vector, the subscript c indicates collision terms, ϵ is the dielectric constant, ϕ is the electric potential, N_D is the density of donors, and N_A is the density of acceptors. Eq. (2) expresses conservation of electron number, Eq. (3) expresses conservation of momentum, Eq. (4) expresses conservation of energy, and Eq. (5) is Poisson's equation for the electric potential.

The transport equations (2)–(4) were derived by Bløtekjær [4] as the first three moments of the Boltzmann equation. The moment expansion is closed at three moments by assuming the Fourier law for heat conduction

$$(6) \qquad \mathbf{q} = -\kappa\nabla T$$

where for semiconductors the thermal conductivity $\kappa = \kappa(n, \ T, \ T_0, \ N_D + N_A)$.

Eqs. (2)–(4) are in conservation form, and may be written in terms of the variables $n, \mathbf{v}, T$, and ϕ. These variables represent the simplest choice for upwind methods.

In the absence of heat conduction ($\kappa = 0$ in Eq. (4)), Eqs. (2)–(4) are the Euler equations [5] of gas dynamics with source terms due to the collision terms and the electric field. Eqs. (2)–(4) are hyperbolic (5 modes) in this case, and the soundspeed $c = \sqrt{\gamma T/m}$. The polytropic gas constant $\gamma = \frac{5}{3}$ for the "monatomic" electron gas. There are five nonlinear waves in the model: two shock waves (with characteristic speeds $v \pm c$, where v is the velocity normal to the wave) and three contact waves (with characteristic speed v). Two contact waves may be labeled by a jump in the tangential velocity $\mathbf{v}_T$ across the wave, and one wave by a jump in the temperature T. With heat conduction (nonzero κ), Eqs. (2)–(4) are hyperbolic (4 modes) plus parabolic (1 mode). The contact wave corresponding to a discontinuity in T has disappeared due to the parabolic heat conduction term $\nabla \cdot (\kappa\nabla T)$ in Eq. (4), and the soundspeed $c = \sqrt{T/m}$. The limit $\kappa \to 0$ is a singular limit, with a discontinuous change in the soundspeed and the number of nonlinear waves [2].

3. The $n^+ - n - n^+$ diode problem. The $n^+ - n - n^+$ diode models electron flow in the channel of a MOSFET, and exhibits hot electron effects at scales on the order of a micron. The diode consists of an n^+ "source" region, an n "channel" region, and an n^+ "drain" region (see Fig. 1).

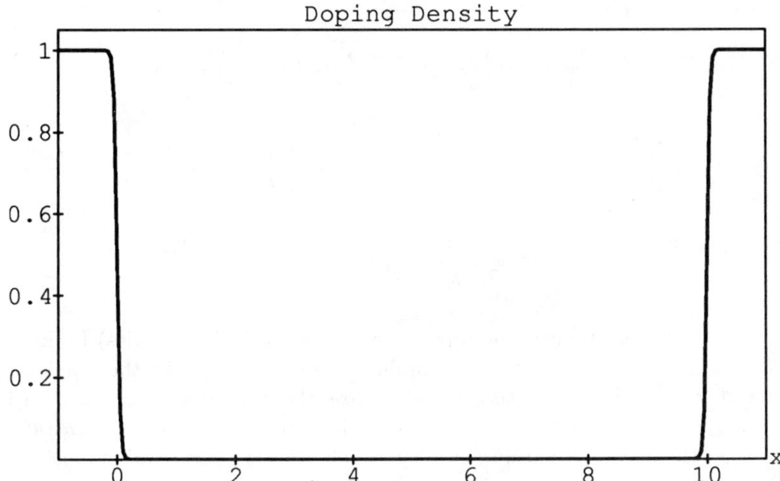

FIG. 1. Doping profile in 10^{18} cm^{-3} for a 1 micron channel device at 77 K. x is in 0.1 microns for all figures.

Since the effects of holes may be neglected for the $n^+ - n - n^+$ diode problem, in Eq. (2)

$$\left(\frac{\partial n}{\partial t}\right)_c = 0. \tag{7}$$

The collision terms in Eqs. (3) and (4) are approximated in terms of momentum and energy relaxation times. For the relaxation times and the thermal conductivity, I take the Baccarani-Wordeman models [6], with τ_w modified from the summary in Ref. [7]:

$$\left(\frac{\partial \mathbf{p}}{\partial t}\right)_c = \frac{-\mathbf{p}}{\tau_p}, \quad \tau_p = m\frac{\mu_{n0}}{e}\frac{T_0}{T} \tag{8}$$

$$\left(\frac{\partial W}{\partial t}\right)_c = \frac{-\left(W - \frac{3}{2}nT_0\right)}{\tau_w}, \quad \tau_w = \frac{m}{2}\frac{\mu_{n0}}{e}\frac{T_0}{T} + \frac{3}{2}\frac{\mu_{n0}}{ev_s^2}T_0 \tag{9}$$

where T_0 is the ambient device temperature, $\mu_{n0} = \mu_{n0}(T_0, N_D + N_A)$ is the low field electron mobility, and $v_s = v_s(T_0)$ is the saturation velocity. The models for τ_p and τ_w include the effects of electron–phonon and electron–impurity collisions. The thermal conductivity κ is specified by the Wiedemann-Franz law

$$\kappa = \kappa_0 \frac{\mu_{n0}}{e} n T_0 \tag{10}$$

where κ_0 is a positive constant.

In one dimension, the hydrodynamic model consists of three nonlinear conservation laws (for electron number, momentum, and energy), plus Poisson's equation for the electric potential:

$$(11) \qquad f_n = \frac{d}{dx}(nv) = 0$$

$$(12) \qquad f_v = \frac{d}{dx}(mnv^2) - en\frac{d\phi}{dx} + \frac{d}{dx}(nT) + \frac{mnv}{\tau_p} = 0$$

$$(13) \qquad f_T = \frac{d}{dx}\left(\frac{5}{2}nvT + \frac{1}{2}mnv^3 - env\phi\right) - \frac{d}{dx}\left(\kappa\frac{dT}{dx}\right) + \frac{\frac{1}{2}mnv^2 + \frac{3}{2}n(T - T_0)}{\tau_w} = 0$$

$$(14) \qquad f_\phi = \epsilon\frac{d^2\phi}{dx^2} + e(N - n) = 0.$$

For boundary conditions (assuming subsonic flow at the boundaries) I take charge neutral contacts ($n = N$) in thermal equilibrium ($T = T_0$) with the ambient temperature at x_{min} and x_{max}, with a bias V across the device: $e\phi(x_{min}) = T\ln(n/n_i)$ and $e\phi(x_{max}) = T\ln(n/n_i) + eV$, where n_i is the intrinsic electron concentration. If $\kappa_0 = 0$, then the energy equation (4) is hyperbolic rather than parabolic, and I drop one boundary condition and specify $n = N$ only at x_{min}.

The variables n, T, and ϕ are defined at the grid points $i = 0, 1, \cdots, N-1, N$, while the velocity v is defined at the midpoints of the elements l_i ($i = 1, \cdots, N$) connecting grid points $i - 1$ and i. The boundary conditions specify n, T, and ϕ at $i = 0$ and $i = N$. Eqs. (11), (13), and (14) are enforced at the interior grid points $i = 1, \cdots, N - 1$, while Eq. (12) is enforced at the midpoints of the elements l_i, $i = 1, \cdots, N$. The discrete equations are then derived by using the second upwind method.

Eqs. (11)–(14) have the form

$$(15) \qquad \begin{bmatrix} f_n \\ f_v \\ f_T \\ f_\phi \end{bmatrix} = \frac{d}{dx}\begin{bmatrix} vg_n \\ vg_v \\ vg_T \\ 0 \end{bmatrix} + \begin{bmatrix} 0 \\ h_v \\ h_T \\ h_\phi \end{bmatrix} + \begin{bmatrix} 0 \\ s_v \\ s_T \\ s_\phi \end{bmatrix} = 0$$

where

$$(16) \qquad g_n = n, \quad g_v = mnv, \quad g_T = \frac{5}{2}nT + \frac{1}{2}mnv^2 - en\phi$$

$$(17) \qquad h_v = -en\frac{d\phi}{dx} + \frac{d}{dx}(nT), \quad h_T = -\frac{d}{dx}\left(\kappa\frac{dT}{dx}\right), \quad h_\phi = \epsilon\frac{d^2\phi}{dx^2}$$

and where the source terms s_v, s_T, and s_ϕ depend only on n, v, and T. In the second upwind method, the advection terms $d(vg)/dx$ in Eq. (15) are discretized using second upwind differences

$$(18) \qquad \frac{d}{dx}(vg)_i \approx (v_{i+1}g_R - v_ig_L)/\Delta x$$

where

$$(19) \qquad g_R = \begin{cases} g_i & (v_{i+1} > 0) \\ g_{i+1} & (v_{i+1} < 0) \end{cases}, \quad g_L = \begin{cases} g_{i-1} & (v_i > 0) \\ g_i & (v_i < 0) \end{cases}$$

and central differences are used for h_v, h_T, and h_ϕ. Recall that the velocity v_{i+1} is located at the midpoint of element l_{i+1}.

To linearize the discretized version of Eqs. (11)–(14), I use Newton's method:

$$(20) \qquad J \begin{bmatrix} \delta n \\ \delta v \\ \delta T \\ \delta \phi \end{bmatrix} = - \begin{bmatrix} f_n \\ f_v \\ f_T \\ f_\phi \end{bmatrix} = -f, \quad \begin{bmatrix} n \\ v \\ T \\ \phi \end{bmatrix} \leftarrow \begin{bmatrix} n \\ v \\ T \\ \phi \end{bmatrix} + t \begin{bmatrix} \delta n \\ \delta v \\ \delta T \\ \delta \phi \end{bmatrix}$$

where J is the Jacobian and t is a damping factor [8] between 0 and 1, chosen to insure that the norm of the residual f decreases monotonically. The Newton method converges quadratically.

Physical parameters. In silicon, the effective electron mass $m = 0.24\ m_e$ at 77 K, where m_e is the electron mass, $\epsilon = 11.7$, and $n_i = 2.84 \times 10^{-20}$ cm^{-3} at 77 K. I use the following model for μ_{n0} [6, 7] at 77 K:

$$(21) \qquad \mu_{n0}(T0 = 77\ \text{K},\ N_i) = \frac{\Delta\mu}{1 + (N_i/N_{ref})^\alpha}$$

$$(22) \qquad \Delta\mu = 18000\ \frac{\text{cm}^2}{\text{Vs}}$$

$$(23) \qquad N_{ref} = 1.44 \times 10^{15}\ \text{cm}^{-3}$$

$$(24) \qquad \alpha = 0.659$$

where $N_i = N_D + N_A$ is the total impurity concentration. The value for $\Delta\mu$ is taken from Monte Carlo simulations of pure Si at 77 K [9]. I take $v_s = 1.25 \times 10^7$ cm/s.

4. Hydrodynamic computations of the electron shock waves. I will present simulations of the $n^+ - n - n^+$ diode in which a shock profile develops in the channel as the supersonic flow on entering the channel breaks to a subsonic flow, in analogy with gas dynamical flow in a Laval nozzle[1]. The $n^+ - n - n^+$ doping of the diode corresponds to the converging/diverging geometry of the Laval nozzle.

The hydrodynamic shocks simulated below have a more complicated structure than shocks supported by the Euler equations of gas dynamics, due to the heat conduction term $\nabla \cdot (\kappa \nabla T)$ in Eq. (4), the relaxation time source terms in Eqs. (3) and (4), and the coupling of the electron gas to the electric field. A shock wave for the inviscid Euler equations is an exact discontinuity in n, v, and T, where v is the normal velocity of the gas [5]. In gas dynamics with heat conduction and viscosity, a shock wave varies rapidly but smoothly over a short viscous length scale [5, 11, 12].

---

[1] See Ref. [10] for an analysis of a steady-state shock wave in a Laval nozzle.

With heat conduction and no viscosity, the situation is more complex [11, 12]. In the frame of the shock, the flow ahead of the shock is supersonic and behind the shock is subsonic. If the Mach number M of the flow ahead of the shock is greater than one but less than a critical Mach number M_c, then the shock is a spread-out smooth profile. If $M > M_c$, then there is an outer profile to the shock wave with an inner exact discontinuity [11]. ($M_c = v_c/c \approx 1.74$ for a polytropic gas with $\gamma = 5/3$, where $c = \sqrt{T/m}$.)

Thus the traditional hyperbolic discontinuous shock wave of gas dynamics ($\kappa_0 = 0$) is spread out due to the parabolic heat conduction term. In other words, the electron shock waves in the hydrodynamic model have both parabolic and hyperbolic aspects.

I will present numerical evidence that the width of the outer shock profile scales linearly with κ_0 in Eq. (10). Thus in the limit of vanishing heat conduction ($\kappa_0 \to 0$), I obtain the traditional discontinuous shock wave of gas dynamics.

For the shock computations, I take a diode consisting of a 0.1 micron source, a 1.0 micron channel, and a 0.1 micron drain. In the n^+ region, the doping density $N = 10^{18}$ cm^{-3} at 77 K, while in the n region $N = 10^{15}$ cm^{-3} at 77 K (see Fig. 1). (I use a hyperbolic tangent fitting over $\pm$ 0.05 microns at the junctions.) The ambient device temperature $T_0 = 77$ K $= 0.00665$ eV.

Figs. 2–5 present a family of electron shock waves at $V = 1$ volt with 240 grid intervals[2]. The shock profile is most clearly visible in the velocity plots (Fig. 2). For comparison's sake, the DAMOCLES velocity is also plotted. The flow is supersonic at the velocity peaks just inside the channel, and subsonic at the end of the waves where the velocity makes a "bend" to the plateau in the channel. The width of the outer shock profile is roughly defined as the distance between where the wave "breaks" after the velocity peak and the beginning of the velocity plateau where v falls below the soundspeed c (see Figs. 3 and 4). To automate the calculation of the shock widths, I will define the shock width as

$$(25) \qquad \text{width} \equiv (v_{max} - v_{min})/\max\{|\Delta v/\Delta x|\}$$

over the wave, where Δv is the change in v over Δx. The Mach numbers and widths of the shock waves are presented in Table 1. The width of the shock wave shows an approximately linear scaling with κ_0 in the range $0 \le \kappa_0 \le 0.5$. For $0.1 > \kappa_0 \to 0$, the width of the computed shock wave is dominated by numerical and artificial viscosity effects. On the other hand, as κ_0 increases beyond 0.5, the width of the shock wave becomes harder and harder to define, due to the spreading out of the wave.

Note that there is an unphysical velocity "spike" near the channel–drain junction for $\kappa_0 \ge 0.25$. However all the "spikes" are subsonic.

Fig. 4 is a detail of the velocity plot for $\kappa_0 = 0$, showing the resolution of a discontinuous shock wave over $\sim 4\Delta x$. A small amount of explicit artificial viscosity of the form

$$(26) \qquad \nu \max\{|v| + c\} \, \Delta x \, m \frac{d^2(nv)}{dx^2}$$

---

[2] The simulation of the 77 K shock wave has converged under mesh refinement with 240 grid intervals. See Ref. [1].

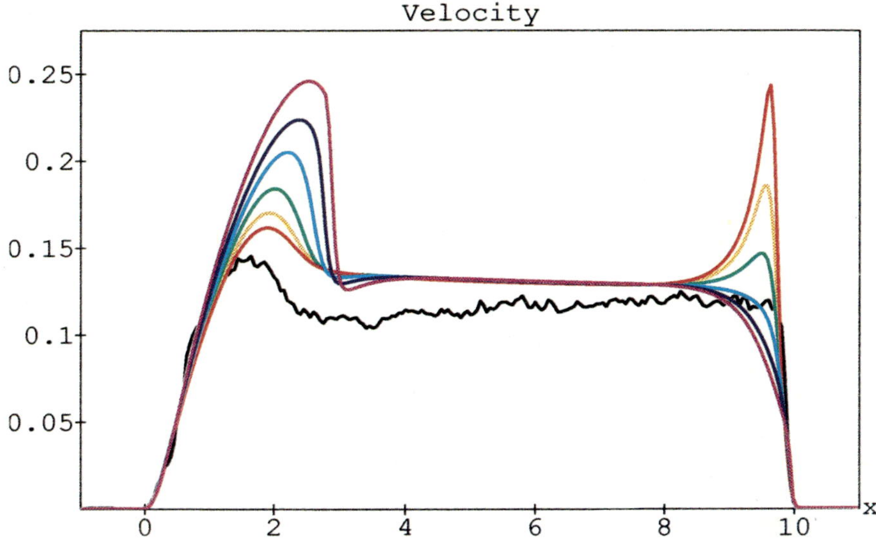

FIG. 2. Electron velocities in 10^8 cm/s for $V = 1$ volt, 1 micron channel, 77 K, $\kappa_0 = 0$ (violet), 0.1 (dark blue), 0.25 (light blue), 0.5 (green), 0.75 (orange), and 1.0 (red). The black jagged curve is the DAMOCLES result.

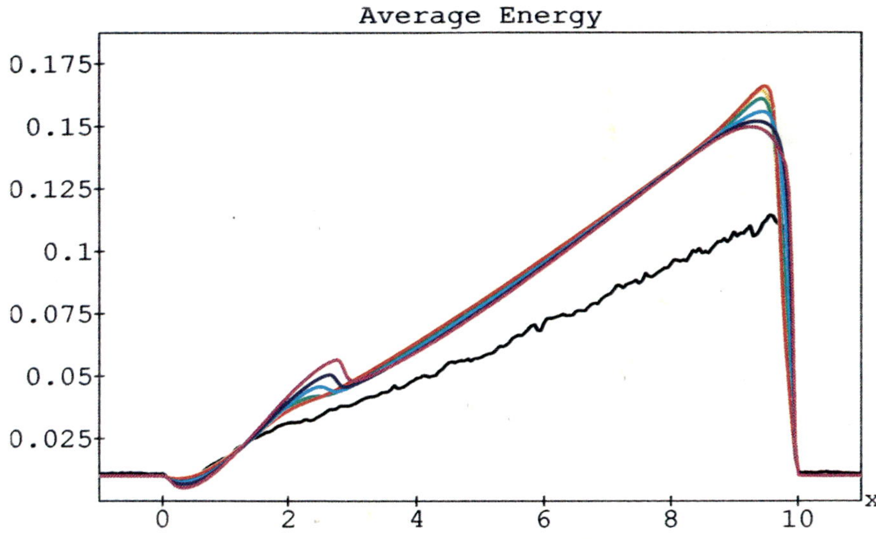

FIG. 5. Average electron energy in eV for $V = 1$ volt, 1 micron channel, 77 K, $\kappa_0 = 0$ (violet), 0.1 (dark blue), 0.25 (light blue), 0.5 (green), 0.75 (orange), and 1.0 (red). The black jagged curve is the DAMOCLES result.

TABLE 1

Mach numbers and widths (in 0.1 microns) of the electron shock waves vs. the amount of heat conduction κ_0.

κ_0	Mach number	width
0	3.8	0.16
0.1	3.2	0.20
0.25	2.8	0.31
0.5	2.4	0.45
0.75	2.2	0.51
1	2.0	0.55

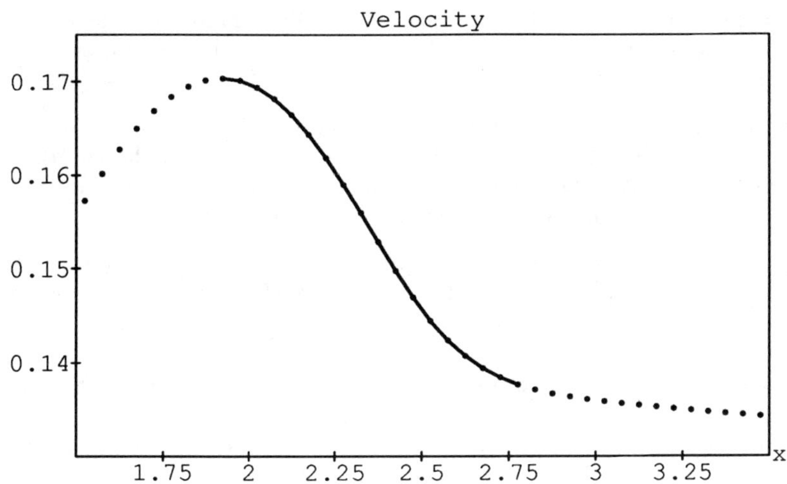

FIG. 3. Detail of electron velocity in 10^8 cm/s for $\kappa_0 = 0.75$, $V = 1$ volt, 1 micron channel, 77 K, illustrating the "width" of the hyperbolic/parabolic shock wave (solid curve) spread out by the large amount of heat conduction in the hydrodynamic model.

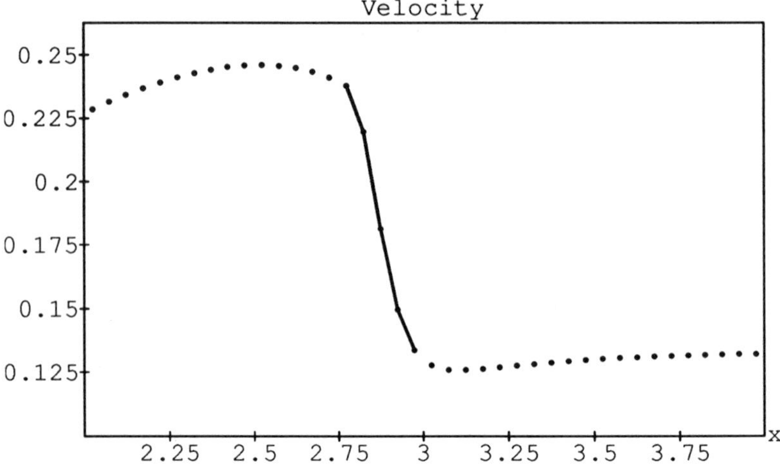

FIG. 4. Detail of electron velocity in 10^8 cm/s for $\kappa_0 = 0$, $V = 1$ volt, 1 micron channel, 77 K, showing the traditional discontinuous shock wave of gas dynamics.

with $\nu = 0.03$ is necessary on the right hand side of Eq. (12) to eliminate oscillations ahead of the shock wave in this case. The shock wave can be resolved over $\sim 2\Delta x$ using a high-order ENO upwind scheme [13].

Fig. 5 demonstrates that the average electron energy is not very sensitive to the amount of heat conduction in the hydrodynamic model. Again the DAMOCLES average energy is shown for the sake of comparison. Note that the hydrodynamic energy with the Baccarani-Wordeman models is consistently too high.

Also note the evidence for an inner discontinuous shock in the average energy curves for $\kappa = 0.1$ and 0.25, since a discontinuity in dT/dx implies a discontinuous jump in v for the inner shock wave[3].

5. Comparison of hydrodynamic and Monte Carlo computations. The "best fit" hydrodynamic simulation uses a two-parameter fit on the amount of heat conduction, with $\kappa_{0L} = 0.75$ at the source–channel junction and $\kappa_{0R} = 0.3$ at the channel-drain junction. In the channel, κ_0 is linearly interpolated between κ_{0L} and κ_{0R}. Fig. 6 shows the DAMOCLES vs. "best fit" hydrodynamic electron velocity. The DAMOCLES velocity exhibits a Mach 2.2 shock profile (spread out due to heat conduction) based on both internal evidence[4] and comparison with the hydrodynamic simulations.

The average electron energies are plotted in Fig. 7. The DAMOCLES simulation indicates two electron temperatures near the channel–drain junction: "hot" 928 K

---

[3] Integrating Eq. (13) over a wave yields $1/2mnv[v^2] = \kappa[dT/dx]$, where $[\psi]$ indicates the jump $\psi_+ - \psi_-$ across the wave [1].

[4] The electron temperature $T \approx 77$ K at the shock wave. Using the effective electron mass approximation, $M = v/c = v/\sqrt{T/m} \approx 2.2$.

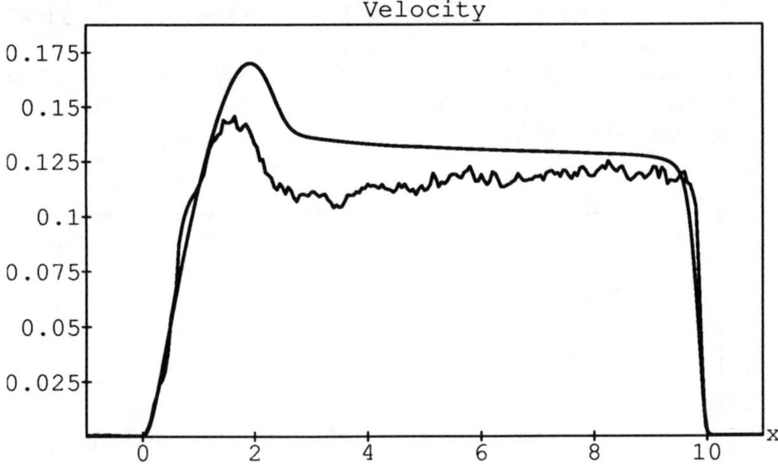

FIG. 6. DAMOCLES vs. "best fit" hydrodynamic electron velocity in 10^8 cm/s for $V = 1$ volt, 1 micron channel, 77 K.

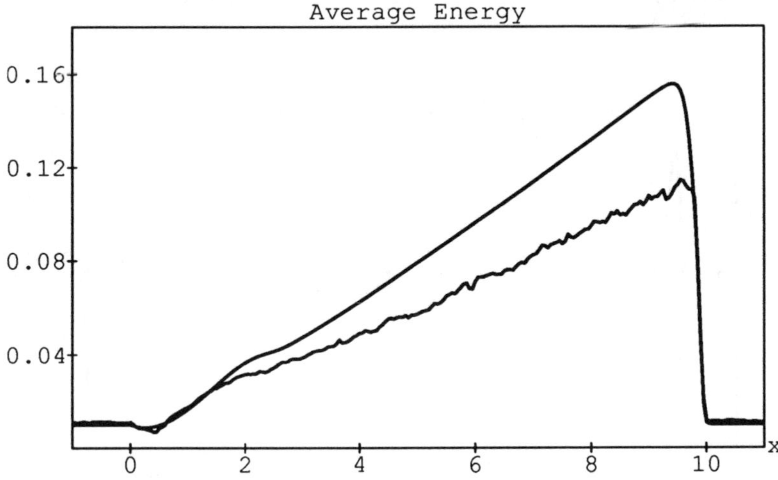

FIG. 7. DAMOCLES vs. "best fit" hydrodynamic average electron energy in eV for $V = 1$ volt, 1 micron channel, 77 K.

electrons flowing from the channel, and "cold" 77 K electrons diffusing from the drain. This feature of the $n^+ - n - n^+$ diode was pointed out in Ref. [14]. The "best fit" hydrodynamic temperature with the Baccarani-Wordeman models again is too high as shown in Fig. 8. The peak electron temperature with the hydrodynamic model is approximately 1100 K.

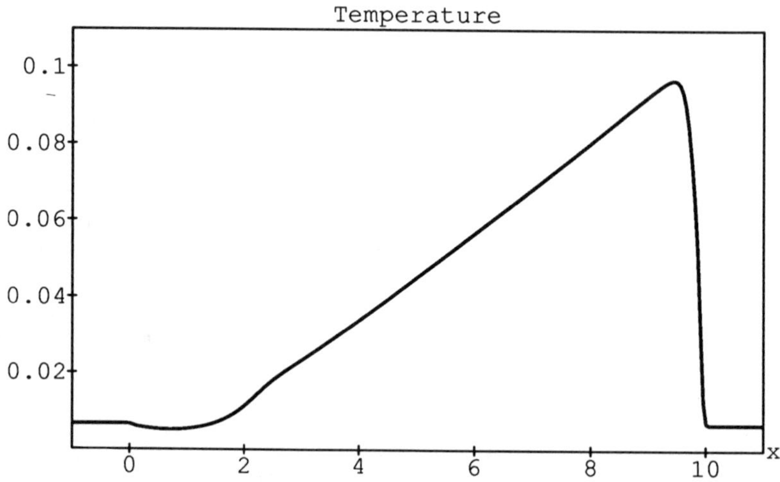

FIG. 8. Electron temperature in eV for the "best fit" hydrodynamic simulation for $V = 1$ volt, 1 micron channel, 77 K.

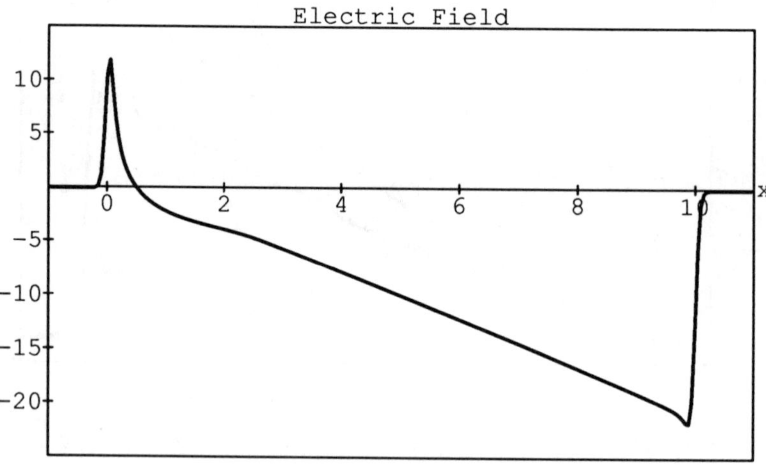

FIG. 9. Electric field in kV/cm for the "best fit" hydrodynamic simulation for $V = 1$ volt, 1 micron channel, 77 K.

The electric field is shown in Fig. 9. The hydrodynamic and DAMOCLES electric potentials basically agree to within the plotting line width.

6. Conclusion. Good agreement between the hydrodynamic and Monte Carlo methods for simulating the electron shock wave can be obtained at a cost of adjusting the amount of heat conduction in the hydrodynamic model with two parameters κ_{0L} and κ_{0R}. This indicates that the Wiedemann-Franz law (10) or even the Fourier law (6) is incorrect, and a better model for heat conduction is essential for making quantitative predictions using the hydrodynamic model.

By using different models for τ_p and τ_w, I expect that better quantitative agreement between the DAMOCLES and hydrodynamic results can be obtained.

The velocity is the variable most sensitive to the amount of heat conduction in the hydrodynamic model. Thus if the average electron energy is physically the most important variable, then the exact value of κ_0 is not very important.

Both internal evidence and comparison with hydrodynamic simulations suggest that the "velocity overshoot" in the DAMOCLES velocity plot should be interpreted as a spread-out Mach 2.2 shock wave. Thus the hydrodynamic prediction of an electron shock wave [1] in Si at 77 K has been confirmed by Monte Carlo simulation of the Boltzmann equation.

Acknowledgement. I would like to thank Steven Laux of the IBM Thomas J. Watson Research Center for providing the DAMOCLES results.

REFERENCES

[1] C. L. Gardner, "Numerical simulation of a steady-state electron shock wave in a submicrometer semiconductor device," *IEEE Transactions on Electron Devices*, vol. 38, pp. 392–398, 1991.

[2] C. L. Gardner, J. W. Jerome, and D. J. Rose, "Numerical methods for the hydrodynamic device model: Subsonic flow," *IEEE Transactions on Computer-Aided Design of Integrated Circuits and Systems*, vol. 8, pp. 501–507, 1989.

[3] M. V. Fischetti and S. E. Laux, "Monte Carlo analysis of electron transport in small semiconductor devices including band-structure and space-charge effects," *Physical Review B*, vol. 38, pp. 9721–9745, 1988.

[4] K. Bløtekjær, "Transport equations for electrons in two-valley semiconductors," *IEEE Transactions on Electron Devices*, vol. ED-17, pp. 38–47, 1970.

[5] R. Courant and K. O. Friedrichs, *Supersonic Flow and Shock Waves.* New York: Springer-Verlag, 1948.

[6] G. Baccarani and M. R. Wordeman, "An investigation of steady-state velocity overshoot effects in Si and GaAs devices," *Solid State Electronics*, vol. 28, pp. 407–416, 1985.

[7] S. Selberherr, "MOS device modeling at 77 K," *IEEE Transactions on Electron Devices*, vol. 36, pp. 1464–1474, 1989.

[8] R. E. Bank and D. J. Rose, "Global approximate Newton methods," *Numerische Mathematik*, vol. 37, pp. 279–295, 1981.

[9] M. V. Fischetti, "Monte Carlo simulation of transport in technologically significant semiconductors of the diamond and zinc-blende structures—Part I: Homogeneous transport," *IEEE Transactions on Electron Devices*, vol. 38, pp. 634–649, 1991.

[10] J. Glimm, G. Marshall, and B. Plohr, "A generalized Riemann problem for quasi-one-dimensional gas flows," *Advances in Applied Mathematics*, vol. 5, pp. 1–30, 1984.

[11] Ya. B. Zel'dovich and Yu. P. Raizer, *Physics of Shock Waves and High-Temperature Hydrodynamic Phenomena.* New York: Academic Press, 1966.

[12] R. Menikoff and B. J. Plohr, "The Riemann problem for fluid flow of real materials," *Reviews of Modern Physics*, vol. 61, pp. 75–130, 1989.

[13] C. L. Gardner and E. Fatemi, "Analysis of the electron shock wave in the hydrodynamic model for semiconductor devices," in preparation.

[14] M. R. Pinto, W. M. Coughran, C. S. Rafferty, R. K. Smith, and E. Sangiorgi, "Device simula-

tion for silicon ULSI," in *Computational Electronics: Semiconductor Transport and Device Simulation*, Boston: Kluwer Academic Publishers, 1991.

MACROSCOPIC AND MICROSCOPIC APPROACH FOR THE SIMULATION OF SHORT DEVICES

A. GNUDI*, D. VENTURA*, G. BACCARANI* AND F. ODEH**

Introduction. The rapid development of semiconductor technology has made possible the industrial fabrication of semiconductor devices with characteristic lengths smaller than $1\,\mu$m. Such devices are characterized by *i)* very large electric fields and *ii)* very rapid spatial variations of the electric field and carrier concentrations. These variations occur over distances comparable with the characteristic lengths of the carrier transport, such as the average momentum and energy relaxation lengths. Because of point *i)* the basic assumption of the drift-diffusion transport model, i.e. small perturbations around equilibrium, breaks down. Usually, empirical corrections to the drift-diffusion model are introduced to account for some off-equilibrium effects. For example, velocity saturation at large electric fields can be achieved by making the mobility field-dependent. This, however, does not solve the problems which arise from point *ii)*. It is therefore necessary to resort to more detailed mathematical models, which are more strongly connected with the basic microscopic transport picture.

From a practical point of view, an important concern of the device designer is to keep under control such effects as MOSFET's substrate currents and hot carrier injection into the gate oxide (which limits the reliability of the device), or the base-collector breakdown voltage in high-speed BJT's (which represents a serious limit to the scaling rules). It is thus desirable to have simulation tools which can take into account and predict with reasonable approximation all such physical effects. Several ways have been explored, ranging from simple modifications of the drift-diffusion model [1], to a full Monte Carlo particle simulation [2].

The purpose of this note is to briefly describe the application of two approaches towards modeling high-energy related effects in small devices. The first approach is based on the *hydrodynamic* model, which is a macroscopic approximation to the Boltzmann Transport Equation (BTE) beyond drift-diffusion, whereas the second one is a low-order representation of the BTE based on projection on an appropriate basis. We also discuss the use of these two techniques for the calculation of MOSFET's substrate currents and BJT's collector current multiplication factor.

In Section I the fundamental features of the hydrodynamic model are recalled and a model for impact ionization consistent with the basic assumptions of hydrodynamic transport is discussed. An application to the computation of MOSFET's substrate currents is presented in Section II. In Section III we describe a microscopic transport model based on spherical harmonics expansion of the BTE. This technique is applied in Section IV to the simulation of a thin-base BJT and to the computation of the collector current multiplication factor.

---

*University of Bologna, Department of Electronics, viale Risorgimento 2, 40136 Bologna, ITALY.

**IBM T.J. Watson Research Center, Box 218 Yorktown Heights, NY.

I. The hydrodynamic model and impact ionization. For the sake of completeness, we start by recalling the equations of the hydrodynamic model for electrons in steady-state

$$(\text{I.1,a}) \qquad U - \frac{1}{q}\,\text{div}(\mathbf{J}_n) = 0$$

$$(\text{I.1,b}) \quad \mathbf{J}_n - \frac{\tau_{pn}}{q}\,(\mathbf{J}_n \bullet \text{grad})\,\frac{\mathbf{J}_n}{n} = q\,\mu_n\left[\frac{k_B T_n}{q}\,\text{grad}(n) + n\,\text{grad}\left(\frac{k_B T_n}{q} - \varphi\right)\right]$$

$$(\text{I.1,c}) \qquad -\,\text{div}\left[\kappa_n\,\text{grad}(T_n) + \frac{\mathbf{J}_n}{q}\,(w + k_B T_n)\right] = \mathbf{E} \bullet \mathbf{J}_n - n\,\frac{w - w_0}{\tau_{wn}} - U\,w$$

In the above, n and $\mathbf{J}_n$ denote the electron and the current densities resp., w is the electron average energy, related to the mean velocity v and to the effective electron temperature T_n by the expression

$$(\text{I.2}) \qquad w = \frac{1}{2}mv^2 + \frac{3}{2}k_B T_n\,,$$

w_0 is the equilibrium energy, φ and $\mathbf{E}$ are the electrostatic potential and field, which are governed by Poisson equation. The expression (I.2) interpretes the average energy as the sum of the contributions from the average drift velocity and from the random motion around that average.

The equations are derived by the usual moments' method applied to the BTE under the following simplifying assumptions *i)* the temperature tensor (defined as the second moment of the particle distribution function around its mean) reduces to a scalar, *ii)* the hierarchy of moments is closed by relating the *heat flow* of the electron gas $\mathbf{Q}$ to the electron temperature via $\mathbf{Q} = -\kappa_n\,\text{grad}(T_n)$, where κ_n is the *thermal conductivity*, given by Wiedmann-Franz law $\kappa_n = (5/2 + c)\,(k_B/q)^2\,\sigma_n\,T_n$, *iii)* the effects of collisions on momentum and energy are modelled by the relaxation time approximation. In the fundamental equations τ_{pn} and τ_{wn} denote the momentum and energy relaxation times and $\mu_n = (q/m)\,\tau_{pn}$. Usually the mobility μ_n is modelled as a decreasing function of the average energy, due to the increase of the scattering rates with the particle energy.

We now mention some of the basic features which distinguish the model from the standard drift-diffusion. First is the introduction of the energy balance equation (I.1,c). At the right-hand-side we recognize *i)* the forcing term $\mathbf{E} \bullet \mathbf{J}_n$, which represents the energy absorbed by the electrons from the electric field per unit time and volume, *ii)* the energy $n\,(w - w_0)/\tau_{wn}$ delivered by the electrons to the lattice per unit time and volume due to collisions with optical phonons, *iii)* the energy lost in the recombination processes $U\,w$. The algebraic sum of the three terms equals the spatial variation of the energy flow at the left-hand-side, which is the sum of the heat flow and the transported energy. Due to the non-zero energy-relaxation time, the mean energy w and the electron temperature T_n turns out to be larger, in regions of large electric fields, than their equilibrium values. Second, in the momentum balance equation, and apart from the non-linear convective term $(\tau_{pn}/q)(\mathbf{J}_n \bullet \mathrm{grad})(\mathbf{J}_n/n)$, there are two distinguishing features from the drift-diffusion model. One is the existence of a *thermoelectric field*, the gradient of $k_B T_n/q$, which provides an additional driving force that accounts for the tendency of electrons to diffuse from hot toward cold regions. The other is the relation between diffusion coefficient and mobility given by

$$\text{(I.3)} \qquad\qquad D_n = \frac{k_B T_n}{q}\,\mu_n$$

which involves the electron temperature T_n rather than the lattice temperature. Thus in hot regions, where T_n is large, the hydrodynamic model predicts a larger diffusion than the drift-diffusion, reflecting the increased random velocity over the equilibrium condition.

Another important feature stems from the energy equation and from the above mentioned assumptions on the mobility. The combination of these two facts is responsible for *velocity overshoot*. This can be most easily seen in a one-dimensional setting where the electric field is increasing in the direction of motion. A simple consequence of the energy balance equation is that the average energy w is less than the value corresponding to the local field in homogeneous conditions. Thus, since the mobility is a decreasing function of w, it follows that the hydrodynamic velocity is larger than the velocity based on a mobility which is a function of the local electric field, as is commonly assumed in the drift-diffusion model (for details see [3]).

A final remark: the numerical aspects of the hydrodynamic model are somewhat more delicate than those of the drift-diffusion, because of the *i)* increased stiffness of the momentum equation with respect to the drift-diffusion model, *ii)* the complications introduced by the convective term which allows for extreme gradients in velocity and electron density, *iii)* the roughness of the source term $\mathbf{E} \bullet \mathbf{J}$ in the energy equation, due to the difficulty of obtaining a smooth representation of the current density vector $\mathbf{J}$. These issues have been addressed in some specialized papers (e.g. [4]), which show that the above difficulties can be taken care of by means of a generalized Scharfetter-Gummel scheme coupled with careful iteration methods.

We now discuss the modeling of MOSFET's substrate currents due to impact ionization; an issue which is quite important for device designers. Several works have recently appeared on the subject [e.g. 5,6], offering a variety of methods, which range from the use of semi-analytical models, to the direct application of the energy balance equation, to, finally, the combination of the drift-diffusion or the energy balance equations with Monte Carlo, in order to extract more detailed information on the high-energy tail of the distribution.

In the drift-diffusion codes, impact ionization is usually treated in the spirit of the so called *lucky-electron* model, which derives its main concept from the original work of Shockley [7]; there are obvious limitations in short devices due to the neglect of effects caused by fast variations of the electric field. Several proposals were made for incorporating impact ionization in the hydrodynamic equations in a physically sound way; for example, in [5] the ionization coefficient is given by an exponential in the homogeneous electric field associated with the electron temperature, while [6] uses a generalization of the lucky electron concept coupled to the hydrodynamic model.

Another approach, proposed in [8], was adopted in the simulations presented in the next section. A short description follows: one assumes that the even part of the electron energy distribution can be described by a *heated maxwellian* function of the form $f(E) \propto \exp(-E/k_B T_n)$. The microscopic ionization scattering rate is then calculated assuming a Coulomb interaction potential between two colliding electrons, one in the conduction band and the other in the valence band. The resulting generation rate turns out to be a function of the electron temperature, of the form

$$(I.4) \qquad G_{II} = n\, Y(u) = \frac{1}{2\tau_o} \left[\sqrt{u/\pi}\, \exp\left(-1/u\right) - \mathrm{erfc}\left(1/\sqrt{u}\right) \right]$$

where τ_o is a normalization time constant, and $u = (k_B T_n)/E_{th}$ is proportional to the electron mean energy normalized by the ionization threshold energy E_{th}. The effects due to fast spatially varying electric fields are incorporated by means of the T_n dependence, via the solution of the energy balance equation.

The above generation rate enters the continuity equation

$$(I.5) \qquad U - \frac{1}{q}\,\mathrm{div}\left(\mathbf{J}_n\right) = G_{II}$$

and the energy-balance equation

$$(I.6) \quad -\mathrm{div}\left[\kappa\,\mathrm{grad}\left(T_n\right) + \frac{\mathbf{J}_n}{q}(w + k_B T_n) \right] = \mathbf{E} \bullet \mathbf{J}_n - n\,\frac{w - w_0}{\tau_{wn}} - U\,w - E_G\,G_{II}$$

where E_G is the band gap. The term $E_G\,G_{II}$ in (I.6) represents the loss of energy per unit volume and time due to the ionization processes. The coefficients τ_o and τ_{wn} can be considered as adjustable parameters. They may be found by fitting the substrate currents of a particular device, in the way explained in the next section.

II. Simulation of an N-channel MOSFET. We present below the results of the simulation of short n-channel MOSFET's, including impact ionization, and the comparison of the calculated substrate currents with measured data. The simulated structures are the n-channel devices of the $0.25\,\mu$m CMOS technology presented in [9], characterized by an oxide thickness $t_{ox} = 7\,$nm, a junction depth $x_j = 0.1\mu$m and a substrate doping $N_{sub} = 10^{17}\,$cm^{-3}. The devices selected for our comparison have different effective channel lengths, ranging from $0.35\,\mu$m to $4\,\mu$m. The simulations have been carried out with the hydrodynamic version of the two-dimensional device analysis program HFIELDS [10]. The latter solves the electron hydrodynamic equations, including the generation term G_{II} in the form

(I.4) self-consistently with Poisson equation. In an n-channel device, for conditions far from the drain-bulk breakdown, holes are not expected to give any appreciable contribution to the total ionization rate, and hence are treated in the drift-diffusion approximation. We have used the mobility model [11], defined by

$$\text{(II.1,a)} \qquad \mu_n(w) = \frac{\mu_o}{1 + \alpha\,(w - w_o)}$$

$$\text{(II.1,b)} \qquad \alpha = \frac{\mu_o}{q\,\tau_{wn}\,v_{sat}^2}$$

with $\mu_o = \mu_{oo} / (1 + E_\perp / E_{ref})$, where $E_\perp$ is the electric field perpendicular to the Si-SiO$_2$ interface, which accounts for surface mobility degradation. The turn-on characteristics for different channel lengths with $V_{DS} = 50$ mV (Fig. 1) were used to calibrate the parameters μ_{oo} and E_{ref}. The source/drain contact resistances, whose effect is non-negligible in the shortest simulated device ($L = 0.35\,\mu$m), were accounted for by means of external lumped resistances in series to the source/drain contacts.

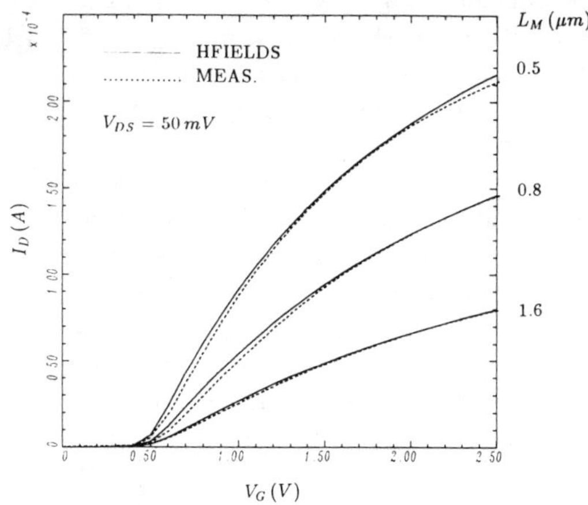

Figure 1. Comparison between measured and simulated turn on characteristics ($V_{DS} = 50\,\text{mV}$) for the n-MOSFET's indicated in the text for three different channel lengths (L_M is the gate mask length).

The high-drain voltage turn on characteristics ($V_{DS} = 2.5\,\text{V}$) show good agreement with the experimental data when $v_{sat} = 9.2 \times 10^6$ cm/sec is used in the expression (II.1,b), in good agreement with the measurements in [12] (Fig. 2). We have observed that the effect of velocity overshoot on the drain current is quite small. In fact, the overshoot is in general quite dependent on the relaxation-time τ_{wn}; however, even in the shortest device, changing τ_{wn} within the typical values in the literature did not significantly affect the drain current. An explanation for this can be found by considering that the velocity overshoot is mainly due to the spatial variation of the electric field, as seen in section I. In MOSFET's, the largest variation of the field occurs near the drain, while the drain current is controlled by the portion of the channel at the source side.

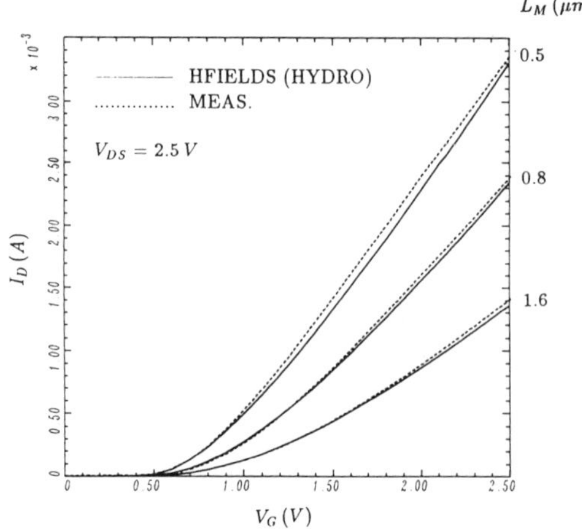

Figure 2. Comparison between measured and simulated high-drain voltage turn on characteristics ($V_{DS} = 2.5\,\text{V}$) for the n-MOSFET's indicated in the text for three different channel lengths (L_M is the gate mask length).

The typical behaviors of the electric potential, electron concentration and normalized electron temperature (T_n/T_L) for the $0.35\,\mu m$ device, ($V_{GS} = 1.5$ V, $V_{DS} = 3.0$ V) are shown in Figs. 3, 4 and 5 respectively. For comparison, we give the corresponding drift-diffusion electron concentration in Fig. 6. The temperature clearly exhibits a ridge all around the high-field region near the drain, and suddenly drops as the neutral drain region is entered. The ripple on the bulk side of the drain junction is due to the aforementioned roughness of the $\mathbf{E} \bullet \mathbf{J}$ term, and to the difficulty of accurately resolving the very rapid variation of temperature.

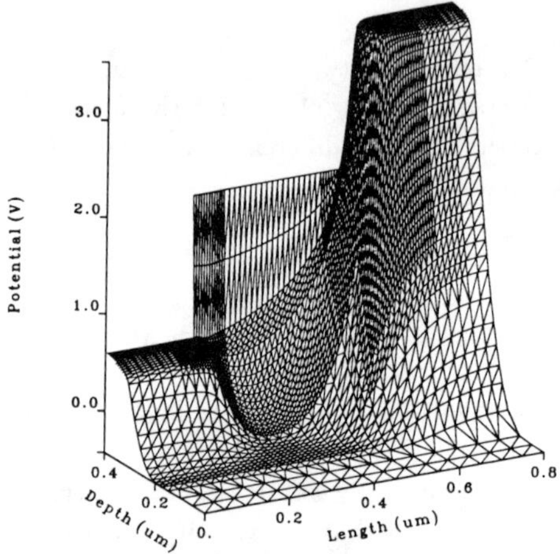

Figure 3. Electric potential in the simulated $0.35\,\mu m$ n-MOSFET ($V_{GS} = 1.5\,\text{V}$, $V_{DS} = 3.0\,\text{V}$).

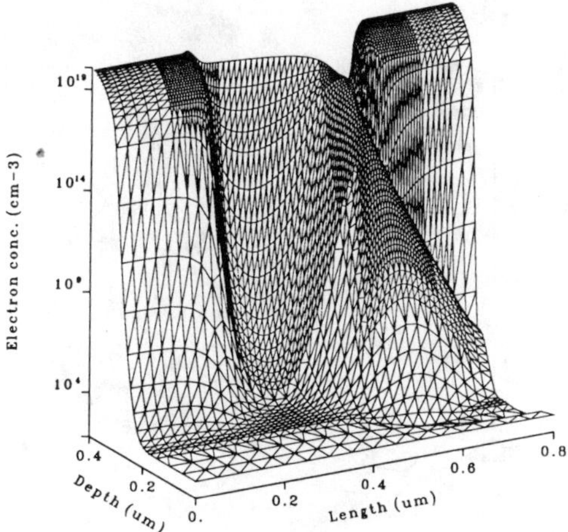

Figure 4. Electron concentration in the simulated $0.35\,\mu$m n-MOSFET ($V_{GS} = 1.5\,$V, $V_{DS} = 3.0\,$V).

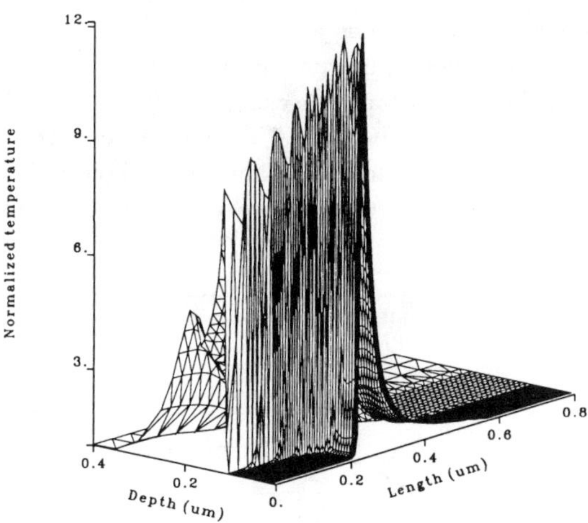

Figure 5. Normalized electron temperature T_n/T_L ($T_L = 300$ K) in the simulated $0.35\,\mu$m n-MOSFET ($V_{GS} = 1.5\,$V, $V_{DS} = 3.0\,$V).

Comparing Figs. 4 and 6 clearly shows the effect of the hydrodynamic increased electron diffusivity, due to the generalized Einstein relation (I.3), which results in an electron concentration around the drain junction much higher than that of the drift-diffusion.

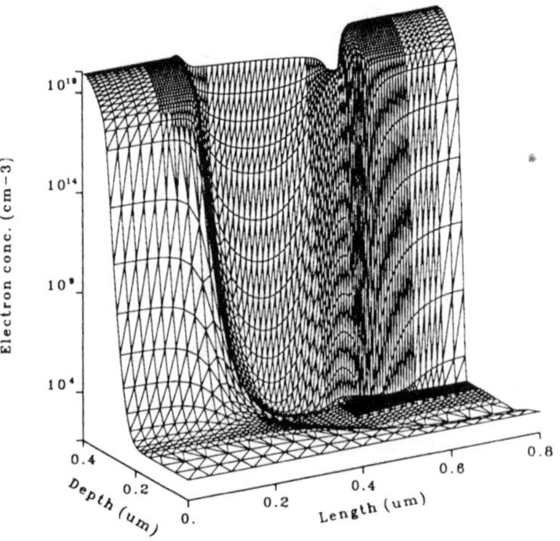

Figure 6. Electron concentration obtained with the drift-diffusion model for the same device and the same bias as for Fig. 4.

A comparison between simulated and measured substrate currents is shown in Fig. 7, for different bias conditions and channel lengths. The two adjustable parameters τ_o and τ_{wn} were found by comparing the measured and the calculated substrate currents for the longest available channel length, for the lowest source-to-drain voltage (Fig. 7,a: $L = 4\,\mu$m, $V_{DS} = 2.5\,$V). The result is $1/\tau_o = 7 \times 10^{15}\,\text{sec}^{-1}$ and $\tau_{wn} = 0.1$ psec. A suitable value for the ionization threshold energy was found to be $E_{th} = 2.0\,$eV, which is quite close to the 1.8 eV extracted experimentally in [13].

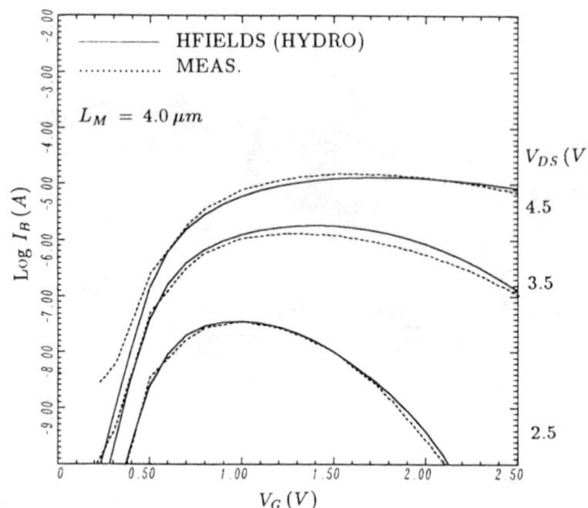

Figure 7,a. Comparison between simulated (hydrodynamic) and measured substrate currents for the n-MOSFET's indicated in the text for different channel lengths (L_M is the gate mask length).

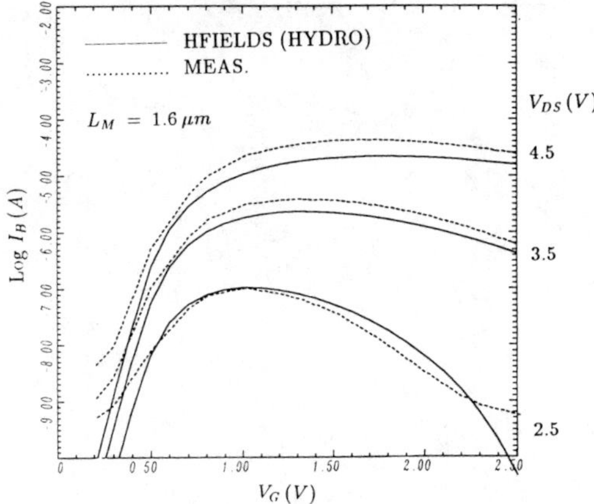

Figure 7,b.

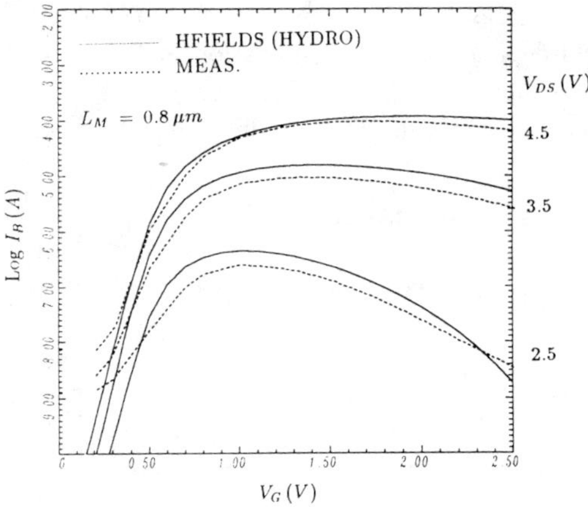

Figure 7,c.

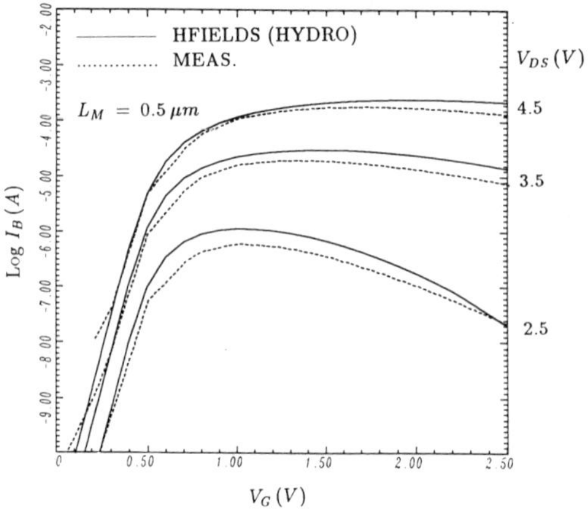

Figure 7,d.

It is seen from Fig. 7 that substrate currents can be predicted with sufficient accuracy for all the considered channel lengths down to $0.35\,\mu m$, for drain voltages up to at least 4.5 V. For comparison, we used the drift-diffusion version of HFIELDS, with an electric-field-dependent impact ionization model which was fitted at small V_{DS}, to compute the substrate current. Fig. 8 shows the main result and exhibits a rather poor agreement with experimental data, especially for high biases.

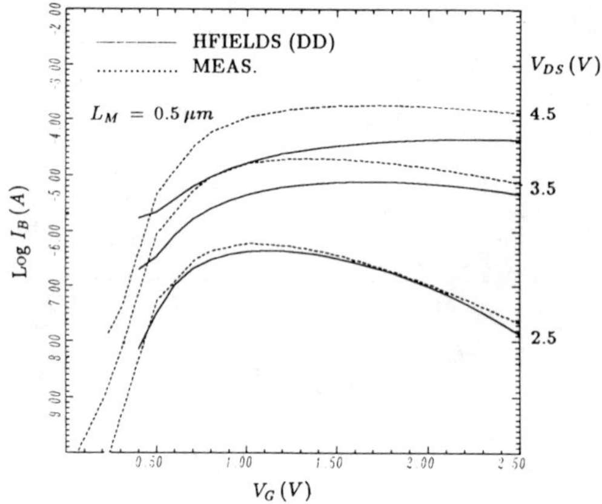

Figure 8. Comparison between simulated (drift-diffusion) and measured substrate currents for the $0.5\,\mu m$ gate mask length.

III. Spherical harmonics and the BTE. Macroscopic models, such as the hydrodynamic model described above, ignore most of the details of the behavior of the electron distribution in the momentum space. In order to regain, at least partially, some of this information - needed for proper modeling of impact ionization - approximate solutions to the BTE have to be obtained. In this section we use a spherical harmonics expansion to reduce the BTE to the energy-space domain (E, x) and show how the resulting system of equations can be solved numerically in one-dimension [14]. A similar approach has been used in [15] for the homogeneous case. In the next section we apply this technique to obtain the electron energy distribution pointwise in a one-dimensional n-p-n BJT as well as the collector current multiplication factor.

Consider the classical stationary, linear BTE for electrons in the conduction band

$$\mathbf{u}_g(\mathbf{k}) \cdot \nabla_{\mathbf{r}} f(\mathbf{r}, \mathbf{k}) - \frac{q}{\hbar} \mathbf{F}(\mathbf{r}) \cdot \nabla_{\mathbf{k}} f(\mathbf{r}, \mathbf{k}) =$$

$$(\text{III.1}) \qquad \int S(\mathbf{k}', \mathbf{k}) \, f(\mathbf{r}, \mathbf{k}') \, \mathrm{d}^3 k' - f(\mathbf{r}, \mathbf{k}) \int S(\mathbf{k}, \mathbf{k}') \, \mathrm{d}^3 k'$$

where $\mathbf{u}_g$ is the group velocity, $S(\mathbf{k}, \mathbf{k}')$ the differential electron scattering probability per unit time from state $\mathbf{k}$ to state $\mathbf{k}'$, and $f(\mathbf{r}, \mathbf{k})$ the distribution function. In (III.1), the exclusion principle and electron-electron scattering, which make the collision operator non linear, are neglected. At any point $\mathbf{r}$ in the position space, we define a local polar coordinate system $(\hat{\mathbf{J}}(\mathbf{r}), \hat{\mathbf{e}}_1(\mathbf{r}), \hat{\mathbf{e}}_2(\mathbf{r}))$, where $\hat{\mathbf{J}}$ is the current-density unit vector, $\hat{\mathbf{e}}_1$ lies in the plane $(\hat{\mathbf{J}}, \mathbf{F})$ and $\hat{\mathbf{e}}_2$ is normal to that plane. Each vector $\mathbf{k}$ is expressed by the local polar coordinates $(k, \theta(\mathbf{r}), \varphi(\mathbf{r}))$, and the occupation probability f is then expanded in spherical harmonics

$$(\text{III.2}) \qquad f(\mathbf{r}, \mathbf{k}) = f(\mathbf{r}, k, \theta, \varphi) = \sum_{l} \sum_{m=-l}^{l} Y_l^m(\theta, \varphi) \, f_l^m(\mathbf{r}, k)$$

where the coefficients f_l^m depend on $\mathbf{k}$ only via the modulus k. We simplify expression (III.2) by assuming f to be symmetric around the polar axis $\hat{\mathbf{J}}$, thus ruling out any φ dependence. Such an assumption is satisfied for a spherically symmetric band in a one-dimensional case, and holds approximately true for a smoothly varying $\hat{\mathbf{J}}$ field. With the above assumption, and truncating the series to the third term, f can now be expressed as

$$(\text{III.3}) \qquad f(\mathbf{r}, \mathbf{k}) = f_0(\mathbf{r}, k) + f_1(\mathbf{r}, k) \cos \theta + f_2(\mathbf{r}, k) \frac{1}{2} (3 \cos^2 \theta - 1) + \dots$$

Equations for the unknown coefficients f_i are derived in a standard way by replacing expression (III.3) into the BTE and balancing the coefficients of the harmonics of the same order. A remark is necessary on the treatment of the collision terms. If we assume that the scattering rate $S(\mathbf{k}, \mathbf{k}')$ depends only on the modulus of $\mathbf{k}$ and

$\mathbf{k}'$ and the angle ξ between them, then, using the addition theorem for Legendre polynomials, one obtains

$$\int S(\mathbf{k}',\mathbf{k})\, f(\mathbf{r},\mathbf{k}')\, \mathrm{d}^3 k' =$$

$$4\pi \int S_0(k',k)\, f_0(\mathbf{r},k')\, k'^2\, \mathrm{d}k' + \frac{4}{3}\pi \cos\theta \int S_1(k',k)\, f_1(\mathbf{r},k')\, k'^2\, \mathrm{d}k' +$$

$$\text{(III.4)} \qquad +\frac{4}{5}\pi \frac{1}{2}(3\cos^2\theta - 1) \int S_2(k',k)\, f_2(\mathbf{r},k')\, k'^2\, \mathrm{d}k' + \dots$$

where $S_i(k',k)$ are the coefficients of the expansion

$$S(\mathbf{k}',\mathbf{k}) = S(k',k,\cos\xi) =$$

$$\text{(III.5)} \qquad S_0(k',k) + S_1(k',k)\cos\xi + S_2(k',k)\frac{1}{2}(3\cos^2\xi - 1) + \dots$$

We point out that only the first two terms of the expansion (III.2) are necessary to fully evaluate the first moments of the distribution. Actually

$$\langle A \rangle \equiv \frac{1}{n} \int f(\mathbf{k}) A(\mathbf{k})\, \mathrm{d}k = \frac{1}{n} \int f(\mathbf{k}) A(\mathbf{k})\, k^2 \,\mathrm{d}k\, \mathrm{d}\Omega =$$

$$= \frac{1}{n} \sum_{i,j,m} \int f_i(k)\, Y_i^0(\theta)\, A_j^m(k)\, Y_j^m(\varphi,\theta)\, \mathrm{d}\Omega\, k^2 \mathrm{d}k =$$

$$\text{(III.6)} \quad = \frac{1}{n} \sum_{i,j,m} \int f_i(k) A_j^m(k) k^2 \mathrm{d}k\, c_i\, c_j\, \delta_i^j\, \delta_0^m = \frac{1}{n} \sum_i \int f_i(k) A_i^0(k)\, c_i^2\, k^2 \mathrm{d}k$$

where the normalization coefficients are defined as $c_i = \sqrt{(4\pi)/(2i+1)}$. The electron concentration n is

$$\text{(III.7)} \qquad n = \int f(\mathbf{k})\mathrm{d}^3 k = 4\pi \int f_0(k)\, k^2\, \mathrm{d}k$$

For the mean energy (assuming spherically symmetric band) we have $A = A_0^0 = E(k)$, and for the mean velocity

$$\text{(III.8)} \qquad \mathbf{A}(\mathbf{k}) = u_g(k)\hat{\mathbf{k}} = u_g(k)(\hat{\mathbf{J}}\cos\theta + \hat{\mathbf{e}}_1 \sin\theta \cos\varphi + \hat{\mathbf{e}}_2 \sin\theta \sin\varphi)$$

Hence only the term $\mathbf{A}_1^0 = u_g(k)\hat{\mathbf{J}}$, involving f_1, differs from zero.

In this paper we shall consider the following scattering mechanisms: elastic acoustic phonons, optical phonons, ionized impurities and impact ionization. The expressions of the scattering rates and the physical parameters are taken from [16]. For acoustic phonons we have

$$\text{(III.9)} \qquad S^{ac}(\mathbf{k}, \mathbf{k}') = S_0^{ac}(k, k') = \frac{2\pi k_B T_0}{\hbar u_l^2 \rho} \varepsilon^2 \delta[E(\mathbf{k}') - E(\mathbf{k})] \equiv c_{ac}\delta(E' - E)$$

where u_l is the longitudinal sound velocity, ρ the semiconductor density and ε the coupling energy.

The optical phonon scattering rate is written as

$$S^{op}(\mathbf{k}, \mathbf{k}') = S_0^{op}(k, k') = \frac{\pi(D_t K)^2}{\rho \omega_{op}} \left[N_{op}^+; N_{op}\right] \delta[E(\mathbf{k}') - E(\mathbf{k}) \pm \hbar\omega_{op}] =$$

$$\text{(III.10)} \qquad\qquad\qquad \equiv c_{op}\left[N_{op}^+; N_{op}\right] \delta(E' - E \pm \hbar\omega_{op})$$

where N_{op} is the phonon occupation number, $N_{op}^+ = N_{op} + 1$, $D_t K$ is the optical coupling energy and $\hbar\omega_{op}$ the optical phonon energy; the upper sign together with N_{op}^+ corresponds to emission, the lower one together with N_{op} corresponds to absorption. We consider impurity scattering according to the Brooks-Herring formulation

$$\text{(III.11)} \qquad S^i(\mathbf{k}, \mathbf{k}') = \frac{Z^2 q^4}{\epsilon_s^2 \hbar} N_i \left[\beta^2 + 2k^2(1 - \cos\xi)\right]^{-2} \delta(E' - E)$$

where N_i is the impurity concentration, β the inverse screening length and ξ the angle between $\mathbf{k}$ and $\mathbf{k}'$. Since impurity scattering is non-isotropic, the function S_i has to be expanded in spherical harmonics as described above to obtain

$$\text{(III.12,a)} \qquad S^i(k, k', \xi) = S_0^i(k, k') + S_1^i(k, k')\cos\xi + \dots$$

where

$$\text{(III.12,b)} \qquad S_0^i(k, k') = \frac{Z^2 q^4}{\epsilon_s^2 \hbar} N_i \left[\beta^2(4k^2 + \beta^2)\right]^{-1} \delta(E' - E)$$

$$\text{(III.12,c)} \quad S_1^i(k, k') = \frac{Z^2 q^4}{\epsilon_s^2 \hbar} N_i \frac{3}{8k^4} \left[\frac{4k^2(\beta^2 + 2k^2)}{\beta^2(\beta^2 + 4k^2)} - \ln\left(\frac{4k^2 + \beta^2}{\beta^2}\right)\right] \delta(E' - E)$$

The scattering matrix for impact ionization, in the Born approximation, is

$$\text{(III.13,a)} \qquad S^{ii}(\mathbf{k}, \mathbf{k}', \mathbf{k}'') = b_{ii}\left[a^2 + (\mathbf{k}' - \mathbf{k})^2\right]^{-2} \delta(E - E' - E'' - E_G)$$

where $\mathbf{k}$ and $\mathbf{k}'$ are the initial and final states of the ionizing electron, $\mathbf{k}''$ is the final state of the electron coming from the valence band, b_{ii} a normalizing constant, E_G

the energy gap and a the inverse screening length. The expansion of (III.13,a) into spherical harmonics analogous to the Eqs. (III.12,b)-(III.12,c) gives

$$(\text{III.13,b}) \quad S_0^{ii}(k, k', k'') = b_{ii} \left[(a^2 + k^2 + k'^2)^2 - 4k'^2 k^2\right]^{-1} \delta(E - E' - E'' - E_G)$$

$$S_1^{ii}(k, k', k'') = \delta(E - E' - E'' - E_G) \frac{3 b_{ii}}{8 k'^2 k^2}$$

$$(\text{III.13,c}) \qquad \times \left[\frac{4 k' k \, (a^2 + k^2 + k'^2)}{(a^2 + k^2 + k'^2)^2 - 4 k'^2 k^2} + \log \frac{a^2 + k^2 + k'^2 - 2 k' k}{a^2 + k^2 + k'^2 + 2 k' k}\right]$$

To account for the two different origins of the incoming electrons, i.e. the valence and the conduction band, we define

$$(\text{III.14,a}) \qquad A(E', E) = \int g(E'') \left[S_0^{ii}(E', E, E'') + S_0^{ii}(E', E'', E)\right] \mathrm{d}E''$$

$$(\text{III.14,b}) \qquad B(E', E) = \int g(E'') S_1^{ii}(E', E, E'') \mathrm{d}E''$$

In this derivation we assumed i) flat valence band, ii) occupation probability of the valence band equal to one.

For a spherically symmetric band structure, the equations for f_0 and f_1 in the (E, x) domain, in one-dimension, then turn out to be

$$(\text{III.15,a}) \qquad \frac{\partial}{\partial x} f_1 - qF \left(\frac{\partial}{\partial E} f_1 + \frac{1}{\gamma} \frac{\mathrm{d}\gamma}{\mathrm{d}E} f_1\right) =$$

$$-3 \, c_{ii} \frac{g(E)}{u_g} f_0(E) + \frac{3}{u_g} \int A(E', E) \, f_0(E') \, g(E') \, \mathrm{d}E' +$$

$$\frac{3 c_{op}}{u_g} \left\{g^+(E) \left[N_{op}^+ f_0^+(E) - N_{op} f_0(E)\right] - g^-(E) \left[N_{op}^+ f_0(E) - N_{op} f_0^-(E)\right]\right\}$$

$$(\text{III.15,b}) \qquad \frac{\partial}{\partial x} f_0 - qF \frac{\partial}{\partial E} f_0 + \frac{1}{\lambda} f_1 - \frac{1}{3 u_g} \int B(E', E) \, f_1(E') \, g(E') \, \mathrm{d}E' = 0$$

In the above equations, $\gamma(E) = (\hbar^2 k^2)/(2 m^*)$ defines the band-shape, $g(E)$ is the density of states defined by $g(E) = 2\pi \, (2 m^*/\hbar^2)^{3/2} \, \gamma^{1/2} \gamma'$, $u_g(E) = (2/m^*)^{1/2} \gamma^{1/2}/\gamma'$ is the group velocity, and $\lambda(E)$ is the carrier mean free path
(III.16)

$$\lambda(E) = \frac{u_g(E)}{c_{ac} \, g(E) + c_{op} \, N_{op}^+ \, g^-(E) + c_{op} \, N_{op} \, g^+(E) + [c_i(E, N_i) + c_{ii}(E)] \, g(E)}$$

where we have used the notation $g^+(E) = g(E + \hbar\omega_{op})$ and a similar notation for the other functions. In (III.15)-(III.16) c_i and c_{ii} are the effective impurity scattering probability and the total impact ionization scattering probability defined by

$$(\text{III.17}) \quad c_i(E) g(E) h(E) = \int \left[S_0^i(E, E') h(E) - \frac{1}{3} S_1^i(E, E') h(E')\right] g(E') \, \mathrm{d}E'$$

$$(\text{III.18}) \qquad c_{ii}(E)\,g(E) = \int S_0^{ii}(E, E', E'')\,g(E')\,g(E'')\,\mathrm{d}E'\mathrm{d}E''$$

where $h(E)$ is an arbitrary function.

The equations showed thus far are valid for any spherically symmetric band shape. We adopted the band model of Ref. [16], which consists of three isotropic parabolic upper bands and one non-parabolic lower band, as shown in Fig. 9. A feature of this band structure is that it reproduces the experimental density of states in silicon up to 3.4 eV. In our actual computations we have considered only the two lowest bands, with which the dynamics of electrons up to 2.6 eV is correctly described. In the first non-parabolic band $\gamma(E) = E(1+\alpha E)$, whereas in the second band $\gamma(E) = E_{max} - E$. It is worth noticing that there is a population for each of the two bands, and therefore two corresponding equations of the form (III.15) with different coefficients have to be solved. The matching conditions at the wave vector $\mathbf{k}_j$ at the band junction are

$$f_0^{(1)}(k_j) = f_0^{(2)}(k_j)$$

$$f_1^{(1)}(k_j) = -f_1^{(2)}(k_j)$$

where the upper indices refer to the two populations. These conditions are derived from the following condition on the total distribution function

$$f^{(1)}(\mathbf{k}_j) = f^{(2)}(-\mathbf{k}_j)$$

Eqs. (III.15) can be expressed in a more compact form using the transformation of variables $H \equiv E - q\phi(x)$, where $\phi(x)$ is the electric potential and H the total energy. By defining the coefficients of the expansion in the new space

$$(\text{III.19}) \qquad f_i(E, x) = f_i(H + q\phi(x), x) \equiv \mathcal{F}_i(H, x)$$

equations (III.15) become

$$(\text{III.20,a}) \qquad \frac{\partial}{\partial x}\left[-g\,u_g\mathcal{F}_1\right] + 3c_{op}\,g\left\{g^+\left[N_{op}^+\mathcal{F}_o^+ - N_{op}\mathcal{F}_o\right] - \right.$$

$$\left. -g^-\left[N_{op}^+\mathcal{F}_0 - N_{op}\mathcal{F}_0^-\right]\right\} - 3\,g^2 c_{ii}\,\mathcal{F}_0 + 3\,g\int A(H', H)\mathcal{F}_0(H')\,g(H')\,\mathrm{d}H' = 0$$

$$(\text{III.20,b}) \qquad \mathcal{F}_1 = -\lambda\frac{\partial \mathcal{F}_0}{\partial x} + \frac{\lambda}{3u_g}\int B(H', H)\,\mathcal{F}_1(H')\,g(H')\,\mathrm{d}H'$$

where the relation $g = (8\pi m^*\gamma)/(\hbar^3 u_g)$ has been applied, along with a slight abuse of notation in using the same name for functions in the original and transformed spaces. Eq. (III.20,a) is the continuity equation in the (H, x) space, which approximates the original BTE in the $(\mathbf{k}, x)$ space.

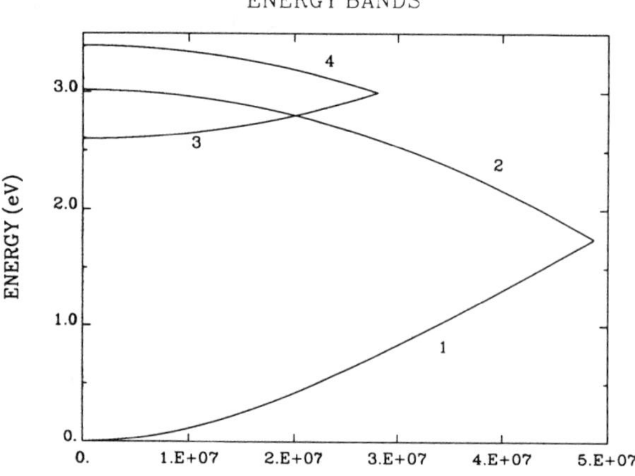

Figure 9. Energy band shape adopted in the model.

From the above equations, neglecting for the moment the integral terms and eliminating $\mathcal{F}_1$, one obtains a second order linear difference-differential equation with non-constant coefficients, defined in the two-dimensional domain (H, x) with curvilinear boundaries $x = 0$, $x = l_{max}$, $H = -q\phi(x)$, $H = E_{max} - q\phi(x)$, where E_{max} is the energy at the top of band 2. Equilibrium distributions for $\mathcal{F}_0$ are assumed at the boundaries $x = 0$, and $x = l_{max}$. At the energy extrema one imposes regularity of the distribution function. For the discretization of the differential operator we used the usual box scheme, with piecewise constant fluxes on every interval. Note that the coefficient $g\lambda u_g$ vanishes at these extrema with order $E^{3/2}$. This, together with the regularity of the solution, implies the vanishing of the fluxes associated with the intervals next to the boundaries.

We have used a grid which is composed of nodes at constant total energy H, uniformly spaced in energy by intervals $\Delta H = \hbar\omega_{op}/n$, with n integer. An important feature of Eqs. (III.20) is the absence of partial derivatives with respect to H. Therefore, each node of coordinates (H, x) is connected along the H direction only with the nodes at $(H + \hbar\omega_{op}, x)$ and $(H - \hbar\omega_{op}, x)$, via the difference operator. The resulting algebraic system is thus decomposed into n decoupled subsystems, which can be solved independently with a considerable speed-up of the computation.

The kernels of the integral operators are not sufficiently peaked to be approximated by δ-functions. A direct iterative approach turned out to be an efficient way of solving the full system.

For completeness we give below the asymptotic behavior of the solution to the homogeneous problem near the origin of $\mathbf{k}$ space for the two bands. These are:

$$f_0(k) = a_0 + O(k^2)$$

(III.21)
$$f_1(k) = O(k)$$

In the energy space this translates to

$$q^2 F^2 \tau(E_{max}) \nabla_E f_0(E_{max}) = c_{op} m^* g(E - \hbar\omega) \left[N^+ f_0(E_{max}) - N f_0(E_{max} - \hbar\omega) \right]$$

$$\text{(III.22)} \qquad\qquad f_1(E_{max}) = 0$$

and similarly for $E = 0$. A convenient normalization condition has also to be added in order to remove the trivial null solution.

IV. Application to the BJT. The simulated structure is a one-dimensional n-p-n BJT similar to the one investigated in [17]. Piecewise uniform impurity concentration has been assumed within the emitter, base, collector and subcollector regions (Fig. 10). Two cases have been considered, namely, n and p collector regions, which are known to provide qualitatively different electric field shapes. Simulations performed with the hydrodynamic version of the code HFIELDS, with a base-emitter voltage $V_{BE} = 0.97\,\text{V}$ and a collector-base voltage $V_{CB} = 3\,\text{V}$, provided the electric-field profiles of Fig. 11. The base-emitter voltage has been selected so that the electron current density turns out to be about $10^5\,\text{A/cm}^2$. Since bandgap narrowing due to heavy doping is neglected in this simulation, the value of V_{BE} is somewhat overestimated. As expected the peak of the electric field, in the n^+-p-n^--n^+ case, is located at the base-collector junction while, in the n^+-p-p^--n^+ case, the field peaks at the subcollector junction [17]. The first set of results described below refer to the computations performed neglecting impact ionization, while the second set takes that effect into consideration.

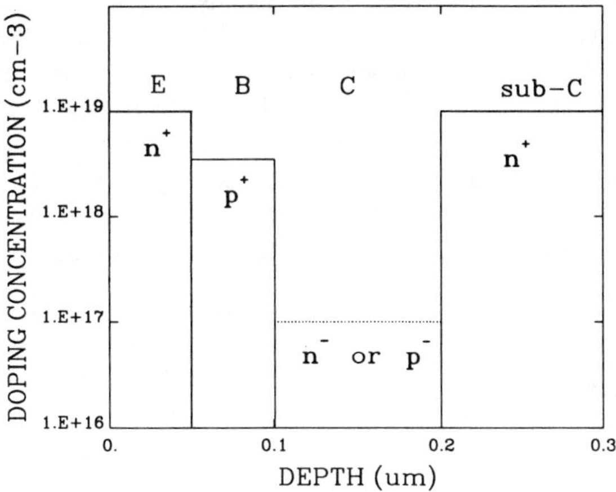

Figure 10. Doping profile of the simulated BJTs.

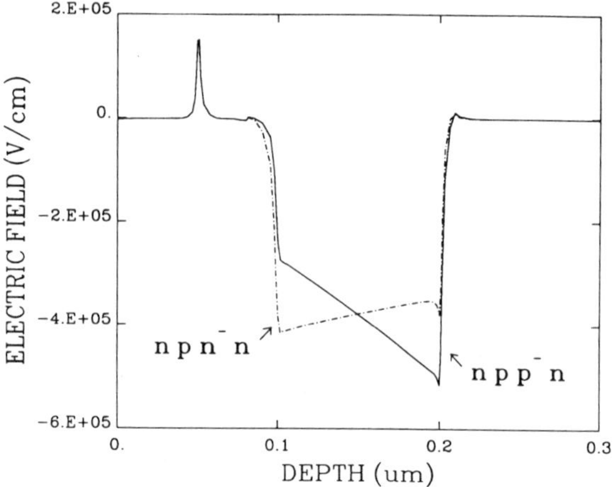

Figure 11. Electric field in the simulated BJTs for the two different doping profiles.

Figs. 12 and 13 compare the electron velocity and normalized energy profiles resulting from the HD program and from the solution of the present model. For consistency, the mobility parameters, saturation velocity, and the energy relaxation time in the HD model were calculated by means of a best fitting procedure of the average velocity and the mean energy vs. field curves obtained with the present approximate solution of the BTE in homogeneous conditions. As seen in Fig. 13, the agreement for the energy is fairly good, even though the HD model tends to provide slightly lower energies in the rising part of the curve. Also, the peak velocities at the base-collector junction are in excellent agreement (Fig. 12). On the other hand, the HD model tends to overestimate the electron velocity at the subcollector edge of the space charge region, where very large negative gradients of the energy occur.

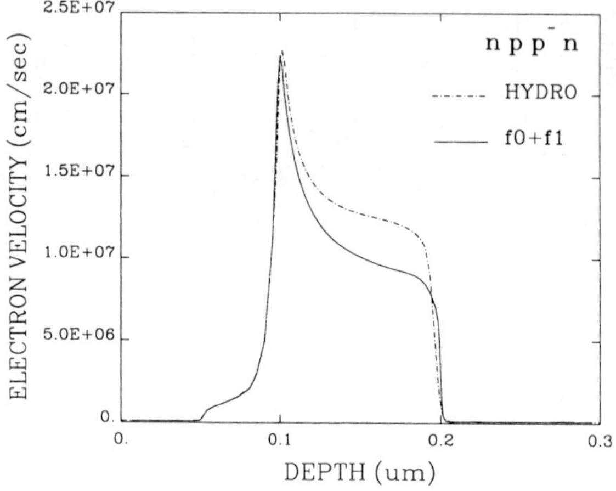

Figure 12. Mean velocity from the spherical harmonics model compared with the hydrodynamic model.

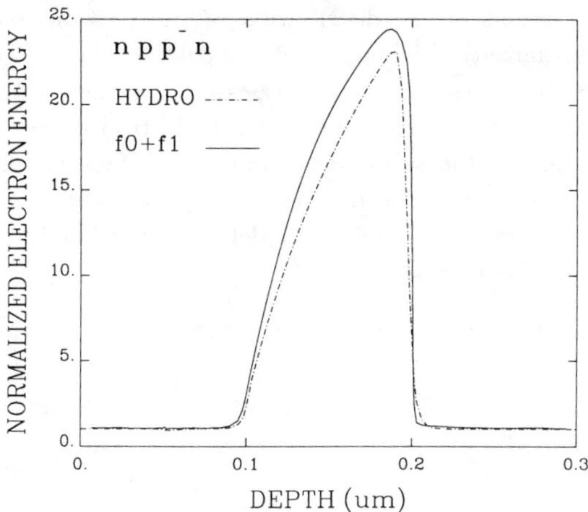

Figure 13. Mean energy from the spherical harmonics model
compared with the hydrodynamic model.

Fig. 14 shows the electron distribution for the n^+-p-p^--n^+ device. In the picture, the numerical solution obtained on the discretization grid has been interpolated in the (E, x) space and exhibited on a coarser grid. One can clearly see the heating of the tail of the distribution as electrons move deeper and deeper into the collector, while quasi-equilibrium distributions are observed in the emitter and in the base regions. Two populations seem to coexist within the subcollector, as revealed by the two different slopes of the distribution for high and low energies: hot electrons travelling across the collector space-charge region, and majority carriers in quasi-equilibrium conditions.

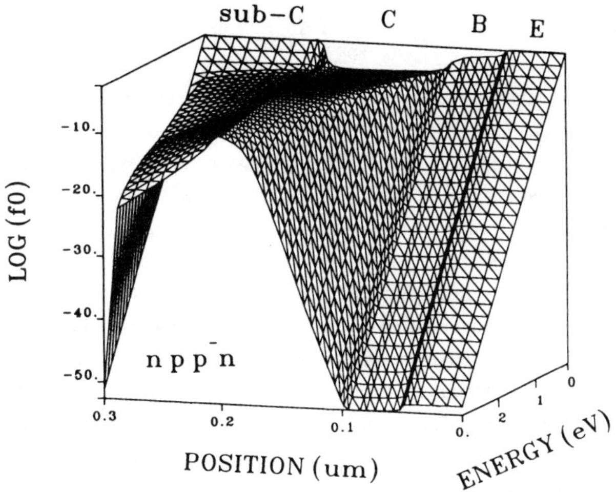

Figure 14. Perspective plot of the electron population f0 as a
function of position and energy.

Two different sections of the distribution of the previous picture are shown in Fig. 15 (the normalization is such that $\int f\,g\,dE = 1$), at two points in the space charge region where the average energies are very nearly the same. Section $x = 0.15\,\mu$m is located in the rising part of the electron energy, whereas section $x = 0.2\,\mu$m is located at the subcollector junction, where the average energy is rapidly decreasing. As can be seen, the high-energy tail behaves very differently in the two cases. Hence, impact-ionization models that are purely based on electron temperature may provide unreliable results.

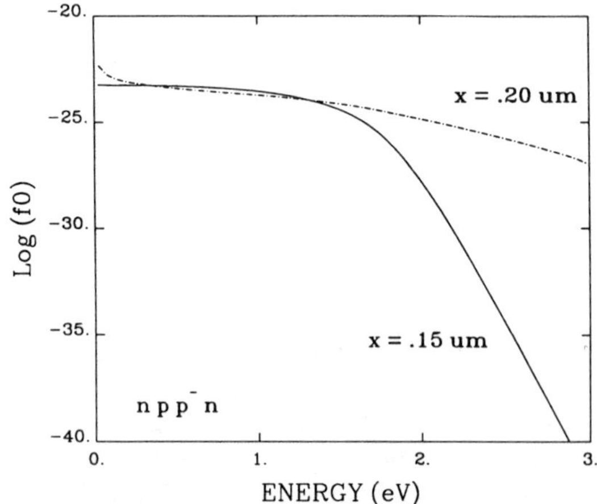

Figure 15. Two sections of Fig. 14 at different positions with the same mean energy.

Fig. 16 shows the impact ionization coefficient α, defined as

$$\text{(IV.1)} \qquad \alpha = \frac{1}{nv}\int f_0 c_{ii} g^2 \mathrm{d}E,$$

as a function of the inverse electric field in a homogeneous medium. In the region of large fields, our results closely follow experimental data by Van Overstraeten, while, in the region of relatively-small fields, they nearly overlap experimental data by Lee. Also, Monte Carlo results [18] are in close agreement with the present model. In Fig. 17 the ionization coefficient within a similar bipolar device is compared against the static-field ionization coefficient, showing large discrepancies due to the non-locality of impact ionization. The multiplication factor $M - 1 = (J_{out} - J_{in})/J_{in}$ is shown in Fig. 18 against V_{CB} for three different values of the collector doping concentration. We want to stress that no fitting parameter has been adjusted, and yet a reasonable qualitative agreement with experiment was obtained. The quantitative differences may be due to a number of reasons, including the uncertainty in the doping profiles. Finally, Fig. 19 shows the effect of impact ionization on the tail of the distribution at some convenient location in the collector.

The computational effort of this one-dimensional model is comparable to that of a two-dimensional hydrodynamic or drift-diffusion model because of the introduction of the energy variable.

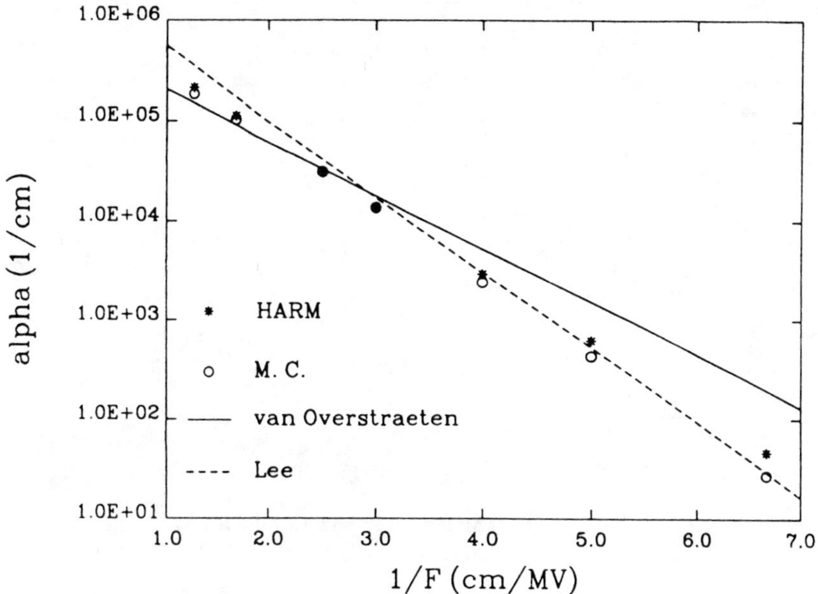

Figure 16. The ionization coefficient as a function of the inverse electric field.

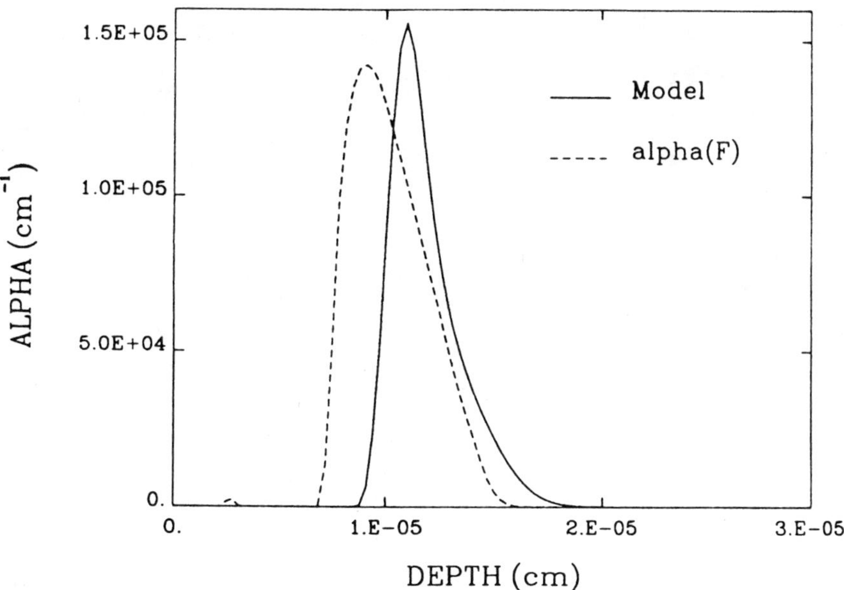

Figure 17. A comparison between ionization coefficients from the spherical harmonics model and from the field-dependent macroscopic model.

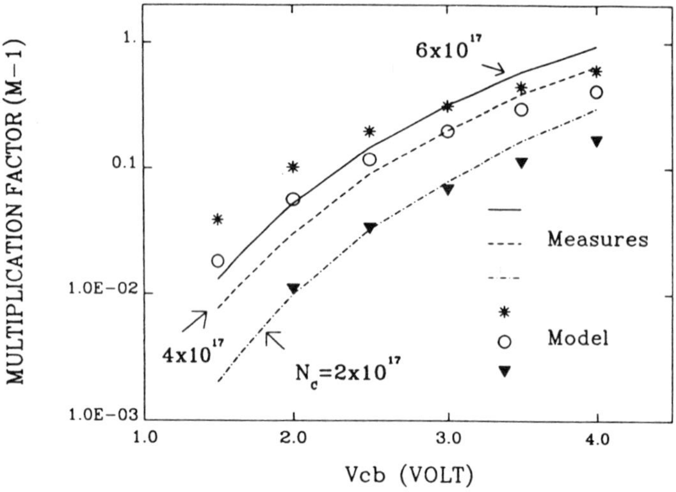

Figure 18. The collector current multiplication factor M-1 compared with experimental data.

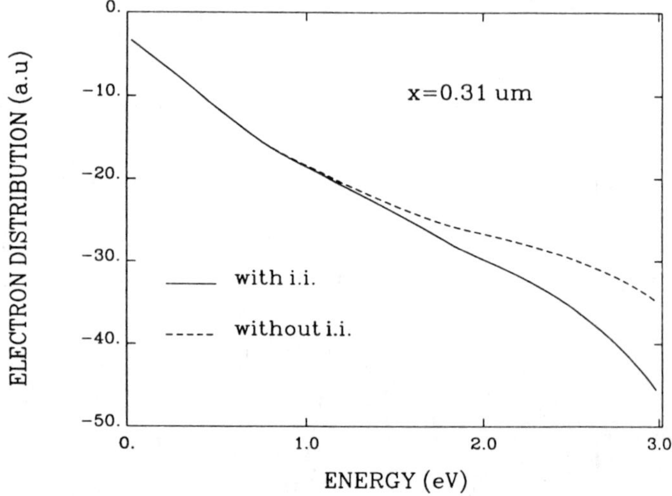

Figure 19. The electron distribution function within the subcollector region with and without impact ionization.

Conclusions. In this paper we described two approaches to the problem of modeling high energy effects in small devices. The macroscopic approach, hydrodynamic model, was applied to the computation of bulk current in MOSFETs and gave reasonable results after some physical parameters were calibrated by experiments on long devices. More detailed information were obtained from a semi-microscopic model approximating the BTE. This approach requires only the correct strength of the scattering constants obtainable from homogeneous conditions in order to achieve good agreement with experiments.

Acknowledgments. This work has been supported by IBM, General-Technology Division, Essex Junction VT, 05452 - USA.

REFERENCES

[1] K. K. Thornber, *Current Equations for Velocity Overshoot*, IEEE Electron Device Letters, vol. EDL-3, no. 3 (1982), p. 69.

[2] M. V. Fischetti, S. E. Laux, *Monte-Carlo Analysis of Electron Transport in Small Semiconductor Devices Including Band-Structure and Space-Charge Effects*, Phys. Rev. B, vol. 38 (1988), p. 9721.

[3] A. Gnudi, F. Odeh, M. Rudan, *Investigation of Non-Local Transport Phenomena in Small Semiconductor Devices*, European Trans. on Tel., vol. 1, (1990), p. 307.

[4] M. Rudan, F. Odeh, *Multi-Dimensional Discretization Scheme for the Hydrodynamic Model of Semiconductor Equations*, COMPEL, vol. 5, (1986), p. 149.

[5] K. Katayama, T. Toyabe, *A New Hot Carrier Simulation Method Based on Full 3D Hydrodynamic Equations*, IEDM-89 Tech. Dig. (1989), p. 135.

[6] B. Meinerzhagen, *Consistent Gate and Substrate Current Modeling based on Energy Transport and the Lucky Electron Concept*, IEDM-88 Tech. Dig. (1988), p. 504.

[7] W. Shockley, *Problems Related to p-n Junctions in Silicon*, Solid-State Elect., vol. 2 (1961), p. 35.

[8] E. Schöll, W. Quade, *Effect of Impact Ionization on Hot-Carrier Energy and Momentum Relaxation in Semiconductors*, J. Phys. C, vol. 20 (1987), p. L861.

[9] B. Davari et al., *A High Performance 0.25 μm CMOS Technology*, IEDM-88 Tech. Dig. (1988), p. 56.

[10] A. Forghieri et al., *A New Discretization Strategy of the Semiconductor Equations Comprising Momentum and Energy Balance*, IEEE Trans. on CAD of ICAS, vol. 7 (1988), p. 231.

[11] W. Haensch, M. Miura-Mattausch, *The Hot-Electron Problem in Small Semiconductor Devices*, J. Appl. Phys., vol. 60 (1986), p. 650.

[12] J. A. Cooper, D. F. Nelson, *High-Field Drift Velocity of Electrons at the $Si - SiO_2$ Interface as determined by a Time-of-Flight Technique*, J. Appl. Phys., vol. 54 (1983), p. 1445.

[13] R. V. Overstraeten, H. De Man, *Measurements of the Ionization Rates in Diffused Silicon p-n Junctions*, Solid-State Elect., vol. 13 (1970), p. 583.

[14,a] D. Ventura, A. Gnudi, G. Baccarani, *An Efficient Method for Evaluating the Energy Distribution of Electrons in Semiconductors Based on Spherical Harmonics Expansion*, to be published on IEICE Trans.

[14,b] A. Gnudi, D. Ventura, G. Baccarani, *One-Dimensional Simulation of a Bipolar Transistor by Means of Spherical Harmonics Expansion of the Boltzmann Transport Equation*, Proc. of the SISDEP '91 Conf., September 1991, Zurich, p. 205.

[15,a] N. Goldsman, Y. Wu, J. Frey, *Efficient Calculation of Ionization Coefficients in Silicon from the Energy Distribution Function*, J. Appl. Phys., vol. 68, no. 3 (1990), p. 1075.

[15,b] N. Goldsman, L. Henrickson, J. Frey, *A Physics-Based Analytical/Numerical Solution to the Boltzmann Transport Equation for Use in Device Simulation*, Solid-St. Electron., vol. 34, no. 4 (1991), p. 389.

[16] R. Brunetti et al., *A many-band silicon model for hot-electron transport at high energies*, Solid-St. Electron., vol. 32, no. 12 (1989), p. 1663.

[17] E. F. Crabbé et al., *The Impact of Non-Equilibrium Transport on Breakdown and Transit Time in Bipolar Transistors*, IEDM-90 Tech. Dig. (1990), p. 463.

[18] R. Thoma et al., *An Improved Impact-Ionization Model for High-Energy Electron Transport in Si with Monte Carlo Simulation*, J. Appl. Phys., vol. 69, no. 4 (1991), p. 2300.

DERIVATION OF THE HIGH FIELD
SEMICONDUCTOR EQUATIONS

P.S. HAGAN*, R.W. COX** AND B.A. WAGNER†

Abstract. Electron and hole densities evolve in $x - z$ phase space according to Boltzmann equations. When the mean free path of the particles is short *and* the electric force on the particles is weak, a well-known expansion (the Hilbert expansion) can be used to solve the Boltzmann equation. This asymptotic solution shows that the spatial density of electrons and holes evolves according to diffusion-drift equations. In fact, the Hilbert expansion leads directly to the Basic Semiconductor (van Roosbroeck) Equations.

As devices become smaller, electric fields become stronger, which renders the Basic Semiconductor Equations increasingly inaccurate. To remedy this problem, we use singular perturbation techniques to obtain a new asymptotic expansion for the Boltzmann equation. Like the Hilbert expansion, the new expansion requires the mean free path to be short compared to all macroscopic length scales. However, it does *not* require the electric forces to be weak. The new expansion shows that spatial densities obey diffusion-drift equations as before, but the diffusivity D and mobility μ turn out to be nonlinear functions of the electric field. In particular, our analysis determines the field-dependent mobilities $\mu(E)$ and diffusivities $D(E)$ directly from the scattering operator. By carrying out this asymptotic expansion to higher order, we obtain the high frequency corrections to the drift velocity and diffusivity, and also the corrections due to gradients in the electric field. Remarkably, we find that Einstein's relation is still satisfied, even with these corrections.

The new diffusion-drift equations, together with Poisson's equation for the electric field, form the *high-field semiconductor equations*, which can be expected to be accurate regardless of the strength of the electric fields within the semiconductor. In addition, our analysis determines the entire momentum distribution of the particles, so we derive a very accurate *first moment model* for semiconductors by substituting the asymptotically-correct distribution back into the Boltzmann equation and taking moments; this model is roughly analogous to a hydrodynamic model without an energy equation. Finally, we present the extension of the high field diffusion-drift equations to three dimensions.

1. Introduction. Particles densities in *phase-space* usually evolve according to Boltzmann equations (also called transport, or kinetic equations). Here we analyze the Boltzmann equations which govern electrons and holes in semiconductors in the "short mean free path" regime.

If the mean free path is short *and* the electric force on the particles is weak, a well-known asymptotic expansions (the Hilbert expansion [1,2]) can be used to reduce Boltzmann equations to radically simpler diffusion-drift equations. When the Hilbert expansion is used for both the electron and hole populations, and when Poisson's equation is used to determine the electric field from the particle densities, the resulting system of equations is the *Basic Semiconductor Equations* (BSEs), or van Roosbroeck equations [3]. In the past, these equations have provided the basis for most VLSI device modeling [2-5]. As semiconductor devices become smaller, however, the strength of their internal electric fields increases dramatically. As a result, the BSEs become increasingly inaccurate as device sizes shrink, and are not

---

*Computer Research Group (C-3), MS B265, Los Alamos National Laboratory, Los Alamos, NM 87545.

**Department of Computer & Information Science, Indiana University – Purdue University, Indianapolis, IN 46205.

†Department of Mathematics, University of Arizona, Tucson, AZ 85721.

useful for predicting the operating characteristics of VLSI devices much smaller than a micron. However, the Boltzmann equation is expected to be valid in VLSI structures down to $0.1\mu m$ in size [6,7].

Here we use singular perturbation techniques to analyze the Boltzmann equation in one spatial dimension. (The corresponding three dimensional analysis is done in [9].) This analysis yields a new asymptotic expansion for the Boltzmann equation which does *not* require the electric forces to be weak. As in the Hilbert expansion, we find that the *spatial densities* (as opposed to *phase-space densities*) evolve according to diffusion-drift equations to leading order. However, the diffusivity D and mobility μ turn out to be nonlinear functions of the electric force on the particles, and our analysis shows how the correct field-dependent mobilities $\mu(E)$ and diffusivities $D(E)$ can be found directly from the scattering operator. By carrying out this asymptotic expansion through $O(\varepsilon)$, we also obtain the corrections to the drift velocity and diffusion coefficient due to gradients in the electric fields, as well as the high frequency corrections to the diffusion-drift equation. Remarkably, Einstein's relation [8] is still satisfied, even with these corrections.

Our analysis yields the entire phase-space distribution of the particles. So in §7 we derive a very accurate *first moment model* by substituting the asymptotically-correct momentum distribution into the Boltzmann equation, and equating the zeroeth and first-order moments. The resulting model is like a diffusion-drift model with some non-locality in the particle currents; equivalently, it is similar to a hydro-dynamic model without an energy equation. However, as we shall see, the nonlinear advection term is exactly canceled by the stress tensor, so this model predicts that electron shock waves will not form. Finally, in Appendix A we apply this work to the Boltzmann equations governing both electrons and holes, and thus derive the correct *high-field semiconductor equations*. In §8 we also present the results for three spatial dimensions, which are obtained in [9].

To simplify the analysis, we assume that only electrons in a single conduction band and holes in a single valence band are important; multiple bands and valleys are considered elsewhere [10]. Since the analysis for electrons and holes is virtually identical, we only present the analysis for electrons; the corresponding results for holes are given as part of the high field semiconductor equations in Appendix A. We use standard notation for general kinetic (transport) theory; the more specialized notation for semiconductors is also given in Appendix A. Define

$$(1.1a) \qquad p(t, x, z) = \text{density of electrons in } x - z \text{ phase space,}$$

$$(1.1b) \qquad v(z) = \text{velocity of an electron with momentum } z.$$

Here $z \equiv \hbar k$ is the *crystal momentum* of the electron, where k is the electron's wavenumber. Thus, z is a periodic variable with a period corresponding to the first Brillouin zone. In addition, the velocity in (1.1b) is given by $v(z) = \mathcal{E}'(z)$, where $\mathcal{E}(z)$ is the appropriate band energy. See Appendix A. Before continuing, let us define the usual inner product

$$(1.2) \qquad \langle f, g \rangle \equiv \int f^*(z) g(z) dz.$$

(Unless stated otherwise, all z integrals are to be taken over the first Brillouin zone). Let us also define the adjoint $\mathcal{L}^+$ of a linear operator $\mathcal{L}$ by

$$(1.3) \qquad \langle f, \mathcal{L}g \rangle \equiv \langle \mathcal{L}^+ f, g \rangle \text{ for all } f(z), g(z).$$

With this notation, the Boltzmann equation for electrons in a semi-conductor is

$$(1.4) \qquad p_t + v(z)p_x + F(x)p_z = \mathcal{L}p,$$

where $F(x) = -qE(x)$ is the force on the electrons. On the left side are the streaming terms, which account for the advection of electrons through phase-space. On the right side is the *scattering operator* $\mathcal{L}$, which accounts for "collisions" of the electrons, both among themselves and with other constituents of the environment, such as phonons and impurities. The scattering operator acts only on the z dependence of p, and determines how the momentum distribution of the electrons evolves due to collisions. In this paper we restrict ourselves to linear scattering operators, which restricts us to applications with moderate electron densities.

To ensure that the scattering operator and Boltzmann equation are physically sensible, we require them to satisfy the following properties:

A. *Symmetry.* We assume that $v(z)$ is odd in z and that $\mathcal{L}$ is even:

$$(1.5) \qquad v(z) = -v(-z); \quad f(z) = \mathcal{L}p(z) \Rightarrow f(-z) = \mathcal{L}p(-z).$$

B. *Conservation of particles.* Scattering processes cannot create or destroy electrons, so we have

$$(1.6) \qquad \langle 1, \mathcal{L}p \rangle \equiv \int \mathcal{L}p \, dz = 0 \quad \text{for any } p(z).$$

C. *Equilibrium.* In the absence of external forces and spatial gradients, the momentum distribution $p(t, z)$ must relax to equilibrium. Thus, the equation

$$(1.7) \qquad p_t = \mathcal{L}p$$

must have a unique (normalized) stable steady solution, namely the *equilibrium distribution* $r(z)$, which satisfies

$$(1.8a) \qquad \mathcal{L}r = 0, \qquad \langle 1, r \rangle = 1.$$

Since the scattering operator $\mathcal{L}$ is linear, to be physically consistent we must take $r(z)$ to be Maxwellian:

$$(1.8b) \qquad r(z) = \frac{1}{N_0} \exp\{-\beta\mathcal{E}(z)\},$$

with

(1.8c)
$$v(z) = \frac{d}{dz}\mathcal{E}(z) \quad \text{and} \quad \beta = 1/k_B T.$$

Here $\mathcal{E}(z)$ is the kinetic energy (band energy) at momentum z, and the normalization constant, N_0, is given by $\langle 1, e^{-\beta\mathcal{E}(z)} \rangle$.

D. *Detailed balance*. The scattering operator must satisfy detailed balance, so

(1.9)
$$\mathcal{L}\{qr\} = r(z)\mathcal{L}^+ q \quad \text{for any } q(z).$$

E. *Fredholm Index 0*. Since $\langle 1, \mathcal{L}p \rangle = 0$ for any $p(z)$, the problem

(1.10a)
$$\mathcal{L}p = f$$

can only have a solution if

(1.10b)
$$\langle 1, f \rangle = 0.$$

We assume that $\mathcal{L}$ is a Fredholm operator with index 0, so that $\langle 1, f \rangle = 0$ is a necessary *and* sufficient condition for $\mathcal{L}p = f$ to be solvable. Similarly, the adjoint problem $\mathcal{L}^+ q = g$ is solvable if and only if the Fredholm condition $\langle g, r \rangle = 0$ is satisfied.

Note that the conservation property, (1.6), implies that

(1.11a)
$$\mathcal{L}^+ 1 = 0,$$

and the symmetry property implies that $r(z)$ and $\mathcal{E}(z)$ are even:

(1.11b)
$$r(z) = r(-z), \quad \mathcal{E}(z) = \mathcal{E}(-z).$$

2. Non-dimensionalization. The first step in analyzing the Boltzmann equation is non-dimensionalization. A typical momentum scale for the scattering process, z_{th}, can be obtained from the equilibrium momentum distribution $r(z)$. (See figure below.) Then $v_{th} \equiv v(z_{th})$ can be used to define a natural velocity scale. The speed at which a typical solution $p(t, z)$ of (1.7) approaches $r(z)$ can be used to define a characteristic timescale for the scattering process, the *thermalization time* τ_{th}. The *mean free length*, also called the *scattering length*, can now be defined as

(2.1)
$$\ell = v_{th}\tau_{th} = v(z_{th})\tau_{th}.$$

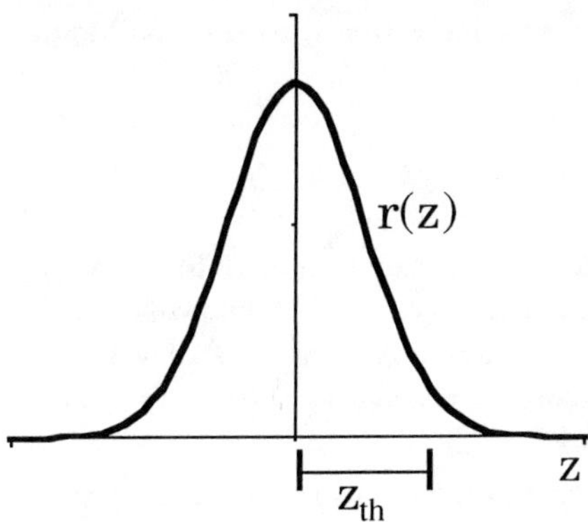

Crudely speaking, the mean free length represents the typical distance a group of electrons can travel (in the absence of any applied forces) before they attain their equilibrium distribution $r(z)$. Let L be a typical macroscopic length scale, and define

$$(2.2) \qquad \varepsilon = \ell/L.$$

Finally, let E_{typ} be the typical size of the electric field, so that $f_{\text{det}} = qE_{\text{typ}}$ is the typical magnitude of the force $F(x)$, and define

$$(2.3) \qquad z_{\text{det}} = f_{\text{det}}\tau_{th}.$$

Crudely speaking, z_{det} is the typical momentum change caused by the force F between significant scattering events; clearly z_{det} represents a balance between the acceleration due to $F(x)$, and the "randomization" of z due to scattering events.

To non-dimensionalize (1.4), let

$$(2.4a) \qquad x_{\text{new}} = x/L, \qquad z_{\text{new}} = z/z_{th}, \qquad v_{\text{new}} = v/v_{th}, \qquad t_{\text{new}} = \varepsilon^2 t/\tau_{th}$$

$$(2.4b) \qquad F_{\text{new}} = F/f_{\text{det}}, \qquad \mathcal{L}_{\text{new}} = \tau_{th}\mathcal{L}, \qquad \mathcal{E}_{\text{new}} = \mathcal{E}/(v_{th}z_{th}), \qquad \beta_{\text{new}} = (v_{th}z_{th})\beta.$$

In terms of the new variables, (1.4) becomes

$$(2.5) \qquad \varepsilon^2 p_t + \varepsilon v(z)p_x + \kappa F(x)p_z = \mathcal{L}p,$$

where $\kappa = z_{\text{det}}/z_{th}$.

In this paper we solve (2.5) using singular perturbation techniques based on $\varepsilon \ll 1$; that is, on the scattering length ℓ being much shorter than any macroscopic length scale. Consequently, this analysis is only valid in regions where the electric field $E(x)$ changes slowly on length scales comparable to ℓ. Similarly, if we were to allow $\mathcal{L}$ to vary with x, then we would have to require $\mathcal{L}$ to vary slowly on length scales comparable to ℓ.

There are two extremely important distinguished limits of (2.5),

$$\kappa = O(\varepsilon) \quad \textit{weak force regime},$$
$$\kappa = O(1) \quad \textit{strong force regime}.$$

Below we obtain the asymptotic solution of (2.5) for both regimes. For the weak force regime, this solution is just the well-known *Hilbert expansion* [1,2], which is presented below for completeness. For the strong force regime, the asymptotic solution of the Boltzmann equation was previously unknown, and represents the key advance in our work.

Both the Hilbert expansion and our new results represent *outer solutions*. Next to all material interfaces and spatial boundaries, there exist extremely thin *kinetic boundary layers*, which are only one or two scattering lengths thick. Both the Hilbert expansion and our results are not valid within such boundary layers. Even though these boundary layers are extremely thin, they can play important roles in semiconductors. We are currently developing analytical and numerical techniques for resolving such layers. Besides these boundary layers, there are usually rapidly-decaying initial layers, which die out on the ultra-short $t \sim O(\varepsilon^2)$ timescale; the results presented here are not valid during this time.

3. Weak force analysis (Hilbert expansion). Here we analyze the weak force regime. Without loss of generality, we take $\kappa = \varepsilon$ and analyze

$$(3.1) \qquad \varepsilon^2 p_t + \varepsilon v(z)p_x + \varepsilon F(x)p_z = \mathcal{L}p.$$

To obtain the outer solution to (3.1), we expand $p(t, x, z)$ as

$$(3.2a) \qquad p(t, x, z) = p^0(t, x, z) + \varepsilon p^1(t, x, z) + \varepsilon^2 p^2(t, x, z) + \dots,$$

where the expansion will be constructed so that

$$(3.2b) \qquad \langle 1, p^i \rangle = 0 \quad \text{for } i = 1, 2, 3, \dots.$$

Hence the spatial density of electrons, $\langle 1, p \rangle$, will be given by the leading term $\langle 1, p^0 \rangle$.

Substituting (3.2) into (3.1), we obtain

$$(3.3) \qquad \mathcal{L}p^0 = 0.$$

Thus,

$$(3.4) \qquad p^0(t, x, z) = n(t, x)r(z),$$

where the coefficient $n(t, x)$ is unknown at this stage of the expansion. Note that

$$n(t, x) = \langle 1, p^0(t, x, z) \rangle = \langle 1, p(t, x, z) \rangle,$$

so $n(t, x)$ is the spatial density of electrons.

At $O(\varepsilon)$ we have

$$(3.5a) \qquad \mathcal{L}p^1 = v(z)p_x^0 + F(x)p_z^0 = (n_x - \beta Fn)v(z)r(z),$$

$$(3.5b) \qquad \langle 1, p^1 \rangle = 0.$$

See equation (1.8). Since $v(z)r(z)$ is odd in z, the Fredholm alternative (1.10b) is satisfied, and the problem

$$(3.6a) \qquad \mathcal{L}\tilde{r} = -v(z)r(z)$$

is solvable. We define $\tilde{r}(z)$ uniquely by requiring it to be odd,

$$(3.6b) \qquad \tilde{r}(z) = -\tilde{r}(-z).$$

Then $\langle 1, \tilde{r} \rangle = 0$, so the solution of (3.5) is

$$(3.7) \qquad p^1(t, x, z) = -(n_x - \beta Fn)\tilde{r}(z).$$

At $O(\varepsilon^2)$, equation (3.1) yields

$$(3.8) \qquad \begin{aligned} \mathcal{L}p^2 &= p_t^0 + vp_x^1 + Fp_z^1 \\ &= n_t r(z) - (n_x - \beta Fn)_x v(z)\tilde{r}(z) - F(n_x - \beta F_n)\tilde{r}'(z). \end{aligned}$$

Since $\tilde{r}'(z)$ is a derivative, clearly $\langle 1, \tilde{r}'(z) \rangle = 0$, so applying the solvability condition to (3.8) yields the diffusion-drift equation

$$(3.9a) \qquad n_t = D_0 n_{xx} - \mu(Fn)_x + \dots \qquad \text{(weak force)},$$

where the diffusivity D_0 and the mobility μ are given by

$$(3.9b) \qquad D_0 \equiv \langle r, \tilde{r} \rangle, \quad \mu = \beta D_0 \equiv D_0/k_B T.$$

Thus, the phase-space density is given by

$$(3.9c) \qquad p(t, x, z) = n \cdot r(z) - \varepsilon\{n_x - \beta Fn\}\tilde{r}(z) + \dots \qquad \text{(weak force)},$$

where the spatial density of electrons, $n(t, x)$, evolves according to (3.9a). Note that (3.9b) shows that the Einstein relation, $D_0 = k_B T \mu$, is satisfied [8].

In Appendix B we carry out this expansion to higher order, and obtain the correct diffusion-drift equation through $O(\varepsilon^2)$. See equations (B.17).

4. Strong force results. Scaling. We now investigate the strong force regime. Without loss of generality, we take $\kappa = 1$ and analyze

$$(4.1) \qquad \varepsilon^2 p_t + \varepsilon v(z)p_x + F(x)p_z = \mathcal{L}p \quad \text{(strong force)}.$$

To obtain the right overall scaling for this case, it is helpful to consider the weak force result (3.9a) with F replaced by F/ε:

$$(4.2) \qquad n_t = D_0 n_{xx} - \mu(Fn)_x/\varepsilon.$$

Even though this equation is *not* correct, it should still enable us to predict the scaling for the strong force regime.

Equation (4.2) shows that the density p depends on the short t/ε timescale. From the steady state of (4.2), we see that p also depends on the short $\tilde{x} = x/\varepsilon$ space scale. Now, at first sight it may appear that since p depends on the x/ε space scale, it must also depend on the very short t/ε^2 timescale. Apart from a rapid initial transient, this is not true. Consider the evolution of a packet of electrons which is initially smooth on the $O(1)$ space scale. Equation (4.2) implies that the packet will advect rapidly with velocity $\mu F(x)/\varepsilon$ until it reaches a stagnation point, where the force $F(x)$ is zero. Near the stagnation point, the much weaker force will slowly compress the packet, while diffusion will tend to spread the packet out. It is only the long-time balance of these two effects that results in the eventual narrow spatial distribution. So after the rapid initial transient dies out, $p(t, x, z)$ will depend on timescales no shorter than t/ε. Consequently, we can drop the transient term $\varepsilon^2 p_t$ in (4.1) to leading order, and study

$$(4.3) \qquad \varepsilon v(z)p_x = \mathcal{L}p - F(x)p_z.$$

Since we expect p to depend on the short x/ε space scale, we cannot drop the $\varepsilon v(z)p_x$ term, and must solve the full steady transport equation. However, even though we expect the solution p to depend on the short x/ε scale, note that equation (4.3) depends explicitly on x only through the force $F(x)$. Thus, the equation itself varies only over the $O(1)$ space scale. Our asymptotic analysis is based on exploiting the disparity between the $O(\varepsilon)$ length scale of the solution, and the $O(1)$ length scale of the equation.

The asymptotic analysis is carried out in §5 and §6. The final result can be expressed simply in terms of eigenvalues and eigenfunctions of (4.3). Separating variables,

$$(4.4) \qquad p = e^{\lambda x/\varepsilon}\Theta(z),$$

leads to an eigenvalue problem parameterized by F:

$$(4.5) \qquad \mathcal{M}(F)\Theta \equiv \mathcal{L}\Theta - F\Theta_z = \lambda v(z)\Theta.$$

Two eigenfunctions turn out to be crucial. First, it is easily seen that $\lambda = 0$ is an eigenvalue at each F. Indeed, the adjoint of $\mathcal{M}$ is

$$(4.6) \qquad \mathcal{M}^+ q = \mathcal{L}^+ q + F q_z,$$

and since $\mathcal{L}^+ 1 = 0$ (see (1.11a)), clearly $\mathcal{M}^+ 1 = 0$. Therefore $\mathcal{M}^+$, and hence $\mathcal{M}$, must have a zero eigenvalue. So at each F there must be a null function $\phi(z; F)$ satisfying

$$(4.7a) \qquad \mathcal{M}\phi \equiv \mathcal{L}\phi - F\phi_z = 0, \quad \langle 1, \phi \rangle = 1.$$

Also, clearly the equilibrium distribution $r(z)$ is an eigenfunction with eigenvalue $\lambda = \beta F$:

$$(4.7b) \qquad \mathcal{M}r \equiv \mathcal{L}r - F r_z = \beta F v(z) r(z), \quad \langle 1, r \rangle = 1,$$

since $\mathcal{L}r = 0$ and $r_z = -\beta v r$.

The asymptotic analysis in §6 shows that the solution of the Boltzmann equation (4.1) is

$$(4.8a) \quad p(t, x, z) = n(t, x) r(z) + \left(n - \frac{\varepsilon n_x}{\beta F} \right) (\phi(z; F) - r(z)) + \ldots \quad \text{(strong force)}$$

to leading order. Here $n(t, x) \equiv \langle 1, p(t, x, z) \rangle$ is the spatial density of electrons, and satisfies the diffusion-drift equation

$$(4.8b) \qquad n_t = (D n_x)_x - (\nu n)_x / \varepsilon + \ldots \quad \text{(strong force)}$$

to leading order, where the diffusivity $D(F)$ and the drift velocity $\nu(F)/\varepsilon$ are non-linear functions of the electric force F:

$$(4.8c) \qquad D(F) \equiv \langle v, \phi \rangle / \beta F, \quad \nu(F) \equiv \langle v, \phi \rangle.$$

Here $\phi(z; F)$ is defined by (4.7a). Despite the βF denominators in (4.8), none of the quantities in (4.8) are singular at $F = 0$. In fact, it turns out that (4.8) reduces exactly to the weak force result (3.9) in the limit $F \to 0$.

Also, to revert back to the original *dimensional* variables (see (2.4)), one simply has to set $\varepsilon = 1$ in (4.8).

5. Eigenvalues and eigenfunctions. In this section we introduce the eigenfunctions and adjoint eigenfunctions needed to solve equation (4.1). Consider the eigenvalue problem parameterized by F,

$$(5.1a) \qquad \mathcal{M}(F)p \equiv \mathcal{L}p - F p_z = \lambda v(z) p,$$

and its adjoint[1]

$$(5.1b) \qquad \mathcal{M}^+(F)q \equiv \mathcal{L}^+ q + F q_z = \lambda^* v(z) q,$$

---

[1] At $F = 0$ the property of detailed balance (and the negative semi-definiteness of $\mathcal{L}$) guarantee that there are no complex eigenvalues λ. However, for $F \neq 0$, pairs of eigenvalues may coalesce and go into the complex plane.

where λ^* denotes the complex conjugate of λ. Suppose that $\Theta(z)$ is an eigenfunction with eigenvalue λ_1, and suppose that $\chi(z)$ is an adjoint eigenfunction with eigenvalue λ_2. Then

$$\lambda_1 \langle v\chi, \Theta \rangle = \langle \chi, \mathcal{M}\Theta \rangle = \langle \mathcal{M}^+\chi, \Theta \rangle = \lambda_2 \langle v\chi, \Theta \rangle,$$

so clearly $\langle v\chi, \Theta \rangle = 0$ if $\lambda_1 \neq \lambda_2$. We conclude that eigenfunctions and adjoint eigenfunctions corresponding to distinct eigenvalues are co-orthogonal in the sense that $\langle v\chi, \Theta \rangle = 0$, regardless of whether $\Theta(z)$ and/or $\chi(z)$ belong to eigenvalues in the continuous or discrete spectrum of $\mathcal{L}$.

Eigenfunctions at $F = 0$. At $F = 0$ the eigenvalue equations are

$$\text{(5.2)} \qquad \mathcal{L}p = \lambda v p, \quad \mathcal{L}^+ q = \lambda v q.$$

Clearly $\lambda = 0$ is an eigenvalue with eigenfunction $r(z)$ and adjoint eigenfunction 1:

$$\text{(5.3)} \qquad \mathcal{L}r = 0, \quad \mathcal{L}^+ 1 = 0.$$

To determine if this eigenvalue is simple, we differentiate (5.2) with respect to λ, and set $\lambda = 0$:

$$\text{(5.4)} \qquad \mathcal{L}p_\lambda = vr, \quad \mathcal{L}^+ q_\lambda = v.$$

Since $v(z)$ is odd and $r(z)$ is even, clearly $\langle v, r \rangle = 0$. Hence (5.4) is solvable, and $\lambda = 0$ has a multiplicity of at least two. Define $\tilde{r}(z)$ and $\ell(z)$ as the unique solutions of

$$\text{(5.5a)} \qquad \mathcal{L}\tilde{r} = -v(z)r(z), \quad \mathcal{L}^+ \ell = -v(z),$$

with

$$\text{(5.5b)} \qquad \tilde{r}(-z) = -\tilde{r}(z), \quad \ell(-z) = -\ell(z);$$

then $-\tilde{r}(z)$ and $-\ell(z)$ are the generalized eigenfunctions p_λ and q_λ. Note that the property of detailed balance implies that

$$\text{(5.5c)} \qquad \tilde{r}(z) = r(z)\ell(z),$$

and that the orthogonality relations

$$\text{(5.6a)} \qquad \langle v, r \rangle = \langle v\ell, \tilde{r} \rangle = 0$$

follow from symmetry. It is also easily shown (from the property of detailed balance and the negative semi-definiteness of $\mathcal{L}$) that

$$\text{(5.6b)} \qquad D_0 \equiv \langle v, \tilde{r} \rangle \equiv \langle v\ell, r \rangle > 0.$$

To determine whether the multiplicity of $\lambda = 0$ is greater than two, we differentiate (5.2) again, obtaining

$$(5.7) \qquad \mathcal{L}p_{\lambda\lambda} = 2vp_\lambda = -2v\tilde{r}, \quad \mathcal{L}^+ q_{\lambda\lambda} = 2vq_\lambda = -2v\ell.$$

These equations are solvable, and the multiplicity of $\lambda = 0$ is greater than two, if and only if $D_0 = \langle v, \tilde{r} \rangle = \langle v\ell, r \rangle = 0$. Since $D_0 > 0$, the multiplicity of $\lambda = 0$ is exactly two.

Besides the double eigenvalue at $\lambda = 0$, $\mathcal{L}$ will have other discrete eigenvalues and/or a continuous spectrum. In particular, if $\Theta(z)$ is an eigenfunction with eigenvalue λ, then $\Theta(-z)$ is an eigenfunction with eigenvalue $-\lambda$. So the spectrum of $\mathcal{L}$ is symmetric about $\lambda = 0$, and contains both positive and negative eigenvalues. In addition, the adjoint eigenfunction with eigenvalue λ is $\chi(z) = \Theta(z)/r(z)$. Finally, since the eigenfunctions at $\lambda = 0$ must be orthogonal to all other eigenfunctions, we have

$$(5.8) \qquad \langle v, \Theta \rangle = 0, \quad \langle v\ell, \Theta \rangle = 0, \quad \langle v\chi, r \rangle = 0, \quad \langle v\chi, \tilde{r} \rangle = 0.$$

Eigenfunctions at $F \neq 0$. Now consider eigenvalues and eigenfunctions for $F \neq 0$. As mentioned earlier, one eigenvalue is zero for all F. Moreover, the eigenvalue $\lambda = 0$ has multiplicity two at $F = 0$, so there must be exactly one other eigenvalue which goes through zero at $F = 0$. (See figure below). Clearly this other eigenvalue is $\lambda = \beta F$. Let $\phi(z; F)$ and $r(z)$ be the eigenfunctions corresponding to these eigenvalues, and let 1 and $\psi(z; F)$ be the adjoint eigenfunctions:

$$(5.9a) \qquad \mathcal{M}\phi \equiv \mathcal{L}\phi - F\phi_z = 0$$

$$(5.9b) \qquad \mathcal{M}^+ 1 = 0$$

$$(5.9c) \qquad \mathcal{M}r \equiv \mathcal{L}r - Fr_z = \beta Fvr$$

$$(5.9d) \qquad \mathcal{M}^+ \psi \equiv \mathcal{L}^+\psi + F\psi_z = \beta Fv\psi.$$

We normalize ϕ and ψ so that

$$(5.10) \qquad \langle 1, \phi \rangle = 1, \quad \langle 1, r \rangle = 1, \quad \langle \psi, r \rangle = 1.$$

Note that these eigenfunctions satisfy the symmetry relations

$$(5.11a) \qquad \phi(z; -F) = \phi(-z; F),$$

$$(5.11b) \qquad \psi(z; -F) = \psi(-z; F),$$

and that the property of detailed balance yields

$$(5.11c) \qquad \phi(z; F) = r(z)\psi(z; -F) = r(z)\psi(-z; F).$$

Together, (5.11a) and (5.11b) imply that these functions satisfy the orthogonality relations

$$(5.12a) \qquad \langle v, r \rangle = \langle v\psi, \phi \rangle = 0.$$

In addition, we define $\nu(F) = \langle v, \phi \rangle$, and note that

$$(5.12b) \qquad \nu(F) = \langle v, \phi \rangle, \quad \nu(-F) = -\nu(F) = \langle v\psi, r \rangle.$$

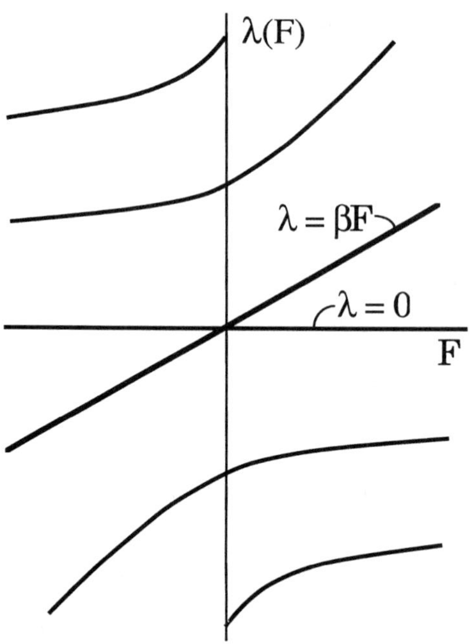

Besides the eigenvalues $\lambda = 0$ and $\lambda = \beta F$, $\mathcal{M}(F)$ will have other discrete eigenvalues and/or a continuous spectrum. In particular, the property of detailed balance shows that if $\Theta(z)$ is an eigenfunction of $\mathcal{M}(F)$ with eigenvalue λ, then $\chi(z) \equiv \Theta(-z)/r(z)$ is an adjoint eigenfunction of $\mathcal{M}(F)$ with eigenvalue $-(\lambda - \beta F)$. Thus, if λ belongs to the spectrum, then $-(\lambda - \beta F)$ does also, so the spectrum of $\mathcal{M}(F)$ is symmetric about $\lambda = \beta F/2$ and contains both positive and negative eigenvalues. Moreover, suppose that λ belongs to the spectrum of $\mathcal{M}(F)$, and let $\Theta(z)$ and $\chi(z)$ be the corresponding eigenfunction and adjoint eigenfunction. Then if $\lambda \neq 0$ and $\lambda \neq \beta F$, the orthogonality conditions yield

$$(5.13) \qquad \langle v, \Theta \rangle = 0, \quad \langle v\psi, \Theta \rangle = 0, \quad \langle v\chi, \phi \rangle = 0, \quad \langle v\chi, r \rangle = 0.$$

Eigenfunctions near $F = 0$. Analogous to "passage through resonance" phenomena, special care must be taken when F is near zero: Because of the nearness of the eigenvalues $\lambda = \beta F$ and $\lambda = 0$, the two eigenspaces $\phi(z; F)$ and $r(z)$ interact very strongly as F goes through zero. To help understand this interaction, $\phi(z; F)$ and $\psi(z; F)$ are expanded in powers of F in Appendix D. In particular, these expansions show that

$$(5.14) \qquad \nu(F) \equiv \langle v, \phi \rangle = \beta F \langle v, \tilde{r} \rangle + O(\beta^3 F^3) \quad \text{for } F \text{ small.}$$

6. Analysis of the strong force case. Since the density p evolves on the t/ε time scale, let

$$(6.1) \qquad t^{\text{new}} \equiv t/\varepsilon.$$

Then the Boltzmann equation (4.1) becomes

$$(6.2) \qquad \varepsilon p_t + \varepsilon v(z) p_x = \mathcal{M}(F) p \quad \text{(strong force)}.$$

To motivate our asymptotic solution, note that (6.2) is

$$(6.3) \qquad \varepsilon v(z) p_x = \mathcal{M}(F) p \equiv \mathcal{L} p - F(x) p_z$$

to leading order. Now, we can easily construct the solution to (6.3) corresponding to the eigenvalue $\lambda(F) = 0$,

$$(6.4\text{a}) \qquad h^0(x, z) = \frac{1}{\nu(F)} \left\{ \phi(z; F) + \varepsilon h_1^0(x, z) + \varepsilon^2 h_2^0(x, z) + \dots \right\},$$

via a power series. Corresponding to the eigenvalue $\lambda = \beta F$, we have the exact solution

$$(6.4\text{b}) \qquad h^1(x, z) = \exp \left\{ \frac{\beta}{\varepsilon} \int_0^x F(x) dx \right\} r(z)$$

to (6.3). Moreover, suppose that $\Theta(z; F)$ is an eigenfunction whose eigenvalue $\lambda(F)$ is part of the continuous or discrete spectrum of $\mathcal{M}(F)$, and suppose that $\lambda \neq 0$ and $\lambda \neq \beta F$. Then we can easily construct the solution of (6.3) corresponding to the eigenvalue $\lambda(F)$,

$$(6.4\text{c}) \qquad h(\lambda, x, z) = \exp \left\{ \frac{1}{\varepsilon} \int_0^x \{\lambda(F) + \dots \} dx \right\} \{\theta(z; F) + \varepsilon h_1(\lambda, x, z) + \dots \},$$

by using geometrical optics.

The general solution of (6.3) can be obtained as a superposition of (6.4a), (6.4b), and components of the form (6.4c). However, suppose that the solution contained components of the form (6.4c) at leading order. Then the solution would become transcendentally large, either as x increased (if $\lambda > 0$) or as x decreased (if $\lambda < 0$). Thus, such components *cannot be part of the leading order solution* except very near to spatial boundaries, where either x cannot increase or x cannot decrease. That is, components of the form (6.4c) can only exist at leading order in the *kinetic boundary layers* adjacent to spatial boundaries and material interfaces. Beyond these boundary layers, in the *outer region*, the solution can only contain components of the form (6.4a) and (6.4b)[2]. Thus we conclude that the *outer solution* of (6.3) must be of the form

$$(6.5\text{a}) \qquad p(t, x, z) = a(t, x) \phi(z; F) + b(t, x) r(z) + \varepsilon R(t, x, z),$$

---

[2] The same reasoning allows us to eliminate terms of the form (6.4b) at leading order *only* if $F(x)$ has one sign. In general, $F(x)$ will be positive in some regions and negative in others, so we expect the solution to contain a term of the form (6.4b).

where, without loss of generality, we can take $R(t, x, z)$ to be orthogonal to both $\phi(z; F)$ and $r(z)$:

$$(6.5b) \qquad \langle v, R \rangle = 0, \quad \langle v\psi, R \rangle = 0.$$

Since the full Boltzmann equation (6.2) only differs from (6.3) by $O(\varepsilon)$, clearly the outer solution of (6.2) is also of the form (6.5). Thus, the outer solution to equation (6.2) can be obtained by assuming that $p(t, x, z)$ is of the form (6.5), and by carefully treating

$$(6.5c) \quad \varepsilon a_x = O(1), \; \varepsilon b_x = O(1), \; \varepsilon R_x = O(1), \; a_t = O(1), \; b_t = O(1), \; R_t = O(1)$$

See the scaling discussion in §4.

Substituting (6.5) into (6.2) yields the exact equation

$$(6.6) \quad \mathcal{M}(F)R - \varepsilon v R_x = \frac{1}{\varepsilon}\left(\varepsilon b_x - \beta F b\right) vr + a_x v\phi + aF_x v\phi_F + a_t\phi + b_t r + \varepsilon R_t.$$

Defining $n(t, x)$ to be the spatial density of electrons,

$$(6.7) \qquad n(t, x) \equiv \langle 1, p \rangle = a + b + \varepsilon\langle 1, R \rangle,$$

and taking the inner product of (6.6) with 1, we obtain

$$(6.8a) \qquad n_t = -\{a\nu(F)\}_x.$$

Here we have used $\langle 1, \mathcal{M}R \rangle = \langle v, R \rangle = 0$ and $\nu(F) \equiv \langle v, \phi \rangle$. Similarly, taking the inner product of (6.6) with $\psi(z; F)$ yields

$$(6.8b) \quad (\varepsilon b_x - \beta F b)\nu(F) = \varepsilon aF_x\langle v\psi, \phi_F \rangle + \varepsilon a_t\langle \psi, \phi \rangle + \varepsilon b_t - \varepsilon^2 F_x\langle v\psi_F, R \rangle + \varepsilon^2\langle \psi, R_t \rangle.$$

Finally, from (6.8a), (6.8b), and (6.6) we obtain

$$(6.8c) \quad \mathcal{M}(F)R - \varepsilon v R_x = aF_x\left\{v\phi_F - \frac{\nu'(F)}{\nu(F)}v\phi + \frac{\langle v\psi, \phi_F \rangle}{\nu(F)}vr\right\} - \varepsilon F_x\frac{\langle v\psi_F, R \rangle}{\nu(F)}vr$$

$$+ a_t\left\{\phi - \frac{v\phi}{\nu(F)} + \frac{\langle \psi, \phi \rangle}{\nu(F)}vr\right\} + b_t\left\{r - \frac{v(\phi - r)}{\nu(F)}\right\}$$

$$+ \varepsilon\left\{R_t - \frac{\langle 1, R_t \rangle}{\nu(F)}v\phi + \frac{\langle \psi, R_t \rangle}{\nu(F)}vr\right\}.$$

Note that (6.8a)-(6.8c) are exact.

6.1 Leading order solution. At leading order, (6.7) and (6.8) yield

$$(6.9) \qquad n = a + b, \quad n_t = -\{a\nu(F)\}_x, \quad \varepsilon b_x = \beta F b.$$

The second equation implies that $a_x = O(1)$, so we can re-write the last equation as $\varepsilon n_x = \beta F(n - a)$ to leading order. Consequently,

$$(6.10) \qquad a = n - \frac{\varepsilon n_x}{\beta F} + O(\varepsilon), \quad b = \frac{\varepsilon n_x}{\beta F} + O(\varepsilon).$$

The second equation now shows that the spatial density evolves according to the diffusion-drift equation

(6.11a) $$n_t = (\varepsilon D n_x - \nu n)_x + O(\varepsilon) \quad \text{(strong force)},$$

to leading order, where the drift velocity $\nu(F)$ and the diffusivity $D(F)$ are given by

(6.11b) $$\nu(F) = \langle v, \phi \rangle, \quad D = \nu(F)/\beta F \quad \text{(strong force)}.$$

Once equations (6.11) have been solved for the spatial density $n(t,x)$, then equations (6.5) and (6.10) provide the leading order phase-space distribution:

(6.11c) $$p(t,x,z) = nr(z) + \left(n - \frac{\varepsilon n_x}{\beta F} \right) (\phi(z;F) - r(z)) + \cdots .$$

6.2. Higher order analysis. In Appendix C we carry out the asymptotic analysis of the strong force regime to higher order. There we find that the spatial density of electrons evolves according to

(6.12a) $$\frac{\partial n}{\partial t} = \frac{\partial}{\partial x} \left\{ \left(1 - \varepsilon \gamma(F) \frac{\partial}{\partial t} \right) \left(\varepsilon \widetilde{D} \frac{\partial n}{\partial x} - \tilde{\nu} n \right) \right\} + O(\varepsilon^2)$$

where the drift velocity $\tilde{\nu}$ and diffusivity $\widetilde{D}$ are now given by

(6.12b) $$\tilde{\nu} = \langle v, \phi \rangle \{ 1 - \varepsilon \alpha(F) F_x + \cdots \}, \quad \widetilde{D} = \tilde{\nu}/\beta F = k_B T \tilde{\nu}/F.$$

Here $\alpha(F)$ and $\gamma(F)$ are defined by

(6.13a) $$\alpha(F) = \langle 1, S^1 \rangle, \quad \gamma(F) = \langle 1, S^2 \rangle,$$

where $S^1(z;F)$ and $S^2(z;F)$ are defined as the solutions of

(6.13b) $$\mathcal{M} S^1 = v \phi_F - \frac{\nu'(F)}{\nu(F)} v(\phi - r), \quad \langle v, S^1 \rangle = 0,$$

(6.13c) $$\mathcal{M} S^2 = \phi - \frac{v(\phi - r)}{\nu(F)}, \quad \langle v, S^2 \rangle = 0.$$

Remarkably, (6.12b) shows that Einstein's relation is still satisfied, even though the drift velocity and diffusion coefficient depend on the gradient of the force at this order.

Appendix C demonstrates that, with sufficient patience, we can analyze the strong force case to as high of order as we wish. However, the asymptotic analysis is valid *only* in regions where the electric field changes gradually over a single mean free path, and the domain-of-validity of higher order results is usually more restricted than that of lower order results [14]. One often expects there to be thin layers (e.g., next to the source and drain, as well as other junctions) in which the electric field changes to rapidly for asymptotic solutions to be very accurate. In these regions non-local effects should be significant, since the electron will experience significantly different environments before scattering. Below we account for such effects in the simplest possible way, by using a Karman-Pohlhausen procedure to derive a first moment model.

7. First moment model. whenever the electric field varies gradually over a mean free path, the momentum distribution of the electrons is of the form

$$(7.1) \qquad p(t, x, z) = a(t, x)\phi(z; F) + b(t, x)r(z)$$

to leading order. See equation (6.5). This distribution should also be fairly realistic in any thin layers where the electric field varies more rapidly. So consider the following procedure: Suppose we demand that the momentum distribution always be of the form (7.1); then we cannot satisfy the Boltzmann equation

$$(7.2) \qquad \varepsilon p_t + \varepsilon v(z)p_x + F(x)p_z = \mathcal{L}p$$

exactly, so instead we demand that the zeroeth and first-order moments of (7.2) be satisfied exactly; this then gives us two equations to determine $a(t, x)$ and $b(t, x)$.

The upshop of this Karman-Pohlhausen procedure is a *first moment model* for semiconductors. In regions where the electric field *does* vary gradually over a mean free path, this model automatically reduces to the asymptotically correct solution, (6.11). In any thin regions where the electric field changes more rapidly, at least the first moments of the Boltzmann equation are satisfied exactly. Consequently, we expect the first moment model to be very accurate, although we have not yet made extensive comparisons between the first moment model and direct (numerical) solutions of the Boltzmann equations.

To make the derivation more transparent, let us re-write (7.1) as

$$(7.3a) \qquad p(t, x, z) = n(t, x)r(z) + J(t, x)\frac{\phi(z; F) - r(z)}{\nu(F)}.$$

Then

$$(7.3b) \qquad n(t, x) \equiv \langle 1, p \rangle, \quad J(t, x) \equiv \langle v, p \rangle,$$

so $n(t, x)$ is the spatial density and $J(t, x)$ is the flux of electrons. Taking the inner product of (7.2) with respect to 1 yields

$$(7.4a) \qquad n_t + J_x = 0,$$

and taking the inner product of (7.2) with respect to $v(z)$ yields

$$(7.4b) \qquad \varepsilon J_t + \varepsilon \langle v^2, p \rangle_x + F\langle v, p_z \rangle = \langle v, \mathcal{L}p \rangle.$$

Substituting the presumed distribution (7.3a) into (7.4b), and using

$$(7.5) \qquad \mathcal{L}r = 0, \quad \mathcal{L}\phi = F\phi_z, \quad r_z = -\beta vr,$$

we obtain the *first moment model*

$$(7.6a) \qquad n_t + J_x = 0,$$

$$(7.6b) \qquad J_t + \{c(F)J\}_x = \frac{\sigma}{\varepsilon D}\left(\nu(F)n - \varepsilon D(F)n_x - J\right),$$

where

$$(7.6c) \qquad c(F) = \frac{\langle v^2, \phi - r\rangle}{\nu(F)}, \qquad \sigma = \langle v^2, r\rangle,$$

and where the drift velocity and diffusivity are defined by

$$(7.6b) \qquad \nu(F) = \langle v, \phi\rangle, \quad D(F) = \nu(F)/\beta F,$$

as before. Note that whenever the force F changes slowly on the x/ε space scale, equation (7.6b) yields

$$(7.7) \qquad J \sim \nu(F)n - \varepsilon D(F)n_x.$$

So whenever the electric field *does* vary gradually, the first moment model reduces to the asymptotically-correct diffusion-drift model (6.11), as expected.

Surprisingly, there is no nonlinear advection term in (7.6b). Unlike the hydrodynamic model, the first moment model predicts that there will *not* be any electron shock waves. Further investigation shows that the nonlinear advection term is missing because the stress-tensor cancels it out exactly.

8. Three dimensions. Future work. The new field-dependent diffusion-drift equation (6.11), and the first moment model (7.6), are the key results of our analysis. Along with the equivalent results for holes and Poisson's equation, they constitute the high field semiconductor equations in Appendix A.

We are currently analyzing the Boltzmann equation in three dimensions [9,10]. Although this analysis is substantially more difficult [15], arguments similar to the ones presented here shows that the spatial density $n(t, \mathbf{x})$ satisfies

$$(8.1a) \qquad n_t = \nabla \cdot (\varepsilon D \nabla n - \nu n) + \varepsilon^2 \nabla \cdot M \cdot \nabla \left(\frac{\boldsymbol{\nu} \cdot \nabla n}{\beta \mathbf{F} \cdot \boldsymbol{\nu}}\right) + \cdots$$

to leading order, where the drift velocity $\boldsymbol{\nu}(\mathbf{F})$, diffusivity tensor D, and dispersion tensor M are defined by

$$(8.1b) \qquad \boldsymbol{\nu}(\mathbf{F}) \equiv \langle \mathbf{v}(\mathbf{z}), \phi\rangle, \quad D \equiv \frac{\boldsymbol{\nu} \otimes \boldsymbol{\nu}}{\beta \mathbf{F} \cdot \boldsymbol{\nu}} + \langle 1, \mathbf{v} \otimes \boldsymbol{\Phi}\rangle, \quad M = \langle 1, \mathbf{w} \otimes (\widetilde{\boldsymbol{\Phi}} - \boldsymbol{\Phi})\rangle.$$

Here $\mathbf{w}(\mathbf{z}; \mathbf{F})$ is orthogonal to the force $\mathbf{F}$, and is related to the velocity $\mathbf{v}(\mathbf{z})$ by

$$(8.1c) \qquad \mathbf{w}(\mathbf{z}; \mathbf{F}) = \mathbf{v}(\mathbf{z}) - \frac{\mathbf{v} \cdot \mathbf{F}}{\boldsymbol{\nu} \cdot \mathbf{F}}\boldsymbol{\nu}(\mathbf{F}),$$

and $\phi(\mathbf{z}; \mathbf{F})$, $\boldsymbol{\Phi}(\mathbf{z}; \mathbf{F})$, and $\widetilde{\boldsymbol{\Phi}}(\mathbf{z}; \mathbf{F})$ are defined by

$$(8.1d) \qquad \mathcal{M}(\mathbf{F})\phi \equiv \mathcal{L}\phi - \mathbf{F} \cdot \nabla_z \phi = 0, \qquad \text{with } \langle 1, \phi\rangle = 1,$$

$$(8.1e) \qquad \mathcal{M}(\mathbf{F})\boldsymbol{\Phi} = -\mathbf{w}\phi, \qquad\qquad\quad \text{with } \langle 1, \boldsymbol{\Phi}\rangle = \mathbf{0},$$

$$(8.1f) \qquad \mathcal{M}(\mathbf{F})\widetilde{\boldsymbol{\Phi}} - \beta \mathbf{F} \cdot \mathbf{v}\widetilde{\boldsymbol{\Phi}} = -\mathbf{w}r, \qquad \text{with } \langle 1, \widetilde{\boldsymbol{\Phi}}\rangle = \mathbf{0}.$$

Since $\boldsymbol{\Phi} \cdot \mathbf{F} \equiv 0$, we see that $\beta D_M \mathbf{F} \equiv \boldsymbol{\nu}(\mathbf{F})$. Thus, Einstein's relation is once again satisfied, at least if we disregard the dispersion term. (The dispersion term does not arise in one dimension because $M\mathbf{F} = \mathbf{0}$; in 1-d this implies that $M = 0$.) We are currently using this asymptotic analysis to develop a three dimensional first moment model analogous to the one in §7.

In the immediate future, we intend to compare direct (numerical) solutions of the Boltzmann equation to the high-field diffusion-drift equations (6.11) and (6.12), and to the first moment model (7.4). Preliminary results show very good agreement between the analytical predictions and the numerical solutions. We also intend to evaluate our results with realistic scattering operators for semiconductors, and then compare the theoretically-predicted drift velocities and diffusivity tensors with experimental values. This work should be completed shortly.

Next to all material interfaces and spatial boundaries, there exist extremely thin *kinetic boundary layers*, which are only one or two mean free lengths thick. The high-field diffusion-drift equations represent *outer solutions*, which are not valid within such boundary layers, while the first moment model is only approximately correct in these boundary layers. Even though these boundary layers are extremely thin, they can play critical roles in semiconductors. We are currently developing analytical and numerical techniques analogous to the ones in [15-17] for resolving such layers.

Appendix A. Other notation. The high field semiconductor equations. Let $\mathcal{E}_n(\mathbf{k})$ be the energy of the relevant conduction band, where $\mathbf{k}$ is the electronic wavenumber, and recall that $\mathcal{E}_n(\mathbf{k})$ is periodic with a period corresponding to the first Brillouin zone. Also recall that the velocity of an electron with wavenumber $\mathbf{k}$ is

$$(A.1) \qquad \mathbf{v}^e(\mathbf{k}) \equiv \frac{1}{\hbar}\boldsymbol{\nabla}_{\mathbf{k}}\mathcal{E}_n(\mathbf{k}).$$

Let $f^e(t, \mathbf{x}, \mathbf{k})$ be the probability density that state $\mathbf{k}$ in band n is occupied at position $\mathbf{x}$. Then, neglecting inter-band transitions, the Boltzmann equation for the electrons in band n is

$$(A.2) \qquad f_t^e + \frac{1}{\hbar}\boldsymbol{\nabla}_{\mathbf{k}}\mathcal{E}_n(\mathbf{k}) \cdot \boldsymbol{\nabla}_{\mathbf{x}} f^e - \frac{q\mathbf{E}}{\hbar} \cdot \boldsymbol{\nabla}_{\mathbf{k}} f^e = \mathcal{L}^{(e)} f^e.$$

Here we are assuming that the nonlinear scattering terms can be omitted; this implies that the scattering operator $\mathcal{L}^{(e)}$ has a Maxwellian equilibrium distribution [2,5,13],

$$(A.3) \qquad r^e(\mathbf{k}) = \frac{1}{N_e} \exp\{-\beta\mathcal{E}_n(\mathbf{k})\}, \quad \text{with } \beta = k_B T,$$

where the normalization N_e will be chosen later.

Before continuing, let us connect this notation with the notation used previously. Recall that $\mathbf{z} = \hbar\mathbf{k}$ is the crystal momentum of an electron, and that

$$(A.4) \qquad p(t, \mathbf{x}, \mathbf{z})d^3 z = \frac{1}{r\pi^3} f^e(t, \mathbf{x}, \mathbf{k})d^3 k$$

is the phase-space density for electrons in a single band [5,13]. Using $\mathcal{E}(\mathbf{z})$ to denote $\mathcal{E}_n(\mathbf{z}/\hbar)$, the Boltzmann equation (A.2) can be written as

$$(A.5) \qquad p_t + \mathbf{v}(\mathbf{z}) \cdot \nabla_{\mathbf{x}} p + \mathbf{F}(\mathbf{x}) \cdot \nabla_{\mathbf{z}} p = \mathcal{L}^{(e)} p.$$

Here $\mathbf{v}(\mathbf{z}) = \nabla_{\mathbf{z}} \mathcal{E}(\mathbf{z})$, and $\mathbf{F} = -q\mathbf{E}$ is the force on the electrons.

For holes, we let $\mathcal{E}_m(\mathbf{k})$ be the energy of the relevant valence band, and let $f^h(t, \mathbf{x}, \mathbf{k})$ be the probability density that state $\mathbf{k}$ in band m is *unoccupied* at position $\mathbf{x}$. Since the deterministic orbits of electrons and holes are exactly the same, the Boltzmann equation for holes is

$$(A.6) \qquad f_t^h + \frac{1}{\hbar} \nabla_{\mathbf{k}} \mathcal{E}_m(\mathbf{k}) \cdot \nabla_{\mathbf{x}} f^h - \frac{q\mathbf{E}}{\hbar} \cdot \nabla_{\mathbf{k}} f^h = \mathcal{L}^{(h)} f^h.$$

As before, we are assuming that the density of holes is not so large that the nonlinear scattering terms need to be included; this implies that the scattering operator $\mathcal{L}^{(h)}$ has the equilibrium distribution

$$(A.7) \qquad r^h(\mathbf{k}) = \frac{1}{N_h} \exp\{+\beta \mathcal{E}_m(\mathbf{k})\}.$$

High field equations. In one spatial dimension, the phase-space densities are

$$(A.8) \qquad p^e(t, x, z)dz = \frac{1}{\pi} f^e(t, x, k)dk, \quad p^h(t, x, z)dz = \frac{1}{\pi} f^h(t, x, k)dk,$$

so the spatial density of electrons and holes is

$$(A.9) \qquad n(t, x) = \langle 1, f^e \rangle, \quad p(t, x) = \langle 1, f^h \rangle,$$

where the inner product is defined as

$$(A.10) \qquad \langle f, g \rangle \equiv \frac{1}{\pi} \int f^*(k) g(k) dk.$$

Accordingly, we normalize r^e and r^h so that $\langle 1, r^e \rangle = \langle 1, r^h \rangle = 1$.

To express the high field equations, define $\phi^e(k; E)$ and $\phi^h(k; E)$ by

$$(A.11a) \qquad \mathcal{L}^{(e)} \phi^e = -\frac{qE}{\hbar} \phi_k^e, \quad \langle 1, \phi^3 \rangle = 1,$$

$$(A.11b) \qquad \mathcal{L}^{(h)} \phi^h = -\frac{qE}{\hbar} \phi_k^h, \quad \langle 1, \phi^h \rangle = 1,$$

and define the drift velocities and diffusivities

$$(A.12a) \qquad \nu_e(E) = -\langle v^3, \phi^e \rangle, \quad D_e(E) = k_B T \nu_e(E)/qE,$$

$$(A.12b) \qquad \nu_h(E) = +\langle v^h, \phi^h \rangle, \quad D_h(E) = k_B T \nu_h(E)/qE,$$

where $v^e(k)$ and $v^h(k)$ are the electron and hole velocities:

$$\text{(A.13)} \qquad v^e(k) = \frac{1}{\hbar}\mathcal{E}'_n(k), \quad v^h(k) = \frac{1}{\hbar}\mathcal{E}'_m(k).$$

Adapting the moment model of §7 to the Boltzmann equations (A.2) and (A.6), we find that the spatial densities of electrons and holes evolve according to

$$\text{(A.14a)} \qquad n_t + J^e_x = 0, \quad p_t + J^h_x = 0,$$

where the electron and hole particle fluxes are given by

(A.14b)
$$J^e_t = \{c^e(E)J^e\}_x = \frac{\alpha^e}{D_e(E)}\left(-\nu_e(E)n - D_e(E)n_x - J^e\right),$$

(A.14c)
$$J^h_t + \{c^h(E)J^h\}_x = \frac{\alpha^h}{D_h(E)}\left(\nu_h(E)p - D_h(E)p_x - J^h\right).$$

Here the electric field is determined by Poisson's equation,

$$\text{(A.14d)} \qquad E = -\Phi_x, \quad -(\varepsilon\Phi_x)_x = q(p - n + C),$$

where C is the charge concentration of ionized impurities, and

$$\text{(A.15a)} \qquad c^e(E) = \langle (v^e)^2, (\phi^3 - r^3)\rangle/\nu_e(E), \quad \alpha^e = \langle (v^e)^2, r^e\rangle,$$

$$\text{(A.15b)} \qquad c^h(E) = \langle (v^h)^2, (\phi^h - r^h)\rangle/\nu_h(E), \quad \alpha^h = \langle (v^h)^2, r^h\rangle,$$

Equations (A.14a)-(A.14d) form the high field semiconductor equations. Note that wherever the electric field varies slowly over a mean free path, we can replace (A.14b) and (A.14c) by their "local equilibrium values"

$$\text{(A.16)} \qquad J^e = -\nu_e(E)n - D_e(E)n_x, \quad J^h = \nu_h(E)p - D_h(E)p_x,$$

and obtain the asymptotically-correct field-dependent diffusion-drift model.

Appendix B. Higher order solution for the weak force regime. Here we carry out the asymptotic expansion

$$\text{(B.1a)} \qquad p(t, x, z) = p^0(t, x, z) + \varepsilon p^1(t, x, z) + \varepsilon^2 p^2(t, x, z) + \ldots,$$

$$\text{(B.1b)} \qquad \langle 1, p^i\rangle = 0 \quad \text{for } i = 1, 2, 3, \ldots$$

for the solution of the weak force Boltzmann equation

$$\text{(B.2)} \qquad \varepsilon^2 p_t + \varepsilon v(z)p_x + \varepsilon F(x)p_z = \mathcal{L}p$$

through $O(\varepsilon^2)$. In §3 we found that

(B.3a) $$p^0(t,x,z) = n(t,x)r(z), \quad p^1(t,x,z) = -(n_x - \beta Fn)\tilde{r}(z),$$

where the spatial density $n(t,x)$ satisfies

(B.3b) $$n_t = \langle v, \tilde{r}\rangle(n_x - \beta Fn)_x + \ldots$$

to leading order. Define $\alpha_1(t,x)$ and $\alpha_2(t,x)$ by

(B.4) $$n_t = \langle v, \tilde{r}\rangle(n_x - \beta Fn)_x + \varepsilon\alpha_1 + \varepsilon^2\alpha_2 + \ldots .$$

Then expanding the Boltzmann equation yields

(5.5a) $$\mathcal{L}p^2 = vp_x^1 + Fp_z^1 + \langle v, \tilde{r}\rangle(n_x - \beta Fn)_x r(z) \quad \langle 1, p^2\rangle = 0$$
(5.5b) $$\mathcal{L}p^3 = vp_x^2 + Fp_z^2 + p_t^1 + \alpha_1 r(z), \quad \langle 1, p^3\rangle = 0$$
(B.5c) $$\mathcal{L}p^4 = vp_x^3 + Fp_z^3 + p_t^2 + \alpha_2 r(z), \quad \langle 1, p^4\rangle = 0$$

through $O(\varepsilon^4)$.

Substituting (B.3a) into (B.5a), we obtain

(B.5) $$\mathcal{L}p^2 = -(n_x - \beta F_n)_x[v\tilde{r} - \langle v, \tilde{r}\rangle r] - F(n_x - \beta Fn)\tilde{r}_z, \quad \langle 1, p^2\rangle = 0.$$

Define $\Phi(z)$ and $\chi(z)$ to be the unique solutions of

(B.6a) $$\mathcal{L}\Phi = r - v\tilde{r}/\langle v, \tilde{r}\rangle, \quad \langle 1, \Phi\rangle = 0,$$
(B.6b) $$\mathcal{L}\chi = \tilde{r}_z, \quad \langle 1, \chi\rangle = 0;$$

then the solution of (B.5) is

(B.7) $$p^2 = \langle v, \tilde{r}\rangle(n_x - \beta Fn)_x\Phi(z) - F(n_x - \beta Fn)\chi(z).$$

Since

(B.8) $$\Phi(z) = \Phi(-z), \quad \chi(z) = \chi(-z),$$

p^0 is even in z, p^1 is odd in z, and p^2 is even in z. Consequently, integrating (B.5b) and (B.5c) over z yields the solvability conditions

(B.9a) $$\alpha_1 = 0,$$

(B.9b) $$\alpha_2\langle 1, r\rangle + F\langle 1, p_z^3\rangle + \langle v, p^3\rangle_x + \langle 1, p^2\rangle_t = 0.$$

Since $\langle 1, r\rangle = 1$ and $\langle 1, p^2\rangle = 0$, we obtain

(B.10) $$\alpha_1 = 0, \quad \alpha_2 = -\langle v, p^3\rangle_x.$$

180

To simplify (B.10), define ℓ by

(B.11a)
$$\mathcal{L}^+\ell = -v, \quad \ell(z) = -\ell(-z),$$

and note that

(B.11b)
$$\tilde{r}(z) = r(z)\ell(z).$$

Then

(B.12)
$$\alpha_2 = \langle \mathcal{L}^+\ell, p^3 \rangle_x = \langle \ell, p^1 \rangle_{tx} + \langle v\ell, p^2 \rangle_{xx} + (F\langle \ell, p_z^2 \rangle)_x,$$

and since

(B.13)
$$p^1 = -(n_x - \beta Fn)\tilde{r}(z), \quad p^2 = n_t\Phi(z) - F(n_x - \beta Fn)\chi(z) + O(\varepsilon^2),$$

we have

(B.14)
$$\begin{aligned}
\alpha_2 = &-\langle \ell, \tilde{r} \rangle (n_x - \beta Fn)_{xt} + [\langle v\ell, \Phi \rangle n_x + \langle \ell, \Phi_z \rangle Fn]_{xt} \\
&- \langle v\ell, \chi \rangle [F(n_x - \beta Fn)]_{xx} - \langle \ell, \chi_z \rangle [F^2(n_x - \beta Fn)]_x .
\end{aligned}$$

Equation (B.14) is the $O(\varepsilon^2)$ correction to the diffusion drift equation, equation (B.3b). It can be simplified by using the property of detailed balance to derive the identity

(B.15)
$$\langle \ell, \Phi_z \rangle = -\beta \langle v\ell, \Phi \rangle + \langle v\ell, \chi \rangle / \langle v, \tilde{r} \rangle.$$

From this identity and (B.3b), we obtain

(B.16)
$$\alpha_2 = \frac{\langle v\ell, \Phi \rangle - \langle \ell, \tilde{r} \rangle}{\langle v, \tilde{r} \rangle} n_{tt} - \left\{ (\langle v\ell, \chi \rangle F_x + \langle \ell, \chi_z \rangle F^2)(n_x - \beta Fn) \right\}_x .$$

Hence, the spatial density $n(t, x)$ evolves according to the equation

(B.17a)
$$n_t + \varepsilon^2 \gamma n_{tt} = (\overline{D}n_x - \bar{\mu}Fn)_x$$

through $O(\varepsilon^2)$, where the diffusion coefficient $\overline{D}$ and mobility $\bar{\mu}$ are given by

(B.17b)
$$\overline{D} = \langle v, \tilde{r} \rangle - \varepsilon^2 F_x \langle v\ell, \chi \rangle - \varepsilon^2 F^2 \langle \ell, \chi_2 \rangle + \dots,$$

(B.17c)
$$\bar{\mu} = \beta\overline{D} = \overline{D}/k_BT,$$

and

(B.17d)
$$\gamma = \langle \ell, \tilde{r} - v\Phi \rangle / \langle v, \tilde{r} \rangle.$$

Moreover, the phase-space density is

(B.18)
$$p(t, x, z) = n(t, x)r(z) - \varepsilon(n_x - \beta Fn)\tilde{r}(z) + \varepsilon^2 n_t \Phi(z) - \varepsilon^2 F(n_x - \beta Fn)\chi(z) + \dots$$

Appendix C. Higher order analysis of the strong force regime. Here we carry out the strong force expansion to higher order. At leading order, (6.8c) is

(C.1)
$$\mathcal{M}(F)R - \varepsilon v R_x = aF_x \left\{ v\phi_F - \frac{\nu'(F)}{\nu(F)}v\phi + \frac{\langle v\psi, \phi_F \rangle}{\nu(F)}vr \right\}$$
$$+ a_t \left\{ \phi - \frac{v\phi}{\nu(F)} + \frac{\langle \psi, \phi \rangle}{\nu(F)}vr \right\} + b_t \left\{ r - \frac{v(\phi - r)}{\nu(F)} \right\}.$$

to solve (C.1), define $R^1(z; F)$, $R^2(z; F)$, and $R^3(z; F)$ as the unique solutions to

(C.2a)
$$\mathcal{M}R^1 = v\phi_F - \frac{\nu'(F)}{\nu(F)}v\phi + \frac{\langle v\psi, \phi_F \rangle}{\nu(F)}vr, \qquad \langle v, R^1 \rangle = 0,$$

(C.2b)
$$\mathcal{M}R^2 = \phi - \frac{v\phi}{\nu(F)} + \frac{\langle \psi, \phi \rangle}{\nu(F)}vr, \qquad\qquad \langle v, R^2 \rangle = 0,$$

(C.2c)
$$\mathcal{M}R^3 - \beta FvR^3 = r - \frac{v(\phi - r)}{\nu(F)}, \qquad\qquad \langle v\psi, R^3 \rangle = 0.$$

Then

(C.3)
$$R(t, x, z) = aF_x R^1 + a_t R^2 + b_t R^3 + O(\varepsilon),$$

and so

(C.4a)
$$n = a + b + \varepsilon aF_x \langle 1, R^1 \rangle + \varepsilon a_t \langle 1, R^2 \rangle + \varepsilon b_t \langle 1, R^3 \rangle + \dots,$$

(C.4b)
$$n_t = -(a\nu(F))_x,$$

(C.4c)
$$b - \frac{\varepsilon}{\beta F}b_x = -\varepsilon aF_x \frac{\langle v\psi, \phi_F \rangle}{\beta F\nu} - \varepsilon a_t \frac{\langle \psi, \phi \rangle}{\beta F\nu} - \frac{\varepsilon b_t}{\beta F\nu} + \dots,$$

through $O(\varepsilon)$. From (C.4a) and (C.4b), we now obtain

(C.5a)
$$b = n - a - \varepsilon aF_x \langle 1, R^1 \rangle - \varepsilon a_t \langle 1, R^2 \rangle - \varepsilon b_t \langle 1, R^3 \rangle + O(\varepsilon^2),$$

and

(C.5b)
$$a_x = -(n_t + \nu' aF_x)/\nu, \qquad b - \frac{\varepsilon}{\beta F}b_x = O(\varepsilon),$$

which yields

(C.6)
$$n - \frac{\varepsilon n_x}{\beta F} - \varepsilon a_t \left(\langle 1, R^2 \rangle - \frac{\langle (\psi - 1), \phi \rangle}{\beta F\nu} \right) = a \left(1 + \varepsilon F_x \langle 1, R^1 \rangle - \varepsilon F_x \frac{\langle v(\psi - 1), \phi_F \rangle}{\beta F\nu} \right)$$

through $O(\varepsilon)$. That is,

$$(\text{C.7}) \qquad n - \frac{\varepsilon n_x}{\beta F} - \varepsilon a_t \langle 1, S^2 \rangle = a(1 + \varepsilon F_x \langle 1, S^1 \rangle),$$

where $S^1(z; F)$ and $S^2(a; F)$ are the solutions of

$$(\text{C.8a}) \qquad \mathcal{M}S^1 = v\phi_F - \frac{\nu'(F)}{\nu(F)} v(\phi - r), \qquad \langle v, S^1 \rangle = 0,$$

$$(\text{C.8b}) \qquad \mathcal{M}S^2 = \phi - \frac{v(\phi - r)}{\nu(F)}, \qquad \langle v, S^2 \rangle = 0.$$

From (C.4b) and (C.7), we now discover that the spatial density satisfies

$$(\text{C.9a}) \qquad \frac{\partial n}{\partial t} = \frac{\partial}{\partial x} \left\{ \left(1 - \varepsilon \langle 1, S^2 \rangle \frac{\partial}{\partial t} \right) \left(\varepsilon \tilde{D} \frac{\partial n}{\partial x} - \tilde{\nu} n \right) \right\} + O(\varepsilon^2),$$

where the drift velocity $\tilde{\nu}$ and diffusivity $\tilde{D}$ are given by

$$(\text{C.9b}) \qquad \tilde{\nu} = \langle v, \phi \rangle \left\{ 1 - \varepsilon F_x \langle 1, S^1 \rangle + \ldots \right\}, \qquad \tilde{D} = \tilde{\nu}/\beta F = k_B T \tilde{\nu}/F.$$

Once (C.9) has been solved for $n(t, x)$, the phase space distribution is given by

$$(\text{C.9c})$$

$$p(t, x, z) = nr + \left(n - \frac{\varepsilon n_x}{\beta F} \right) \left(\phi - r + \varepsilon F_x \left[S^1 - \langle 1, S^1 \rangle \phi \right] \right)$$

$$+ \varepsilon \frac{\partial}{\partial t} \left(n - \frac{\varepsilon n_x}{\beta F} \right) \left(S^2 - \langle 1, S^2 \rangle \phi \right) + \frac{\varepsilon^2}{\beta F} n_{xt} \left(R^3 - \langle 1, R^3 \rangle r \right) + O(\varepsilon^2).$$

Appendix D. Eigenfunctions near $F = 0$. Here we expand $\phi(z; F)$ and $\psi(z; F)$ in powers of F to help understand the interaction of the two eigenspaces $\phi(z; F)$ and $r(z)$ as F goes through zero. Define

$$(\text{D.1a}) \qquad \mathcal{L}r^{(k)} = \frac{1}{\beta} r_z^{(k-1)}, \qquad \langle 1, r^{(k)} \rangle = 0, \quad k = 2, 3, \ldots,$$

$$(\text{D.1b}) \qquad \mathcal{L}^+ \ell^{(k)} = \frac{1}{\beta} \ell_z^{(k-1)} - v \ell^{(k-1)}, \qquad \langle \ell^{(k)}, r \rangle = 0 \quad k = 2, 3, \ldots,$$

with

$$(\text{D.1c}) \qquad r^{(0)} = r(z), \quad \ell^{(0)} = 1, \quad r^{(1)} = \tilde{r}(z), \quad \ell^{(1)} = \ell,$$

and note that $r^{(k)}(z) = r(z)\ell^{(k)}(z)$. Clearly,

$$(\text{D.2a}) \qquad \phi(z; F) = r(z) + \beta F \tilde{r}(z) + \beta^2 F^2 r^{(2)}(z) + \ldots,$$

$$(\text{D.2b}) \qquad \psi(z; F) = 1 - \beta F \ell(z) + \beta^2 F^2 \ell^{(2)}(z) - \ldots,$$

so $\nu(F)$ can be expanded as

$$(\text{D.3}) \qquad \nu(F) \equiv \langle v, \phi \rangle = \beta F \langle v, \tilde{r} \rangle + \beta^3 F^3 \langle v, r^{(3)} \rangle + \ldots .$$

REFERENCES

[1] D. HILBERT, *Begründung der kinetischen Gastheorie*, Math. Ann. 72 (1912), pp. 562–577.

[2] P.A. MARKOWICH, C.A. RINGHOFER, AND C. SCHMEISER, *Semiconductor Equations*, Springer-Verlag, New York (1990).

[3] W.V. VAN ROOSBROECK, *Theory of flow of electrons and holes in germanium and other semiconductors*, Bell Syst. Tech. J. 29 (1950), pp. 560–607.

[4] P.A. MARKOWICH, *The Stationary Semiconductor Device Equations*, Springer-Verlag, New York (1986).

[5] MICHAEL SHUR, *Physics of Semiconductor Devices*, Prentice-Hall, Englewood Cliffs (1990).

[6] W. HÄNSCH AND M. MIURA-MATTAUSCH, *The hot electron problem in small semiconductor devices*, J. Appl. Phys. 60 (1986), pp. 650–655.

[7] J.R. BARKER AND D.K. FERRY, *On the physics and modeling of small semiconductor devices*, Solid State Elect. 23 (1980), pp. 519–530.

[8] A. EINSTEIN, *Über die von der molekularkinetischen Theorie der Wärme geforderte Bewegung von in ruhenden Flüssigkeiten suspendierten Teilchen*, Annalen de Physik 17 (1905), pp. 549–560.

[9] P.S. HAGAN, R.W. COX, AND B.A. WAGNER, *Boltzmann and Fökker-Planck equations in the strong force regime*, in preparation.

[10] P.S. HAGAN, R.W. COX AND B.A. WAGNER, *High field semiconductor equations*, in preparation.

[11] R.W. COX, P.S. HAGAN, AND B.A. WAGNER, *Boltzmann equations and the strong force generalization of the Hilbert expansion*, submitted.

[12] B.A. WAGNER, R.W. COX, AND P.S. HAGAN, *Extension of the drift-diffusion equation for semiconductors to the high field regime*, submitted.

[13] N.C. ASHCROFT AND N.D. MERMIN, *Solid State Physics*, Holt-Sounders, New York (1976).

[14] J. KEVORKIAN AND J.D. COLE, *Perturbation Methods in Applied Mathematics*, Springer-Verlag, New York (1981).

[15] D.S. COHEN AND T.S.M. POPEYE, *On the influence of leafy vegetables on physical exertion*, in preparation.

[16] P.S. HAGAN AND M.M. KLOSEK, *Half-range expansions for kinetic problems with Stürm-Liouville scattering operators*, in preparation.

[17] P.S. HAGAN, C.R. DOERING, AND C.D. LEVERMORE, *Mean exit times for particles driven by weakly colored noise*, SIAM J. Appl. Math. 49 (1989), pp. 1480–1513.

ENERGY MODELS FOR ONE-CARRIER TRANSPORT
IN SEMICONDUCTOR DEVICES

JOSEPH W. JEROME* AND CHI-WANG SHU†

Abstract. Moment models of carrier transport, derived from the Boltzmann equation, have made possible the simulation of certain key effects through such realistic assumptions as energy dependent mobility functions. This type of global dependence permits the observation of velocity overshoot in the vicinity of device junctions, not discerned via classical drift-diffusion models, which are primarily local in nature. It has been found that a critical role is played in the hydrodynamic model by the heat conduction term. When ignored, the overshoot is inappropriately damped. When the standard choice of the Wiedemann-Franz law is made for the conductivity, spurious overshoot is observed. Agreement with Monte-Carlo simulation in this regime has required empirical modification of this law, as observed by IBM researchers, or nonstandard choices. In this paper, simulations of the hydrodynamic model in one and two dimensions, as well as simulations of a newly developed energy model, the RT model, will be presented. The RT model, intermediate between the hydrodynamic and drift-diffusion model, was developed at the University of Illinois to eliminate the parabolic energy band and Maxwellian distribution assumptions, and to reduce the spurious overshoot with physically consistent assumptions. The algorithms employed for both models are the essentially non-oscillatory shock capturing algorithms, developed at UCLA during the last decade. Some mathematical results will be presented, and contrasted with the highly developed state of the drift-diffusion model.

Key words. Energy models, shock capturing algorithms, conservation laws, velocity overshoot, parabolic and nonparabolic energy bands

AMS(MOS) subject classifications. 76N10, 82A70, 35L65, 65C20

1. Introduction.

During the last decade, device modeling has attempted to incorporate general carrier heating, velocity overshoot, and various small device features into carrier simulation. The popular wisdom emerging from such concentrated study holds that global dependence of critical quantities, such as mobilities, on energy and/or temperature, is essential if such phenomena are to be modeled adequately. In this paper, we examine in detail the simulation of two such energy models, including the hydrodynamic model and the RT model. We describe the models, summarize some associated mathematical results, as well as the basic features of the numerical algorithm, and then present the results of extensive numerical simulations for two-dimensional MESFET devices, and for one-dimensional diodes. Both models represent one carrier flow. The hydrodynamic model contains hyperbolic modes related to the momentum equations, while the RT model does not possess such modes. In both cases, however, we employ a conservation law format, and numerical methods suitable for such systems. The ENO (essentially non-oscillatory) method employed makes use of adaptive stencils, and is particularly adept at shock capturing if the parameter regime crosses from supersonic to subsonic. Even if this does not occur, the convective terms are effectively discretized, via this procedure,

*Department of Mathematics, Northwestern University, Evanston, IL 60208.
†Division of Applied Mathematics, Brown University, Providence, RI 02912.

in both models. The first use of such methods in device simulation was in [7], followed by the study [6], in which shocks were detected in micron devices at liquid Nitrogen temperatures, and at room temperature in shorter devices, by independent numerical techniques.

Our development of the RT model follows that of [5]. These researchers attempted to utilize a microscopic relaxation time approximation, which would allow for nonparabolic energy bands and non-Maxwellian distribution functions. The approach allows for parameter fitting of certain key quantities via Monte-Carlo simulation.

One of the principal conclusions of the paper is the essential dependence of the hydrodynamic model upon the heat conduction term. Standard choices lead to numerically detected spurious overshoot at the drain junction of an $n^+ - n - n^+$ diode, while other choices significantly damp this overshoot. Monte-Carlo simulations show that substantial underestimation occurs when the heat conduction term is neglected. We refer the reader to [10], and to the simulation results of this paper for amplification. The RT model was developed, partly in response to the continuing debate concerning heat conduction processes in the hydrodynamic model.

The status of mathematical results differs sharply between the hydrodynamic model, on the one hand, and the drift-diffusion model on the other. For the former, we have summarized two results, one by Gamba (cf. [8]) for an idealized model, in which the adiabatic relation is employed, and another by Gardner, Jerome, and Rose (cf. [9]) in which a Newton-Kantorovich theorem is developed for the $n^+ - n - n^+$ diode, yielding existence and convergence in a specialized subsonic regime. The drift-diffusion model, on the other hand, has been widely studied. Existence and approximation results have been carefully developed, although uniqueness is still not well understood for this model. Existence for the steady-state model is due in varying degrees of generality to many authors, including Mock ([17]), Seidman ([21]), and the first author ([13]). A convergence theory, based upon a calculus due to Krasnosel'skii, was presented in ([15]). Mathematical results have not yet been developed for the strongly nonlinear RT model.

2. Hydrodynamic and drift-diffusion models.

2.1. Mass, momentum and energy transport equations. The equations as presented here are discussed in references [3], [20], and [4]. They are derived as the first three moments of the Boltzmann equation, with the latter written for electrons moving in an electric field as

$$(2.1) \qquad \frac{\partial f}{\partial t} + u \cdot \nabla_x f - \frac{e}{m} F \cdot \nabla_u f = C.$$

Here, $f = f(x, u, t)$ is the numerical distribution function of a carrier species, x is the position vector, u is the species' group velocity vector, $F = F(x, t)$ is the electric field, e is the electron charge modulus, m is the effective electron mass, and C is the time rate of change of f due to collisions. In the Boltzmann equation above, it has been assumed that the traditional Lorentz force field does not have a component induced by an external magnetic field. The moment equations, which will be derived subsequently, are expressed in terms of certain dependent variables, where n is the electron concentration, v is the average velocity, p is the momentum density, P is the symmetric pressure tensor, q is the the heat flux, e_I is the internal energy, and C_n, C_p, and C_W represent moments of C, taken with respect to the functions

$$h_0(u) \equiv 1,$$
$$h_1(u) = mu,$$
$$h_2(u) = \frac{m}{2} \mid u \mid^2 .$$

The equations are given by:

$$(2.2) \qquad \frac{\partial n}{\partial t} + \nabla \cdot (nv) = C_n,$$

$$(2.3) \qquad \frac{\partial p}{\partial t} + v(\nabla \cdot p) + (p \cdot \nabla)v = -enF - \nabla \cdot P + C_p,$$

$$\frac{\partial}{\partial t}\left(\frac{mn}{2} \mid v \mid^2 + mne_I\right) + \nabla \cdot \left(v\left\{\frac{mn}{2} \mid v \mid^2 + mne_I\right\}\right) =$$
$$(2.4) \qquad -env \cdot F - \nabla \cdot (vP) - \nabla \cdot q + C_W.$$

The first Maxwell equation for the electric potential must be adjoined; each species contributes a corresponding moment subsystem, with appropriately signed charge. We begin the derivation with the definitions and assumptions. The concentration is given by $n := \int f\, du$; the average velocity by $v := \frac{1}{n}\int uf\, du$; the momentum by $p := mnv$; the random velocity by $c := u - v$; the pressure tensor by $P_{ij} := m \int c_i c_j f\, du$; and the internal energy density by $e_I := \frac{1}{2n}\int \mid c \mid^2 f\, du$. This function represents energy/unit mass/unit concentration. The heat flux q is given by $q_i := \frac{m}{2}\int c_i \mid c \mid^2 f\, du$. Finally, for reference in subsequent subsections, the electron current density is given by $J := -env$, and the energy flux is given by $S := \int u\{\frac{m}{2} \mid u \mid^2\}f\, du$. The assumptions on f are now stated. The function f is assumed to decrease sufficiently rapidly at infinity:

$$\lim_{|u| \to \infty} h_i(u)f(u) = 0, \ i = 0,1,2.$$

The derivation of (2.2), (2.3), (2.4) proceeds by multiplying the Boltzmann equation (2.1) by h_0, h_1, and h_2, respectively, and integrating over group velocity space. With the application of certain standard identities ([16]), the mass/momentum/energy system is obtained. In addition to these transport equations, we have Poisson's equation for the electric field, where $n_d :=$ doping and $\epsilon :=$ dielectric:

$$(2.5) \qquad F = -\nabla\phi,$$

$$(2.6) \qquad \nabla \cdot (\epsilon\nabla\phi) = -\sum e_i n_i - n_d.$$

Here, we have used the convention that there are different species, each of concentration n_i and charge e_i. The entire system consists of equations (2.2), (2.3), (2.4), repeated according to species, and (2.5), (2.6).

2.2. Moment closure and relaxation relations. The system derived in the preceding subsection has fifteen dependent variables in the case of one species, determined by ϕ, n, v, P, e_I, and q. By moment closure is meant the selection of compatible relations among these variables, so that the number of equations is equal in number to the remaining primitive variables selected. The relations to follow are characterized by the isotropic/parabolic energy band assumption. We begin by introducing a new tensor variable T, the carrier temperature, defined by

$$P_{ij} = nkT_{ij}$$

where k is Boltzmann's constant, and a scalar variable W, the total carrier energy. A program of reduction to a set of basic variables, n, v, W, and ϕ, or a set equivalent to these, can be implemented by the following assumptions:

- The pressure tensor is isotropic, with diagonal entries P_s and off-diagonal entries zero, for a suitable scalar function, P_s. P_s is related to e_I via $mne_I = \frac{3}{2}P_s$.

- It follows from the previous assumption that temperature may be represented by a scalar quantity T, and that the internal energy is represented in terms of T by

$$me_I = \frac{3}{2}kT.$$

- The total energy density (per unit concentration) w is given by combining internal energy and parabolic energy bands with m assumed constant:

$$w = me_I + \frac{1}{2}m \mid v \mid^2,$$

and the total energy (per unit volume) W is the product, $W = nw$.

- The heat flux is obtained by a differential expression involving the temperature:

$$q = -\kappa \nabla T.$$

Here, κ is the thermal conductivity governed by the Wiedemann-Franz law (cf. [2]), described by

$$(2.7) \qquad \kappa = \left(\frac{5}{2} + r\right) n \frac{k^2 \mu_0}{e} T \left(\frac{T}{T_0}\right)^r.$$

The standard choice for r is $r = -1$, but this has some associated difficulties. This will be amplified later in the paper. Here we simply remark that the term raised to the exponent r in (2.7) is proportional to the mobility, which in turn is proportional to the momentum relaxation time.

In the case of N species, the closure relations determine $(d + 2)N + 1$ variables in d spatial dimensions. It is possible to rewrite the system (2.2, 2.3, 2.4) with the closure assumptions incorporated. We have the following.

$$(2.8) \qquad \frac{\partial n}{\partial t} + \nabla \cdot (nv) = C_n,$$

$$(2.9) \qquad \frac{\partial p}{\partial t} + v(\nabla \cdot p) + (p \cdot \nabla)v = -enF - \nabla(nkT) + C_p,$$

$$\frac{\partial W}{\partial t} + \nabla \cdot (v\,W) = -env \cdot F - \nabla \cdot (vnkT)$$
$$(2.10) \qquad\qquad\qquad + \nabla \cdot (\kappa \nabla T) + C_W.$$

The final step deals with the replacement of the collision moments. Motivated by the approach of [18], [1], [20], and [11], we define the recombination rate R and the momentum and energy relaxation times, τ_p and τ_w, respectively, in terms of averaged collision moments as follows.

1. The particle recombination rate R is given by

$$R := -C_n := -\int C\,du.$$

2. The momentum relaxation time τ_p is given via

$$\frac{p}{\tau_p} := -m \int uC\,du := -C_p.$$

3. The energy relaxation time τ_w is given via

$$-\frac{W - W_0}{\tau_w} := \frac{m}{2} \int |u|^2 \, f \, du := C_W.$$

Here, W_0 denotes the rest energy, $\frac{3}{2}kT_0$, where T_0 is the lattice temperature.

The forms for the relaxation times used in [1] and retained by subsequent authors are:

$$(2.11) \qquad \tau_p = c_p \left(\frac{T}{T_0}\right)^r,$$

$$(2.12) \qquad \tau_w = c_w \frac{T}{T + T_0} + \frac{1}{2}\tau_p.$$

Here, c_p and c_w are physical constants, and the standard choice for r, just as in (2.7), is -1.

2.3. Drift-diffusion model. The drift-diffusion model may be obtained by taking zeroth order moments of the BTE and adjoining the Poisson equation. Thus, one obtains the system for N carriers with concentrations n_i and charge e_i, $i = 1, \cdots, N$:

$$(2.13) \qquad \frac{\partial n_i}{\partial t} + \nabla \cdot J_i = -R_i,$$

$$(2.14) \qquad F = -\nabla \phi,$$

$$(2.15) \qquad \nabla \cdot (\epsilon \nabla \phi) = -\sum e_i n_i - n_d.$$

There still remains the issue of determining the constitutive current relations. Classical drift-diffusion theory gives, for $N = 2$, $n_1 = n$, and $n_2 = p$,

$$
(2.16) \qquad J_n = -e\mu_n n \nabla\phi + e D_n \nabla n,
$$

$$
(2.17) \qquad J_p = -e\mu_p p \nabla\phi - e D_p \nabla p.
$$

The introduction of exponential relations for n and p is also common, as is the use of the Einstein relations linking the mobilities, μ_n, μ_p, and the diffusion coefficients D_n, D_p. These relations are specified by

$$
(2.18) \qquad D_n = (kT/e)\mu_n,
$$

$$
(2.19) \qquad D_p = (kT/e)\mu_p.
$$

It is also possible to derive the constitutive relations (2.16), (2.17), from the first order moment relations under the assumption that the momentum relaxation times tend to zero. The details are given in [20]. In fact, the constitutive relations include a heat flux term as well, which is suppressed at constant temperature. If it is not suppressed, one has an energy drift-diffusion model. In this derivation, one uses the definition of mobility in terms of relaxation time.

3. RT models.

In this section, we shall employ a microscopic assumption upon the momentum relaxation time, viz. , we shall assume that the collision term C in (2.1) is of the form,

$$
(3.1) \qquad C = -\frac{f_1}{\tau_p},
$$

where f_1 is the odd part of f. Note that this contrasts with the macroscopic assumption on τ_p, employed in the hydrodynamic model as described in Section 2. There, the representation defining τ_p was a post averaged expression. Here, the expression is employed in the averaging. In this case, we may obtain an expression for the energy flux S:

$$
(3.2) \qquad S = -[n\mu^E F + \nabla(n D^E)],
$$

where μ^E and D^E are tensor expressions for mobility and diffusion, defined in terms of moments, and E represents average energy per unit concentration. The details are furnished in [5]. It is also shown there that the current density has the usual drift-diffusion form, with tensor expressions for mobility and diffusion. The RT model makes the following microscopic assumptions, with distinction between $\mathcal{E}$ and its average, E.

    1. The even part of f is isotropic, and a function of $\mathcal{E}$ alone, and the relaxation time is an inverse power function of $\mathcal{E}$:

$$
f_0 = f_0(\mathcal{E}),
$$

$$
\tau_p = \tau_p(\mathcal{E}) = C\mathcal{E}^{-r}.
$$

2. The microscopic kinetic energy is a quadratic function of $\mathcal{E}$, and the mass is not assumed constant:

$$(3.3) \qquad G(\mathcal{E}) := \frac{m^*}{2} \mid u \mid^2 \; = \mathcal{E} + \alpha \mathcal{E}^2,$$

where α is an appropriate fitting parameter.

3. The temperature is a modified variable in terms of which the following constitutive relation holds for E:

$$(3.4) \qquad E = \frac{3}{2} kT \left(1 + \frac{5}{2}\alpha kT\right).$$

Equation (3.4) allows for nonparabolic energy bands as well as non-Maxwellian distributions. Altogether, the model may be written in terms of the Poisson equation, (2.6), in conjunction with the system,

$$(3.5) \qquad \nabla \cdot J = 0,$$

$$(3.6) \qquad \nabla \cdot S = J \cdot F - n \left\langle \frac{\partial E}{\partial t} \mid_{coll} \right\rangle.$$

Here, J and S have been described previously, the latter in (3.2). In the expressions for J and S, the assumptions made for the model lead to scalar representations for the mobility and diffusion coefficients. For example, the choice made in [5] leads to

$$(3.7) \qquad \mu = \mu_0 T_0/T,$$

$$(3.8) \qquad \mu^E = \frac{3}{2} \left(1 - \frac{\alpha}{2} kT\right) \mu kT.$$

The diffusion coefficients are defined by Einstein's relations. The collision term in (3.6) is a specified quadratic function of T. One significant advantage of the microscopic relaxation time (RT) assumption is that certain key parameters may be fitted via Monte-Carlo preprocessing, ensuring reliability of their values.

4. Mathematical results for the hydrodynamic model.

In this section, we shall describe some recent mathematical results, obtained in one spatial dimension. In the first subsection, we shall present existence and boundary layer results for a simplified version of the steady state hydrodynamic model. This will be followed by a convergence analysis for Newton's method in the subsonic case.

4.1. Existence and boundary layer theory. We first write down the one dimensional evolution system in the case of a single carrier, in the absence of recombination.

$$(4.1) \qquad \frac{\partial n}{\partial t} + \frac{\partial(nv)}{\partial x} = 0,$$

$$(4.2) \qquad \frac{\partial p}{\partial t} + \frac{\partial(pv + knT)}{\partial x} = -enF - \frac{p}{\tau_p},$$

$$(4.3) \qquad \frac{\partial W}{\partial t} + \frac{\partial(vW + vknT)}{\partial x} = -envF + \frac{\partial(\kappa(\partial T)/(\partial x))}{\partial x} - \frac{W - W_0}{\tau_w},$$

$$(4.4) \qquad \epsilon\frac{\partial F}{\partial x} = -en - n_d.$$

The corresponding steady state system is obtained by setting the time derivatives equal to zero.

If we rewrite the second steady state equation by use of the pressure, P, we obtain the equation,

$$(4.5) \qquad \frac{\partial(pv + P)}{\partial x} = -enF - \frac{p}{\tau_p}.$$

The approach of [8] is to eliminate the energy equation (4.3) from the system, and replace its role by a relation in the spirit of gas dynamics, i. e. by the constitutive relation,

$$(4.6) \qquad P(n) = Kn^\gamma, \ \gamma > 1.$$

When units are selected in which $e = 1$, $\epsilon = 1$, $m = 1$, and $K = 1$, we obtain the system, in which n and ϕ are the only dependent variables,

$$(4.7) \qquad j := nv \equiv \text{constant} ,$$

$$(4.8) \qquad (F(n))_x := \left(\frac{j}{n} + n^\gamma\right)_x = -n\phi_x - \frac{j}{\tau_p} := -S(\phi_x, n),$$

$$(4.9) \qquad \phi_{xx} = n - n_d.$$

One can nominally specify boundary conditions on n and ϕ at the endpoints of the device, taken here as the interval, $[0, 1]$. If the doping is such that the built-in potential is the same at both ends of the device, then we may take,

$$(4.10) \qquad \phi(0) = 0, \ \phi(1) = \phi_1,$$

for ϕ and

$$(4.11) \qquad n(0) = n_0, \ n(1) = n_1,$$

for n, where ϕ_1 must satisfy the following consistency relation with respect to j, n_0, and n_1:

$$(4.12) \qquad \phi_1 = f(n_1, j) - f(n_0, j) + j \int_0^1 \frac{dx}{\tau(n, j)n(x)}.$$

Here,

$$(4.13) \qquad f(n, j) = \frac{j^2}{2n^2} + \gamma n^{\gamma - 2}.$$

It is shown in [8] that a weak solution exists for the system (4.7), (4.8), (4.9), satisfying the boundary conditions exactly, or, in lieu of this, satisfying precise limiting relationships. The result for ϕ is classical, because the equation is elliptic.

The result for n is provisional, and is detailed now. If $j > 0$, there is a weak solution n such that the relation,

$$(4.14) \qquad n = G + \alpha,$$

holds, where G is Hölder $\left(\frac{1}{2}\right)$ continuous, and α is monotone increasing. Although n need not be of bounded variation, it is an entropy solution, in the sense that the function of x,

$$(4.14) \qquad H(n(x)) = (F(n(x)) - F(n_{\min})) \,\text{signum}\, (n(x) - n_{\min}) + Cx,$$

is monotone increasing. Here, $n_{\min}$ is a minimum (location) for F, and $C = \sup S$ over relevant arguments. The following precise statement is available for subsonic boundary conditions. If $n_0, n_1 > n_{\min}$, then the following holds.

- Either
$$n(1) = n_1,$$

- or
$$\lim_{x \to 1^-} n(x)$$

  exists, and is a supersonic value, i. e. , is less than $n_{\min}$, and even less than the conjugate value of n_1.

- Similarly, either
$$n(0) = n_0,$$

- or
$$\lim_{x \to 0^+} n(x)$$

  exists and is not less than $n_{\min}$.

In the second instance of both cases above, boundary layers occur, the one on the right involving transition through the supersonic regime.

4.2. Newton convergence theory. In this subsection, we order the basic variables as v, n, T, and ϕ, because of symmetry considerations. Dirichlet boundary conditions are imposed, on n, T, ϕ, with $n(x_{\min}) = n(x_{\max})$. In one dimension, the steady state equations are obtained from (4.1), (4.2), (4.3), and (4.4) by setting the time derivatives equal to zero. Since, by the conservation of mass equation, $nv = j$, it follows that the boundary conditions for v are periodic. The map defined by bringing all terms to the left hand side of the steady state system is called Φ. The linearized equations thus assume the form, where the boundary conditions are homogeneous Dirichlet conditions for δn, δT, and $\delta \phi$, and the boundary conditions on δv are prescribed to be periodic,

$$(4.16) \qquad \begin{bmatrix} 0 \\ 0 \\ -\frac{1}{n}\frac{d}{dx}\left(\kappa\frac{d\delta T}{dx}\right) \\ -\epsilon\frac{d^2\delta\phi}{dx^2} \end{bmatrix} + \begin{bmatrix} A & B \\ C & D \end{bmatrix} \frac{d}{dx} \begin{bmatrix} \delta v \\ \delta n \\ \delta T \\ \delta \phi \end{bmatrix} + \begin{bmatrix} E & F \\ G & H \end{bmatrix} \begin{bmatrix} \delta v \\ \delta n \\ \delta T \\ \delta \phi \end{bmatrix} = f.$$

The (spatially) dependent eigenvalues of the *symmetric* matrix A are calculated to be

$$(4.17) \qquad \lambda = \frac{1}{2}\left(n + \frac{T}{mn}\right) \pm \frac{1}{2}\sqrt{\left(n + \frac{T}{mn}\right)^2 - 4\left(\frac{T}{m} - v^2\right)}.$$

Here,

$$(4.18) \qquad A = \begin{bmatrix} n & v \\ v & \frac{T}{mn} \end{bmatrix},$$

and the smaller eigenvalue is positive if n and T are strictly positive, and if

$$(4.19) \qquad v^2 < \frac{T}{m} = c^2.$$

This type of point in function space is termed a subsonic point. This case was first considered in [9], where damped Newton/standard finite difference methods were presented. When Newton's method is employed in this way, it is essential to determine conditions under which the linear increments are appropriately bounded. This is equivalent to uniform bounds for the operator derivative inverse maps, and represents one of the three properties for an (exact) operator Newton method to yield existence of a root and R-quadratic convergence. The other two are sufficient regularity, and a sufficiently small starting residual, as measured in the range space norm. Explicit representations of B, C, D and of F, G, H are given in [14]. Moreover, if the system map Φ, subject to appropriate Dirichlet boundary conditions on n, T, and ϕ, and periodic boundary conditions on v, accordingly has the domain $D_\Phi \subset X = \prod_1^2 W^{1,\infty} \times \prod_1^2 W^{2,\infty}$, and range in $Y = \prod_1^4 L^\infty$, then (4.19) will hold for every element in a closed ball $B_{r_0} \subset X$, centered at a subsonic element $u_0 \in X$, such that n and T are strictly positive, if r_0 is sufficiently small. It is appropriate to assume at the outset, then, that $D_\Phi \subset B_{r_0}$, so that every function point in D_Φ satisfies (4.19); we may also assume that n and T are uniformly bounded away from zero in this set.

The Lipschitz property of the map Φ', where here we view the system (4.16) as the representation for $\Phi'(v, n, T, \phi)(z, \omega) = f$, with

$$(4.20) \qquad z = (\delta v, \delta n), \qquad \omega = (\delta T, \delta \phi), \qquad f = (f_1, f_2),$$

is evident from the representation for Φ'.

The uniform inverse bounding proceeds as follows. As shown in [9], with the function spaces selected in this paper, the H^1 product norm of z can be estimated in terms of the L^2 norms of ω, ω', and f_1, under the conjunction of the hypothesis (4.19) and the $L^2 \times L^2$ coerciveness assumption,

$$(4.21) \qquad \Lambda = E + E^* - \frac{dA}{dx} \text{ is uniformly positive definite.}$$

Here, E^* is the matrix transpose of E and the latter is defined by

$$(4.22) \qquad E = \begin{bmatrix} \frac{dn}{dx} & \frac{dv}{dx} \\ \frac{dv}{dx} + \frac{1}{\tau_p} & \left(\frac{-1}{mn^2}\right) T \frac{dn}{dx} \end{bmatrix}.$$

A final calculation, making use of the inner product of $[0, \omega]$ with (4.16), and the hypothesis,

$$(4.23) \qquad h_{11} \text{ is positive, and sufficiently large,}$$

where h_{11} is the nonzero entry of H, given by

$$(4.24) \qquad h_{11} = \frac{dv}{dx} + \frac{\frac{dW}{dT} \tau_w - (W - W_0)\tau_w'}{\tau_w^2},$$

with $\tau_w' = \frac{d\tau_w}{dT}$, shows that the H^1 product norm of $[z, \omega]$ is estimated in terms of the L^2 norm of f. This series of calculations controls the L^∞ norm of $[z, \omega]$; the L^∞ norm of $[z', \omega'']$ is now estimated by direct use of the system (4.16), making use of the fact that $[v, n, T, \phi] \in B_{r_0}$. We have now outlined the proof of the following.

THEOREM 4.1. *Let the function spaces X and Y be selected as above, let the steady state system map Φ be given with Dirichlet boundary conditions on n, T, ϕ, and periodic boundary conditions on v, such that $D_\Phi \subset B_{r_0}$, where every point in B_{r_0} is a subsonic point, with uniform positivity bounds. If (4.21) and (4.24) hold, and x_0 is such that $\Phi(x_0)$ is sufficiently small, then an R-quadratically convergent Newton sequence $\{x_n\}$ may be defined in the standard way, with limit x, satisfying $\Phi(x) = 0$.*

5. Discrete schemes based on adaptive stencils: ENO. In this section, we shall briefly describe the ENO schemes as developed in [24] and [25]. Consider a system of hyperbolic conservation laws of the form

$$(5.1) \qquad u_t + \sum_{1}^{d} f_i(u)_{x_i} = g(u, x, t),$$

where

$$u = (u_1, \cdots, u_m)^T, \ x = (x_1, \cdots, x_m),$$

and the hyperbolicity condition,

$$\sum_{1}^{d} \xi_i \frac{\partial f_i}{\partial u} \quad \text{is diagonalizable, with real eigenvalues ,}$$

holds for any real $\xi = (\xi_1, \cdots, \xi_d)$. An initial condition is adjoined to (5.1).

For systems of conservation laws, local field by field decomposition is used, to resolve waves in different characteristic directions. Analytical expressions are employed for the eigenvalues and eigenvectors of an averaged Jacobian matrix. Typically, the Roe average [19] is employed. One feature of the ENO schemes in [24] and [25], which is distinct from the original ENO schemes of Harten et al [12], is that multidimensional regions are treated dimension by dimension: when computing $f_i(u)_{x_i}$ in any particular direction, variables in all other directions are kept constant, and the Jacobians are treated in this direction. This, in essence, reduces the determination of the scheme to the case of a single conservation law in one spatial dimension. Thus, to describe the schemes, consider the scalar one dimensional problem, and a conservative approximation of the spatial operator given by

$$(5.2) \qquad L(u)_j = -\frac{1}{\triangle x}(\hat{f}_{j+\frac{1}{2}} - \hat{f}_{j-\frac{1}{2}}).$$

Here, the numerical flux $\hat{f}$ is assumed consistent:

$$(5.3) \qquad \hat{f}_{j+\frac{1}{2}} = \hat{f}(u_{j-l}, \cdots, u_{j+k}); \qquad \hat{f}(u, \cdots, u) = f(u).$$

The conservative scheme (5.2), which characterizes the $\hat{f}$ divided difference as an approximation to $f(u)_x$, suggests that $\hat{f}$ can be identified with an appropriate function h satisfying

$$(5.4) \qquad f(u(x)) = \int_{x-\frac{\triangle x}{2}}^{x+\frac{\triangle x}{2}} h(\xi)\, d\xi.$$

If H is any primitive of h, then h can be computed from H'. H itself can be constructed by Newton's divided difference method, beginning with differences of order one, since the constant term is arbitrary. The necessary divided differences of H, of a given order, are expressed as constant multiples of those of f of order one lower. After the polynomial Q of degree $r+1$ has been constructed, set

$$(5.5) \qquad \hat{f}_{j+\frac{1}{2}} = \frac{d}{dx}Q(x)_{|x=x_{j+\frac{1}{2}}},$$

to obtain an rth order method. The construction is based on an adaptive stencil in the following sense:

- One begins with an appropriate starting point to the left or right of the current "cell" by means of upwinding as determined by the sign of the derivative of a selected flux.
- As the order of the divided differences is increased, the divided differences themselves determine the stencil: the "smaller" divided difference is chosen from two possible choices at each stage, ensuring a smoothest fit.
- Lax-Friedrichs building blocks or Roe building blocks can both be used. For the latter, in cells with sonic points, a local Lax-Friedrichs building block is used to avoid expansion shocks.

Steady states are reached by explicit time stepping of arbitrary order; nonstandard high order Runge-Kutta methods exist [24] which preserve nonlinear stability of the first order Euler forward version under suitable CFL time step restrictions. The computer program is fully vectorized for computations on Cray supercomputers. For details of the efficient implementation, see [23].

6. Conservation law format for hydrodynamic and RT models.

In this section, we shall specify the conservation law format for the two dimensional hydrodynamic model, and for the one dimensional RT model.

6.1. Hydrodynamic model conservation format. Define the vector of dependent variables as

$$(6.1) \qquad u = (n, \sigma, \tau, W),$$

where $p = (\sigma, \tau)$. The system (2.8), (2.9), (2.10) can be written in the concise form, in two dimensions, as

$$(6.2) \qquad u_t + f_1(u)_x + f_2(u)_y = c(u) + G(u, \phi) + (0, 0, 0, \nabla \cdot (\kappa \nabla T)).$$

The following identifications have been made in (6.2).

$$(6.3) \qquad f_1(u) = \left(\frac{\sigma}{m}, \ \frac{2}{3} \left(\frac{\sigma^2}{mn} + W - \frac{\tau^2}{2mn} \right), \ \frac{\sigma\tau}{mn}, \ \frac{5\sigma W}{3mn} - \sigma \frac{\sigma^2 + \tau^2}{3m^2 n^2} \right),$$

$$(6.4) \qquad f_2(u) = \left(\frac{\tau}{m}, \ \frac{\sigma\tau}{mn}, \ \frac{2}{3} \left(\frac{\tau^2}{mn} + W - \frac{\sigma^2}{2mn} \right), \ \frac{5\tau W}{3mn} - \tau \frac{\sigma^2 + \tau^2}{3m^2 n^2} \right),$$

$$(6.5) \qquad c(u) = \left(0, -\frac{\sigma}{\tau_p}, -\frac{\tau}{\tau_p}, -\frac{W - W_0}{\tau_w} \right),$$

$$(6.6) \qquad G(u) = (0, -enF_1, -enF_2, -enF \cdot v).$$

The eigenvalues and eigenvectors of f_1' and f_2' are known (cf. [23]), and are readily incorporated into the field by field decomposition required for the implementation of ENO.

6.2. RT conservation format. We shall present the conservation law form of the RT model. We begin with the vector form,

$$(6.7) \qquad u_t + f(u)_x = g(u)_{xx} + h(u).$$

In equation (6.7),

$$(6.8) \qquad u = (en, \frac{nE}{m}),$$

$$(6.9) \qquad f(u) = \phi' n \left(e\mu(E), \ \mu^E(E) + D(E) \right),$$

$$(6.10)$$
$$g(u) = (nD(E), \ nD^E(E)),$$

$$(6.11)$$
$$h(u) = \left(0, \ en\mu(E)(\phi')^2 + \frac{e}{\epsilon}(n - n_d)nD(E) - n \left\langle \frac{\partial E}{\partial t} \bigg|_{coll} \right\rangle \right).$$

It can be shown that the left hand side defines a hyperbolic system, since the eigenvalues of $f'(u)$ are real, for all positive n and T.

7. Numerical simulation results.

We now present numerical simulation results for one carrier, two dimensional MESFET devices and one dimensional diodes. The third order ENO shock-capturing algorithm with Lax-Friedrichs building blocks, as described briefly in Section 5 and in more detail in [25], is applied to the hyperbolic part (the left hand side) of Equations (6.2) and (6.7). A nonlinearly stable third order Runge-Kutta time discretization [24] is used for the time evolution towards steady states. The forcing terms on the right hand side of (6.2) and (6.7) are treated in a time consistent way in the Runge-Kutta time stepping. The double derivative terms on the right hand side of (6.2) and (6.7) are approximated by standard central differences owing to their dissipative nature. The Poisson equation (2.6) is solved by direct Gauss elimination for one spatial dimension and by Successive Over-Relaxation (SOR) or the Conjugate Gradient (CG) method for two spatial dimensions. Initial conditions are chosen as $n = n_d$ for the concentration, $T = T_0$ for the temperature, and $u = v = 0$ (two spatial dimensions) or $u = 0$ (one spatial dimension) for the velocities. A continuation method is used to reach the steady state: the voltage bias is taken initially as zero and is gradually increased to the required value, with the steady state solution of a lower biased case used as the initial condition for a higher one.

7.1. Two dimensional MESFET. We simulate, using the Hydrodynamic model (6.2)-(2.6), a two dimensional MESFET of the size $0.6 \times 0.2 \mu m^2$. The source and the drain each occupies $0.1 \mu m$ at the upper left and the upper right, respectively, with the gate occupying $0.2 \mu m$ at the upper middle (Figure 1, left). The doping is defined by $n_d = 3 \times 10^{17} cm^{-3}$ in $[0, 0.1] \times [0.15, 0.2]$ and in $[0.5, 0.6] \times [0.15, 0.2]$, and $n_d = 1 \times 10^{17} cm^{-3}$ elsewhere, with abrupt junctions (Figure 1, right). A uniform grid of 96×32 points is used. Notice that even if we may not have shocks in the solution, the initial condition $n = n_d$ is discontinuous, and the final steady state solution has a sharp transition around the junction. With the relatively coarse grid we use, the non-oscillatory shock capturing feature of the ENO algorithm is essential for the stability of the numerical procedure.

We apply, at the source and drain, a voltage bias $vbias = 2V$. The gate is a Schottky contact, with a negative voltage bias $vgate = -0.8V$ and a very low concentration value $n = 3.9 \times 10^5 cm^{-3}$ obtained from Equation (5.1-19) of [22]. The lattice temperature is taken as $T_0 = 300°K$. The numerical boundary conditions are summarized as follows (where $\Phi_0 = \frac{k_b T}{e} \ln\left(\frac{n_d}{n_i}\right)$ with $k_b = 0.138 \times 10^{-4}$, $e = 0.1602$, and $n_i = 1.4 \times 10^{10} cm^{-3}$ in our units):

- At the source ($0 \leq x \leq 0.1, y = 0.2$): $\Phi = \Phi_0$ for the potential; $n = 3 \times 10^{17} cm^{-3}$ for the concentration; $T = 300°K$ for the temperature; $u = 0\mu m/ps$ for the horizontal velocity; and Neumann boundary condition for the vertical velocity v (i.e. $\frac{\partial v}{\partial \vec{n}} = 0$ where $\vec{n}$ is the normal direction of the boundary).

- At the drain ($0.5 \leq x \leq 0.6, y = 0.2$): $\Phi = \Phi_0 + vbias = \Phi_0 + 2$ for the potential; $n = 3 \times 10^{17} cm^{-3}$ for the concentration; $T = 300°K$ for the temperature; $u = 0\mu m/ps$ for the horizontal velocity; and Neumann boundary condition for the vertical velocity v.

- At the gate $(0.2 \leq x \leq 0.4, y = 0.2)$: $\Phi = \Phi_0 + vgate = \Phi_0 - 0.8$ for the potential; $n = 3.9 \times 10^5 cm^{-3}$ for the concentration; $T = 300°K$ for the temperature; $u = 0 \mu m/ps$ for the horizontal velocity; and Neumann boundary condition for the vertical velocity v.

- At all other parts of the boundary $(0.1 \leq x \leq 0.2, y = 0.2; 0.4 \leq x \leq 0.5, y = 0.2; x = 0, 0 \leq y \leq 0.2; x = 0.6, 0 \leq y \leq 0.2;$ and $0 \leq x \leq 0.6, y = 0)$, all variables are equipped with Neumann boundary conditions.

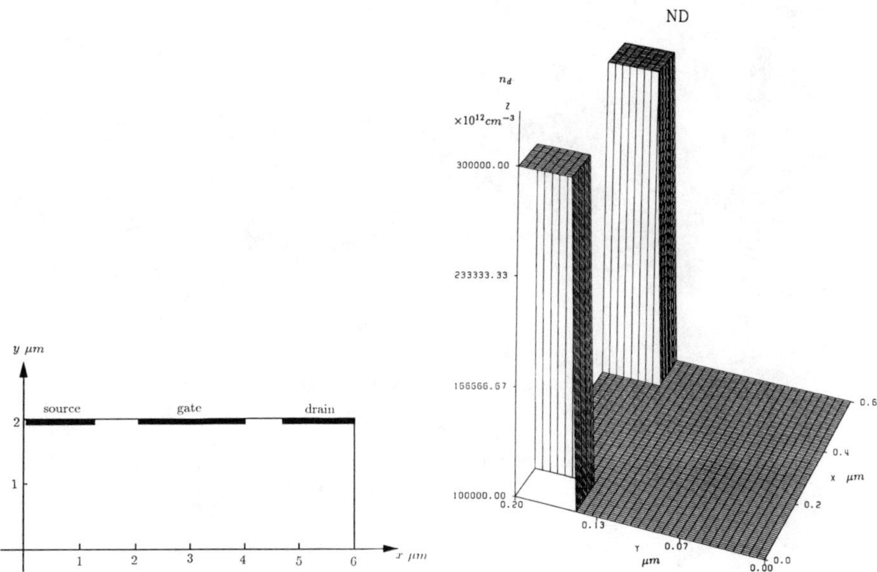

Figure 1: Two dimensional MESFET. Left: the geometry; Right: the doping n_d.

The boundary conditions chosen are based upon physical and numerical considerations. They may not be adequate mathematically, as is evident from some serious boundary layers observable in Figures 2 through 6. ENO methods, owing to their upwind nature, are robust to different boundary conditions (including over-specified boundary conditions) and do not exhibit numerical difficulties in the presence of such boundary layers, even with the extremely low concentration prescribed at the gate (around 10^{-12} relative to the high doping). We point out, however, that boundary conditions affect the global solution significantly. We have also simulated the same problem with different boundary conditions, for example with Dirichlet boundary conditions everywhere for the temperature, or with Neumann boundary conditions for all variables except for the potential at the contacts. The numerical results (not shown in this paper) are noticeably different. This indicates the importance of studying adequate boundary conditions, from both a physical and a mathematical point of view.

In Figures 2 through 6, we show pictures of the concentration n, temperature T, horizontal velocity u, vertical velocity v, and the potential Φ. Surfaces of the solution are shown at the left, and cuts at $y = 0.175$, which cut through the middle of the high doping "blobs" horizontally, are shown at the right.

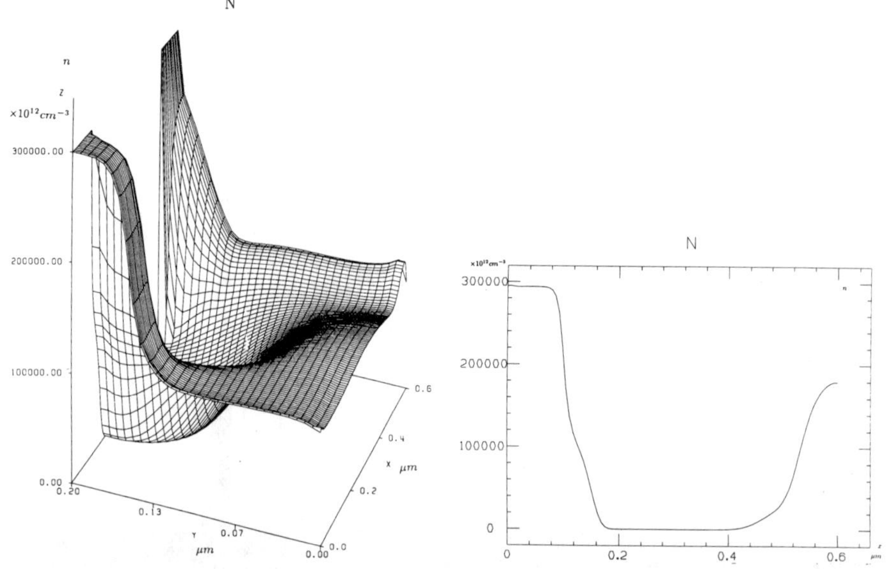

Figure 2: Two dimensional MESFET, concentration n. Left: surface of the solution; Right: cut at $y = 0.175$.

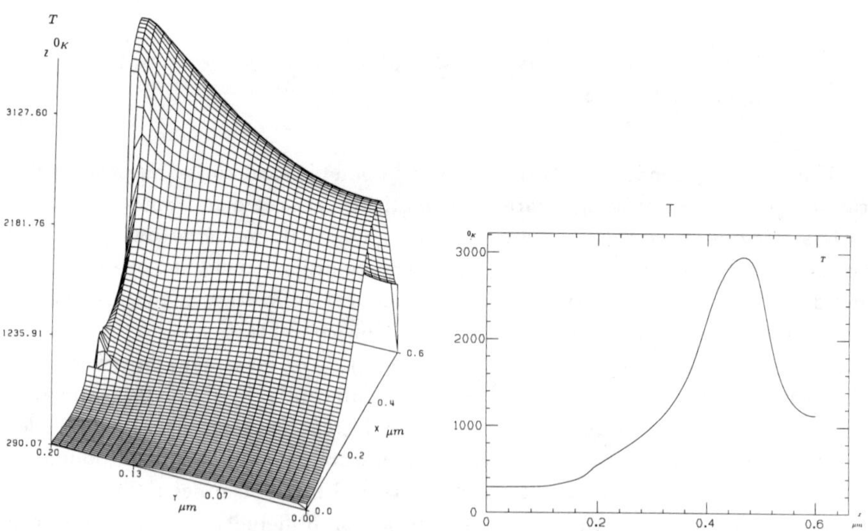

Figure 3: Two dimensional MESFET, temperature T. Left: surface of the solution; Right: cut at $y = 0.175$.

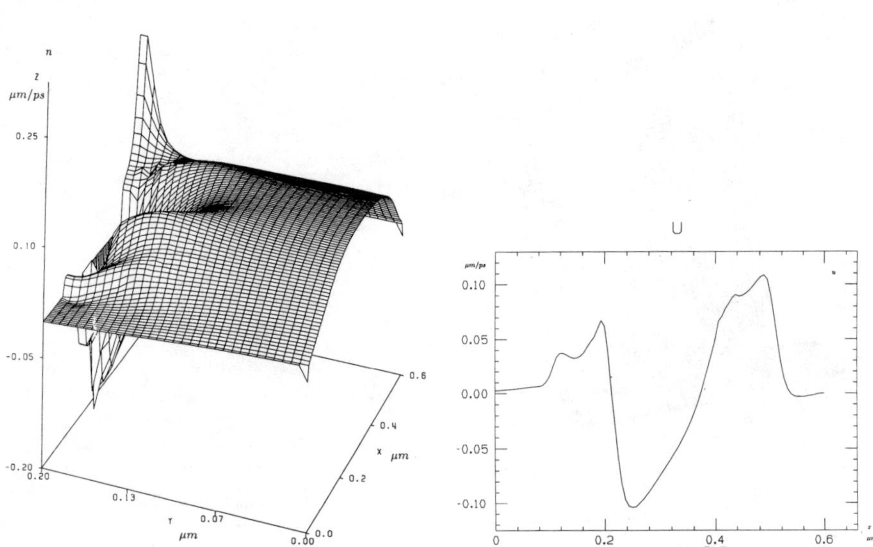

Figure 4: Two dimensional MESFET, horizontal velocity u. Left: surface of the solution; Right: cut at $y = 0.175$.

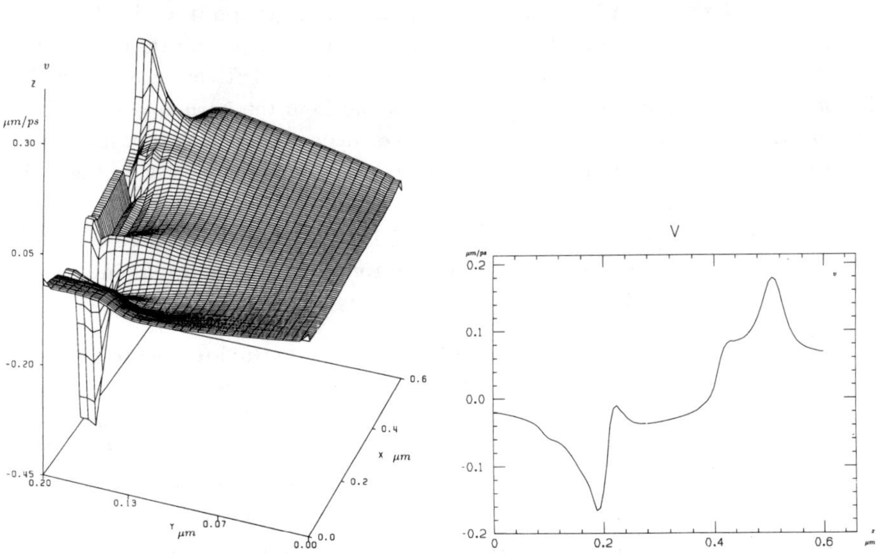

Figure 5: Two dimensional MESFET, vertical velocity v. Left: surface of the solution; Right: cut at $y = 0.175$.

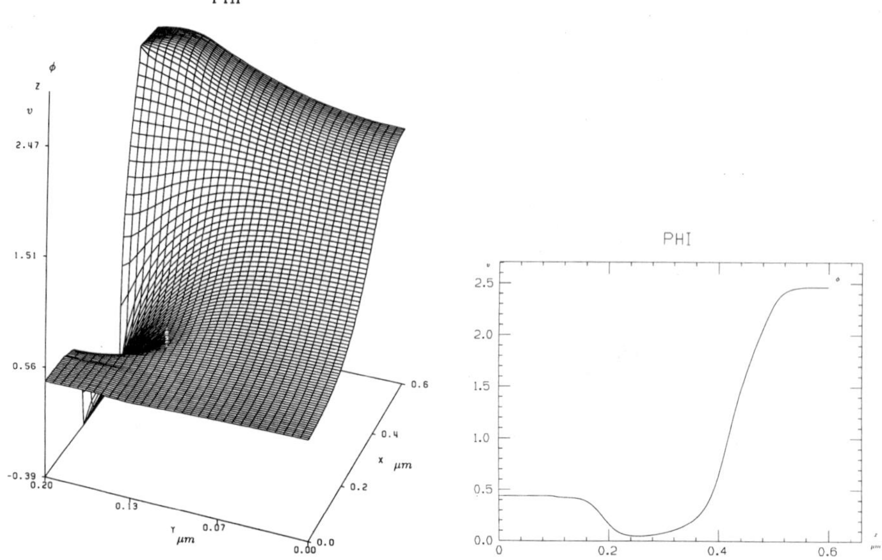

Figure 6: Two dimensional MESFET, potential Φ. Left: surface of the solution; Right: cut at $y = 0.175$.

Notice that there is a boundary layer for the concentration n at the drain but not at the source. Also notice the rapid drop of n at the depletion region near the gate. The temperature achieves its maximum around the left corner of the drain. The leakage current at the gate appears negligible from the normal velocity component, while the horizontal component shows evidence of strong carrier movement toward the source beneath the left gate area, and strong movement toward the drain immediately to the left of the drain junction.

We have also simulated the same MESFET with a higher doping ratio: $3 \times 10^{17} cm^{-3}$ in the high doping region versus $1 \times 10^{15} cm^{-3}$ in the low doping region. We observe similar results (pictures not shown here).

7.2. HD model for a one dimensional diode — spurious velocity overshoot. A notorious phenomenon of HD models is that spurious velocity overshoot occurs at the drain junction of an n^{+}-n-n^{+} diode. It is intrinsic to the model and is not a numerical artifact, as is verified by our grid refinement study and by using different numerical algorithms. This phenomenon is closely related to the physical assumption governing the heat conduction term. Gnudi, Odeh and Rudan [10] observed that the spurious overshoot can be greatly reduced by an empirical modification of the Wiedemann-Franz law for the thermal conductivity.

In this subsection we perform an extensive numerical study of the dependency of the spurious velocity overshoot upon the heat conduction term. The n^{+}-n-n^{+} diode we simulate has a length $0.6 \mu m$, with a doping defined by $n_d = 5 \times 10^{17} cm^{-3}$ in $[0, 0.1]$ and in $[0.5, 0.6]$, and $n_d = 2 \times 10^{15} cm^{-3}$ in $[0.15, 0.45]$, with smooth junctions (Figure 7). The lattice temperature is taken as $T_0 = 296.21^{\circ} K$. We apply a voltage $vbias = 1.5V$, as is the case in [10].

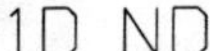

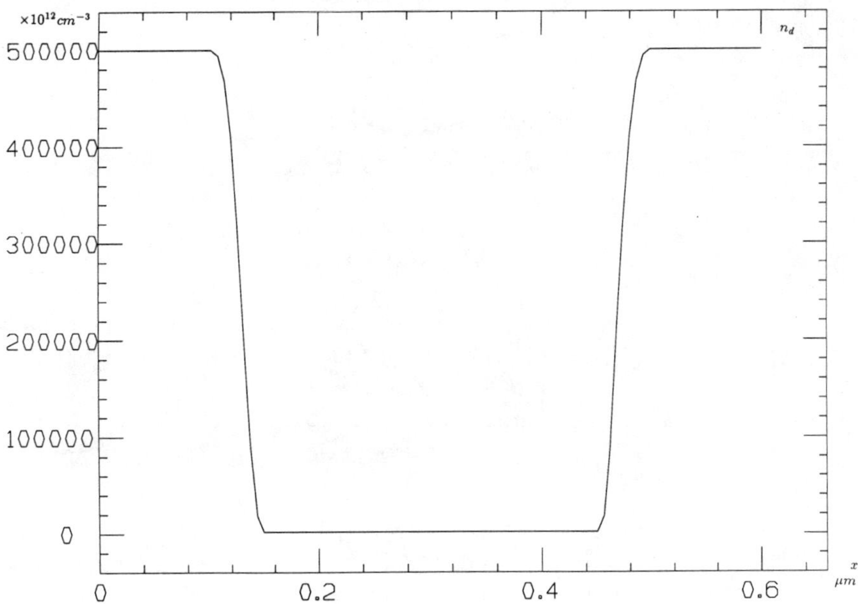

Figure 7: The doping n_d for the one dimensional n^+-n-n^+ diode.

The standard HD model uses $r = -1$ in the Wiedemann-Franz law (2.7) and the relaxation times (2.11). The numerical solution of this is shown as solid lines in Figure 8. We can clearly observe the spurious velocity overshoot at the right junction, but otherwise the solution is basically correct comparing with direct Monto-Carlo simulations (not shown). When r is taken as -2 in (2.7) and (2.11), the solution is completely wrong (dashed line in Figures 8). However, when one takes $r = -2$ only in the coefficient of κ in (2.7) but leaves $r = -1$ in the power of κ in (2.7) and in (2.11), i.e., when one uses

$$(7.1) \qquad \kappa = \frac{1}{2} n \frac{k^2 \mu_0}{e} T \left(\frac{T}{T_0} \right)^{-1},$$

in the place of (2.7) and leaves $r = -1$ in (2.11) unchanged, as was done in [10], one obtains a greatly reduced spurious overshoot (the circles in Figures 8). Finally, the result with $r = -2$ in κ in (2.7) but with $r = -1$ in (2.11) unchanged, is shown by pluses in Figure 8. We can see that the spurious overshoot also disappears.

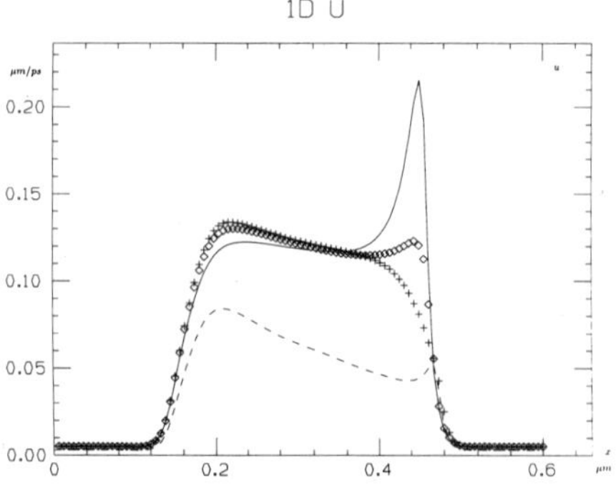

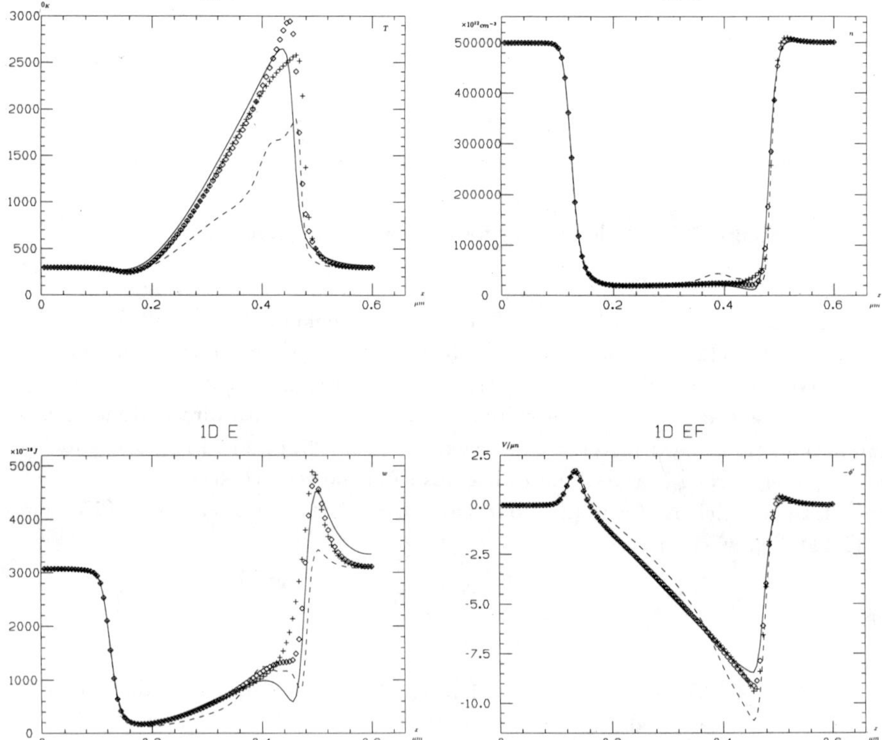

Figure 8: HD for one dimensional n^+-n-n^+ diode. Velocity u (upper), temperature T (middle left), concentration n (middle right), total energy W (lower left) and electric field $-\Phi'$ (lower right). Solid line: r=-1; dashed line: r=-2; circles: r=-2 in the coefficient of (2.7); pluses: r=-2 in (2.7).

7.3. RT model for a one dimensional diode. We present numerical simulation results for the RT model, described in Section 3, for the same one dimensional diode used in Subsection 7.2. Although the RT model is a parabolic system with two equations, the existence of sharp transition regions near the junctions justifies the usage of ENO shock capturing algorithms for the hyperbolic part.

In Figure 9, we show the results of velocity u, temperature T, concentration n, total energy W, and electric field $-\Phi'$ of the RT simulation, in circles, in a background of standard HD results ($r = -1$) in solid lines, and of HD results with $r = -2$ in the coefficient of κ in (2.7) but with $r = -1$ in the power (i.e., (2.7) is replaced by (7.1)), and $r = -1$ in (2.11), in dashed lines. We can see that the RT model greatly reduces the spurious velocity overshoot and is comparable with the result of the empirically modified HD result in dashed lines.

Extensive numerical tests about the RT model, as well as comparisons between the RT and HD models, constitute ongoing research, jointly with U. Ravaioli, E. Kan and D. Chen at the University of Illinois.

Acknowledgements. We would like to thank Edwin Kan, Umberto Ravaioli and Stanley Osher for helpful discussions. The first author is supported by the National Science Foundation under grant DMS-8922398. The second author is supported by the Army Research Office under grant DAAL03-91-G-0123 and by the National Aeronautics and Space Administration under grant NAG1-1145 and under contract NAS1-18605 while in residence at ICASE, NASA Langley Research Center, Hampton, VA 23665. Computations were performed on the Cray YMP at the Pittsburgh Supercomputing Center and on the Cray YMP at the University of Illinois.

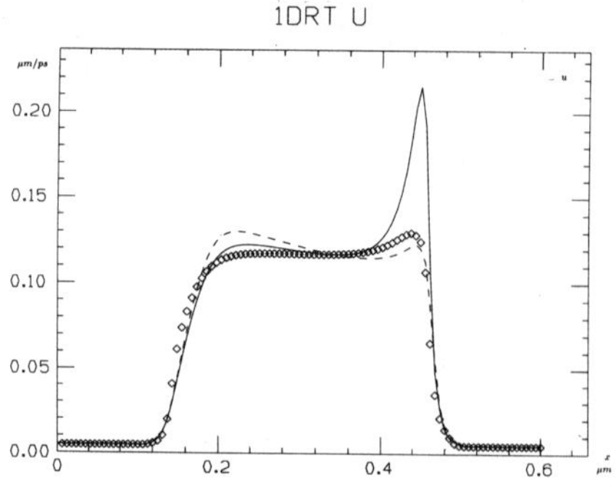

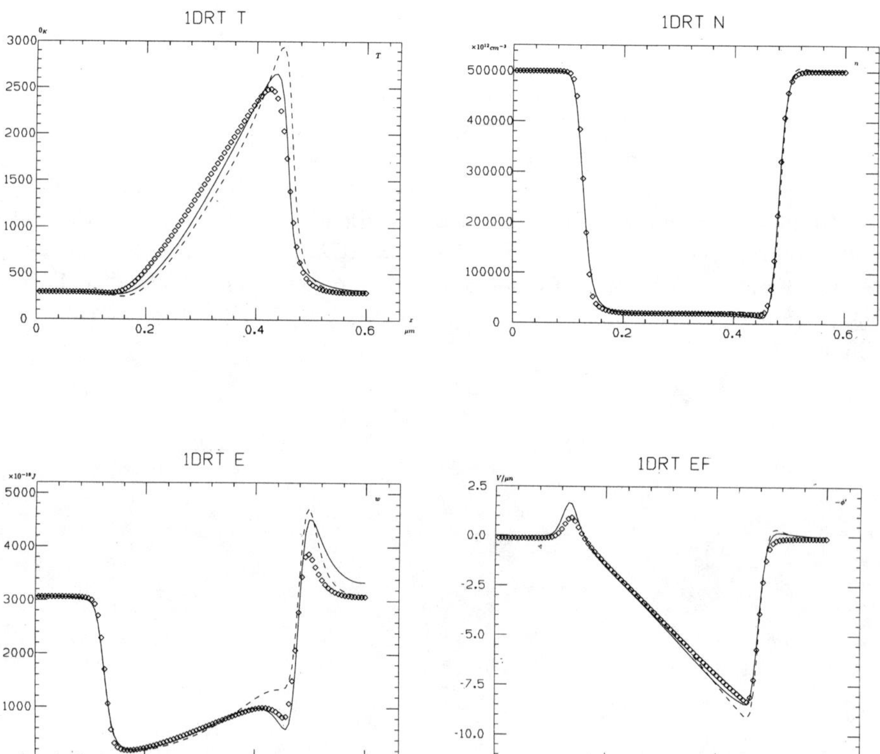

Figure 9: One dimensional n^+-n-n^+ diode. Velocity u (upper), temperature T (middle left), concentration n (middle right), total energy W (lower left), and electric field $-\Phi'$ (lower right). Solid line: standard HD with r=-1; circles: RT; dashed lines: HD with r=-2 in the coefficient of (2.7).

REFERENCES

[1] G. BACCARANI AND M.R. WORDEMAN, *An investigation of steady-state velocity overshoot effects in Si and GaAs devices*, Solid State Electr., 28 (1985), pp. 407–416.

[2] F.J. BLATT, *Physics of Electric Conduction in Solids*, McGraw Hill, New York, 1968.

[3] K. BLOTEKJAER, *Transport equations for electrons in two-valley semiconductors*, IEEE Trans. Electron Devices, 17 (1970), pp. 38–47.

[4] C. CERCIGNANI, *The Boltzmann Equation and its Application*, Springer-Verlag, New York, 1987.

[5] D. CHEN, E. KAN, K. HESS, AND U. RAVAIOLI, *Steady-state macroscopic transport equations and coefficients for submicron device modeling*, to appear.

[6] E. FATEMI, C. GARDNER, J. JEROME, S. OSHER, AND D. ROSE, *Simulation of a steady-state electron shock wave in a submicron semiconductor device using high order upwind methods*, in K. Hess, J. P. Leburton, and U. Ravaioli, editors, Computational Electronics, Kluwer Academic Publishers (1991), pp. 27–32.

[7] E. FATEMI, J. JEROME, AND S. OSHER, *Solution of the hydrodynamic device model using high-order nonoscillatory shock capturing algorithms*, IEEE Transactions on Computer-Aided Design of Integrated Circuits and Systems, CAD-10 (1991), pp. 232–244.

[8] IRENE M. GAMBA, *Stationary transonic solutions for a one-dimensional hydrodynamic model for semiconductors*, Communications in P.D.E, 17 (1992), pp. 553–577.

[9] C.L. GARDNER, J.W. JEROME, AND D.J. ROSE, *Numerical methods for the hydrodynamic device model: Subsonic flow*, IEEE Transactions on Computer-Aided Design of Integrated Circuits and Systems, CAD-8 (1989), pp. 501–507.

[10] A. GNUDI, F. ODEH, AND M. RUDAN, *An efficient discretization scheme for the energy continuity equation in semiconductors*, in Proceedings of SISDP (1988), pp. 387–390.

[11] W. HÄNSCH AND M. MIURA-MATTAUSCH, *The hot-electron problem in small semiconductor devices*, J. Appl. Physics 60 (1986), pp. 650–656.

[12] A. HARTEN, B. ENGQUIST, S. OSHER AND S. CHAKRAVARTHY, *Uniformly high order accurate essentially non-oscillatory schemes, III*, J. Comp. Phys., 71 (1987), pp. 231–303.

[13] JOSEPH W. JEROME, *Consistency of semiconductor modelling: An existence/stability analysis for the stationary van Roosbroeck system*, SIAM J. Appl. Math., 45(4), August 1985, pp. 565–590.

[14] JOSEPH W. JEROME, *Algorithmic aspects of the hydrodynamic and drift-diffusion models*, in Mathematical Modelling and Simulation of Electrical Circuits and Semiconductor Devices (R.E. Bank, R. Bulirsch, and K. Merten, editors), Birkhäuser Verlag (1990), pp. 217–236.

[15] JOSEPH W. JEROME AND THOMAS KERKHOVEN, *A finite element approximation theory for the drift-diffusion semiconductor model*, SIAM J. Num. Anal., 28 (1991), pp. 403–422.

[16] JOSEPH W. JEROME, *Mathematical Theory and Approximation of Semiconductor Models*, SIAM, 1994.

[17] M.S. MOCK, *On Equations Describing Steady-State Carrier Distributions in a Semiconductor Device*, Comm. Pure Appl. Math., 25 (1972), pp. 781–792.

[18] J.P. NOUGIER, J. VAISSIERE, D. GASQUET, J. ZIMMERMANN, AND E. CONSTANT, *Determination of the transient regime in semiconductor devices using relaxation time approximations*, J. Appl. Phys., 52 (1981), pp. 825–832.

[19] P. ROE, *Approximate Riemann solvers, parameter vectors, and difference schemes*, J. Comp. Phys., 27 (1978), pp. 1–31.

[20] M. RUDAN AND F. ODEH, *Multi-dimensional discretization scheme for the hydrodynamic model of semiconductor devices*, COMPEL, 5 (1986), pp. 149–183.

[21] T. SEIDMAN, *Steady state solutions of diffusion reaction systems with electrostatic convection*, Nonlinear Anal., 4 (1980), pp. 623–637.

[22] S. SELBERHERR, *Analysis and Simulation of Semiconductor Devices*, Springer-Verlag, Wien - New York, 1984.

[23] C.-W. SHU, G. ERLEBACHER, T. ZANG, D. WHITAKER, AND S. OSHER, *High-order ENO schemes applied to two- and three-dimensional compressible flow*, J. Appl. Numer. Math., 9 (1992), pp. 45–71.

[24] C.-W. SHU AND S.J. OSHER, *Efficient implementation of essentially non-oscillatory shock capturing schemes*, J. Comp. Phys., 77 (1988), pp. 439–471.

[25] C.-W. SHU AND S.J. OSHER, *Efficient implementation of essentially non-oscillatory shock capturing schemes, II*, J. Comp. Phys., 83 (1989), pp. 32–78.

SOME APPLICATIONS OF ASYMPTOTIC METHODS IN SEMICONDUCTOR DEVICE MODELING*

LEONID V. KALACHEV**

Abstract. This short survey of results concerning the applications of perturbation analysis in semiconductor device modeling is devoted mostly to problems that were solved using the method of composite asymptotic expansions or the, so-called, boundary function method. Thorough description of this approach can be found in Vasil'eva and Butuzov [17], [18], [19] and in O'Malley [13], [14]. The main ideas of the method are illustrated below on the example of the singularly perturbed problem for the Gunn diode. Here the construction of the leading order terms of the asymptotic solution is discussed. This gives the opportunity to obtain the main characteristics of the device to the zeroth order. More detailed analysis of the asymptotic approximation for the solution of the Gunn diode problem, including the construction of higher order terms, will be published later. To make the presentation more compact some cumbersome details of the solution algorithm have been omitted.

1. The statement of the problem for the Gunn diode. For the Gunn diode, consisting of a homogeneously doped piece of semiconductor (typically, gallium arsenide (GaAs)), we consider a spatially one-dimensional model for which the nondimensionalized system of equations can be written in a form:

$$\frac{\partial E}{\partial x} = n - 1 \qquad \text{(Poisson equation)}, \tag{1.1}$$

$$\frac{\partial n}{\partial t} = -\frac{\partial}{\partial x} J_n \qquad \text{(continuity equation)}, \tag{1.2}$$

$$J = J_n + \frac{\partial E}{\partial t} \qquad \text{(total current density)}, \tag{1.3}$$

$$J_n = nv(E) - \epsilon \frac{\partial n}{\partial x} \qquad \text{(electron current density)}. \tag{1.4}$$

Here $v(E)$, the charge carriers velocity, is represented in Figure 1; E_{cr}, a critical value of the field, is such that for $E > E_{\mathrm{cr}}$, the bulk differential conductivity of the device becomes negative; the saturation velocity v_{sat} and E_{min} are defined so that $v_{\mathrm{sat}} = v(\infty) = v(E_{\mathrm{min}})$, v is scaled by v_{sat}; the electric field density E, the charge carriers density n and the currents J and J_n are scaled by E_{cr}, constant donor concentration n_0 and $v_{\mathrm{sat}} \cdot n_0$ respectively; the characteristic time is given by ℓ/v_{sat}, where ℓ is a characteristic length.

The following additional conditions are imposed:

$$n(0) = n(1) = 1 \quad \text{(ohmic contacts)}, \tag{1.5}$$

$$\int_0^1 E(x)dx = U \quad \text{(bias voltage)}. \tag{1.6}$$

---

*This research was supported in part by the Institute for Mathematics and its Applications with funds provided by the National Science Foundation.

**Department of Applied Mathematics, FS-20, University of Washington, Seattle, WA 98195.

The applied voltage U is scaled by $E_{cr} \cdot \ell$.

There are two characteristic parameters in the problem:

$$\lambda = \sqrt{\frac{\epsilon_s e_{cr}}{q n_0 \ell}} \quad \text{(the scaled Debye length)},$$

and

$$\gamma = \frac{D}{v_{sat} \cdot \ell},$$

where ϵ_s is the permittivity of the semiconductor, q is the charge of electron, and D is a diffusivity. The case when both parameters λ and γ are small is considered in Markowich et al. [12], while discussion of the case where the diffusion term is omitted can be found in Shaw et al. [15]. We consider only the case when $\lambda \sim 0(1)$, $0 < \gamma \ll 1$. Without loss of generality we assume that $\lambda = 1$, and $\gamma = \epsilon$, where $0 < \epsilon \ll 1$ is a small parameter.

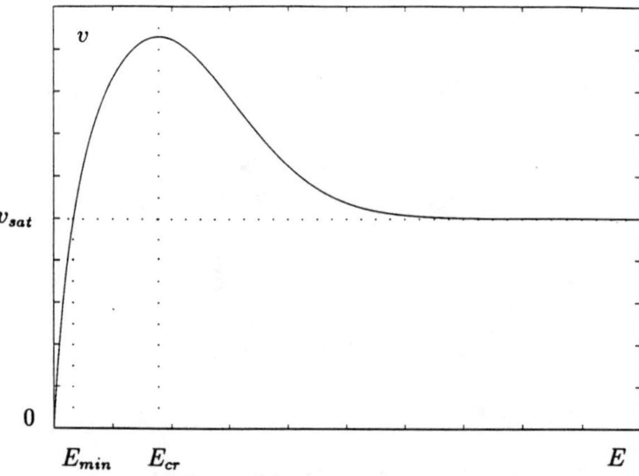

Figure 1. Dependence $v = v(E)$ for GaAs.

It is known that for applied voltages exceeding some threshold value, two working regimes with different currents exist for the Gunn diode, with the regime corresponding to the larger current being unstable (see, e.g. Shaw et al. [15], Szmolyan [16]). The so-called, *trivial* solution of the problem (1.1)-(1.6) corresponding to this unstable regime can be easily written out as

$$n = 1, \quad E_{triv} = U = \text{const},$$
$$J_n = n \cdot v(E_{triv}) = v(U),$$
$$J = J_n.$$

The other (stable) solution is known to have a pulse-like form (see Figure 2), with the pulses for E (and n) moving with velocity v. We will construct the asymptotic approximation for this solution when the pulse lies entirely within the domain $x \in (0, 1)$ (we will not here discuss the transition processes of formation or disappearance of such pulses).

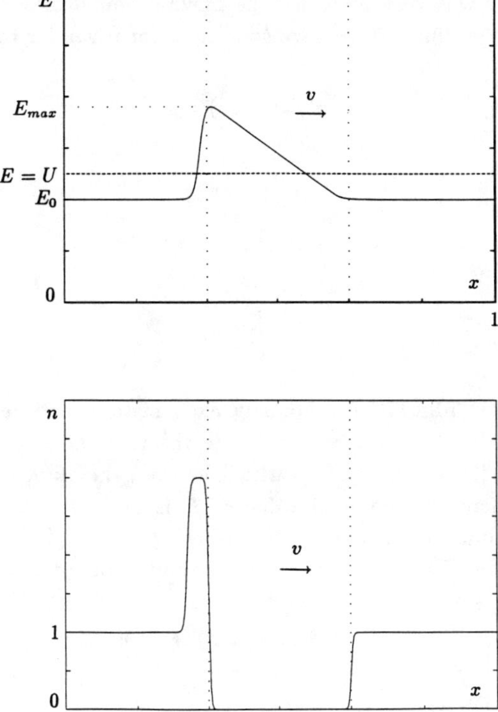

Figure 2. The structure of the solution.

Let us introduce the new independent variable z associated with the moving structure:

$$(1.7) \qquad z = x - ct,$$

where c is unknown velocity of the structure. We seek functions E and n depending on variable z:

$$(1.8) \qquad E = E(z), \quad n = n(z).$$

Taking into account (1.7), (1.8), the system (1.1)-(1.4) can be rewritten in a form:

$$(1.8) \qquad c\frac{\partial n}{\partial z} = \frac{\partial}{\partial z}\left(nv(E) - \epsilon\frac{\partial n}{\partial z}\right)$$

$$(1.9) \qquad J = nv(E) - \epsilon\frac{\partial n}{\partial z} - c\frac{\partial E}{\partial z},$$

$$(1.10) \qquad \frac{\partial E}{\partial z} = n - 1.$$

Condition (1.6) changes very little; on the moving boundaries $z = z'(t)$ and $z = z''(t)$ (in the new coordinates) we have conditions for n similar to (1.5):

$$(1.11) \qquad n(z') = n(z'') = 1,$$

$$(1.12) \qquad \int_{z'}^{z''} E(x)\,dx = U,$$

where $z' = z'(t)$, $z'' = z''(t)$, $z'' - z' = 1$.

2. Asymptotic algorithm. For any fixed instant of time (when the pulse is entirely within the domain), we subdivide the interval $z \in [z', z'']$ into three subintervals $[z', 0]$, $[0, \Delta z]$, $[\Delta z, z'']$ (without loss of generality, we associate $z = 0$ with the point where the maximal value of E is observed) and seek a uniform asymptotic approximation for the solution of the problem (1.8)-(1.12) in the form (cf. the notations for E-functions in Figure 3; similar for n-functions):

$$(2.1) \qquad E(z) = \begin{cases} \overline{E}^1(z) + \Pi^* E(\xi), z \in [z', 0], \xi \le 0; \\ \overline{E}^2(z) + \Pi E(\xi) + Q^* E(\eta), z \in [0, \Delta z], \xi \ge 0, \eta \le 0; \\ \overline{E}^3(z) + Q E(\eta), z \in [\Delta z, z''], \eta \ge 0; \end{cases}$$

$$(2.2) \qquad n(z) = \begin{cases} \overline{n}^1(z) + \tfrac{1}{\varepsilon}\Pi^* n(\xi), z \in [z', 0], \xi \ge 0; \\ \overline{n}^2(z) + \tfrac{1}{\varepsilon}\Pi n(\xi) + Q^* n(\eta), z \in [0, \Delta z], \xi \ge 0, \eta \le 0; \\ \overline{n}^3(z) + Q n(\eta), z \in [\Delta z, z''], \eta \ge 0. \end{cases}$$

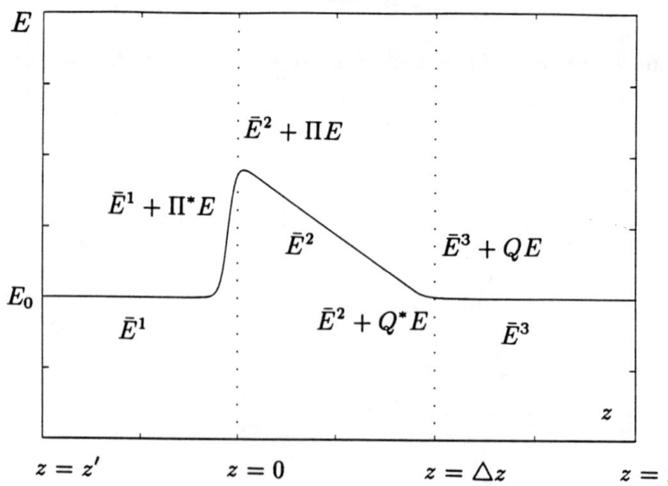

Figure 3. Notations for E-functions.

$$(2.3) \qquad\qquad J = J_0 + \sqrt{\epsilon}J_1 + \epsilon J_2 + \ldots,$$

$$(2.4) \qquad\qquad c = c_0 + \sqrt{\epsilon}c_1 + \epsilon c_2 + \ldots.$$

Here $\xi = z/\epsilon$ and $\eta = (z - \Delta z)/\sqrt{\epsilon}$ are stretched variables; $\overline{E}^i$, $\overline{n}^i (i = 1, 2, 3)$ are regular functions; boundary functions Π^*, Π and Q^*, Q depend on the variables ξ and η respectively. Each term in the sums (2.1), (2.2) is, in turn, a power series expansion in powers of $\sqrt{\epsilon}$ (the appearance of such an asymptotic sequence is connected with the construction of higher order terms of the asymptotic solution in the vicinity of the point $z = \Delta z$); for example

$$\overline{E}^1(z, \epsilon) = \sum_{i=0}^{\infty}(\sqrt{\epsilon})^i \overline{E}^1_i(z), \quad \Pi^* E(\xi, \epsilon) = \sum_{i=0}^{\infty}(\sqrt{\epsilon})^i \Pi^*_i E(\xi), \text{ etc.}$$

We require that boundary functions decay when corresponding stretched variables tend to $+\infty$ or $-\infty$; for example

$$\Pi^*_i E(-\infty) = 0, \quad \Pi_i E(+\infty) = 0, \text{ etc.}$$

The expansion (2.3) for J does not contain any boundary terms or any dependence on z. This simply reflects the fact that for a one-dimensional device without internal sources and drains the current is constant. For a nonlinear function $v(E)$ we must use an asymptotic representation similar to (2.1), (2.2):

$$(2.5) \qquad v(E) = \begin{cases} \overline{v}^1(\overline{E}^1) + \Pi^* v(\xi), z \in [z'_1 0), \xi \leq 0; \\ \overline{v}^2(\overline{E}^2) + \Pi v(\xi) + Q^* v(\eta), z \in [0, \Delta z], \xi \geq 0, \eta \leq 0; \\ \overline{v}^3(\overline{E}^3) + Q v(\eta), z \in [\Delta z, z''], \eta \geq 0; \end{cases}$$

where

$$\Pi^* v(\xi) = v(\overline{E}^1(\epsilon\xi) + \Pi^* E(\xi)) - v(\overline{E}^1(\epsilon\xi)), \xi \leq 0;$$

$$\Pi v(\xi) = v(\overline{E}^2(\epsilon\xi) + \Pi E(\xi)) - v(\overline{E}^2(\epsilon\xi)), \xi \geq 0;$$

and analogous expressions hold for $Q^* v(\eta)$, $Q v(\eta)$. It can be easily shown that for exponentially decaying $\Pi^*_i E$-functions the functions $\Pi^*_i v(\xi)$ will also be exponentially decaying.

Substituting (2.1)-(2.5) into (1.8)-(1.12) we can determine the terms of the asymptotic approximation by a standard procedure (note that Δz must be determined along with the construction of the asymptotic solution).

In the present discussion, the most important aspect is the construction of the zeroth order terms for the function E, because they define at the zeroth order the main characteristics of the Gunn diode: the velocity of the structure and therefore the current, the amplitude of the pulse, etc. In the following we will take into account that the density of electrons satisfies $n \geq 0$, and that for applied voltages satisfying $U \geq 1$, the difference between $E = U$ and E_0 (see Figure 2) is of the order $0(1)$.

The determination of the regular functions to the zeroth order can be easily obtained by putting $\epsilon = 0$ in the original system (1.8)-(1.10). For the solutions on the subintervals $[z', 0]$ and $[\Delta z, z'']$ (the boundary conditions for $\overline{n}_0^i$ are $\overline{n}_0^1(z') = 1$, $\overline{n}_0^3(z'') = 1$) we write ($i = 1, 3$):

$$\overline{n}_0^i = 1, \frac{\partial \overline{E}_0^i}{\partial z} = \overline{n}_0^i - 1 = 0 \quad \text{and hence } \overline{E}_0^i = \text{const},$$

$$(2.6) \qquad J_0 = \overline{n}_0^i \cdot v(\overline{E}_0^i) - c_0 \frac{\partial \overline{E}_0^i}{\partial z} = v(\overline{E}_0^i) = \text{const}.$$

By virtue of (2.6) and (1.12)

$$(2.7) \qquad \overline{E}_0^1 = \overline{E}_0^3 \equiv E_0,$$

where E_0 is some unknown constant (see Figure 2). (the discussion of equality (2.7) can be also found in Markowich et al. [12].)

For the subinterval $[0, \Delta z]$ the only other solution of the degenerate system that will make it possible to satisfy the integral condition (1.12) is

$$\overline{n}_0^2 = 0,$$

$$\frac{\partial \overline{E}_0^2}{\partial z} = \overline{n}_0^2 - 1 = -1 \text{ and hence } \overline{E}_0^2 = -z + K,$$

$$(2.8) \qquad J_0 = \overline{n}_0^2 \cdot v(\overline{E}_0^2) - c_0 \frac{\partial \overline{E}_0^2}{\partial z} = c_0,$$

(here, K is some unknown constant).

From (2.6), (2.7), (2.8) (with $J_0 = \text{const}$ throughout the device) it follows that

$$(2.9) \qquad c_0 = v(E_0).$$

It can be easily shown that the boundary functions $\Pi_0 E$, $Q_0^* E$, $Q_0 E \equiv 0$. In the subinterval $[z', 0]$ we must construct the boundary function $\Pi_0^* E$ to give us the transition from $E_{\max} = \max_{[0, \Delta z]} \overline{E}_0^2$ to the solution $\overline{E}_0^1 = E_0$ (see Figure 4). The equations for $\Pi_0^* E$, $\Pi_0^* n$ can be written out as follows ($\xi \leq 0$):

$$(2.10) \qquad 0 = \Pi_0^* n \cdot v(E_0 + \Pi_0^* E) - \frac{\partial \Pi_0^* n}{\partial \xi} - c_0 \frac{\partial \Pi_0^* E}{\partial \xi},$$

$$(2.11) \qquad \frac{\partial \Pi_0^* E}{\partial \xi} = \Pi_0^* n.$$

Eliminating $\Pi_0^* n$ from (2.10) and adding the decay conditions at $-\infty$, we obtain

$$(2.12) \qquad \frac{\partial^2 \Pi_0^* E}{\partial \xi^2} = [v(E_0 + \Pi_0^* E) - v(E_0)] \frac{\partial \Pi_0^* E}{\partial \xi},$$

$$(2.13) \qquad \Pi_0^* E(-\infty) = \frac{\partial \Pi_0^* E}{\partial \xi}(-\infty) = 0.$$

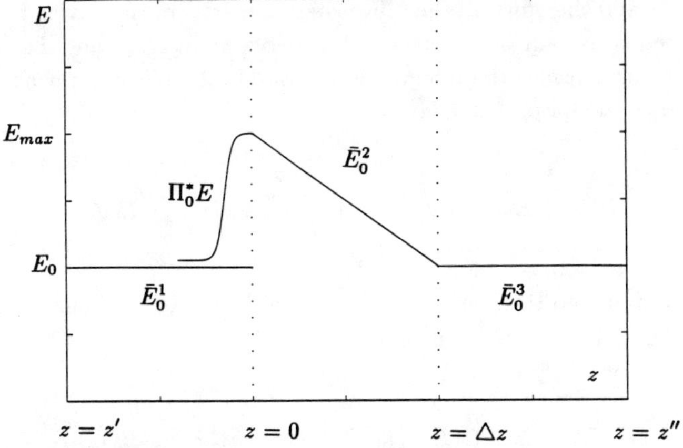

Figure 4. The structure of the zeroth order approximation for function E.

Taking into account (2.13) we can integrate (2.12) once to obtain

$$(2.14) \qquad \frac{\partial \Pi_0^* E}{\partial \xi} = \int_0^{\Pi_0^* E} [v(E_0 + s) - v(E_0)]ds.$$

The value $\Pi_0^* E_{\max} = \max_{\xi \leq 0}(\Pi_0^* E)$ is then defined by the well-known "equal area rule" (see Markowich et al. [12], etc.) following from (2.14) (Figure 5):

$$(2.15) \qquad 0 = \int_0^{\Pi_0^* E_{\max}} [v(E_0 + s) - v(E_0)]ds, \quad \Pi_0^* E_{\max} \neq 0.$$

this relation gives the implicit function $\Pi_0^* E_{\max}(E_0)$.

At the point $z = 0$ equality takes place (see Figure 4):

$$(2.16) \qquad \max(\overline{E}_0^2) = E_0 + \Pi_0^* E_{\max}(E_0)$$

As soon as $\Pi_0^* E_{\max}$ is known (it depends on the still unknown constant E_0) the expression for the implicit function $\Pi_0^* E(\xi)$ can be easily written out:

$$(2.17) \qquad \int_{\Pi_0^* E_{\max}}^{\Pi_0^* E} \frac{dz}{\left[\int_0^z (v(E_0 + s) - v(E_0))ds\right]} = -\xi.$$

It follows from (2.11) that the expression for $\Pi_0^* n$ is given by (2.14). It can be easily shown that $\Pi_0^* E(\xi)$ and $\Pi_0^* n(\xi)$ decay exponentially as $\xi \to -\infty$.

216

To find E_0 and therefore all the other characteristics of the device to the zeroth order, we need to obtain one more relation between $\Pi_0^* E_{\max}$ and E_0 in addition to (2.15). Let us consider the integral condition (1.12). We can rewrite it (to the zeroth order) in the form:

$$(2.18) \qquad U = \int_{z'}^{z''} \overline{E}_0 \, dz = \int_{z'}^{0} E_0 \, dz + \int_{0}^{\Delta z} \overline{E}_0^2(z) \, dz + \int_{\Delta z}^{z''} E_0 \, dz.$$

The boundary function $\Pi_0^* E$ does not enter (2.18) because its impact to the integral is of the order $O(\epsilon)$. From (2.16) and the relation $\overline{E}_0^2 = -z + K$, the following expressions can be easily derived:

$$K = E_0 + \Pi_0^* E_{\max}(E_0), \quad \Delta z = \Pi_0^* E_{\max}(E_0).$$

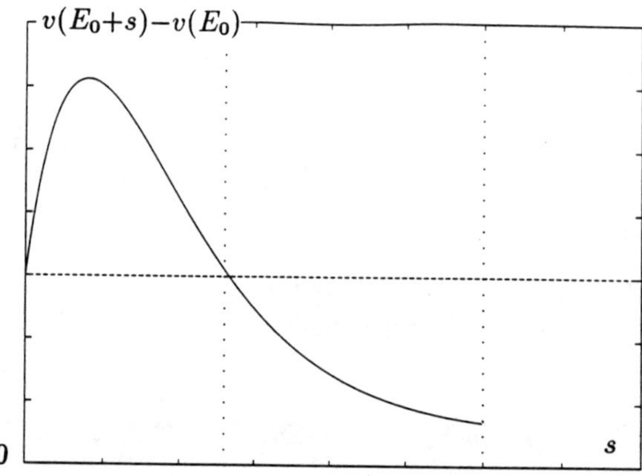

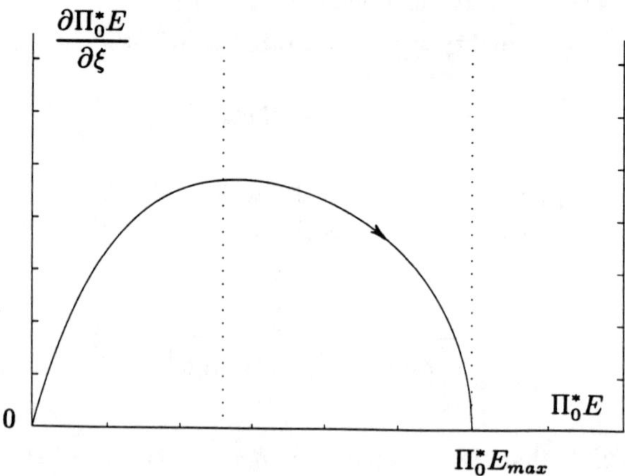

Figure 5. Equal area rule.

Substituting $\overline{E}_0^2$, K and Δz into (2.18) we get

$$(2.19) \qquad \Pi_0^* E_{\max} = +\sqrt{2(U - E_0)}.$$

Substituting (2.19) into (2.15) we obtain

$$(2.20) \qquad \int_0^{\sqrt{2(U-E_0)}} (v(E_0 + s) - v(E_0))ds = 0.$$

The solution E_0 of the equation (2.20) can be found numerically. For known E_0 the values $J_0 = v(E_0)$, $c_0 = v(E_0)$, $\Pi_0^* E_{\max}(E_0)$, $\Pi_0^* E(\xi)$, $\Pi_0^* n(\xi)$, $\Delta z(E_0)$ will also be known.

Let us consider different possibilities that might occur for the solution of (2.20). The trivial solution $E_0 = U$ always exists, it corresponds to the trivial solution of the whole problem that is stable for $U < 1$ and unstable for $U > 1$. For $U > 1$ the nontrivial solution E_0' corresponds to the point A in the Figure 6 where the curves F and G intersect ($F = \Pi_0^* E_{\max}(E_0)$ is defined implicitly by (2.14) and $G = \Pi_0^* E_{\max}(E_0)$ is defined by (2.19)). This solution is known to be stable. For $U < \alpha < 1$, where $\alpha \sim 0(1)$ is a constant that can be found, no nontrivial solution exists (Figure 7). For $\alpha < U \lesssim 1$ the situation shown in Figure 8 is possible (we take into account that $\frac{\partial F}{\partial E_0}(1) = -1$, $\frac{\partial G}{\partial E_0}(U) = -\infty$): two nontrivial solutions E_0' and E_0'' of (2.20) exist corresponding to the intersection points A and B respectively. The trivial solution of the original problem with $U \lesssim 1$ is stable (Szmolyan [16]), the solution of the problem with $E_0 = E_0'$ is expected to be stable and the solution with $E_0 = E_0''$ to be unstable. The fact that three solutions exist (including the trivial one) can be used to explain the hysteresis effects that were experimentally observed for the Gunn diode: for increasing and decreasing applied bias voltages, different paths of voltage-current characteristics were obtained (Shaw et al. [15]). Other terms of the asymptotic solution (including higher order terms) can be constructed likewise.

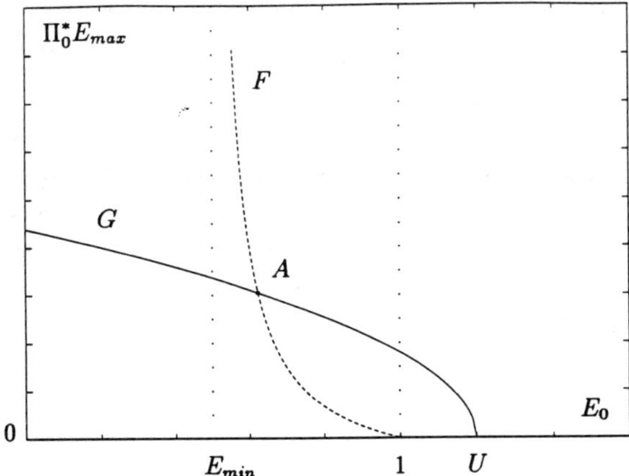

Figure 6. Unique solution exists for $U > 1$.

218

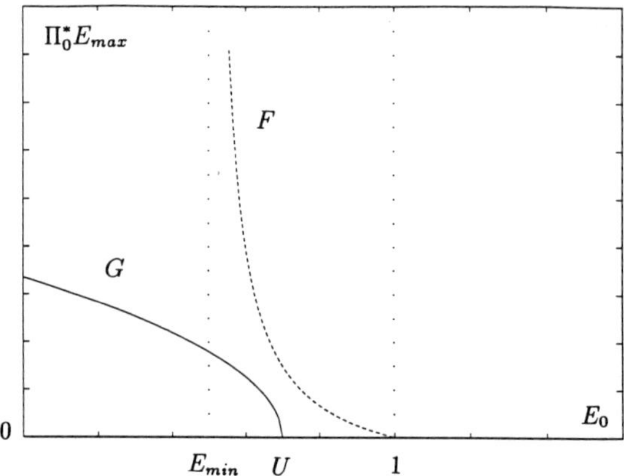

Figure 7. No solutions for $U < \alpha < 1$.

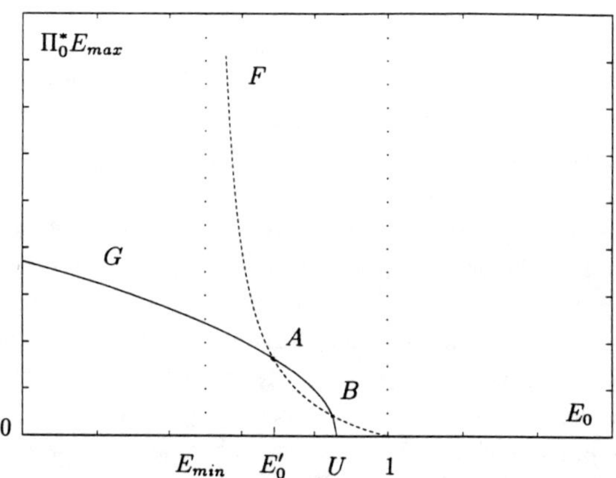

Figure 8. The case when $\alpha < U \lesssim 1$.

3. Other problems. In this section some other applications of the asymptotic analysis to semiconductor device modeling will be briefly discussed.

1. The most widespread model that is now used for numerical simulations of the processes in the semiconductor devices is still a drift-diffusion model. When Gummel-type iteration schemes are applied to solve the drift-diffusion equations numerically the speed of convergence and sometimes the convergence itself depend crucially on the successfully chosen initial iterate. In Kalachev and Obukhov [9] the singularly perturbed Poisson equation (one of the drift-diffusion equations) was considered (in dimensionless form) as:

$$(3.1) \qquad \alpha^2 \Delta \Psi = n - p - N,$$

$$(3.2) \qquad n = \exp(\Psi - \varphi_n), \quad p = \exp(\varphi_p - \Psi).$$

Here $\alpha = L_0/L$, where L_0 is the Debye length, L is characteristic length; the electrostatic potential Ψ and the Fermi quasilevels φ_n, φ_p are measured in units of kT/q (k is Boltzmann's constant, T is the absolute temperature, q is the charge of the electron); n, p and N are the concentrations of electrons, holes, and the dopant concentration in units of intrinsic concentration n_i. The small parameter $\epsilon = \alpha/\sqrt{m}$, for $m = \max|N|$, enters the equation (3.1) making the problem for the Poisson equation singularly perturbed. For contemporary semiconductor devices, $\epsilon \sim 10^{-1} - 10^{-4}$. This problem has to be solved by successive approximations at each step of the iterative Gummel-type process of obtaining the solution for the full drift-diffusion model. In [9] the boundary function method was used to construct the initial iterate to solve the Poisson equation in a rectangular domain modeling the two-dimensional semiconductor structure when some voltages were applied to the contacts. Some segments of the boundary modeled the ohmic contacts, while on the rest of the boundary the homogeneous Neumann conditions were prescribed.

The solution of (3.1), (3.2) is conveniently sought in the form

$$\Psi = \Psi_0 + \varphi,$$

where Ψ_0 is the solution of the quasineutrality equation

$$n - p - N = 0$$

satisfying the boundary conditions at the ohmic contacts. Then we have the singularly perturbed boundary value problem for the potential φ:

$$(3.3) \qquad \epsilon^2 \Delta\varphi = A(x,z)\sinh\varphi + B(x,z)(\cosh\varphi - 1) - g(x,z,\epsilon),$$

$$(3.4) \qquad \varphi|_{\Gamma_i} = 0 \quad \text{(ohmic contacts)},$$

$$(3.5) \qquad \frac{\partial\varphi}{\partial\nu}\Big|_{\gamma_i} = -\frac{\partial\Psi_0}{\partial\nu}\Big|_{\gamma_i} \sim 0(1) \quad \text{(rest of the boundary)}.$$

Here x, z are spatial coordinates; $A(x,z)$, $B(x,z)$, $g(x,z,\epsilon)$ are known, sufficiently smooth functions; $A > B$ for all (x,z); and $\partial/\partial\nu$ is the outward normal derivative.

Under certain conditions the full asymptotic approximation for the solution of (3.3)-(3.5) is constructed. It happens that the boundary functions appear only to the order $O(\epsilon)$ and the explicit expressions for them can be easily obtained. This fact simplifies the use of the asymptotic solution in a numerical algorithm. Numerical computations have shown that, when the asymptotic solution is used as the initial iterate, the convergence of the numerical process for the drift-diffusion model is speeded up by a factor of 5-10. In Kalachev et al. [8] the case of a gate contact is considered and the estimation of the asymptotic remainder is presented. Problems for the singularly perturbed Poisson equation in a three-dimensional semiconductor structure and in the case of large outer electric field are solved in [10], [11] respectively.

2. The boundary function method was used to construct asymptotic solutions of the full drift-diffusion models for one-dimensional devices by Belyanin [1], [2],

[3], etc., in these papers the ratio of Debye length to the length of the device was considered as small parameter. In [1] the asymptotic approximation is constructed for the solution of the system modeling a diode in a nonstationary case. The nonlinear parabolic equation for the electron concentration $\overline{n}_0$ (a regular function to the zeroth order) is solved numerically, hole concentration $\overline{p}_0$ and electric field $\overline{E}_0$ are expressed algebraically through $\overline{n}_0$, and the formulae for the boundary functions and higher order terms are written out explicitly. In [2] the stationary problems for the diode in the cases of moderate and large applied currents are considered. In [3] the stationary problem is solved for a one-dimensional device containing an arbitrary number of $p-n$ junctions and bias contacts, and a theorem on estimating of the remainder is proved.

It is worthwhile to mention the paper by Vasil'eva et al. [6], where the one dimensional problem for the diode (thyristor structure) is posed as the optimal control problem. For given voltage-current characteristics, the synthesis of the device with such characteristics is discussed, when a doping level $N(-1 \leq N \leq 1)$ is considered to be the control function.

3. Some other asymptotic problems of the semiconductor device modeling concerning the asymptotic solution of the stationary drift-diffusion model in the case of large generation-recombination terms in a two-dimensional domain, the internal transition layers in a thin semiconductor films, the asymptotic derivation of the ambipolar diffusion equation for the intrinsic semiconductors with the discussion of the correct boundary conditions for this equation, are presented in [4], [5], [7].

It is my pleasure to thank Robert O'Malley, Harold Grubin, Christian Schmeiser and Peter Szmolyan for numerous and fruitful discussions concerning the Gunn diode problem during the IMA Workshops on Semiconductors, July 15 – August 9, 1991, where this paper was written.

REFERENCES

[1] M.P. BELYANIN, *Numerical-asymptotic solution of a nonstationary singularly perturbed problem from the theory of semiconductor devices (Russian)*, Diff. Uravneniya 21, No. 8, (1985), pp. 1436–1440.

[2] M.P. BELYANIN, *Asymptotic solution of one model for $p-n$ junction (Russian)*, Zh. Vichislit. Matem. i Matem. Fiziki 26, No. 2, translated into English in U.S.S.R. Comput. Math. and Math. Phys. (1986), pp. 306–311.

[3] M.P. BELYANIN, *On the asymptotics in a one-dimensional model of some semiconductor devices*, U.S.S.R. Comput. Math. and Math. Phys. 28, (1988), pp. 21–34.

[4] M.P. BELYANIN, L.V. KALACHEV, E.V. MAMONTOV, *Application of the boundary function method for the simulation of some semiconductor devices*, to appear in Mat. Model.

[5] M.P. BELYANIN, A.B. VASIL'EVA, *On an inner transition layer in a problem of the theory of semiconductor films (Russian)*, Zh. tVichislit. Matem. i Matem. Fiziki 28, No. 2, translated into English in U.S.S.R. Comput. Math. and Math. Phys. (1988), pp. 223–236.

[6] M.P. BELYANIN, A.B. VASIL'EVA, A.V. VORONOV, A.V. TIKHONRAVOV, *An asymptotic approach to the problem of designing a semiconductor device (Russian)*, Mat. Model. 1, No. 9 (1989), pp. 43–63.

[7] V.F. BUTUZOV, L.V. KALACHEV, *Asymptotic derivation of the ambipolar diffusion equation in the physics of semiconductors*, submitted for publication in U.S.S.R. Comput. Math. and Math. Phys..

[8] L.V. KALACHEV, S.V. KRUCHKOV, I.A. OBUKHOV, *Asymptotic analysis of the Poisson equation in semiconductors (Russian)*, Mat. Model. 1, No. 9 (1989), pp. 129–140.

[9] L.V. KALACHEV, I.A. OBUKHOV, *Approximate solution of the Poisson equation for a model of a two-dimensional semiconductor structure*, Vestnik Mosk. Universiteta, Fizika, 44, No. 3, translated into English (1989), pp. 63–68.

[10] L.V. KALACHEV, I.A. OBUKHOV, *Asymptotic solution of the Poisson equation in a three-dimensional semiconductor structure*, to appear in U.S.S.R. Comp. Math. and Math. Phys..

[11] L.V. KALACHEV, I.A. OBUKHOV, *Asymptotic solution of the Poisson equation in the case of a large outer field*, submitted for publication in Mat. Model.

[12] P.A. MARKOWICH, C.A. RINGHOFER, C. SCHMEISER, *Semiconductor Equations*, Springer-Verlag, Wien-New York (1990).

[13] R.E. O'MALLEY, JR., *Introduction to Singular Perturbations*, Academic Press, New York (1974).

[14] R.E. O'MALLEY, JR., *Singular Perturbations Methods for Ordinary Differential Equations*, Springer-Verlag, New York (1991).

[15] M.P. SHAW, H.L. GRUBIN, P.R. SOLOMON, *Gunn-Hilsum effect*, Academic Press, New York (1979).

[16] P. SZMOLYAN, *Asymptotic analysis of the Gunn effect*, IMA Preprint, Univ. of Minnesota (1989).

[17] A.B. VASIL'EVA, V.F. BUTUZOV, *Asymptotic Expansions of the Solutions of Singularly Perturbed Equations (Russian)*, Nauka, Moscow (1979).

[18] A.B. VASIL'EVA, V.F. BUTUZOV, *Singularly Perturbed Equations in the Critical Case (Russian)*, Moscow State University, Moscow (1978); translation into English (1980).

[19] A.B. VASIL'EVA, V.F. BUTUZOV, *Asymptotic Methods in Singular Perturbation Theory*, Visshaya Shkola, Moscow (1990).

DISCRETIZATION OF THREE DIMENSIONAL DRIFT-DIFFUSION EQUATIONS BY NUMERICALLY STABLE FINITE ELEMENTS

THOMAS KERKHOVEN*

Abstract. Many of the commonly employed discretizations of the drift-diffusion current continuity equations (including the Scharfetter-Gummel perpendicular bisector box-method discretization) can be expressed in terms of the Slotboom variables ν and ω. The exponential upwinding techniques, which are usually included, bring the drift-diffusion current continuity equations in terms of the Slotboom variables in self-adjoint form.

The piecewise linear finite element approximation which solves Galerkin's equations is shown to be maximum stable for mixed Neumann-Dirichlet boundary conditions in two and three dimensions, even if no discrete extrema principles prevail. For the Delaunay triangulation in two dimensions, and the Delaunay tetrahedryzation in three dimensions, the box method discretization based on perpendicular bisectors yields a discretized system with extrema principles for the Slotboom variables. Box methods can be considered as Petrov-Galerkin methods for a piecewise linear approximation with piecewise constant test-functions.

Petrov-Galerkin finite element error analysis requires demonstration of an appropriate inf-sup condition. Complications arise in demonstrating this condition because piecewise constant test-functions are not of H^1 regularity. These complications can be circumvented by approximating the piecewise constant test functions by continuous piecewise polynomial test functions. The box-method for a piecewise linear approximation function need not generate a system of linear equations which is identical to Galerkin's equations. Nevertheless, under rather general conditions, the piecewise linear approximation that is obtained by the box-method realizes the same order of accuracy as the solution to Galerkin equations.

Key words. semiconductor simulation, convergence, finite elements, maximum stability

AMS(MOS) subject classifications. 65C20, 35J60, 47H17

1. Introduction. In the analytical and numerical modeling of semiconductor devices by the drift-diffusion model, a priori bounds on the extrema of the solution and charge conservation are of prime importance [16, 14, 24, 22, 4]. A priori bounds on the extrema of the solution to a discretized model are an essential ingredient in the finite element approximation theory for the drift-diffusion semiconductor model in [19, 15] as well. Existence and stability of solutions to the discretized model is demonstrated analogously to the proof for the original model in [14]. In this paper, we examine both Galerkin's equations for a piecewise linear approximation and the box method discretization. The solution to Galerkin's equations for the current-continuity equations can be shown to be maximum stable. However, this stability is subject to rather stringent meshwidth restrictions. Alternatively, we present a three dimensional finite element approximation theory for the finite difference box-method discretizations that are commonly used [4]. For these box-method equations, L_∞ stability and existence of solutions can be asserted at arbitrary meshwidth. Moreover, the box-method [26, 4] realizes conservation of charge for the discretized equations at arbitrary meshwidth.

In [3], the two dimensionsional box-integration discretization was analyzed as a Petrov-Galerkin method. Moreover, under the Assumption (2.4) in that paper, an inf-sup condition was shown to hold, which allowed the demonstration that this box method realizes the same order of accuracy as the Galerkin approximation by

---

* Department of Computer Science, University of Illinois, Urbana, IL 61801

piecewise linear finite elements. In this paper we examine the questions of finite element convergence, charge conservation, and a priori maximum stability in two and three dimensions. The Assumption (2.4) in [3] is circumvented, and we emphasize the computational practicality of the approaches. In the two dimensional analysis in [12], Assumption (2.4) of Bank and Rose was removed by a different technique than employed in this paper. Furthermore, the assumptions on the variable coefficient in the self-adjoint elliptic operator are somewhat different.

The electrostatic potential and the quasi-Fermi levels are scaled by the thermal voltage, $U_T = k_B T/q$, where q denotes the size of the electron charge, T is the temperature which is assumed to be constant, k_B is Boltzmann's constant. The concentrations n and p of the electron and hole carriers are scaled by the intrinsic concentration n_i and represented in terms of the quasi-Fermi levels v and w and the dimensionless electrostatic potential u through the relations $n = e^{u-v}$, $p = e^{w-u}$. As usual, the Slotboom variables $\nu = e^{-v}$ and $\omega = e^{w}$ are obtained from the quasi-Fermi level through exponentiation. The current continuity equations in the steady-state semiconductor model are then given by

$$
\begin{aligned}
-\nabla \cdot (D_n e^{u-v} \nabla v) + (G - R)(u, v, w) &= 0, \\
-\nabla \cdot (D_p e^{w-u} \nabla w) - (G - R)(u, v, w) &= 0.
\end{aligned}
$$

Here, Einstein's relations (see e.g. [25]) have been employed, and $G - R$ denotes a scaled generation-recombination term. For clarity, we shall partially restrict ourselves to the case of vanishing generation-recombination terms. The system can then be written

$$
\begin{aligned}
(1.1) && -\nabla \cdot (D_n e^{u} \nabla e^{-v}) &= 0, \\
(1.2) && -\nabla \cdot (D_p e^{-u} \nabla e^{w}) &= 0,
\end{aligned}
$$

subject to mixed Dirichlet/homogeneous Neumann boundary conditions, taken on the Dirichlet part, Σ_D, of the device boundary, and on its complement, Σ_N, respectively.

In terms of the Slotboom variables ν and ω, the current continuity equations are self-adjoint. The equations for the Slotboom variables ν and ω are discretized by a Petrov-Galerkin method [2] for a piecewise linear approximation with more general test functions. The order of convergence for the Petrov-Galerkin discretization of the single boundary value problems follows from the precise interpolation space in which the solution to these boundary value problems lies. This rather technical question is discussed in §**3.3** of [15]. This latter paper establishes the convergence of the finite element solution of the entire drift-diffusion semiconductor model by an analysis which is based on the formalism of Krasnosel'skii and his co-workers in [20]. Although the analysis in [15] deals explicitly with a piecewise linear Galerkin method, the results for the coupled system may be adapted to the more general Petrov-Galerkin framework which is presented here as well.

2. Approximation by piecewise linear finite elements. We begin the discretization of the drift-diffusion equations (1.1) and (1.2) by approximating the electrostatic potential function u, and the Slotboom variables ν and ω, from the space of piecewise linear finite element functions, S_h, on a simplicial mesh. The functions of

S_h are continuous and are linear on each simplex S. As usual, $h = \max_S\{diam\ S\}$. The Petrov-Galerkin discretization of the current continuity equations for piecewise linear Slotboom variables ν and ω is given by,

$$(2.1) \qquad \langle \mu_n e^{U_h} \nabla V_h, \nabla \phi_i \rangle = \langle G - R, \phi_i \rangle, \quad \text{for} \quad i = 1, \cdots M,$$

and

$$(2.2) \qquad \langle \mu_p e^{-U_h} \nabla W_h, \nabla \phi_i \rangle = \langle G - R, \phi_i \rangle, \quad \text{for} \quad i = 1, \cdots M.$$

In these latter equations $V_h \in \bar{V}_I + S_h$, $W_h \in \bar{W}_I + S_h$, where we have selected piecewise linear interpolants of the boundary data for $\bar{V}_I$ and $\bar{W}_I$. The integrals here can be evaluated in closed form.

Fox box-method discretizations, charge conservation may be concluded with respect to a suitably chosen grid of boxes that is dual to the finite element grid. Such box-methods may be regarded as Petrov-Galerkin methods with piecewise constant test functions ϕ_{B_j}. However, because such piecewise constant test functions are not of H^1 regularity, the demonstration of an inf-sup condition is complicated. Therefore, Petrov-Galerkin finite element error analysis becomes problematic. We propose to circumvent the complications which are introduced by the piecewise constant test functions, by employing Lemma 4.1. This lemma allows us to approximate the box-method finite element equations by Petrov-Galerkin equations in which the test functions are of H^1 regularity.

3. L_∞ stability. In the analyses of the drift-diffusion semiconductor model in [16, 14, 24, 22], stability is employed in the form of the a priori L_∞ bounds on the components u, v, and w. These a priori extrema bounds for the coupled system are obtained from the extrema principles for the constituent boundary value problems. A piecewise linear finite element function inherits an extrema principle from the solution to the original elliptic boundary value problem, if the finite element discretization yields a stress matrix which has a positive inverse. For both Galerkin's equations for piecewise linear finite elements, and the box-method, geometric restrictions have to be imposed on the mesh so that the stress matrix has a positive inverse. It is shown in §3.1 that the conditions for the box-method based on perpendicular bisectors are satisfied by N-dimensional Delaunay triangulations [6], which can be generated efficiently.

The regularity of the solutions to the mixed Dirichlet–Neuman boundary value problems in (1.1) and (1.2) is limited. Therefore, standard finite element error analysis yields convergence in the energy norm only to the order h^q, where $q = 1$ in one dimension, and $q \leq 1/2$ in two and three dimensions (see [15]. This again implies that piecewise linear finite elements already recover the optimal order of convergence, both in energy and in displacement.

3.1. Stress Matrix, Mesh geometry, and Maximum Principles. The solution u to the differential equation

$$(3.1) \qquad -\nabla \cdot [a(x)\nabla u] = 0,$$

(where $0 < a_{min} \leq a(x) \leq a_{max} < \infty$) on an N dimensional domain G, assumes its extrema on the boundary ∂G, (see e.g. [10], §**2.2**). Extrema principles for solutions to discretizations of (3.1) depend on the discretization procedure and the mesh geometry. A solution U_h to a finite element discretization of (3.1) satisfies a discrete extrema principle if the inverse of the finite element matrix is positive. We consider first Galerkin's equations for a piecewise linear finite element approximation to (3.1). In this case the inverse of the stress matrix is positive if the off-diagonal elements a_{ij} in the stress matrix are non positive. We express the entries of the element stress matrix for this piecewise linear finite element discretization of (3.1) on a simplicial mesh in terms of the geometric properties of the element S as stated below.

DEFINITION 3.1. *Let S be an N-dimensional simplicial finite element for which*

- *V_S is the volume,*
- *$\vec{v}_i$ is a vertex,*
- *e_{ij} is the edge connecting vertices $\vec{v}_i$ and $\vec{v}_j$,*
- *F_k is the face opposite to the vertex k, with measure $|F_k|$. In the global finite element mesh we will also employ the notation F_{rs}, meaning the face in between the finite elements S_s and S_r,*
- *n_i is the normal distance of v_i to F_i,*
- *γ_{ij} is the angle between the inward normal vectors to the faces F_i and F_j,*
- *ϕ_l is the piecewise linear nodal basis function which is 1 at vertex $\vec{v}_l$,*
-

$$\alpha_{ij} \equiv \int_S a(x)\nabla\phi_i \cdot \nabla\phi_j dx,$$

is the ij-th entry of the <u>element</u> stiffness matrix,

- *$\langle a(x)\rangle \equiv \int_S a(x)dx/V_S$, the average of $a(x)$ over the element S,*
- *a_{ij} is the ij-th element of the assembled stiffness matrix.*

It follows through elementary calculations, as shown in the appendix of [19], that

$$\alpha_{ij} \equiv \int_S a(x)\nabla\phi_i \cdot \nabla\phi_j dx = \langle a(x)\rangle \cos(\gamma_{ij})\frac{1}{n_i n_j}V,$$

or

$$\alpha_{ij} = \langle a(x)\rangle \cos(\gamma_{ij})\frac{|F_i||F_j|}{N^2 V}.$$

In the application to this paper, the role of a is played by $\mu_n e^u$, $\mu_p e^{-u}$, respectively. The off-diagonal elements in the global stress matrix are obtained by summation over elements S_i, adjacent to edge e_{jk} as stated in Corollary **A.2** from the appendix of [19], quoted next.

COROLLARY 3.1. *Let U_h be a piecewise linear finite element function which satisfies Galerkin's equations for (3.1). Let vertex v_k for $k \neq j$ be "adjacent" to vertex v_j, and let the edge e_{jk} belong to the elements S_i for $i = 1, \cdots, p$. Then, the coefficient a_{jk} in the stress matrix is given by*

$$(3.2) \qquad a_{jk} = \sum_{\substack{adjacent \\ elements\ S_i}} \langle a(x)\rangle_{S_i} \cos(\gamma_{jk})\frac{|F_{j,S_i}||F_{k,S_i}|}{N^2 V_{S_i}}.$$

In two dimensions the expression for the off diagonal entries reduces to the well known form

$$\frac{1}{2}[\langle a(x)\rangle_{T_1} \cot(\phi_1) + \langle a(x)\rangle_{T_2} \cot(\phi_2)],$$

where the T_i are the two triangles adjacent to edge jk and the ϕ_i are the two angles opposite to the edge jk.

For varying $a(x)$, it follows straightforwardly that no off-diagonal elements are positive if all angles in the mesh are bounded from above by $\pi/2$. If $a(x)$ is a constant in 2 dimensions, the sum of every two angles opposite to the same edge should be no larger than π. This condition in two dimensions is satisfied by the Delaunay triangulation (see [6]), and can be generated either by "flipping" of triangles as proposed by Lawson in [21], with worst case computational complexity $O(n^2)$ [8], or more efficiently with computational complexity $O(n \log n)$ [11]. However, for piecewise linear finite elements in three dimensions, the condition that $a_{jk} \leq 0$ in (3.2) is not related to the Delaunay tetrahedryzation [9].

For the piecewise linear Slotboom variable ν_h, let $\mathbf{J}_h = \mu_n e^u \nabla \nu_h$ be the finite element current density. For zero generation-recombination terms, $\mathbf{J}_h$ satisfies the M equations $\int \mathbf{J}_h \cdot \nabla \phi_j dx = 0$. Because exactly one piecewise linear finite element equation corresponds to every vertex v_j in the mesh, a discretization of the current continuity equations can be defined with respect to a dual grid of boxes, in which every vertex in the finite element mesh corresponds to a box in the dual grid.

DEFINITION 3.2. *Let S be an N-dimensional simplicial finite element as defined in definition 3.1. Then, on the element S the mesh of dual boxes is defined such that*

- *To every vertex v_j in the finite element mesh corresponds exactly one box B_j in the mesh of boxes.*
- *To every simplex S_l in the finite element grid corresponds exactly one vertex c_l in the mesh of boxes.*
- *To every edge e_{jk} in the finite element grid corresponds exactly one common (not necessarily planar) face f_{jk} in between the two boxes B_j and B_k.*
- *Every face F_{rs} in the finite element mesh is divided in exactly N connected parts $p_{rs,j}$, where $p_{rs,j}$ contains the vertex v_j in the finite element mesh.*

It is *not* required that the face f_{jk} is a single $N-1$ dimensional plane, and it can be an arbitrary surface. However, computations are simplified by such a choice.

As in [3], the box-method with respect to the mesh of boxes B_j is defined as a finite element method for piecewise linear approximations ν_h and ω_h, with in (2.1) and (2.2) nodal piecewise constant test functions ϕ_{B_j}. The test function ϕ_{B_j} is equal to the constant 1 inside B_j, and 0 elsewhere.

The box-method can be implemented efficiently if the flux through the box face f_{jk}, between the boxes B_j and B_k, depends only on the nodal values U_j and U_k. As is asserted in the following lemma from [18], this situation pertains if every face f_{jk} is normal to its corresponding edge e_{jk}.

LEMMA 3.2. *Let the faces f_{jk} of the boxes B_j be normal to their corresponding edges e_{jk} in the simplicial mesh. Then, the flux through face f_{jk} from box B_j into box*

B_k is equal to

$$\langle a(x)\rangle_{f_{jk}}(|f_{jk}|/|e_{jk}|)[U_j - U_k].$$

In the stress matrix (a_{jk}) the off-diagonal elements are given by.

(3.3) $$- \langle a(x)\rangle_{f_{jk}}(|f_{jk}|/|e_{jk}|).$$

In the Petrov-Galerkin formulation of the box-method that is employed here, the approximating function and the test function are different. Nevertheless, the Lemma 3.2 implies immediately that if the boxes B_j are defined by the perpendicular bisecting planes, then the bilinear form $a(u, v)$ is symmetric.

COROLLARY 3.3. *The stress matrix defined by the box-method in Lemma 3.2 is symmetric.*

Notice that in the box-method with perpendicular bisectors the sign of the off-diagonal elements in the matrix depends only on the geometry of the mesh and not on the smoothness of $a(x)$ in (3.1). This implies that with the box-method discretization a discrete extrema principle is obtained for variable, positive $a(x)$, whenever such a principle holds for Laplace's equation. All off-diagonal elements in the stress matrix for the box-method will be non positive if for all vertices v_j and v_k, connected by an edge e_{jk}, the face f_{jk} has a surface area $|f_{jk}|$, which is not smaller than 0. Non-negativity of the surface areas f_{jk}, results from Delaunay's condition [6] that the circumscribed sphere C_{S_i} of each tetrahedron (or triangle in two dimensions) S_i contains, in its interior, no vertices v_l which do not belong to S_i.

The coefficients a_{ij} for the piecewise linear finite element method in (3.2) and those for the perpendicular bisector box-method, are in general not equal. However, in two dimensions, for constant $a(x)$, the elements a_{jk} of the stress matrix for the piecewise linear finite element method in (3.2), and those for the box-method with perpendicular bisectors given by (3.3) are identical. This point was first observed by Bank and Rose in [3], and is a special case of Corollary 4.3 to Lemma 4.1 in this paper. In three dimensions, the box-method that employs perpendicular bisecting planes, gives rise to a stress matrix with positive inverse on the Delaunay tetrahedryzation, whereas Galerkin's equations for piecewise linear finite elements need not. Given a set of vertices v_i in three dimensions, a Delaunay tetrahedryzation can be constructed with computational complexity $O(n^2)$ [17], an efficient implementation is described in [7]. Moreover, in the box method a discretized form of charge conservation is valid. Because the box method, based on perpendicular bisectors, combines charge conservation with maximum stability on Delaunay tetrahedryzations, it appears to have certain advantages over Galerkin's equations for piecewise linear finite elements. However, in those cases that the stress matrix for the box-method discretization of the Laplacean differs from the stress matrix for Galerkin's equations, it becomes more problematic to present, for the box-method, a finite element error analysis that includes demonstration of the inf-sup condition.

Alternatively, stability of a finite element discretization may be obtained from finite element convergence. In §3.2, we show that the finite element projection is indeed stable for the mixed Dirichlet-Neumann boundary value problem in two dimensions, and in three dimensions for sufficient regularity of the boundary contacts.

3.2. L_∞ stability and finite element convergence. Let the piecewise linear function U_h solve Galerkin's equations for (3.1). We show how finite element convergence in the energy norm, and Sobolev's imbedding theorems, can be employed to obtain bounds on the extrema of U_h. Finite element convergence in the maximum norm is derived directly from finite element convergence in the energy because Galerkin's equations realize the best possible accuracy in the energy norm. The results generalize to solutions with less regularity an earlier analysis, for Laplace's equation in two dimensions only, of Schatz in [23]. Furthermore, significantly simpler techniques are employed.

Because Sobolev's imbedding theorems are dependent on the dimensionality of the domain G, the results are weaker for higher dimensions. In one dimension the error satisfies

$$e_{max} = e(x^*) = (1/2)[\int_0^L e_x dx - \int_{x^*}^L e_x dx] \leq (1/2)\|e_x\|_{L_1}.$$

L_∞ stability follows, therefore, straightforwardly from finite element convergence. In fact, if $u \in H^2$, then this analysis yields an order of 1 for convergence in the maximum norm.

To obtain maximum convergence in N dimensions, where $N \geq 2$, we require finite element convergence in the energy norm of the form

$$\int_G a(x)|\nabla(U_h - u)|^2 dx \leq Ch^{N/2-1+q},$$

for any $q > 0$. This implies that, in two dimensions, finite element convergence in the energy to any power of h implies convergence to a marginally smaller power of h of the extrema. In three dimensions, however, we need convergence in the energy to the order $1/2 + q$. This order of convergence appears to be marginally larger than can be inferred from the mixed Dirichlet-Neumann boundary conditions, provided that lightning rod singularities with field sizes $\|\nabla u\| \approx cr^{-1+\epsilon}$ (see e.g. p.97 of Jackson [13]) are excluded. We surmise that in three dimensions for rounded contact edges, that look locally esentially two dimensional, the same regularity results hold as in two dimensions.

We consider the equation (3.1) in N dimensions, subject to general mixed Dirichlet-Neumann boundary conditions, assuming only that $0 < a_{\min} \leq a(x) \leq a_{\max}$. This generality is required for the application of our results to the drift-diffusion semiconductor model, which is our target. We employ the following. By Sobolev's embedding theorems, in N dimensions, where $N \geq 2$, for an arbitrary function u,

$$\|u\|_{L_\infty} \leq C * \|\nabla u\|_{L_q},$$

if $q > N/2$. For a finite element function U_h, $\|\nabla U_h\|_{L_q}$ can be bounded by $\|\nabla U_h\|_{L_2} * h^{-t}$ for a suitable $t > 0$. The finite element method minimizes the energy $\int_G a(x)|\nabla(U_h - u)|^2 dx$ of the difference between the finite element solution U_h, and the solution u, to equation (3.1).

In §3.3 and §1 of [15] it is discussed that for the drift-diffusion semiconductor model the solution u in (3.1) should lie in $H^{1+\theta}$, where $\theta < 1/2$ in 2 dimensions, and $\theta \leq 1/2$ in three dimensions. Next, it is shown in §3.4 of [15] that this again implies that $\|U_h - u\|_{H^1} \leq Ch^\theta$, where the value of θ depends on the dimensionality of the domain G as stated above. In two and more dimensions a bound on $\|U_h - u\|_{H^1}$ does not suffice to obtain a bound on $\max_G |U_h - u|$ through direct application of Sobolev's theorems. For dimensionality of more than one we need an *inverse assumption* on the simplicial mesh (see equation (3.2.28) on p.140 of [5]) which asserts that the diameter h_{S_r} of the smallest simplex S_r in the mesh is bounded from below by a constant times the largest diameter h, i.e.

$$(3.4) \qquad \forall\, S_r \in G \quad h/h_{S_r} \leq c_\nu$$

With the inverse assumption Theorem **(3.2.6)** in [5]) becomes applicable. We restate this theorem below, adapted to the application in this paper. (Here ρ_{S_r} is the diameter of the incribed sphere in S_r, and .)

THEOREM 3.4. *Let there be given an N-dimensional simplicial mesh. Let the simplices S_r satisfy that, (i) there is a constant σ such that $\forall\, S_r, h_{S_r}/\rho_{S_r} \leq \sigma$, (ii) as the meshes are refined, $h \to 0$, (iii) the inverse assumption (3.4) holds. Let $l, m \geq 0$ and $(r, q) \in [1, \infty]$ such that $l \leq m$, and that the piecewise linears are in $W^{l,r}(\hat{S}) \cap W^{m,q}(\hat{S})$. Then, there exists a constant $C = C(c_\sigma, c_\nu, l, r, m, q)$ such that*

$$\forall\, \text{piecewiselinear } v_h, \left(\sum_{S_r \in G} |v_h|^q_{m,q,S_r} \right)^{1/q} \leq$$

$$(3.5) \qquad \frac{C}{(h^N)^{\max\{0,(1/r)-(1/q)\}} h^{m-l}} \left(\sum_{S_r \in G} |v_h|^r_{l,r,S_r} \right)^{1/r}$$

In [18] it is shown that, under the assumption that the function $u \in W^{1+\theta,2}$, the piecewise linear solution to Galerkin's equations converges to the piecewise linear interpolant U_i of u in the maximum norm. Here the Sobolev space of fractional order $W^{1+\theta,2}$ is defined as on page 204 of [1]. We quote the result.

THEOREM 3.5. *On the N dimensional domain G, let the error in the approximation of the function u by the piecewise linear finite element function U_h be bounded by*

$$\|\nabla(U_h - u)\|_{L_2} \leq C \|u\|_{W^{1+\theta,2}} * h^\theta,$$

where $\theta \geq N[1/2 - 1/p]$, for $p > N$. Let U_i be the piecewise linear interpolant of the function u. Then

$$(3.6) \qquad \max_{x \in G} |U_h - U_i| \leq C * \|U\|_{W^{1+\theta,2}} * h^{\theta - N[1/2 - 1/p]}.$$

Theorem 3.5 implies directly that, up to the accuracy of the finite element approximation, an extrema principle can be inferred for piecewise linear solution to Galerkin's equations from an extrema principle for the solution to the original problem. Again from [18] we quote

COROLLARY 3.6. *Let the extrema of the function u on the domain G satisfy*

$$\gamma \leq u(x) \leq \delta.$$

Let U_h be the piecewise linear finite element function that minimizes

$$\int a(x)|\nabla(U_h - u)|^2 dx,$$

and let $u \in W^{1+\theta,2}$ for $\theta \geq N[1/2 - 1/p]$, where $p > N$. Then

$$\max_{x \in G} U_h \leq \delta + ch^s,$$

and

$$\min_{x \in G} U_h \leq \gamma - ch^s,$$

where $s = \theta - N(1/2 - 1/p)$ The current continuity equations are discretized in terms of the Slotboom variables $\nu = e^{-v}$ and $\omega = e^w$. The approximate upper bounds on ν and ω suffice to yield an approximate lower bound on the quasi-Fermi level v and an approximate upper bound on the quasi-Fermi level w. However, because of the singularity in the logarithm at 0, the approximate lower bounds on ν and ω yield an upper bound on v and a lower bound on w only if $\gamma - ch^s > 0$. This requires an upper bound on the meshwidth h unless the discretization yields a discrete extrema principle for the Slotboom variables and, hence, for the quasi-Fermi levels.

4. Piecewise linear finite elements and the box-method. In [3] it was observed that the box-method can be interpreted as a Petrov-Galerkin method for a piecewise linear approximation with piecewise constant test functions. Finite element convergence can then be proven under the condition that the finite element inner product satisfies an appropriate inf-sup condition. Demonstration of the inf-sup condition for this Petrov-Galerkin method is non-trivial, because the piecewise constant test functions are not in the space H^1. Therefore, in [3] the Assumption (2.4) (in [3]) was made. We will circumvent the Assumption (2.4) as made by Bank and Rose as follows: In Lemma 4.1 and Theorem 4.2 we estimate for a piecewise linear approximation U_h, the effect on the Petrov-Galerkin equations of changing test functions ϕ, if the integrals $\int a(x)\phi(x)dx$ of the test function ϕ on finite element faces F_{rs} and over finite element volumes S_r are left unmodified. The results allow us to replace the piecewise constant test functions ϕ_B, which are not in H^1, by smoother test functions ϕ_{mod}, which are in H^1 and which assume equal integrals over finite element faces and volumes as the functions ϕ_B. With the smoothed test functions, the bilinear form $a(u,v)$ is continuous on $H^1 \times H^1$. Therefore, establishing the inf-sup condition for the finite element inner-product $a(u,v)$ is less problematic. The approximation of the piecewise constant ϕ_B by $\phi_{mod} \in H^1$ requires first the following lemma from [18].

LEMMA 4.1. *Let U_h be a piecewise linear finite element function on a simplicial mesh on an N dimensional domain G that satisfies the Petrov-Galerkin finite element equation*

$$(4.1) \qquad \int a(x)\nabla U_h \cdot \nabla \phi^{(1)} dx = \int f \phi^{(1)} dx,$$

Let $\phi^{(2)}$ be a second test function such that the weighted integrals of $\phi^{(1)}$ and $\phi^{(2)}$ are equal over all element faces F_{rs}, i.e.

$$(4.2) \qquad \int_{F_{rs}} a\phi^{(1)} dx = \int_{F_{rs}} a\phi^{(2)} dx \;\; \forall F_{rs}.$$

Then, the piecewise linear finite element function U_h satisfies the modification of equation (4.1) in which $\phi^{(1)}$ is replaced by $\phi^{(2)}$ as stated in

$$\int a\nabla U_h \cdot \nabla\phi^{(2)}dx - \int f\phi^{(2)}dx =$$

$$(4.3) \qquad \int [\phi^{(1)} - \phi^{(2)}]\nabla a \cdot \nabla U_h dx + \int f[\phi^{(2)} - \phi^{(1)}]dx.$$

This lemma allows us to bound the modification of the finite element inner products for a Petrov-Galerkin method for a piecewise linear approximation U_h if the test functions ϕ are modified without changing integrals over faces and elements. From [18] we quote.

THEOREM 4.2. *Let U_h be a piecewise linear finite element function on a simplicial mesh on an N dimensional domain G that solves Petrov-Galerkin finite element equation (4.1) Let $\phi^{(2)}$ be a second test function such that the weighted integrals of $\phi^{(1)}$ and $\phi^{(2)}$) are equal on all element faces F_{rs}. Moreover, let the integrals of $\phi^{(1)}$ and ϕ^2 over elements S_r be equal, i.e.*

$$(4.4) \qquad \int_{S_r} \phi^{(1)}dx = \int_{S_r} \phi^{(2)}dx \ \ \forall S_r,$$

Then, U_h satisfies the approximate finite element equations

$$|\int a\nabla U_h \cdot \nabla\phi^{(2)}dx - \int f\phi^{(2)}dx| \leq$$

$$(4.5) \qquad [\|\phi^{(1)} - \phi^{(2)}\|_{L_p}\|\nabla U_h\|_{L_r}|a|_{2,q} + \|\phi^{(2)} - \phi^{(1)}\|_{L_t}|f|_{1,s}]h,$$

where $1/p + 1/q + 1/r = 1$ and $1/s + 1/t = 1$.

As stated in the following corollary from [18], Theorem 4.2 implies that for the Laplacean in N dimensions Galerkin's equations for a piecewise linear approximation are identical to box-method equations if the boxes B_j equidivide the faces F_{rs} in the finite element mesh.

COROLLARY 4.3. *For the Laplacean in N dimensions the piecewise linear finite element method on N-simplices is equivalent to the box-method if, (i) the boxes B_j are dual to the finite element mesh, (ii) the boxes equidivide all faces F_{rs} in the finite element mesh.*

It follows immediately that in two dimensions, the classical box-method based on a dual mesh of perpendicular bisectors is equivalent to the piecewise linear finite element method up to numerical integration of the variable coefficient $a(x)$. In more than two dimensions the box-method in which the face f_{jk} is equal to the perpendicular bisecting plane of edge e_{jk} is in general *not* equivalent to the piecewise linear finite element method. For instance, let, in three dimensions, the boxes B_j intersect with inter element faces F_{rs} along straight lines. Then, the condition that these lines partition the triangular faces F_{rs} into three parts of equal surface area amounts to the requirement that the boxes B_j intersect with the inter element faces F_{rs} along the line pieces that connect the midpoints of the edges e_{jk} in the finite element grid with the centroids of the inter element faces F_{rs}. These linepieces are perpendicular to the edges e_{jk} only if all faces are equilateral. In other words, all tetrahedral finite elements have to be equilateral tetrahedra. But three dimensional space cannot be tetrahedrized with equilateral tetrahedra without leaving holes. Hence, in three dimensions, the box-method based on perpendicular bisecting planes does not coincide with Galerkin's method for piecewise linear finite elements.

We proceed to develop a finite element convergence theory for the box-method, for those cases that the box-method generates a different stress matrix than Galerkin's method. The box-method is interpreted as an approximate Petrov-Galerkin method for a piecewise linear approximation U_h with polynomial test functions ϕ_j.

As observed in [3], the box-method can be considered as a finite element method which employs nodal step functions as test functions, with the step at the location of the faces f_{jk} of the boxes. However, Lemma 4.1 implies that substitution for the box functions ϕ_{B_j} of test functions ϕ_j, with face integrals $\int_{S_{rs}} a\phi_j dx$ equal to the face integrals $\int_{S_{rs}} a\phi_{B_j} dx$ of the nodal box function ϕ_{B_j}, results only in a limited perturbation of the finite element equations. We employ this observation to obtain a finite element error analysis of the box-method. The observation is important because the box functions ϕ_{B_j} are not of H^1 regularity. However, by replacing the box test functions ϕ_{B_j} that do not lie in H^1, by test functions ϕ_j, that do lie in H^1, an inf-sup condition for the Petrov-Galerkin equations (2.1) and (2.2) becomes provable by relatively simple means.

The following lemma from [18] shows that the solution U_B to the finite element box-method equations for (3.1) attains the same order of accuracy as piecewise linear finite elements, provided that the finite element inner product for the smoothed test functions approximates the box-method inner product sufficiently closely, and that this inner product satisfies an appropriate inf-sup condition,

$$(4.6) \qquad \inf_{u \in U_n} \sup_{v \in V_n} \frac{a(u,v)}{\|u\|_U \|v\|_V} \geq \gamma > 0.$$

Substituting smoother test function for the box functions ϕ_B can be viewed as solving approximate finite element equations. Therefore, the following lemma can be viewed as an equivalent of Strang's first Lemma (see e.g. Theorem **4.1.1** in [5]) for the Petrov-Galerkin method.

LEMMA 4.4. *Let U_B solve the Petrov-Galerkin equations (4.1) with box test functions ϕ_{B_j}. Let ψ_j be test functions with equal face and volume integrals as the ϕ_{B_j} as in (4.2) and (4.4). Let $a(u,v)$, with a piecewise linear approximation function U_h and test functions ψ_j, satisfy the inf-sup condition (4.6). Let $a(u,v)$ be continuous in both u and v on H^1, with Lipschitz constant C. Then, for all piecewise linear v_h*

$$(4.7) \qquad \|U_B - u\| \leq (C/\gamma)\|v_h - u\| + (c/\gamma)[\|\nabla U_h\|_{L_r} |a|_{2,q} + |f|_{1,s}]h,$$

where $1/q + 1/r < 1$ and $1/t < 1$ in two dimensions. while $1/q + 1/r = 5/6$ and $1/t = 5/6$ in three dimensions. Hence, U_B realizes the same order of accuracy as the piecewise linear finite element method, provided that $a(x)$ and $f(x)$ are sufficiently regular.

Hence, a box-method yields the same order of accuracy as the piecewise linear finite element method if the following sufficient conditions are satisfied: (i) The piecewise constant box test functions ϕ_{B_j} can be approximated by test functions ϕ_j which satisfy (4.2) and (4.4), and which are continuously imbedded in H^1. (ii) The bilinear form $a(u,v)$ satisfies an inf-sup condition (4.6).

If the face integrals of the box functions ϕ_B are equal to those of piecewise linear test functions, the inf-sup condition (4.6) for the piecewise linear finite element approximation with modified test functions ϕ_j is readily established as in the following lemma from [18].

LEMMA 4.5. *Let the piecewise linear finite element approximation U_h solve the box method equations corresponding to (3.1) on a simplicial mesh in N dimensions. Let the face and element integrals (4.2) and (4.4) of the nodal box functions ϕ_{B_j} be equal to those of the nodal piecewise linear test functions ϕ_j. Then, the bilinear form $a(u,v)$ satisfies the inf-sup condition.*

$$a_{min} \leq \inf_{u \in U_n} \sup_{v \in V_n} \frac{a(u,v)}{\|u\|_U \|v\|_V}$$

Moreover, the Lipschitz constant of $a(u,v)$ on $H^1 \times H^1$ is given by

$$a(u,v) \leq a_{max} * \|u\|\|v\|.$$

For the case that the box functions ϕ_{B_j} do not equidivide the faces of the element, we follow the next procedure. The bilinear form $a(U_h, \psi)$ is continuous from $H^1 \times H^1$ to $\mathbb{R}$. A Petrov-Galerkin test function ϕ_j of H^1 regularity which satisfies (i) above is constructed as follows: We modify the piecewise linear test function ϕ_j by adding to ϕ_j, for each finite element face F_{rs} on which $\phi_j \neq 0$, a piecewise polynomial function μ_{rs} which is nonzero only in the interior of elements S_r and S_s, on both sides of face F_{sr}. Moreover, on each element S_r we add to each nodal basis function a polynomial function that is 0 on the entire surface of S_r. Formally we define:

DEFINITION 4.1. *Let S_r and S_s be adjacent simplices in N dimensions. Let F_{rs} be the face common to S_r, and S_s. Let $J_{F_{rs}}$ be the set of indices of vertices v_i of F_{rs}. Let the nodal piecewise linear basis function $\phi_j(x)$ satisfy $\phi_j(v_i) = \delta_{ij}$. Then the facial correction function μ_{rs}, corresponding to the face F_{rs} is defined:*

$$(4.8) \qquad \mu_{rs}(x) = \prod_{j \in J_{F_{rs}}} \phi_j(x).$$

Let J_{S_s} be the set of indices of the vertices v_j of element S_s. Then, the elemental correction function σ_s, corresponding to element S_s, is defined:

$$(4.9) \qquad \sigma_{S_r}(x) = \prod_{j \in J_{S_{S_r}}} \phi_j(x).$$

Integrals of the μ_{rs} and their products over F_{rs} and S_r, or integrals of σ_s over S_s can be determined analytically [18].

The modified test functions $\psi_j(x)$ are formally defined in the following

DEFINITION 4.2. *On simplicial element S_r, the nodal test function $\psi_j(x)$ is defined in terms of the nodal piecewise linear basis functions $\phi_j(x)$ as follows:*

Let μ_k be the corresponding to face F_k. Let σ_{S_r} correspond to element S_r. Then

$$(4.10) \qquad \psi_j(x) = \phi_j(x) + \sum_{k \neq j} \alpha_{j,k} \mu_k(x) + \beta_j \sigma_{S_r}(x)$$

Where for every face F_k the sum

$$(4.11) \qquad \sum_{j \neq k} \alpha_{j,k} = 0,$$

and for every element S_r

$$(4.12) \qquad \sum_{j} \beta_j = 0,$$

These modified test functions allow us to apply inf-sup error analysis with both approximation and test functions in H^1. Some procedures to establish the inf-sup condition in these, more general, cases are presented in [18].

REFERENCES

[1] Robert A. Adams. *Sobolev Spaces*. Academic Press, New York, San Fransisco, London, 1975.

[2] I. Babuška and A.K. Aziz. *The Mathematical Foundations of the Finite Element Method with Applications to Partial Differentiqal Equations*. Academic Press, New York, San Fransisco, London, 1972.

[3] Randolph E. Bank and Donald J. Rose. Some Error Estimates for the Box Method. *SIAM J. on Numer. Anal.*, 24:777–787, 1987.

[4] Randolph E. Bank, Donald J. Rose, and Wolfgang Fichtner. Numerical Methods for Semiconductor Device Simulation. *SIAM J. on Scient. and Statist. Comp.*, 4(3):416–435, September 1983.

[5] Philippe G. Ciarlet. *The Finite Element Method for Elliptic Problems*. North Holland, Amsterdam, New York, Oxford, 1978.

[6] B. Delaunay. Sur La Sphére Vide. *Izvestia Akademii Nauk SSSR, Otdelenie Matematicheskii i Estestvennyka Nauk*, 7(6):792–800, 1932.

[7] David P. Dobkin and Michael J. Laszlo. Primitives for the manipulation of three-dimensional subdivisions. *Algorithmica*, 4:3–32, 1989.

[8] Herbert Edelsbrunner. The Computational Geometry Column. *Buletin of the EATCS*, 37:109–116, 1988.

[9] Herbert Edelsbrunner. Spatial Triangulations with Dihedral Angle Conditions. In *Proc. Int. Workshop on Discrete Algorithms and Complexity*, pages 83–89, 1989.

[10] David Gilbarg and Neil S. Trudinger. *Elliptic Partial Differential Equations of Second Order*. Springer-Verlag, New York, Berlin, 1983.

[11] L.J. Guibas and J. Stolfi. Primitives for Manipulation of General Subdivisions and the Computation of Voronoi Diagrams. *ACM Transactions on Graphics*, 4(2):74–123, April 1985.

[12] W. Hackbusch. On First and Second Order Box Schemes. *Computing*, 41:277–296, 1989.

[13] J.D. Jackson. *Classical Electrodynamics 2nd edition*. John Wiley and Sons, New York, 1974.

[14] Joseph W. Jerome. Consistency of Semiconductor Modelling: An Existence/Stability Analysis for the Stationary van Roosbroeck System. *SIAM J. Appl. Math.*, 45(4):565–590, August 1985.

[15] Joseph W. Jerome and Thomas Kerkhoven. A Finite Element Approximation Theory For The Drift Diffusion Semiconductor Model. *SIAM J. on Appl. Math.*, 28:403–422, April 1991.

[16] Joseph W. Jerome and Thomas Kerkhoven. *The Steady State Drift-Diffusion Semiconductor Model*. SIAM, Philadelphia, 1992.

[17] Barry Joe. Construction of Three-Dimensional Delaunay Triangulations Using Local Transformations. *Computer Aided Geometric Design*, 8:123–142, 1991.

[18] Thomas Kerkhoven. Piecewise Linear Finite Element Approximations And The Box Method. Technical report, University of Illinois, in preparation.

[19] Thomas Kerkhoven and Joseph W. Jerome. L_∞ Stability of Finite Element Approximations to Elliptic Gradient Equations. *Numerische Mathematik*, 57:561–575, 1990.

[20] M.A. Krasnosel'skii, G.M. Vainikko, P.P. Zabreiko, Ya.B. Rutitskii, and V.Ya. Stetsenko. *Approximate Solution of Operator Equations*. Wolters-Noordhoff, Groningen, 1972.

[21] C.L. Lawson. Software for C^1 Function Interpolation. In J.R. Rice, editor, *Mathematical Software III*, pages 161–194, New York, 1977. Academic Press.

[22] Peter A. Markowich. *The Stationary Semiconductor Device Equations*. Springer-Verlag, Wien, New York, 1986.

[23] Alfred Schatz and Lars B. Wahlbin. On the Quasi Optimality in L_∞ of the H_1-projection into Finite Element Spaces. *Match. Comp.*, 38:1–22, 1982.

[24] Thomas I. Seidman. Steady State Solutions of Diffusion-Reaction Systems with Electrostatic Convection. *Nonlinear Analysis. Theory, Methods and Applications*, 4:623–637, 1980.

[25] S.M. Sze. *Physics of Semiconductor Devices*. Wiley-Interscience, New York, second edition, 1981.

[26] R.S. Varga. *Matrix Iterative Analysis*. Prentice-Hall, Englewood Cliffs, 1962.

MATHEMATICAL MODELING OF QUANTUM WIRES IN PERIODIC HETEROJUNCTION STRUCTURES

THOMAS KERKHOVEN*

Abstract. The confinement of electrons in quantum wires in the corners of a periodic sawtooth potential well of microscopic size is modeled numerically. The potential well is created in a semiconductor nanostructure consisting of alternating sawtooth layers of differently doped GaAs and AlGaAs. The mathematical model is two dimensional and consists of an eigenvalue problem for Schrödinger's equation coupled with Poisson's equation for the electrostatic potential ϕ. This system is solved by an algorithm that employs an underrelaxed fixed point iteration to generate an approximation which is reasonably close to the solution. Subsequently, this approximate solution is employed as an initial guess for a Jacobian-free implementation of an approximate Newton method.

Confinement of quantum wires of electrons to corners of such a sawtooth structure results only if the quantum well is sufficiently deep. To obtain that only the ground state in the corner of the sawtooth well is substantially occupied, the energy levels in the well have to be positioned correctly with respect to the Fermi level E_F in the semiconductor. The energy levels E_i in the well are set correctly with respect to the Fermi level E_F by bending the conduction band upward in the neighborhood of the quantum well. This band bending is brought about by the addition of depleted PN juctions on both sides of the sawtooth structure.

The band bending brought about by the depleted PN juctions does not result in a uniform upward shift in energy of the entire sawtooth structure. The corners of the well, where confinement is desired are shifted upward too far due to local excess of negative charge. By choosing the doping profile of the AlGaAs layers on the two sides of the sawtooth structure asymmetrically, potential wells with bound states can be created in half of the corners of the sawtooth, resulting simultaneously in the absence of bound states in the other half.

Key Words. semiconductor simulation, eigenvalue problems, Quantum mechanics

AMS(MOS) subject classification. 65B, 65H10, 65N

1. Introduction. Mathematical modeling of quantum wires in corners of a jagged periodic structure requires inclusion of minute physical detail. A numerical model of such quantum wires has been formulated which includes two dimensional self-consistent solution of both Poisson's equation for the electrostatic potential, and of an eigenvalue problem for Schrödinger's equation. Numerical simulations with this detailed model allow us to proceed beyond earlier results which only included the solution of an eigenvalue problem for Schrödinger's equation in [11]. The latter simulation suggested that quantum mechanical electrons can be confined in the corner of a periodic jagged potential well, defined by a fixed discontinuity or "step" in the potential only. This jagged well may be created by depositing $Al_xGa_{1-x}As$ and $GaAs$ in alternating layers, each several nanometers thick. The discontinuity in the conduction band energy levels of these two materials would create a quantum potential well in the $GaAs$ region in which electrons may be confined.

Our more elaborate model includes the actual charge density in the quantum well, and allows us to ascertain that only the ground state is significantly occupied. To determine the occupancies of the eigenstates in the well, the location of the energy levels E_l of these eigenstates with respect to the Fermi level E_F must be found. The position of the energy levels E_l in the well with respect to E_F sensitively depends on the extent of the region which is depleted of classical electrons in the semiconductor

---

* Department of Computer Science, University of Illinois, Urbana, IL 61801

TABLE 1

Saw-tooth device dimensions used to generate Table 3.

x-direction		y-direction	
Dimension	Width (Å)	Dimension	Width (Å)
x_1	625	y_1	625
x_2	75	y_2	75
x_3	100	y_3	100
x_4	75	y_4	75
x_5	625	y_5	625

around the energy well. Therefore, the classical electron density has to be included in the model. Moreover, we found that the simplified assumption of an infinitely deep potential well in [11] had to be abandoned, and that spreading of the wavefunctions outside the well significantly affects the solution.

A test quantum wire nanostructure as depicted in Fig. 1 is modeled on the computational domain presented in Fig. 2. A typical example of the dimensions in Fig. 2 at which the model is operated, is presented in Table 1. The concentration x of Al is assumed to be constant throughout the $Al_{1-x}Ga_xAs$ layers. The simulations with the model were carried out with a modified version of the algorithm for the numerical computation of electron states in the cross-section of a quantum wire that was presented in [7].

The two-dimensional physical model consists of Poisson's equation for the electrostatic potential ϕ, coupled with an eigenvalue problem for Schrödinger's equation:

$$-\nabla \cdot [\epsilon \nabla \phi] = \rho,$$
$$-(\hbar^2/2)\nabla \cdot [(1/m)\nabla \psi_l] + [V - E_l]\psi_l = 0. \tag{1.1}$$

Here m is the effective mass, ϵ is the dielectric constant, and ρ is the charge density expressed in terms of the magnitude of the elementary charge e, the electron density n, the hole density p, and the total density of ionized dopants $N_D^+ - N_A^-$ by

$$\rho = e[-n + p + N_D^+ - N_A^-]. \tag{1.2}$$

In Eq. (1.2) the total electron density, $n = n_{qm} + n_{cl}$, is the sum of the quantum mechanical electron density n_{qm}, and the classical electron density n_{cl}. By a suitable choice of doping density $N_D - N_A$, the active region of the device where the quantum mechanical quantum wire is constructed, is depleted of classical electrons $n_{cl}(\vec{x})$. Furthermore, the hole density $p(\vec{x})$ is negligible throughout the device. The ambient temperature of the simulated devices is $T = 4.2\,K$, the temperature of liquid Helium.

The quantum mechanical electron density n_{qm} is obtained from the eigenpairs (E_l, ψ_l) of Schrödinger's equation through

$$n_{qm} = \sum_l N_l \psi_l^2. \tag{1.3}$$

The occupancy level N_l of the l-th eigenstate for the one-dimensional electron gas is expressed in terms of the Fermi level E_F, the effective mass along the wire axis for

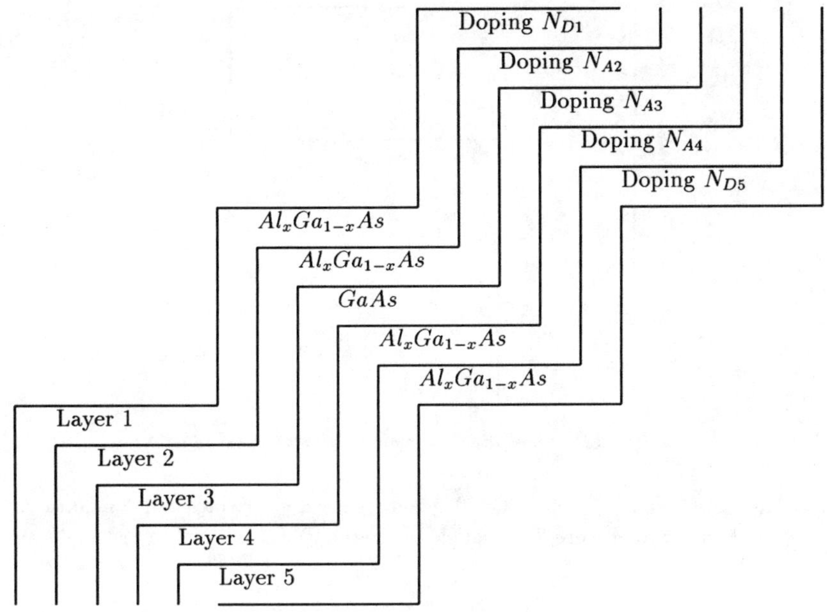

FIG. 1. *Periodic saw-tooth device structure.*

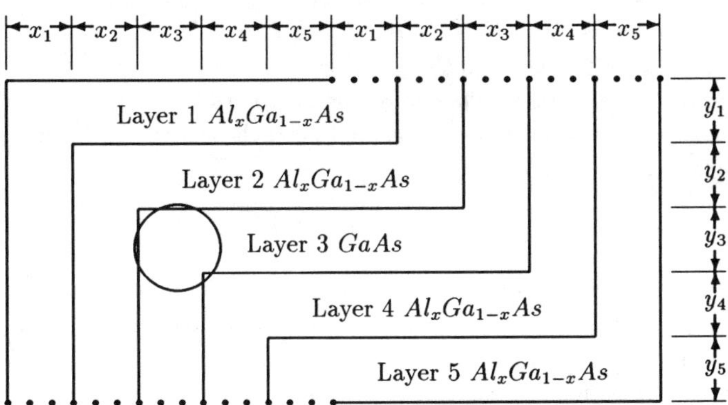

FIG. 2. *Computational domain Ω of the periodic saw-tooth structure with periodic grid points and physical dimensions illustrated.*

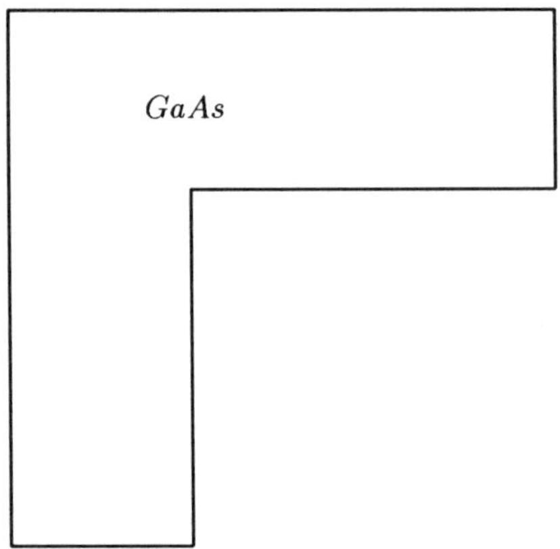

FIG. 3. *Computational domain employed in [11].*

the l-th ladder of eigenstates $m_p^{(l)}$, the degeneracy level $g_\nu^{(l)}$ (which is 1 for GaAs and 2 for Si), and the temperature T, through the Fermi-Dirac integral of order $-\frac{1}{2}$ as

$$(1.4) \qquad N_l = \frac{g_\nu^{(l)}}{\pi} \Big(\frac{2m_p^{(l)} k_B T}{\hbar^2} \Big)^{\frac{1}{2}} F_{-\frac{1}{2}} \Big(\frac{E_F - E_l}{k_B T} \Big).$$

Here, k_B is Boltzmann's constant, and $\hbar = h/(2\pi)$, where h is Planck's constant.

The location of the Fermi level is set at the boundary of the computational domain in the bulk from the assumptions of charge neutrality (including "freeze-out") and thermodynamic equilibrium. The energy levels of the bound state in the corner, which we desire to be occupied, and the first non-bound state, which should remain relatively empty, have to be situated appropriately on both sides of E_F. The position of the energy levels in the quantum well with respect to E_F, is determined by the charge distribution in between the well and the point where thermodynamic equilibrium and charge neutrality are assumed. Therefore, the position of the eigenvalues E_l with respect to the Fermi level in the quantum well depends sensitively on the depletion width. The classical electron density,

$$(1.5) \qquad n_{cl} = [N_{D,bndy} - N_{A,bndy}] \exp[e(\phi - \phi_{bndy})/(k_B T)],$$

is described by the Boltzmann distribution described in [15]. It is included in Poisson's equation so that the precise depletion width of free carriers surrounding the quantum well may be computed. In this equation, $N_{D,bndy}$, $N_{A,bndy}$, and ϕ_{bndy} are the values of the electron density, hole density, and electrostatic potential, respectively, at the boundary of the simulated domain, far from the active regions of the device. Further details of the calculation of the depletion width from n_{cl} are presented in § 4.2.

The potential energy of the conduction band E_C is related to the electrostatic potential $\phi(\vec{x})$ by $E_C(\vec{x}) = -e\phi(\vec{x}) + V_h(\vec{x})$, where $V_h(\vec{x})$ takes into account heterojunction discontinuities and is equal to 0 in the $Al_{1-x}Ga_x As$ surrounding the quantum well. The potential V in Schrödinger's equation is equal to the conduction band energy E_C.

The parameters of the problem have to be adjusted carefully to create an effective quantum wire. These are: (i) The depth ΔE_C of the potential well formed in the jagged periodic structure (This depth ΔE_C is created as a discontinuity in the conduction band between $Al_{1-x}Ga_xAs$ and $GaAs$). (ii) The width w of the potential trough. (iii) The doping profile inside and outside the potential trough.

2. Numerical solution. Self consistency of the numerical model is achieved with a slightly modified version of the iterative algorithm presented in [7]. This algorithm accelerates a nonlinear block Gauß-Seidel relaxation method in which Poisson's equation for the electrostatic potential and the eigenvalue problem for Schrödinger's equation are solved successively. Iterations of this algorithm are represented by application of a fixed point mapping $T_\eta : \eta \to \bar{\eta}$, which maps a modified logarithm of the quantum mechanical electron density n_{qm} to itself. T_η is presented below:

- $n_{qm} = e^\eta - \delta$.
- Employ a damped Newton method to solve a nonlinear version of Poisson's equation for an intermediate estimate to the potential $\phi(n_{qm}, N_D^+ - N_A^-)$. Obtain n_{cl} as well.
- Obtain $V(x) = -e\phi + V_h(x)$ from ϕ.
- Given V, solve the eigenvalue problem for Schrödinger's equation

$$-\frac{\hbar^2}{2m}\nabla^2\psi_l + [V - E_l]\psi_l = 0,$$

- Compute the quantum mechanical electron density

$$n_{qm}(x) = \sum_l N_l\psi_l^2(x).$$

- $\bar{\eta} = \log(n_{qm} + \delta)$.

Here. $\delta = 10^{-14}$ is added to avoid the singularity of the logarithm at zero. The inclusion of the logarithm in the definition of the mapping T_η increases the smoothness of the mapping and facilitates the convergence of the numerical algorithm.

In the iterative algorithm for the solution of the coupled system (1.1), the fixed point iteration T_η is accelerated and stabilized. Initially, stabilization of the fixed point iteration is implemented through underrelaxation as in the following adaptive approach:

- Set $\omega = 1$, choose η_0.
- **iterate on** i
    - $\eta_{i+1} = \omega T_\eta[\eta_i] + (1 - \omega)\eta_i$
    - **if** $\|\eta_{i+1} - \eta_i\| > \|\eta_i - \eta_{i-1}\|$ **then**
      $\omega = \|\eta_i - \eta_{i-1}\|/\|\eta_{i+1} - \eta_i\|$, $\eta_{i+1} = \omega T_\eta[\eta_i] + (1 - \omega)\eta_i$
    - **if** $\|\eta_{i+1} - \eta_i\|/\|\eta_i - \eta_{i-1}\| > \|\eta_i - \eta_{i-1}\|/\|\eta_{i-1} - \eta_{i-2}\|$
    - **then** $\omega := \omega * .8$
- **until** $\omega < \omega_{min}$.

The heuristic motivation for this approach is an induction on a result from [10] for a fixed point mapping T_G similar to T_η applied to the one-dimensional drift diffusion model. For this model it was shown that the nonlinear block Gauß-Seidel approach (Gummel's method) which can be defined through an iteration with T_G, converges while sufficiently far from the solution and slows down as the solution is approached. This behavior can be understood in terms of the maximum principles which hold for the single elliptic equations which constitute the system. Similar maximum principles hold for each of the elliptic problems for the quantum-mechanical model.

Close to the solution the equation $T_\eta(\eta) = \eta$ is solved through an inexact version of Newton's method which is locally quadratically convergent. This implementation of Newton's method is stabilized through a "damping" strategy [6] so as to make it more reliable further away from the solution. To apply Newton's method, the fixed point iteration,

$$\eta_{i+1} = T_\eta(\eta_i),$$

is rewritten as the nonlinear root finding problem

$$n - T_\eta(\eta) = 0.$$

Newton's method requires at every step the solution of the linear system

$$(2.6) \qquad [I - \nabla_\eta T_\eta(\eta_i)]d\eta = -[\eta_i - T_\eta(\eta_i)],$$

where $\nabla_\eta T_\eta(\eta_i)$ is the Jacobian matrix of the mapping T_η at the point η_i. This means that the typically dense linear system $I - \nabla_\eta T_\eta(\eta_i)$ has to be inverted. However, this presents no obstacle because the system of equations is solved in a Jacobian-free manner by the iterative method GMRES [14] which requires only Jacobian-vector products. The latter can be approximated by finite differences.

The only operations on the Jacobian matrix $\nabla_\eta T_\eta(\eta_i)$ that are required for GM-RES are matrix-vector multiplications $w = \nabla_\eta T_\eta(\eta_i)v$, which can be approximated by

$$\nabla_\eta T_\eta(\eta_i)v \approx \frac{T_\eta(\eta_i + hv) - T_\eta(\eta_i)}{h},$$

where η_i is the point where the Jacobian is being evaluated and h is some carefully chosen small scalar. This approximation to the product of a vector by a Jacobian was successfully used in the context of ordinary differential equations [2, 5, 1] and is quite common in nonlinear equation solution methods and optimization methods (see for example [6, 3, 16, 4]).

In an analysis and set of numerical experiments in [9], it was discussed why a Jacobian-free approach based on GMRES is highly suitable for accelerating the convergence of a nonlinear fixed point mapping T defined through successive solution of coupled elliptic problems. It was shown in [9] that for a mapping T_η as defined above the eigenvalue spectrum of $I - \nabla_\eta T_\eta(\eta_i)$ clusters at 1 only, and that the GMRES based approach can take advantage of this feature.

The method is implemented as follows. Solution of Newton's equations (2.6) is equivalent to minimization over $d\eta$ of

$$(2.7) \qquad \|[I - T_\eta](\eta_i) + [I - \nabla_\eta T_\eta(\eta_i)]d\eta\|.$$

where $\|\cdot\|$ is the Euclidean norm in $\mathbb{R}^N$. Suppose that η_i is the current approximation to the solution η^* of $\eta - T_\eta(\eta) = 0$ and that we wish to find a new approximation of the form $\eta_{i+1} = \eta_i + d\eta$. In the nonlinear version of GMRES [16], to be called henceforth NLGMR, the vector $d\eta$ is written in the form

$$(2.8) \qquad d\eta = \sum_{j=1}^{m} \alpha_j v_j,$$

where the $v_j's$ are m orthonormal vectors that form a basis of the Krylov subspace $K_m = \text{span}\{v_1, [I - \nabla_\eta T_\eta(\eta_i)]v_1, ..., [I - \nabla_\eta T_\eta(\eta_i)]^{m-1}v_1\}$. These vectors are easily determined by an Arnoldi process, provided the operation $v \mapsto w = [I - \nabla_\eta T_\eta(\eta_i)]v$ is available. The coefficients α_j are unknowns to be determined.

Minimization of (2.7) can be achieved by applying the GMRES algorithm [14] to the linear system

$$[I - \nabla_\eta T_\eta(\eta_i)]d\eta = -[\eta_i - T_\eta(\eta_i)],$$

starting with the initial solution $d\eta^{(0)} = 0$. Notice that solving this equation exactly will yield the Newton direction $[I - \nabla_\eta T_\eta(\eta_i)]^{-1}[\eta_i - T_\eta(\eta_i)]$, so this procedure is nothing but an inexact Newton method.

As discussed in [9], the accuracy to which Newton's equations are solved is adjusted adaptively. For every iteration in Newton's method, the maximum number m of steps in the GMRES algorithm is varied according to the level of nonlinearity as determined from the residual in Newton's equations. We start out with the number of GMRES iterations $m = 2$. We double the size of the subspace up to a maximum of 25 whenever we find that the residual for the linearized equations is within a factor of 1.5 of the nonlinear residual. The size of the subspace is kept unchanged if the nonlinear residual was in between 1.5 and 5 times the linear residual. Otherwise the size of the subspace is halved. The algorithm is summarized below.

- Form an initial guess η_0 for the electron density.
- **repeat**
  - Employ m steps of GMRES to solve the system

$$[I - \nabla_\eta T_\eta(\eta_i)]d\eta = -[\eta_i - T_\eta(\eta_i)]$$

    for Newton's direction dn without generating the Jacobian $I - \nabla_\eta T_\eta(\eta_i)$.
  - **if** $2/3 \leq \|res_{nl}\|/\|res_{lin}\| \leq 3/2$
  - **then** $m := m * 2$
  - **else if** $3/2 \leq \|res_{nl}\|/\|res_{lin}\| \leq 5$
  - **then** $m := m$
  - **else** $m := m/2$.
  - Perform an inexact linesearch for the stepsize τ along the direction dn to guarantee a decrease of the nonlinear residual $(\eta_i + \tau d\eta) - T_\eta(\eta_i + \tau d\eta)$.
- **until** convergence.

The ultimate rate of convergence per outer iteration will be quadratic for this algorithm because it is a form of Newton's method.

Because the Fermi-Dirac distribution function $F_{-\frac{1}{2}}$ decreases rapidly if the eigenvalue E_l increases, only the lowest energy eigenfunctions of Schrödinger's equation are needed. The eigenvalue problem for Schrödinger's equations is solved by a version of the Chebyshev accelerated subspace iteration subroutine RITZIT of Rutishauser [13] which was accelerated by a factor of two. Details of the implementation are provided in [7].

3. Saw-tooth quantum wire structure model. In the computations presented in [11], the quantum mechanical wave function $\psi(\vec{x})$ is set equal to 0 at the $Al_xGa_{1-x}As/GaAs$ interface. This approach corresponds to an infinitely deep potential well. For a well of finite depth, the wavefunction $\psi(\vec{x})$ may extend significantly outside the well. Therefore, the computational domain is enlarged to allow the wavefunctions to decay appropriately. Furthermore, a shifted periodic boundary condition avoids imposing the artificial boundary condition $\psi(\vec{x}) = 0$ inside the potential trough.

In Fig. 1, layers 1, 2, 4 and 5 consist of $Al_xGa_{1-x}As$, while layer 3 consists of $GaAs$. The impurity doping in layers 1 and 5 are donor type with concentrations N_{D1} and N_{D5}, respectively, and acceptor type in layers 2, 3 and 4 with concentrations N_{A2}, N_{A3} and N_{A4}, respectively. Also, the molar concentration, x, of Al in $Al_xGa_{1-x}As$ has the same value throughout this device and determines the depth of the resulting quantum well as follows. The dependence of the heterojunction discontinuity in the conduction band ΔE_C of the $Al_xGa_{1-x}As/GaAs$ system, ΔE_C, on the concentration x is discussed in [8].

This periodic semiconductor device is simulated on the computational domain Ω presented in Fig. 2. The domain is reduced to one period by using a shifted periodic grid and periodic boundary conditions. The grid continues from the upper-right-half boundary to the lower-left-half boundary. The equations were discretized on a nonuniform rectangular grid of size 95×48. The computations were implemented on a CRAY Y-MP4/464.

4. The Fermi level E_F and the depletion width. We assume thermodynamic equilibrium throughout the device and charge neutrality in the bulk at the boundaries of the computational domain Ω. These two combined assumptions determine the difference between the conduction band E_C and the Fermi level E_F in the bulk. Because the device is designed for low temperatures, "freeze-out" of part of the donors has to be taken into account.

The classical electron density in the bulk is modeled by the Boltzmann distribution as in Eq. (1.5). In writing down Poisson's equation for the electrostatic potential, it has to be taken into account that in the presence of electric fields, the entire donor density will eventually ionize. The depletion width depends slightly on the quantum mechanical electron density n_{qm}, but predominantly on the doping profile.

To prevent flooding of the quantum well, the classical electron density n_{cl} has to be depleted in the vicinity of the quantum wire. Therefore, the conduction band energy $E_C(\vec{x})$ in the immediate vicinity of the quantum wire has to be larger than $E_C(\vec{x})$ in the bulk. Even more so because the depth ΔE_C of the quantum well has to be much larger than $E_C - E_F$ in the bulk to obtain sufficient confinement.

In [8] it is shown that for a band discontinuity $\Delta E_C \leq 55meV$ confinement is lost. This loss of confinement reflects both the limited curvature of ψ_1 outside the well for E_1 close to 0, and significant occupancy of the second, unbound state because $E_2 - E_1$ decreases. With the difference $E_C - E_F \approx 0.182meV$ at a donor doping of

$6 * 10^{17} cm^{-3}$ or $6 * 10^{16} cm^{-3}$ in the bulk, it is evident that at the location of the quantum well the conduction band level E_C has to be shifted with respect to E_F.

The required shift in the conduction band energy $E_C(\vec{x})$ is brought about by embedding the heterojunctions, which define the potential well, between depleted PN junctions. The PN junctions are created by adding, in Fig. 2, the layers 2 and 4, which consist of $Al_x Ga_{1-x} As$ doped with acceptor impurities with concentrations N_{A2} and N_{A4}, respectively. The addition of layers 2 and 4 allows us to adjust the energy levels E_l in the well with respect to E_F, and to prevent the eigenstates from flooding with electrons.

The correct positioning of the energy levels E_l with respect to the Fermi-level E_F is made possible by the self-consistency of the computations. As a result, it becomes possible to determine the wavefunction occupation levels, N_l, and the quantum mechanical electron density, n_{qm}.

4.1. Calculation of $E_C - E_F$ in the bulk. At liquid Helium temperatures the ionization of donors is only partial in the absence of electric fields. Therefore, the "freeze-out" of ionized donors, and the boundary value for the conduction band energy E_C, and the electrostatic potential ϕ, are computed, as usual, from the assumption of charge neutrality,

$$(4.9) \qquad \rho = e[N_D^+ - N_A^- + p - n] = 0.$$

and thermodynamic equilibrium (as in [15].) The latter assumption implies that
- the ionized donor density is

$$(4.10) \qquad N_D^+ = \frac{N_D}{1 + g_D \exp[-(E_D - E_F)/k_B T]},$$

- and the electron density is

$$(4.11) \qquad n = \frac{4}{\sqrt{\pi}} \Big(\frac{m_{de} k_B T}{2\pi\hbar^2}\Big)^{3/2} F_{\frac{1}{2}}\Big(-\frac{E_C - E_F}{k_B T}\Big).$$

In these equations N_D is the concentration of donor impurities, E_D is the energy level of ionized donor impurities, g_D is the ground state degeneracy factor for donor atoms, and m_{de} is the density-of-state effective mass for electrons. Furthermore, since two electrons of opposite spin may occupy the same energy level in the conduction band, $g_D = 2$. The expressions for electron density, n, Eq. (4.11), and hole density, p, are then determined through the Fermi-Dirac integral of order $\frac{1}{2}$ with the location of the ionized donor impurities at $E_C - E_D = 7.0\ meV$. In this application, the ionized acceptor density N_A^-, and the hole density p are negligible. However, they are given by corresponding expressions and included in the routines which compute the amount of "freeze-out."

4.2. The depletion width determined from the classical electron density. Although in the bulk, at thermodynamic equilibrium, only a fraction of the donor atoms is ionized; in the presence of electric fields, full ionization results. Therefore, the classical electron density Eq. (1.5) is normalized to be equal to the fully ionized doping density at the boundary of the computational domain in the bulk. The inclusion of classical electrons causes Poisson's equation for the electrostatic potential to become nonlinear. It is phrased properly as

$$(4.12) \quad -\nabla \cdot [\epsilon \nabla \phi] = e\{-n_{qm} + (N_D - N_A) - (N_{D,bndy} - N_{A,bndy})\exp[e(\frac{\phi - \phi_{bndy}}{k_B T})]\}.$$

At the outer boundary, where $\phi = \phi_{bndy}$, the exponential term in n_{cl} equals unity and charge neutrality prevails.

The nonlinear Poisson equation Eq. (4.12) for the potential ϕ_n is solved by a globally convergent damped Newton method including inexact line searches [12]. The potential ϕ_n at outer iteration n of the algorithm presented in [7] is employed as initial guess $\phi_{n+1,0}$ for Newton's method at the next iteration.

4.3. The positioning of E_1, E_2, and E_F. The Fermi-Dirac distribution function,

$$(4.13) \qquad F(E_l) = \frac{1}{1 + \exp[(E_l - E_F)/k_B T]},$$

decays from 1 to 0 over an energy range of a few $k_B T$ around the Fermi level E_F. Proper design of a quantum wire commands that n_{qm} be dominated by only the eigenpair (E_1, ψ_1). Therefore, the energy levels in the quantum well have to be positioned such that $E_1 < E_F < E_2$.

The occupancy of the ground state E_1 will be 100 times as large as that of the next state E_2 if

$$(4.14) \qquad (E_F - E_1) = (E_2 - E_F) = 4.6\, k_B T.$$

Hence, it is necessary that

$$(4.15) \qquad (E_2 - E_1) \geq 9.2\, k_B T$$

to create a separation of occupancy levels such that $N_1 \geq 100 N_2$.

5. Energy band bending. The static charge in the depleted PN junctions does not result in a uniform shift of the potential well with repect to the bulk $GaAs$. Symmetric doping on both sides of the saw-tooth structure results in upward energy band bending in the neighborhood of the corners of the saw-tooth well due to a local excess of negative charge. This energy band bending reduces the depth of the potential well at the corners and, therefore, may result in the loss of states localized to these corners. However, this detrimental effect of the negative doping, which is employed to force depletion of classical electrons at the potential well, can be reduced, and even reversed, by employing a different doping profile on the two sides of the jagged well.

By choosing in the saw-tooth structure illustrated in Fig. 1, a doping profile in which $N_{D1} \neq N_{D5}$ and $N_{A2} \neq N_{A4}$, the conduction band energy level may be forced to curve downward sufficiently to confine the bound electron states to a single quantum wire in half of the corners. By creating an asymmetric impurity doping profile, the separation between the lowest energy states may be increased (or decreased) by increasing (decreasing) the ratio $(N_{A2} : N_{A4})$ and/or decreasing (increasing) the ratio $(N_{D1} : N_{D5})$. The model with the asymmetric doping profile now provides sufficient

TABLE 2

Saw-tooth device doping parameters used to generate Table 3, Figures 4-7.

Semiconductor		Doping	
Layer	Material	Type	Concentration (cm^{-3})
1	$Al_xGa_{1-x}As$	N_D	6.0×10^{17}
2	$Al_xGa_{1-x}As$	N_A	8.0×10^{16}
3	$GaAs$	N_A	1.0×10^{14}
4	$Al_xGa_{1-x}As$	N_A	8.0×10^{17}
5	$Al_xGa_{1-x}As$	N_D	6.0×10^{16}

TABLE 3

Energy levels and corresponding occupancy levels obtained from a self-consistent device simulation by using the physical parameters listed in Table 2 with $x = 0.1125$, and Table 1.

State l	$E_l - E_F$ (meV)	N_l (μm^{-1})
1	-0.71551	22.09980
2	3.43813	0.00114
3	3.62361	0.00068
4	3.85803	0.00036
5	4.43989	0.00007

separation, in energy, between the lowest energy level and the next highest energy level in addition to appropriate placement of E_l with respect to E_F. The depth of the potential well, ΔE_C, and the doping parameters, N_{D1}, N_{D5}, N_{A2}, and N_{A4}, require fine tuning in order to obtain the desired energy band bending and energy level placement necessary to create a quantum wire.

The asymmetric doping profile allows the confinement of quantum wires in one half of the corners of the saw-tooth structure in Fig. 1. Numerical results in Figures 4–7 illustrate a quantum mechanically confined electron density in a saw-tooth nanostructure. The potential well is $84.4\,meV$ deep. The input parameters listed in Tables 2 with $x = 0.1125$, and 1 are used in the simulation to calculate the energy levels, E_l, of the eigenstates and their corresponding occupancy levels, N_l, which are listed in Table 3. Furthermore, the numerical results for ψ_1, n_{qm}, $-e\phi$, and n_{cl} are plotted in Figs. 4 - 7, respectively. Notice that the depletion width w in Fig. 7 of the classical electron density n_{cl}, does not extend to the boundary of the computational domain Ω. Hence, the assumption of charge neutrality at the boundary of the computational domain Ω is not violated.

6. Conclusion. A more detailed mathematical model of the confinement of electrons to quantum wires in corners of periodic heterojunction structures as proposed in [11] has been formulated. The additional detail shows: (i) In a well of realistic depth there is significant spreading of wavefunctions outside the well. This spreading causes the spacing of the energy levels of the bound states to decrease. (ii) The difference between the energy levels of the bound states and the conduction band in the neigh-

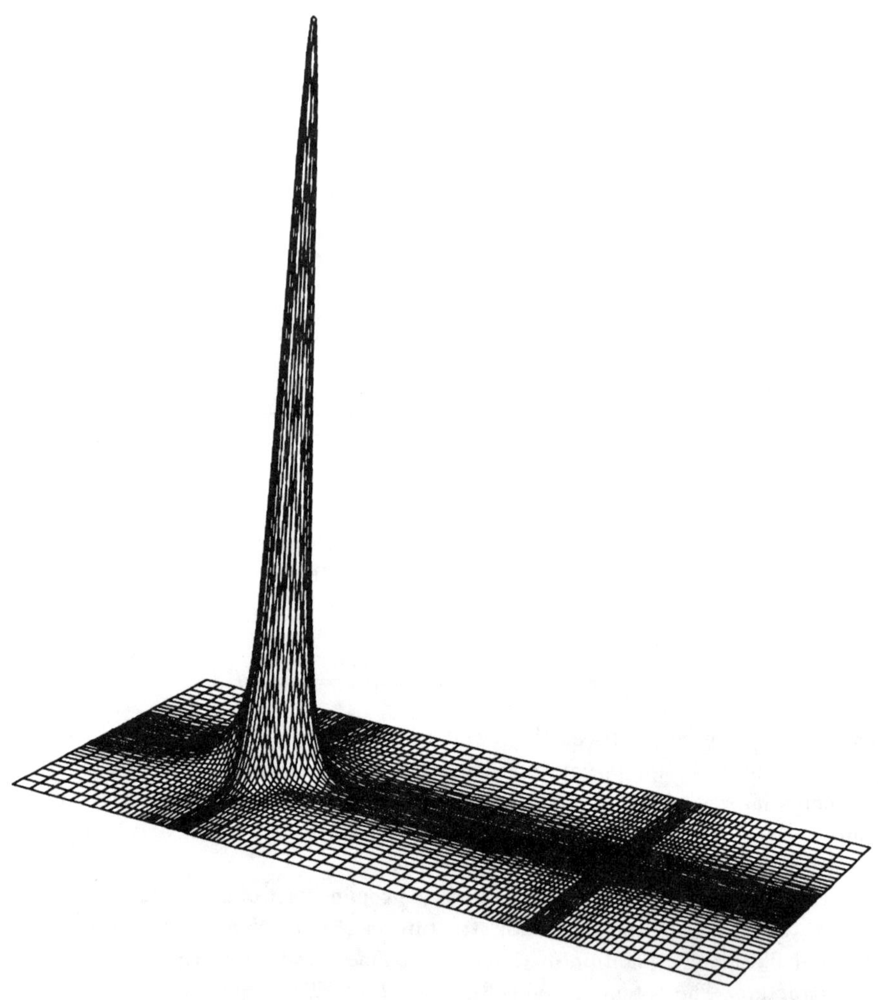

FIG. 4. *Wavefunction, ψ_1, for energy level E_1 listed in Table 3.*

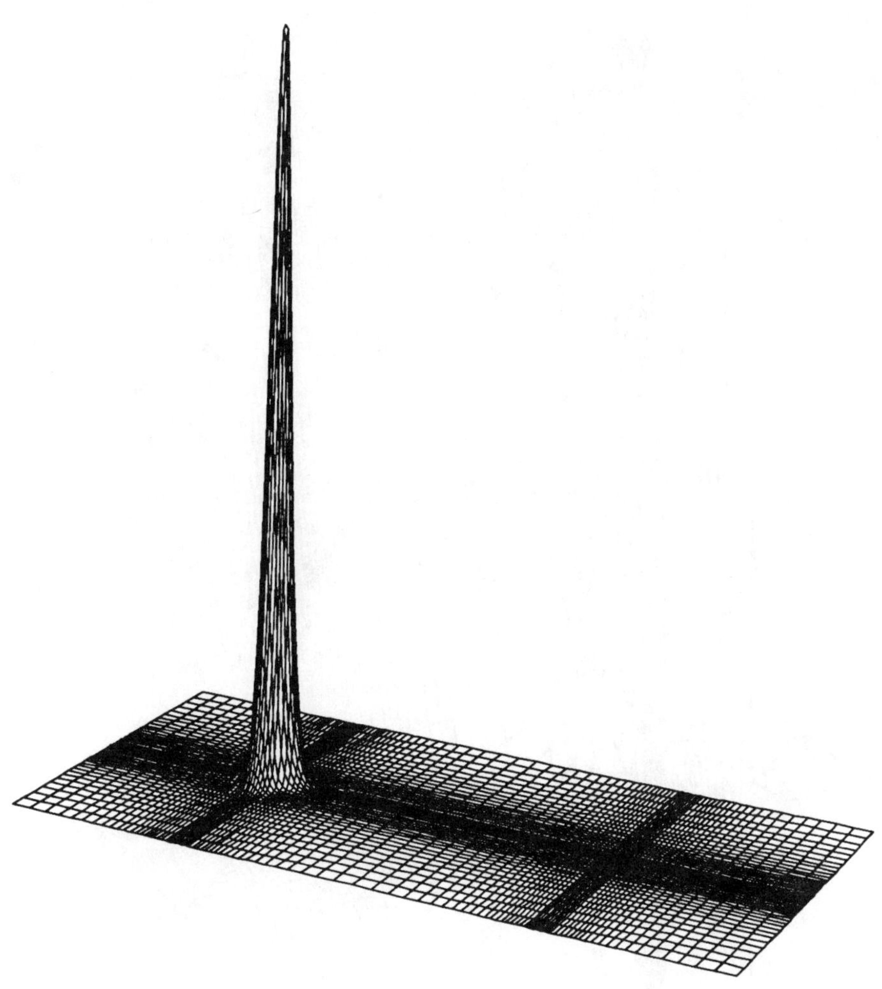

FIG. 5. *Quantum mechanical electron density, n_{qm}, for energy levels listed in Table 3.*

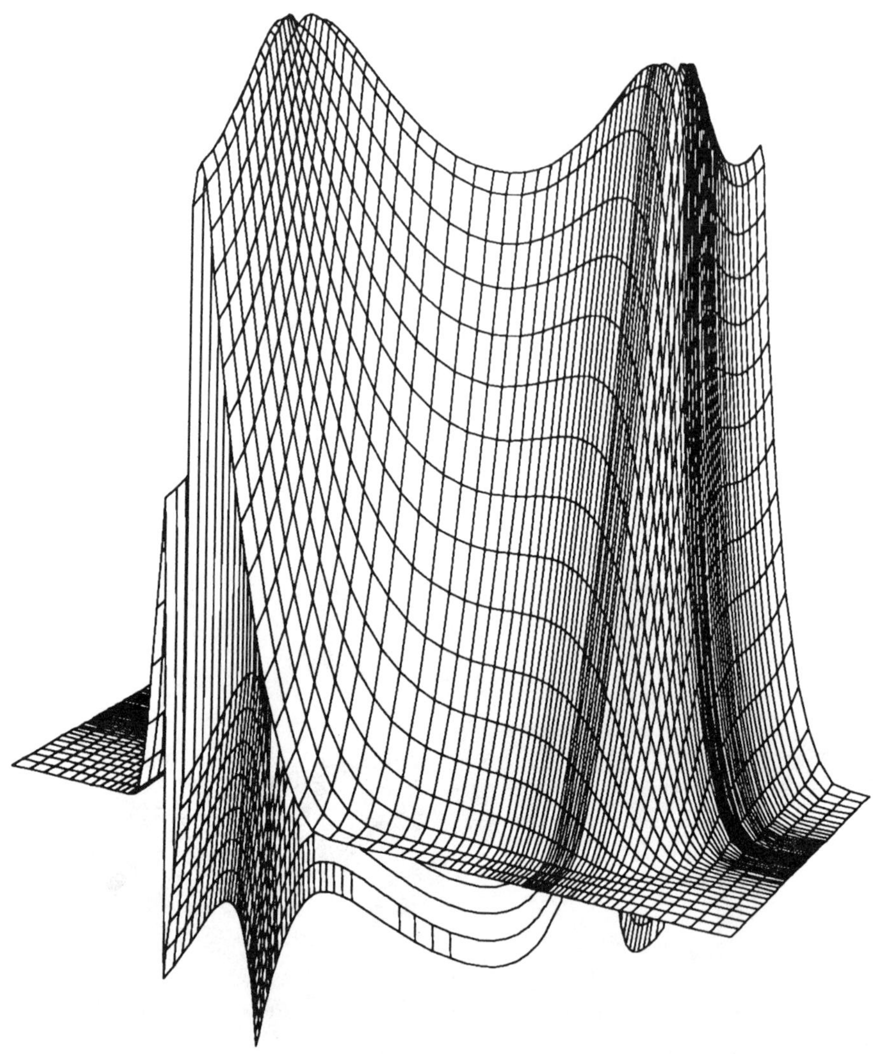

FIG. 6. *Potential energy, $-e\phi$, for energy levels listed in Table 3.*

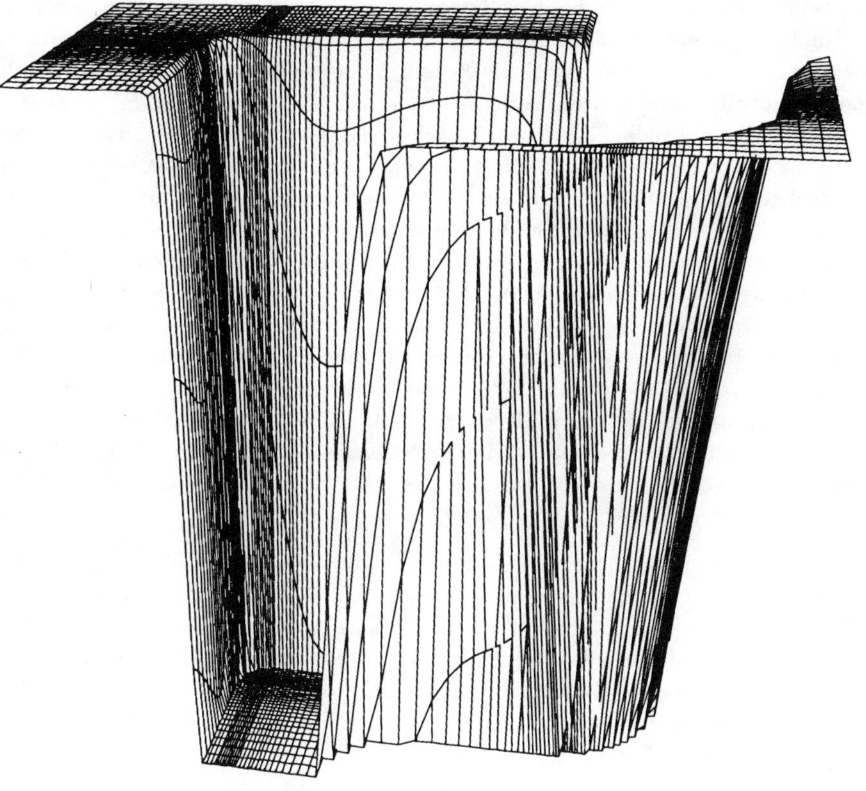

FIG. 7. *Classical electron density n_{cl}, corresponding to Table 3.*

borhood of the well has to be much larger than the difference in energy between the Fermi level E_F and the conduction band in the bulk. However, for proper formation of a quantum wire, the energy levels E_1 and E_2 of the lowest two states have to be positioned with respect to the Fermi level E_F such that $E_1 < E_F < E_2$. Therefore, conduction band bending is required to adjust the energy levels in the quantum well with respect to the Fermi level. (iii) The energy band bending can be brought about by depleted PN junctions on both sides of the periodic sawtooth structure. However, depleted PN junctions formed by layered alternately doped layers of semiconductor materials cause excess negative charge near the corners of the sawtooth structure. As a result the potential energy curves locally upwards and confinement may be lost. (iv) Downward bending of the potential energy can be brought about for one half of the corners of the sawtooth structure by choosing the doping densities on the two sides of the sawtooth structure asymmetrically and different.

The location of the energy levels E_i in the corners of the sawtooth structure depends sensitively on the extent of the depletion around the PN junctions. Accurate determination of this depletion width requires that classical electrons are included in the mathematical model.

7. Acknowledgements. The author gratefully acknowledges the assistance in the generation of the quoted numerical results of: A.T. Galick, M.W. Raschke, U. Ravaioli, and A. Sameh. Support by the National Science Foundation grant EET-8719100 is gratefully acknowledged. The computations on the CRAY Y-MP4/464 of the National Center for Supercomputing Applications (NCSA) were made possible through a block grant of the National Center for Computational Electronics (NCCE) of the University of Illinois.

REFERENCES

[1] Peter Brown and Alan C. Hindmarsh. Matrix-free methods for stiff systems of ODEs. *SIAM J. Numer. Anal.*, 23:610–638, 1986.

[2] I.L. Chern and W.L. Miranker. Dichotomy and conjugate gradients in the stiff initial value problem. Technical Report 8032-34917, IBM, Yorktown Heights, 1980.

[3] T.F. Coleman, B.S. Garbow, and J.J. More. Software for estimating sparse jacobian matrices. Technical Report ANL-MCS-TM-14, Argonne National Laboratory, 1983.

[4] J.E. Dennis, Jr. and Robert B. Schnabel. *Numerical Methods for Unconstrained Optimization and Nonlinear Equations.* Prentice-Hall, Inc., Englewood Cliffs, 1983.

[5] C.W. Gear and Youcef Saad. Iterative solution of linear equations in ODE codes. *SIAM J. Scient. and Statist. Comp.*, 4:583–601, 1983.

[6] Philip E. Gill, Walter Murray, and Margaret H. Wright. *Practical Optimization.* Academic Press, New York, San Fransisco, London, 1981.

[7] T. Kerkhoven, A. T. Galick, U. Ravaioli, J. H. Arends, and Y. Saad. Efficient Numerical Simulation Of Two Dimensional Quantum Wells. *J. Appl. Physics*, pages 3461–3469, October 1990.

[8] T. Kerkhoven, M. Raschke, and U. Ravaioli. Self-Consistent Simulation Of Quantum Wires In Periodic Heterostructure Structures. *to be published*, 1991.

[9] T. Kerkhoven and Y. Saad. On Acceleration Methods For Coupled Nonlinear Elliptic Systems. *Numerische Mathematik*, to appear.

[10] Thomas Kerkhoven. On the Effectiveness of Gummel's Method. *SIAM J. on Scient. & Statist. Comput.*, 9:48–60, January 1988.

[11] Keisuke Kojima, Kazumasa Mitsunaga, and Kazuo Kyuma. Calculation of Two-Dimensional Quantum-Confined Structures Using The Finite Element Method. *Appl. Phys. Lett.*, 55(9):882–884, August 1989.

[12] J.M. Ortega and W.C. Rheinboldt. *Iterative Solution of Nonlinear Equations in Several Variables*. Academic Press, New York, San Fransisco, London, 1970.

[13] H. Rutishauser. Simultaneous Iteration Method for Symmetric Matrices. In *Handbook for Automatic Computation, Vol. II, Linear Algebra*, pages 284–302. Springer Verlag, New York, 1971.

[14] Youcef Saad and Martin H. Schultz. GMRES: A Generalized Minimal Residual Method for solving Nonsymmetric Linear Systems. *SIAM J. Scient. and Statist. Comput.*, 7:856–869, 1986.

[15] S.M. Sze. *Physics of Semiconductor Devices*. Wiley-Interscience, New York, second edition, 1981.

[16] L.B. Wigton, D.P. Yu, and N.L. Young. GMRES acceleration of computational fluid dynamics codes. In *Proc. 7th AIAA conference*, pages 67–74, 370 L'Enfant Promenade,SW Washinton DC 20024, 1985. American Institute of Aeronautics and Astronautics.

NUMERICAL SIMULATION OF MOS TRANSISTORS

ERASMUS LANGER*

Abstract. This contribution is intended to review the international state-of-the-art in numerical simulation of MOS devices. Much emphasis is laid on the discussion of recent refinements to carrier transport models, e.g. drift-diffusion model, enhanced drift-diffusion equations, hydrodynamic model, and Monte Carlo simulation. Adequate models for the physical parameters are reported with suitable parameter values, e.g. carrier mobilities taking into account the various scattering mechanisms, and carrier generation-recombination including impact ionization. Examples are presented for two different types of MOS devices: on the one hand, simulation results of miniaturized MOS transistors are discussed which have been obtained by our simulator MINIMOS 5.0 with additional extensions, and on the other hand, simulation results concerning a power MOS transistor are shown which has been investigated with our device simulation program BAMBI 2.1.

Key words. simulation, MOS transistor, semiconductor equations

1. Introduction. Since 1960 when the demonstration of the practically usable MOS transistor took place [42] its development has shown to evolve dramatically. Today, about thirty years later, integrated circuits with millions of devices per single chip are manufactured. In order to minimize the number of cycles of trial and error in device fabrication improved understanding of basic device operation has attained crucial importance. Thus, numerical modeling of MOS transistors has become a basic requirement for the development of prototype devices although the first modeling attempts were made relatively late.

A two-dimensional solution of the Poisson equation with application to a MOS structure probably was first published by Loeb et al. [50] and Schroeder and Muller [68] in 1968. Since then a lot of work has been contributed dealing with simulation of MOS devices due to their intrinsically two-dimensional nature, e.g.: [90], [52], [38], [58], [87], [73] during the seventies, [14], [77], [88], [46], [69], [92], [55], [22] during the eighties.

Two-dimensional transient simulations of MOSFET's have been carried out in, e.g.: [53], [57], [94], [89], [78], [32], [24], [43]. Three-dimensional static modeling has been published in, e.g.: [7], [11], [39], [80], [49], [13], [86], [76].

The main efforts during the recent years were not focused only on transient simulations and three-dimensional modeling but also on the improvement of the models of the physical parameters, e.g.: [37], [56], [91], [81], on the refinement of the mathematical model, e.g.: [67], [31], [45], on the analysis of the operation of MOSFET's at low temperature, e.g.: [82], [93], [47], [60], [79], [74], and on investigations of special physical effects, e.g.: [30], [62], [61], [28], [75], [34].

Although most activities in MOS transistor simulation are concentrated on the analysis of ultra small devices, there also exist many publications concerning modeling of power MOSFET's, e.g.: [85], [22], [19].

---

*Institute for Microelectronics, Technical University Vienna, A-1040 Wien, AUSTRIA.

2. Transport model. The state-of-the-art in semiconductor technology enables the fabrication of semiconductor devices with characteristic lengths smaller than $1\mu m$ in an industrial environment. Such devices are characterized by very large electric fields and rapid spatial variations of the electric field and carrier concentrations. These variations occur over a distance comparable with the characteristic lengths of carrier transport, i.e. the average momentum relaxation length and the energy relaxation length. Owing to the very large electric fields the assumption of small perturbations around equilibrium breaks down which the drift-diffusion transport model is based on. It was realized long ago that a straightforward extension of the classical semiconductor equations would have to include the energy balance of field-driven carriers [15]. One way for this purpose is an enhanced drift-diffusion model which accounts for the energy balance. The obvious benefit of this approach is that it can be implemented without great efforts in an existing simulation program based on the basic semiconductor equations. Another practicable way to include hot carrier effects is the usage of the hydrodynamic model [65] which implies the energy conservation equation. Last not least, there is to discuss the Monte Carlo method whose most important drawback is – as usual – the enormous amount of required computing resources.

2.1. Drift-diffusion equations. Device Modeling based on the self-consistent solution of fundamental semiconductor equations dates back to the famous work of Gummel in 1964 [25]. Up to now, most of the worldwide used simulation tools for MOS transistors are solving the well known fundamental semiconductor equations consisting of Poisson's equation (1), continuity equations for electrons and holes (2), (3), and the current relations for both carrier types (4), (5).

$$\text{(1)} \qquad \text{div}(\varepsilon \cdot \text{grad}\,\psi) = -\rho$$

$$\text{(2)} \qquad \text{div}\,\vec{J_n} - q \cdot \frac{\partial n}{\partial t} = q \cdot R$$

$$\text{(3)} \qquad \text{div}\,\vec{J_p} + q \cdot \frac{\partial p}{\partial t} = -q \cdot R$$

$$\text{(4)} \qquad \vec{J_n} = q \cdot \mu_n \cdot n \cdot \left(\vec{E} + \frac{1}{n} \cdot \text{grad}\left(n \cdot \frac{k_B \cdot T_n}{q} \right) \right)$$

$$\text{(5)} \qquad \vec{J_p} = q \cdot \mu_p \cdot p \cdot \left(\vec{E} - \frac{1}{p} \cdot \text{grad}\left(p \cdot \frac{k_B \cdot T_p}{q} \right) \right)$$

It should be noted that the current relations in this notation already represent an extension of the classical drift-diffusion equations as there exist different temperatures T_n and T_p within the diffusion terms. The reason for this will be explained in the next section.

As it can be seen from the different activities mentioned above there are ongoing arguments in the scientific community whether these equations are adequate to describe transport in submicron devices. Particularly the current relations (4) and (5) which are the most complex equations out of the set of the basic semiconductor device equations undergo strong criticism in view of, for instance, ballistic transport [64], [35]. Their derivation from more fundamental physical principles is indeed not at all straightforward. They appear therefore with all sorts of slight variations in the specialized literature and a vast number of papers has been published where some of their subtleties are dealt with. Anyway, investigations on ultra short MOSFET's [66] do not give evidence that it is necessary to waive these well established basic equations for silicon devices down to feature sizes in the order of 0.1 microns [79].

2.2. Enhanced drift-diffusion equations. In contrast of the classical drift-diffusion model the solution of the enhanced drift-diffusion equations accounts for hot-carrier transport in semiconductors. An adequate model for already existing simulation programs has been implemented first in MINIMOS [31], our well established tool for the two and three-dimensional analysis of miniaturized MOS transistors.

This model is based on the work of Hänsch and Miura-Mattausch [29] which follows very closely the work of Blotekjaer [6] with the distinction that in the first place a different Ansatz for the distribution function was chosen, and secondly, and more importantly, the relaxation time approximation was not used. This provides a closed set of equations for the electron density n, the electron current density $\vec{J}_n$, energy density and energy current density. From this set of equations a self-consistent mobility and energy relaxation time have been derived rigorously [27]. This formulation covers particle as well as energy balance in a closed form. Unfortunately it would require the development of a new code to be utilized as a simulating tool. Therefore, Hänsch and Miura-Mattausch presented an approximate solution of these equations that could feasibly be used in conventional drift-diffusion codes. Their rational was based on the observation that, strictly speaking, the classical current equation is rigorously valid in the limit of small fields. High-field effects enter through the saturation of the drift velocity and are accounted for using a heuristic field–dependent mobility model, which is a local field-dependent function. Following these classical ideas, Hänsch and Miura-Mattausch introduced a local field-dependent model of the mobilities μ_n and μ_p and the thermal voltages U_{T_n} and U_{T_p} to be used in the modified current relationships (4) and (5).

To model μ and U_T, however, a local solution of the general equations was performed. To this end all spatial derivatives were dropped so that the complicated system of differential equations turned into a simple algebraic equation for the energy density and in turn the mobility. This rather drastic step is justified by the observation that, in conventional simulation, mobility models are inferred from drift velocity measurements in constant electric field. The inhomogeneous situation is then accounted for by an appropriate driving force [71], [72]. Assuming infinite bulk material in the first place and replacing the electric field by the driving force F, the electron mobility reads

$$(6) \qquad \mu_n^{\mathrm{LISF}} = \frac{2\mu_n^{\mathrm{LIS}}}{1 + \sqrt{1 + \left(\dfrac{2 \cdot \mu_n^{\mathrm{LIS}} \cdot F_n}{v_n^{\mathrm{sat}}}\right)^2}}$$

and for the thermal voltage

$$(7) \qquad U_{T_n} = \frac{k_B \cdot T_n}{q} = U_{T_0} + \frac{2}{3} \cdot \tau_n^{\varepsilon} \cdot \left(v_n^{\mathrm{sat}}\right)^2 \cdot \left(\frac{1}{\mu_n^{\mathrm{LISF}}} - \frac{1}{\mu_n^{\mathrm{LIS}}}\right)$$

is obtained. Here, μ_n^{LIS} is the low-field mobility containing lattice, impurity, and surface scattering, τ_n^ε is the energy relaxation time, and v_n^{sat} is the saturation velocity.

Concerning the driving force F, there is some controversy about which one is the most appropriate force in the literature. The problem is that originally the local field-dependent expressions are given with $F = E$. Generalization to the inhomogeneous two-dimensional case is by no means straightforward. There are, however, two limiting cases the driving force has to obey: On the one hand it should give $F = E$ for the homogeneous situation, and on the other hand it has to give a velocity saturation for high-density gradients as well. In this situation carriers do not gain energy from the field, provided it is small, and therefore their mean velocity cannot exceed the thermal velocity, which is not necessarily the same as the high-field saturation velocity v^{sat}. Any physically motivated driving force has to fulfill these requirements. Under the present circumstances a saturation with respect to the gradient of the thermal voltage has to be included too. One appropriate driving force proposed firstly in [31] is

$$(8) \qquad F_n = \left| \operatorname{grad}\psi - \frac{1}{n} \cdot \operatorname{grad}(n \cdot U_{T_n}) \right|.$$

2.3. Hydrodynamic equations. The derivation of the hydrodynamic model describing the carrier transport in semiconductor devices is based on the Boltzmann transport equation

$$(9) \qquad \frac{\partial f}{\partial t} + \vec{u} \cdot \operatorname{grad}_x f \pm \frac{q}{m} \cdot \vec{E} \cdot \operatorname{grad}_u f = C$$

where $f = f(x_j, u_j, t)$ is the distribution function, x_j, $u_j(j = 1, 2, 3)$ are the components of the vector position $\vec{x}$ and group velocity $\vec{u}$, q is the electron charge, m denotes the effective mass of the carriers under investigation, $\vec{E}$ is the electric field, and C denotes the collision integral. In semiconductors it is necessary to write such an equation for each branch of the energy-band function, using the negative sign for electrons and the positive sign for holes. In the above written formulation the constant mass approximation is assumed, and the scalar mass m is a proper average of terms of the mass tensor relative to each extremum of the considered energy-band function. Equation (9) is a partial differential equation in the seven-dimensional phase space $(\vec{x}, \vec{u}, t)$. In order to reduce this equation to a set of equations only in the space and time coordinates usually the so called method of moments is used as described in [65].

The derivation of the hydrodynamic model implies some loss of information on the behavior of the distribution function f. The main reason for this is the absolutely necessary consideration of only a finite number of moments of the Boltzmann transport equation. Additional details can be obtained by introducing more moments, at the expense of an increasing complication of the equations. In order to obtain an equation system comparable to the drift-diffusion model, the following simplifying assumptions have to be taken: the temperature tensor is reduced to a scalar, and the so called relaxation–time approximation is used for the description of the collision terms of the intervalley and intravalley transition type.

After introducing the set of variables commonly used in semiconductor device theory one obtains a current relation

$$(10) \quad \vec{J}_n - \frac{\tau_{pn}}{q}(\vec{J}_n \cdot \mathrm{grad})\frac{\vec{J}_n}{n} = q \cdot \mu_n \left[\frac{k_B \cdot T_n}{q} \mathrm{grad}(n) + n \cdot \mathrm{grad}\left(\frac{k_B \cdot T_n}{q} - \psi \right) \right]$$

and an energy conservation equation

$$(11) \quad -\mathrm{div}\left(\kappa_n \cdot \mathrm{grad}(T_n) + \vec{J}_n \cdot \frac{k_B \cdot T_n}{q} \right) = \vec{J}_n \cdot \mathrm{grad}\left(\frac{w}{q} - \psi \right) - n \cdot \frac{w - w_0}{\tau_{wn}}$$

where τ_{pn} and τ_{wn} are the momentum and energy relaxation times for electrons in the conduction band, respectively, and w_0 is the equilibrium mean energy. These two equations in connection with the continuity equation (2) whose formulation has not changed build the basic equations of the hydrodynamic transport model (for electrons). They have to be solved self-consistently with the corresponding equations for holes and with the Poisson equation.

Regarding the increased expense, the hydrodynamic model has an evident drawback compared to the drift-diffusion model: the hydrodynamic model has two further partial differential equations (the energy conservation equation for both carrier types) which have to be solved self-consistently with the other three (i.e. Poisson equation and both continuity equations). Furthermore, the formulation of the current relations is more complex than in the drift-diffusion model.

But there is no doubt about the fact that in spite of the simplifying assumptions the hydrodynamic model is more rigorous than the classical drift-diffusion model, especially concerning hot carrier transport. Let us compare the two different formulations of the current relations (4) and (10). The current relation of the classical drift-diffusion model, i.e. equation (4) with $T_n = T_p = T_L$, where T_L is the lattice temperature, can be obtained from (10) by neglecting $(\tau_{pn}/q)(\vec{J}_n \cdot \mathrm{grad})(\vec{J}_n/n)$, which is called convective term, and the gradient of $k_B T_n/q$, which is a thermoelectric field. Furthermore, in the classical drift-diffusion model the diffusion coefficient is defined by the Einstein relation

$$(12) \qquad D_{n,p} = \frac{k_B \cdot T_L}{q} \cdot \mu_{n,p}$$

whereas the analogous term in the hydrodynamic model reads

$$(13) \qquad D_{n,p} = \frac{k_B \cdot T_{n,p}}{q} \cdot \mu_{n,p}.$$

If we compare now the current relations of the hydrodynamic model (10) and the enhanced drift-diffusion model (4) we see that the thermoelectric field as well as the so called generalized Einstein relation (13) are included implicitly – the only difference is the neglect of the conductive term. As reported in [65] the influence of the convective term on the solution of the partial differential equation system is very small and is, therefore, commonly neglected even in the actual simulation codes using the hydrodynamic model. Thus, from our point of view, the increased expense of the hydrodynamic equations compared to the enhanced drift-diffusion model briefly described in the previous section is not generally justified.

2.4. Hybrid transport model. Since the very first beginning of Monte Carlo simulations in the field of carrier transport in semiconductor devices [54] the most important drawback up to now is the enormous amount of computing resources required by this method for the solution of the Boltzmann transport equation (9). The main benefits of this method are the facts that it is simple to implement, that sophisticated physical models can be used, and that any desired physical information can easily be extracted [40].

A very promising way is the coupling of the Monte Carlo method and the drift-diffusion equations [45] which is based on the following considerations: The Monte Carlo method allows more accurate physical models and is, therefore, well suited to describe the non-equilibrium transport occurring under conditions appearing in ultra small MOS transistors (i.e. very high electric field in the active region with rapid changes over distances comparable to the mean free path of the carriers). On the other hand, for description of low field transport the drift-diffusion model which uses local transport coefficients provides sufficient accuracy. Moreover, the drift-diffusion model has turned out to be even superior to the Monte Carlo method in regions with retarded fields. Therefore, several attempts were published to combine the drift-diffusion and Monte Carlo technique in order to benefit from the different capabilities of both methods [63], [83]. In the following, a recently performed implementation of the hybrid transport model in the existing simulation tool MINIMOS shall be discussed. It should be noted that up to now this extension has been implemented only in the two-dimensional part of the program package.

Three-dimensional scattering rates in the entire device are assumed, thus neglecting quantization effects eventually occurring in an inversion layer. Multiplication of the Boltzmann transport equation with wave vector component k_i and integration of $\vec{k}$ leads to a momentum balance equation, which reads for electrons

(14)
$$-\left(q \cdot E_i + \frac{1}{n} \sum_{j=1}^{2} \frac{\partial(n \cdot \langle \hbar \cdot k_i \cdot v_j \rangle)}{\partial x_j} \right) = \left\langle \int (\hbar \cdot k_i - \hbar \cdot k_i') \cdot S(\vec{k}, \vec{k}') \cdot d^3 \vec{k}' \right\rangle , i = 1, 2$$

where v denotes the group velocity. (It should be mentioned that in the following the index n for indicating the electrons is omitted as this carrier type is discussed in this section only.) The average operator $\langle A \rangle$ is the mean value of $A(\vec{k})$ weighted by the local distribution function $f(\vec{x}, \vec{k})$. The left hand side can be interpreted as driving force that acts on the electron ensemble, consisting of the electric field plus diffusion term, whereas the right hand side describes the rate of momentum loss due to scatterings. This equation can be expressed in a form similar to the drift-diffusion current relation. The parameters needed in this current relation are derived in the following way.

For band structures with spheric and ellipsoidal energy surfaces the vector valued momentum loss integral is colinear with the momentum $\hbar \vec{k}$

(15)
$$\int (\hbar \cdot \vec{k} - \hbar \cdot \vec{k}') \cdot S(\vec{k}, \vec{k}') \cdot d^3 \vec{k}' = \hbar \vec{k} \cdot \lambda_m(E).$$

Here the proportionality factor $\lambda_m(E)$ is the momentum scattering rate. With the local average velocity the local momentum loss mobility can be defined as

$$\mu = q \cdot \frac{\|\langle \vec{v} \rangle\|}{\|\langle \hbar \cdot \vec{k} \cdot \lambda_m(E) \rangle\|} \tag{16}$$

where $\|B\|$ denotes the maximum norm of B. This definition does not rely on the relaxation time approximation and, since no effective mass occurs in this formula, extension to general bands is straightforward. In the latter case μ would have tensor property. The definition of the thermal voltage tensor (which is proportional to the temperature tensor: $U_{ij} = (k_B/q)T_{ij}$) results directly from the momentum conservation equation (14)

$$U_{ij} = \frac{1}{q} \cdot \langle \hbar \cdot k_i \cdot v_j \rangle. \tag{17}$$

This definition is independent of the underlying band structure model. Inserting these definitions in (14) one obtains a general current relation

$$J_i = q \cdot n \cdot \mu \cdot \left(E_i + \frac{1}{n} \cdot \sum_{j=1}^{2} \frac{\partial(n \cdot U_{ij})}{\partial x_j} \right). \tag{18}$$

The differences between this current relation and the classical one are twofold. Firstly, the diffusion term is due to the tensor property of the thermal voltage more complicated. Secondly, the parameters μ and U_{ij} can no longer be treated simply as parameters depending on electric field or other local quantities, as it is usually done in the conventional drift-diffusion model, because they carry information of the local distribution function. By means of the Monte Carlo method these parameters are evaluated. The conventional simulator using the Monte Carlo parameters $\mu(\vec{x})$ and $U_{ij}(\vec{x})$ in the current relation (18) is then capable of recovering the Monte Carlo results for $n(\vec{x})$ and $\vec{J}_n(\vec{x})$. In this way hot electron effects, such as velocity overshoot and hot carrier diffusion, are consistently included in the conventional simulator. The solution is performed globally in the whole device, but only in the high field region mobility and temperature profiles have to be extracted from the Monte Carlo procedure. In regions with low fields and low spatial inhomogeneities local models can be used thus saving computation time.

The continuity equation in conjunction with a drift-diffusion current relation employing a scalar temperature yields an elliptic partial differential equation which has diagonal form. If an anisotropic temperature is taken into account cross derivatives appear in the elliptic operator. Conventional device simulators solve elliptic systems which are in diagonal form, therefore a scalar temperature is desirable. Neglecting the off diagonal elements in (17) a scalar temperature is used which is the arithmetic mean value of the main diagonal elements.

For the first conduction band of silicon we use a model consisting of six anisotropic valleys with a first order correction for nonparabolicity [41]. Acoustic intravalley scattering in the elastic approximation, intervalley phonon scattering, surface roughness scattering, and coulomb scattering are taken into account. Except of the latter one all mechanisms are isotropic. In the case of surface scattering in the inversion layer the wave vector is redistributed randomly in a plane parallel to the Si-SiO$_2$ interface. For isotropic scattering mechanisms the momentum scattering rate does not differ from the total scattering rate. The following superposition implies independence of all scattering processes

$$(19) \qquad \lambda_m(E) = \lambda_{ac}^{tot} + \lambda_{opt}^{tot} + \lambda_{surf}^{tot} + \lambda_{ion}^{m} .$$

Coulomb interaction is the only one to be treated separately. The momentum loss integral (15) is evaluated by the Brooks-Herring formulation for the transition rate $S(\vec{k}, \vec{k}')$ and

$$(20) \qquad \lambda_{ion}^{m}(E) = \lambda_{ion}^{tot} - \lambda'_{ion}(E)$$

is obtained. The subtraction corresponds to the difference of initial and final wave vector in (15).

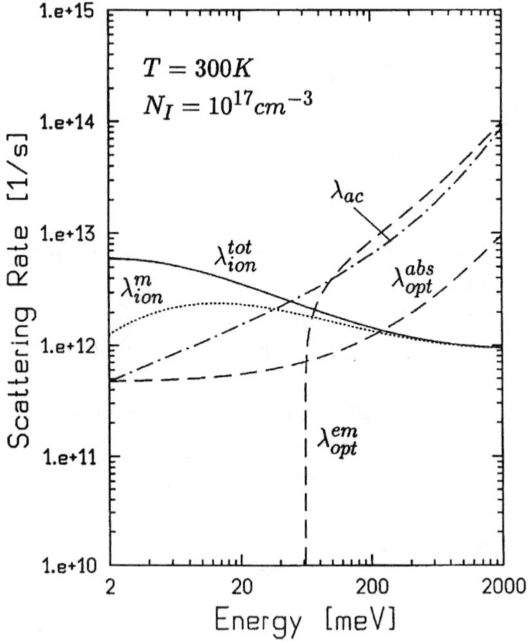

Figure 1: Total and momentum scattering rates used for mobility calculation.

Figure 1 depicts the energy dependencies of total (λ_{ion}^{tot}) and momentum (λ_{ion}^{m}) scattering rate for ionized impurities, and additionally the total scattering rates for acoustic intravalley phonons (λ_{ac}) and one representative intervalley phonon mode (emission and absorption).

3. Models for parameters. The fundamental semiconductor equations include a set of parameters which have to be appropriately modeled in order to describe the various transport phenomena qualitatively and quantitatively correctly. The basic requirement for an estimation of most of these physical parameters is independent more or less of the underlying transport model as the Poisson equation (1) and both continuity equations (2), (3) have to be solved self-consistently in any case.

3.1. Modeling space charge. The Poisson equation (1) requires a model for the space charge ρ which makes use of only the dependent variables ψ, n, p and material properties. The well established approach for this model is to sum up the concentrations with the adequate charge sign multiplied with the elementary charge

$$(21) \qquad \rho = q \cdot (p - n + N_D^+ - N_A^-).$$

The doping concentration is assumed usually to be fully ionized at room temperature. In order to include low temperature phenomena it seems to be necessary to include partial ionization which can be described by

$$(22) \qquad N_D^+ = \frac{N_D}{1 + 2 \cdot \exp\left(\frac{E_{fn} - E_D}{k_B \cdot T}\right)}, \qquad N_A^- = \frac{N_A}{1 + 4 \cdot \exp\left(\frac{E_A - E_{fp}}{k_B \cdot T}\right)}$$

in the classical way. E_D and E_A are the ionization energies of the respective donor and acceptor dopant. Typical values for $E_c - E_D$ and $E_A - E_v$ for the most common dopants in silicon are: 0.054eV for arsenic, 0.045eV for phosphorus, 0.039eV for antimony, and 0.045eV for boron. A quite complete list of values can be found in [84]. Note, that only energy differences can be given (E_c and E_v are the conduction band and the valence band, respectively). The Fermi levels E_{fn} and E_{fp} have to be appropriately related to the dependent variables by making use of Fermi statistics

$$(23) \qquad n = N_c \cdot \frac{2}{\sqrt{\pi}} \cdot F_{1/2}\left(\frac{E_{fn} - E_c}{k_B \cdot T}\right), \qquad p = N_v \cdot \frac{2}{\sqrt{\pi}} \cdot F_{1/2}\left(\frac{E_v - E_{fp}}{k_B \cdot T}\right).$$

N_c and N_v are the density of states in the conduction band and the valence band, respectively. The classical formulae for the density of states are given by

$$(24) \qquad N_c = 2 \cdot \left(\frac{2 \cdot \pi \cdot k_B \cdot T \cdot m_n^*}{h^2}\right)^{3/2}, \qquad N_v = 2 \cdot \left(\frac{2 \cdot \pi \cdot k_B \cdot T \cdot m_p^*}{h^2}\right)^{3/2}.$$

$F_{1/2}$ is the Fermi function of order 1/2 which is defined by

$$(25) \qquad F_{1/2}(x) = \int_0^\infty \frac{\sqrt{y}}{1 + e^{y-x}} \cdot dy.$$

The parameters m_n^* and m_p^* are the effective masses for electrons and holes, respectively. It is worthwhile to note that the ratio of the density of states depends only on the ratio of the effective masses

$$(26) \qquad \frac{N_c}{N_v} = \left(\frac{m_n^*}{m_p^*}\right)^{3/2}.$$

By means of simple algebraic manipulation with the expressions for the carrier concentrations (23) one obtains

$$(27) \qquad \frac{E_{fn} - E_D}{k_B \cdot T} = G_{1/2}\left(\frac{n}{N_c}\right) + \frac{E_c - E_D}{k_B \cdot T}$$

$$(28) \qquad \frac{E_A - E_{fp}}{k_B \cdot T} = G_{1/2}\left(\frac{p}{N_v}\right) + \frac{E_A - E_v}{k_B \cdot T}.$$

$G_{1/2}(x)$ is the inverse Fermi function of order $1/2$ defined by

$$(29) \qquad G_{1/2}\left(\frac{2}{\sqrt{\pi}} \cdot F_{1/2}(x)\right) = x.$$

It is now possible by evaluating the expressions for the density of states (24) with appropriate fits to the effective masses [23] to compute numerical values for the ionized impurity concentrations (22) using only the carrier concentrations as the dependent variables in the basic equations. However, comparisons to experiment indicate that it is better to compute the density of states from relation (26) and a fit to the intrinsic carrier concentration

$$(30) \qquad n_i = \sqrt{N_c \cdot N_v} \cdot \exp\left(-\frac{E_g}{2 \cdot k_B \cdot T}\right)$$

where E_g is the band gap $E_c - E_v$ which can be modeled temperature dependent with the fit provided by Gaensslen et al. [23]. The prefactor in (30) can be fitted to experiments by

$$(31) \qquad \sqrt{N_c \cdot N_v} = \exp\left\{45.13 + 0.75 \cdot \ln\left[\frac{m_n^*}{m_0} \cdot \frac{m_p^*}{m_0} \cdot \left(\frac{T}{300\mathrm{K}}\right)^2\right]\right\} \ \mathrm{cm}^{-3}.$$

With (26) and (31) it is now straightforward to compute the numerical values for the density of states. At room temperature we obtain $N_c = 5.1 \cdot 10^{19}$ cm^{-3} and $N_v = 2.9 \cdot 10^{19}$ cm^{-3}.

It should be noted that equations (22) are valid only for moderate impurity concentrations. For heavy doping the assumption of a localized ionization energy does definitely not hold. Instead an impurity band is formed which may merge with the respective band edge, e.g. [36]. It appears to be appropriate to assume total ionization for concentrations above some threshold value and to account for a suitable functional transition between the classical formulae (22) and total ionization.

3.2. Modeling carrier mobilities. The next set of physical parameters to be considered carefully consists of the carrier mobilities μ_n and μ_p in the current relations (4) and (5) or (10) regarding the hydrodynamic model. The models for the carrier mobilities have to take into account a great variety of scattering mechanisms the most basic one of which is lattice scattering. The lattice mobility in pure silicon can be fitted with simple power laws

$$(32) \qquad \mu_n^{\mathrm{L}} = 1430\frac{\mathrm{cm}^2}{\mathrm{Vs}} \cdot \left(\frac{T}{300\mathrm{K}}\right)^{-2}, \qquad \mu_p^{\mathrm{L}} = 460\frac{\mathrm{cm}^2}{\mathrm{Vs}} \cdot \left(\frac{T}{300\mathrm{K}}\right)^{-2.18}.$$

The expressions (32) fit well experimental data of [1], [8], and [48].

The next effect to be considered is ionized impurity scattering. The best established procedure for this task is to take the functional form (33) of the fit provided by Caughey and Thomas [10] and use temperature dependent coefficients.

$$(33) \qquad \mu_{n,p}^{\mathrm{LI}} = \mu_{n,p}^{\min} + \frac{\mu_{n,p}^{\mathrm{L}} - \mu_{n,p}^{\min}}{1 + \left(\frac{CI}{C_{n,p}^{\mathrm{ref}}}\right)^{\alpha_{n,p}}}$$

$$(34) \qquad \mu_n^{\min} = 80\frac{\mathrm{cm}^2}{\mathrm{Vs}} \cdot \left(\frac{T}{300\mathrm{K}}\right)^{-0.45}, \qquad \mu_p^{\min} = 45\frac{\mathrm{cm}^2}{\mathrm{Vs}} \cdot \left(\frac{T}{300\mathrm{K}}\right)^{-0.45}$$

$$(35) \qquad C_n^{\mathrm{ref}} = 1.12 \cdot 10^{17} \ \mathrm{cm}^{-3} \cdot \left(\frac{T}{300\mathrm{K}}\right)^{3.2}$$

$$C_p^{\mathrm{ref}} = 2.23 \cdot 10^{17} \ \mathrm{cm}^{-3} \cdot \left(\frac{T}{300\mathrm{K}}\right)^{3.2}$$

$$(36) \qquad \alpha_{n,p} = 0.72 \cdot \left(\frac{T}{300\mathrm{K}}\right)^{0.065}$$

The fits (34)-(36) are from [33]. Similar data have been provided in [3] and [20].

In view of partial ionization one should consider neutral impurity scattering. However, in view of the uncertainty of the quantitative values for ionized impurity scattering it seems not to be worthwhile to introduce another scattering mechanism with additional fitting parameters.

Particular emphasis has to be put on surface scattering which is modeled with an expression suggested by Seavey [70]

$$(37) \qquad \mu_{n,p}^{\mathrm{LIS}} = \frac{\mu_{n,p}^{\mathrm{ref}} + (\mu_{n,p}^{\mathrm{LI}} - \mu_{n,p}^{\mathrm{ref}}) \cdot (1 - F(y))}{1 + F(y) \cdot \left(\frac{S_{n,p}}{S_{n,p}^{\mathrm{ref}}}\right)^{\gamma_{n,p}}}$$

with

$$\mu_n^{\mathrm{ref}} = 638\frac{\mathrm{cm}^2}{\mathrm{Vs}} \cdot \left(\frac{T}{300\mathrm{K}}\right)^{-1.19}, \qquad \mu_p^{\mathrm{ref}} = 160\frac{\mathrm{cm}^2}{\mathrm{Vs}} \cdot \left(\frac{T}{300\mathrm{K}}\right)^{-1.09}$$

$$F(y) = \frac{2 \cdot \exp\left(-\left(\frac{y}{y^{\mathrm{ref}}}\right)^2\right)}{1 + \exp\left(-2 \cdot \left(\frac{y}{y^{\mathrm{ref}}}\right)^2\right)}$$

$$S_n = \max\left(0, \frac{\partial\psi}{\partial y}\right), \qquad S_p = \max\left(0, -\frac{\partial\psi}{\partial y}\right).$$

S_n^{ref} is assumed to be $7 \cdot 10^5 \frac{\mathrm{V}}{\mathrm{cm}}$, S_p^{ref} is $2.7 \cdot 10^5 \frac{\mathrm{V}}{\mathrm{cm}}$, γ_n is 1.69, γ_p is 1.0, and y^{ref} is 10nm. The formulae for surface scattering are definitely not the ultimate expressions. They just fit quite reasonably experimental observations.

Velocity saturation can be modeled with formulae (38). The expression for the electron mobility comes directly from the derivation of the enhanced drift-diffusion model (6). For the hole mobility the same functional form has been assumed and has been fitted to experimental data.

$$(38) \qquad \mu_n^{\text{LISF}} = \frac{2\mu_n^{\text{LIS}}}{1 + \sqrt{1 + \left(\frac{2 \cdot \mu_n^{\text{LIS}} \cdot F_n}{v_n^{\text{sat}}}\right)^2}}, \qquad \mu_p^{\text{LISF}} = \frac{\mu_p^{\text{LIS}}}{1 + \frac{\mu_p^{\text{LIS}} \cdot F_p}{v_p^{\text{sat}}}}$$

F_n and F_p are the effective driving forces which read according to (8)

$$(39) \qquad F_n = \left| \text{grad}\,\psi - \frac{1}{n} \cdot \text{grad}(n \cdot U_{T_n}) \right|, \qquad F_p = \left| \text{grad}\,\psi + \frac{1}{p} \cdot \text{grad}(p \cdot U_{T_p}) \right|.$$

The used saturation velocities are given by

$$(40) \qquad \begin{aligned} v_n^{\text{sat}} &= 1.45 \cdot 10^7 \frac{\text{cm}}{\text{s}} \cdot \sqrt{\tanh\left(\frac{155\text{K}}{T}\right)} \\ v_p^{\text{sat}} &= 9.05 \cdot 10^6 \frac{\text{cm}}{\text{s}} \cdot \sqrt{\tanh\left(\frac{312\text{K}}{T}\right)}. \end{aligned}$$

The functional form of these fits is after [1], and the experimental data matched are [1], [9], and [8]. An eventual dependence on the crystallographic orientation which one would deduce from [2], [44] is presently not taken into account.

3.3. Modeling carrier temperatures. To describe carrier heating properly one has to account for local carrier temperatures $T_{n,p}$ in the current relations. As already discussed in the previous chapters this can be achieved by either solving the hydrodynamic equations, or by using a model obtained by series expansions of the solution to the energy conservation equations, e.g. by using the enhanced drift-diffusion model. As already discussed we believe that the latter is sufficient for silicon devices. Applying equation (7) also for holes the enhanced drift-diffusion model yields for the electronic voltages

$$(41) \qquad U_{T_{n,p}} = \frac{k_B \cdot T_{n,p}}{q} = U_{T_0} + \frac{2}{3} \cdot \tau_{n,p}^{\varepsilon} \cdot (v_{n,p}^{\text{sat}})^2 \cdot \left(\frac{1}{\mu_{n,p}^{\text{LISF}}} - \frac{1}{\mu_{n,p}^{\text{LIS}}}\right).$$

The energy relaxation times $\tau_{n,p}^{\varepsilon}$ are in the order of 0.5 picoseconds. They should be modeled as functions of the local doping concentration as motivated by the following reasoning: The product of carrier mobility times electronic voltage which symbolizes a diffusion coefficient must be a decreasing function with increasing carrier voltage [5]. Its maximum is attained at thermal equilibrium. Therefore, the relation

$$(42) \qquad \mu_{n,p}^{\text{LISF}} \cdot U_{T_{n,p}} \leq \mu_{n,p}^{\text{LIS}} \cdot U_{T_0}$$

must hold. Substituting (41) into (42) and rearranging terms one obtains for the energy relaxation times

$$(43) \qquad \tau_{n,p}^{\varepsilon} \leq \frac{3}{2} \cdot U_{T_0} \cdot \frac{\mu_{n,p}^{\text{LIS}}}{(v_{n,p}^{\text{sat}})^2}.$$

In MINIMOS the energy relaxation times are modeled on this basis with a fudge factor γ in the range $[0, 1]$ and a default value of 0.8

$$(44) \qquad \tau_{n,p}^{\varepsilon} = \gamma \cdot \frac{3}{2} \cdot U_{T_0} \cdot \frac{\mu_{n,p}^{\mathrm{LIS}}}{(v_{n,p}^{\mathrm{sat}})^2} \cdot$$

For vanishing doping one obtains the maximum energy relaxation times which are at room temperature $\tau_n^{\varepsilon} = 4.44 \cdot 10^{-13}$s, $\tau_p^{\varepsilon} = 2.24 \cdot 10^{-13}$s.

3.4. Modeling carrier generation/recombination. The right hand sides of the continuity equations are represented by the carrier generation/recombination term which consists of various mechanisms.

An adequate model for thermal generation/recombination is given by the well known Shockley-Read-Hall term [71]

$$(45) \qquad R^{\mathrm{SRH}} = \frac{n \cdot p - n_i^2}{\tau_p \cdot (n + n_1) + \tau_n \cdot (p + p_1)}$$

with

$$(46) \qquad n_1 = n_0 \cdot \frac{1 - f_{t0}}{f_{t0}}, \qquad p_1 = p_0 \cdot \frac{f_{t0}}{1 - f_{t0}}$$

and the doping dependent carrier life times

$$(47) \qquad \tau_n = \frac{\tau_{n0}}{1 + \frac{N_D + N_A}{N_n^{\mathrm{ref}}}}, \qquad \tau_p = \frac{\tau_{p0}}{1 + \frac{N_D + N_A}{N_p^{\mathrm{ref}}}} \cdot$$

f_{t0} in equation (46) denotes the fraction of occupied traps at equilibrium, n_0 and p_0 are the equilibrium carrier densities. The doping dependent empirical model for the carrier life times (47) accounts for additional generation/recombination centers at high doping. The parameters are chosen to fit experimental findings, e.g. $\tau_{n0} = 3.94 \cdot 10^{-4}$s, $N_n^{\mathrm{ref}} = 7.1 \cdot 10^{15}$ cm^{-3}, $\tau_{p0} = 3.94 \cdot 10^{-5}$s, $N_p^{\mathrm{ref}} = 7.1 \cdot 10^{15}$ cm^{-3} [18].

Auger recombination is modeled by

$$(48) \qquad R^{\mathrm{AU}} = (C_{cn} \cdot n + C_{cp} \cdot p) \cdot (n \cdot p - n_i^2)$$

with appropriate temperature dependent Auger coefficients [21]

$$(49) \qquad \begin{aligned} C_{cn} &= 2.8 \cdot 10^{-31} \frac{\mathrm{cm}^6}{\mathrm{s}} \cdot \left(\frac{T}{300\mathrm{K}}\right)^{0.14} \\ C_{cp} &= 9.9 \cdot 10^{-32} \frac{\mathrm{cm}^6}{\mathrm{s}} \cdot \left(\frac{T}{300\mathrm{K}}\right)^{0.2} \end{aligned}$$

As model for impact ionization the old Chynoweth formulation (50) can be used. Although the actuality of this model [12] has been discussed heavily in the scientific community, it still seems that it fulfills the requirements for device simulation quite satisfactorily – at least not less than other models published in literature.

$$(50) \qquad R^{II} = -\alpha_n \cdot \frac{|\vec{J_n}|}{q} - \alpha_p \cdot \frac{|\vec{J_p}|}{q}$$

$$(51) \qquad \alpha_{n,p} = \alpha_{n,p}^{\infty} \cdot \exp\left(-\frac{\beta_{n,p}}{E}\right)$$

The coefficients of (51) can be modeled temperature dependent by (52) and (53) to fit experimental data [16], [17], [59].

$$(52) \qquad \begin{aligned} \alpha_n^{\infty} &= 7 \cdot 10^5 \ \text{cm}^{-1} \cdot \left(0.57 + 0.43 \cdot \left(\frac{T}{300\text{K}}\right)^2\right) \\[2mm] \alpha_p^{\infty} &= 1.58 \cdot 10^6 \ \text{cm}^{-1} \cdot \left(0.58 + 0.42 \cdot \left(\frac{T}{300\text{K}}\right)^2\right) \end{aligned}$$

$$(53) \qquad \begin{aligned} \beta_n &= 1.23 \cdot 10^6 \frac{\text{V}}{\text{cm}} \cdot \left(0.625 + 0.375 \cdot \left(\frac{T}{300\text{K}}\right)\right) \\[2mm] \beta_p &= 2.04 \cdot 10^6 \frac{\text{V}}{\text{cm}} \cdot \left(0.67 + 0.33 \cdot \left(\frac{T}{300\text{K}}\right)\right) \end{aligned}$$

A comparison to other models for impact ionization can be found in [4]. The scatter among the results of the different models indicates that more effort is necessary in this area to better understand impact ionization and to obtain a more rigorous description of this phenomenon.

4. Simulation results. This section deals with recently performed work concerning MOS transistor simulation within the Institute for Microelectronics at the Technical University of Vienna. Two of these topics are new and not yet released implementations in the program package MINIMOS [86], and a third work is occupied with the simulation of power MOSFET's with the more universal usable simulator BAMBI [22].

4.1. The hybrid model applied to a MOSFET. The coupling of Monte Carlo method and drift-diffusion model briefly described in section 2.4 has been implemented in MINIMOS in order to justify the efficiency and accuracy of that compound model with respect to ultra small MOS transistors [45]. An n-channel MOSFET with $L_{\text{gate}} = 0.25\mu$m, $t_{\text{ox}} = 5$ nm was simulated at room temperature using the combined technique. The device has a metallurgical channel length of $L_{\text{eff}} = 0.15\mu$m and exhibits a threshold voltage $U_t = 0.23$V.

For practical simulation of a MOSFET, electrons are injected in source, where they fully thermalize before entering the channel [67]. In the region of interest, usually near the drain, a sufficient large number of particles is supplied by a particle split algorithm, thus reducing the statistical uncertainty of the results. The bias conditions for the following results are $U_{\text{GS}} = U_{\text{DS}} = 2.5$V and $U_{\text{BS}} = 0$V.

The two-dimensional distribution of the main diagonal temperatures in the area near the drain are depicted in Figure 2. The lateral temperature T_{xx} has a maximum value at the surface, while the maximum of T_{yy} is shifted away from the surface. Degradation due to hot electron injection into the oxide can be more accurately modeled by using the spatial distribution of T_{yy} than by using the scalar temperature obtained from average energy. In the performed simulations the off-diagonal temperatures never exceeded 15% of the main-diagonal elements.

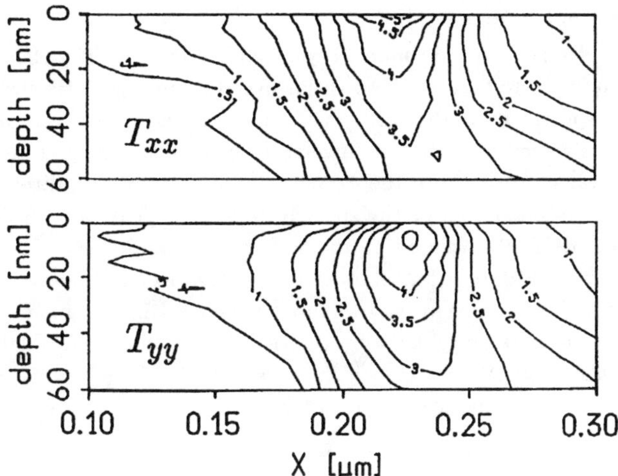

Figure 2: Lateral and transversal temperatures in a quarter micron MOSFET (units [1000K]).

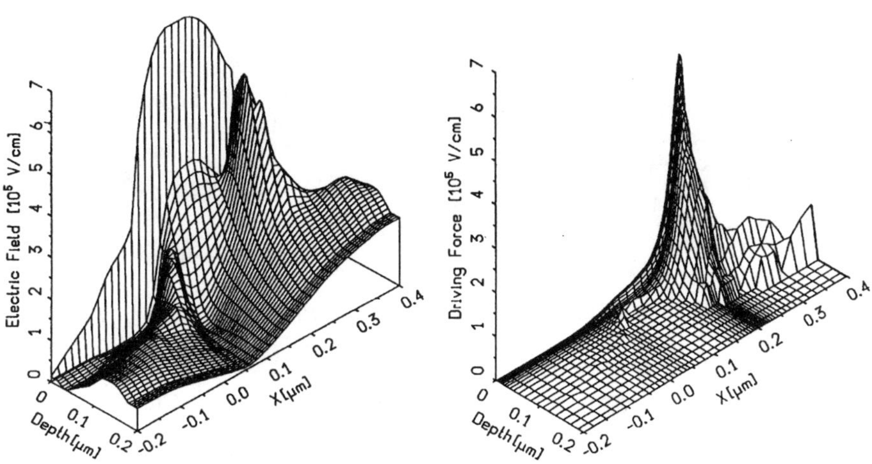

Figure 3: Electric field (left) and driving force (right) calculated by Monte Carlo in the same device.

Comparing the left and the right part of Figure 3 we see that the large normal field within the inversion layer does not appear in the driving force. This is obvious since the normal field is compensated by diffusion. The electric field in the source junction is compensated by diffusion as well, since in this area the driving force calculated by Monte Carlo vanishes. The field peak near the drain edge however appears almost unchanged in the driving force, thus accelerating and heating up the electron gas in this area.

Velocity of electrons in the channel is plotted in Figure 4. The effective channel extends from 0.05μm to 0.2μm, the positions of the junctions of source and drain subdiffusion, respectively. In the first half of this range the surface velocity (curve A) is lower than the velocities within the inversion layer since the electrons are pressed towards the surface. Near drain the pressing force has opposite direction and the electron velocity is maximum at the surface. The field peak near drain induces a velocity overshoot of 90% related to the bulk saturation velocity. A comparison of the two parts of Figure 3 has shown that in the overshoot region diffusion is not important. Therefore, velocity overshoot is treated in this model more like a drift phenomenon. It is reproduced by the drift-diffusion current relation (18) by incorporating the nonlocal mobility which will also be increased compared to a local mobility, which cannot produce velocities larger than the bulk saturation velocity.

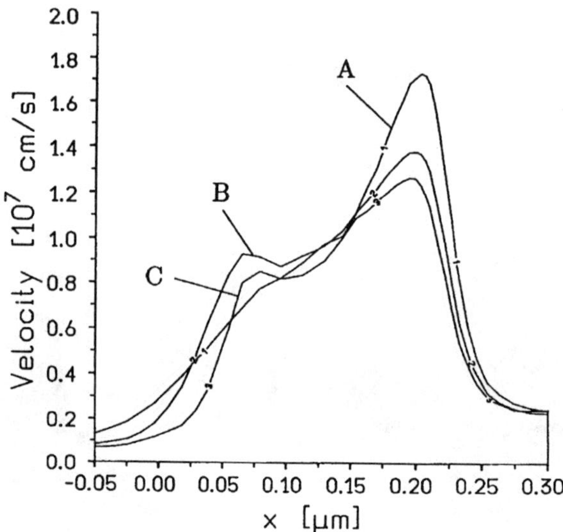

Figure 4: Electron average velocity along the channel. Curve A: at the Si-SiO$_2$ interface. Curves B and C: 5nm and 10nm away from interface.

4.2. Non-degenerate gate effect in MOSFET's. Implanted gate MOS devices have become common in submicron technologies. For that reason the simulator MINIMOS has been extended to solve self-consistently the basic semiconductor equations also in the poly-gate area (fully including non-planar devices) [26]. The Poisson equation is solved in the total simulation area (from y_t until y_B in Figure 5). For the continuity equations two approaches have been implemented. In the first one, both discretized continuity equations are solved in the poly-gate simultaneously with the bulk area (from y_G until y_B in Figure 5). This approach is interesting for the transient simulation, however a proper modeling of mobility and generation-recombination phenomena in polysilicon (e.g. grain boundary recombination) in necessary. The investigations presented here are restricted to steady-state conditions. Then the poly-gate is in thermodynamic equilibrium (net recombination vanishes and leakage currents are negligible). A unique and constant Fermi level exists in the poly-gate, which enables the carrier concentrations to be calculated analytically as a function of the local potential ψ. In such an approach the discretization error of the continuity equation (specially at the non-planar gate/oxide interface — see Figure 5) is avoided. This permits that band gap narrowing and Fermi-Dirac statistics can be implemented in a simpler way than in the first. For a detailed description of the model of heavy doping effects in the gate the interested reader is referred to [26].

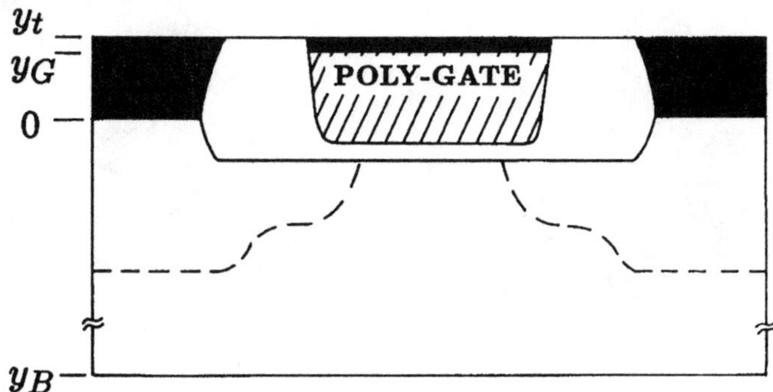

Figure 5: Simulation area of the polysilicon-gate device.

Investigations have been shown that there is a strong influence of the charge at the polysilicon/oxide interface on the field penetration into the gate, and therefore on the flat-band and threshold voltage. There is not much information about the nature of this charge in the literature. Since the polysilicon is deposited over oxide, it is believed that the gate/oxide interface is worse than the interface between thermally grown oxide and bulk silicon. In the actually performed simulations there is a fixed oxide and interface trapped charge (with both donor and acceptor nature) incorporated. The traps at the grain boundaries in the polysilicon presently have not been taken into account. If the doping is several times higher than the equivalent volume trap density in polysilicon N_{tvol} (surface trap density at grain boundary/grain size), the trapped charge is negligible compared to the space charge due to impurity ions.

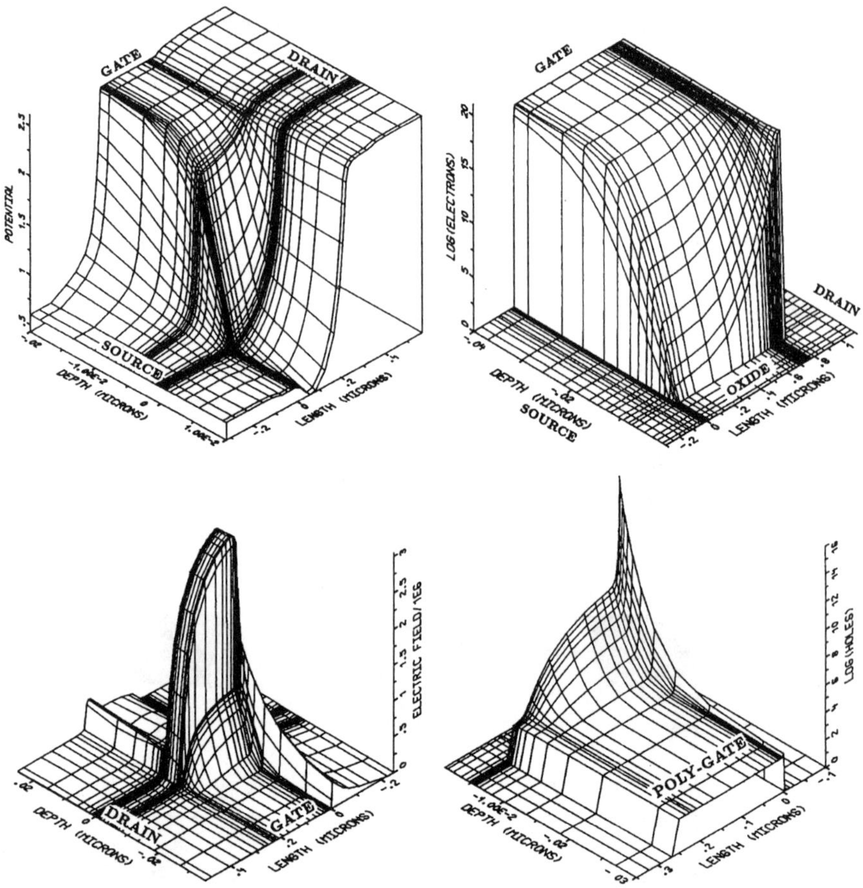

Figure 6: Potential, electric field, electron and hole distributions in an n-gate/n-channel device.

In order to investigate the impact of the poly-gate depletion on the characteristics of thin oxide submicron MOSFET's quarter micron planar devices have been analyzed. The devices have 5nm oxide thickness, threshold voltage ± 0.25V, and are designed for room temperature operation. Multiple implanted source-drain profiles (Figure 5) are reconstructed from the data in literature. For the p-gate device it is assumed that there is no boron penetration. Figure 6 shows the distribution of the potential, field and electron and hole concentrations in the gate of the n-channel/n-gate device with the specifications $t_{ox} = 5$nm, $l = 0.25\mu$m, $N_g = 10^{19}$ cm^{-3}, and the bias conditions $U_{DS} = U_{GS} = 2$V. Due to thin oxide, medium ionized impurity concentration at the gate/oxide interface (10^{19} cm^{-3}) and high gate bias (2V), a remarkable potential drop occurs in the gate as can be seen in the upper left (electric potential) and lower left drawing (electric field) of Figure 6. Note that the view of the upper pictures is another than the view of the lower drawings in Figure 6. These different representations were necessary in order to make the distributions more illustrative. The distributions of the carrier densities (upper right: electrons, lower right: holes) are plotted in logarithmic scale.

The gate-drive is reduced about 20% at the channel end near the source. Note that for this device the inversion in the polygate takes place (beginning at the source channel-end) at $U_{GS} \approx 3.7V$, leading to the recovery of the transconductance (experimental finding in [51]).

The threshold voltage and the potential drop in poly-gate at the threshold versus ionized impurity concentration near the gate/oxide interface N_g are shown for the p-gate/p-channel device in Figure 7. The charge at the gate/oxide interface Q_{go} which is the parameter of the curves in Figure 7 (here assumed as fixed) has a strong influence on the voltage drop in the poly-gate and therefore on the threshold voltage. Assuming Q_{go} to be a positive charge the voltage drop in the gate is increased for a p-gate/p-channel device, while in an n-gate/n-channel device a positive Q_{go} has a screening effect.

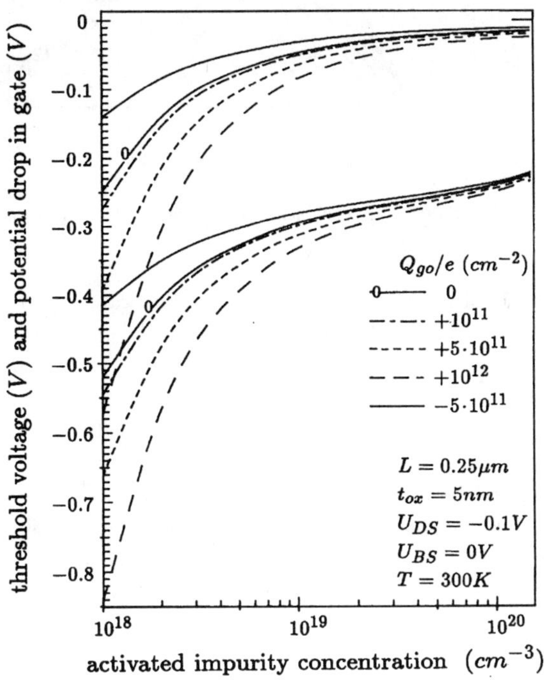

Figure 7: Threshold voltage of p-gate/p-channel device.

The fall-off of the drain current in the saturation region with N_g as parameter is shown in Figure 8. For common values Q_{go} has minor influence, and N_g is the main parameter in determination of the drain current degradation. In order to suppress totally the reduction of the gate drive, the activated impurity concentration near the gate/oxide interface must be at least $4 \cdot 10^{19}$ cm^{-3} for the analyzed 5nm oxide devices.

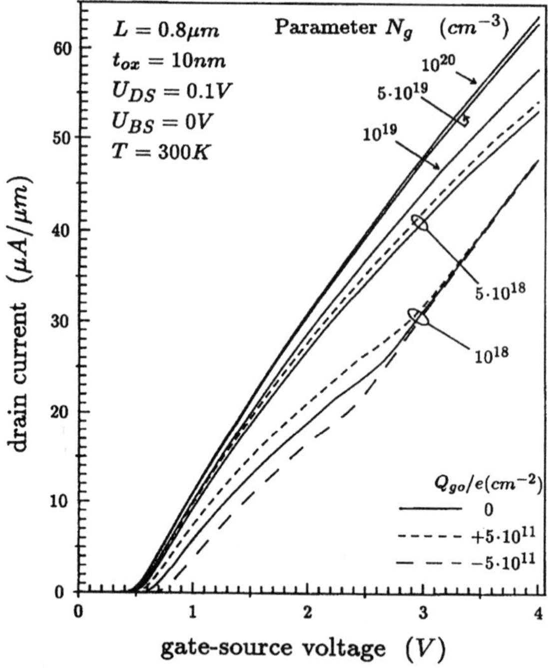

Figure 8: Transfer characteristics in the saturation of p-gate/p-channel device.

The relative ratio of the effective and the terminal gate-source voltage is given roughly by (assuming solely depletion in the gate)

$$(54) \qquad \frac{2}{1 + \sqrt{1 + \frac{2 \cdot \varepsilon_{\text{ox}}^2 \cdot U_{GS}}{1 \cdot \varepsilon_{pg} \cdot t_{\text{ox}}^2 \cdot N_g}}}$$

where ε_{ox} and ε_{pg} are the permittivities in oxide and polysilicon-gate, respectively. Applying different scaling rules on t_{ox} and U_{GS} the poly-gate effect becomes more or less severe by miniaturization. For instance, for the device of Figure 8 the reduction of the current at the gate and drain supply voltage of -2V is 20% at $N_g = 10^{19}$ cm^{-3}, while for a simulated 10nm-oxide (0.5μm) device the corresponding reduction was 12% at -5V.

4.3. Simulation of power MOS transistors. Power devices are characterized by large and often complex geometries and extreme bias conditions. Therefore, simulation tools applicable to the analysis of power devices have to fulfill partially other requirements than simulators for ultra small semiconductor devices. In this section, a typical example for a power MOS transistor simulation performed at our institute will be presented. The simulation has been carried out with the program package BAMBI, a "Basic Analyzer for MOS and Bipolar devices" [22].

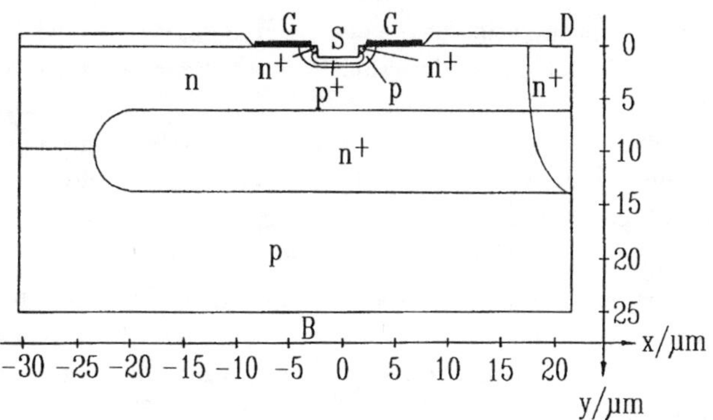

Figure 9: Simulation area of the investigated DMOS-structure.

Figure 9 shows the geometry of an investigated DMOS-structure. The dimensions of that geometry are identical to the used simulation area. Source, gate, drain, and bulk contacts are indicated by S, G, D, and B respectively. Note the very small highly n-doped source regions as well as the small step in the source-gate area of approximately 1μm (small compared to the dimensions of the whole device amounting 25μm $\times$ 55μm). The drain region starts in a depth of approximately 6μm and is connected via a highly n-doped implant to the drain contact at the surface. The bulk contact exists only in this "test transistor" for simplification while in a real device the substrate is contacted at the surface as well.

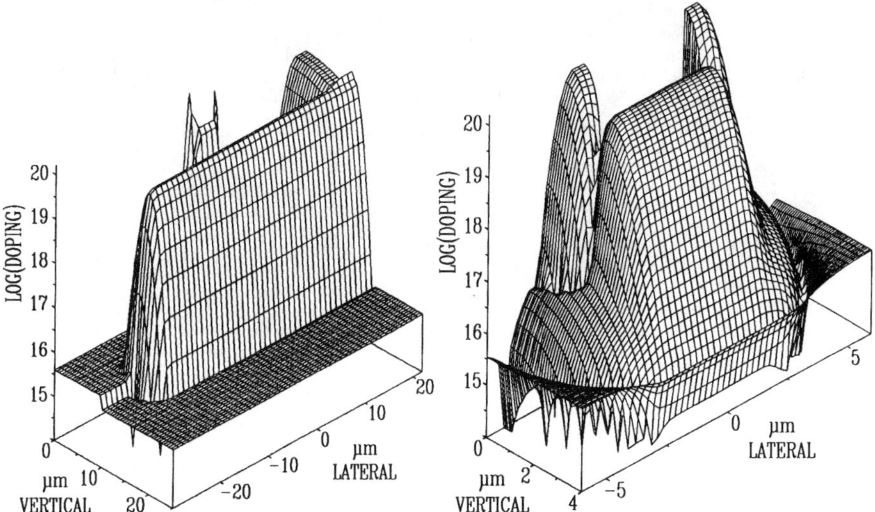

Figure 10: Net doping density in the whole simulation area (left) and in the channel region (right).

The net doping concentration of the investigated device is shown in Figure 10 in a logarithmic scale. The viewing direction for both drawings is given by the lower left edge of the geometry definition within Figure 9. In the left plot of Figure 10 which represents the doping concentration over the whole simulation area one can see the wide drain region in the depth which is connected to the surface. The step at the left vertical axis is coming from the different doping types (donors below the surface and acceptors in the depth as to be seen in Figure 9).

The right drawing of Figure 10 gives a detailed view of the doping profile in the channel and source region which is indicated in the left plot by two peaks only. The two peaks in the right drawing are lying at the surface and represent the source regions which are isolated by a highly doped p-implant. If one follows the doping profile at the surface one sees (from left to right) the n-doped substrate, the np-junction to the channel, and finally the pn^+-junction between channel and source. It should be noted that the doping of this transistor has been approximated simply by superposition of Gaussian profiles.

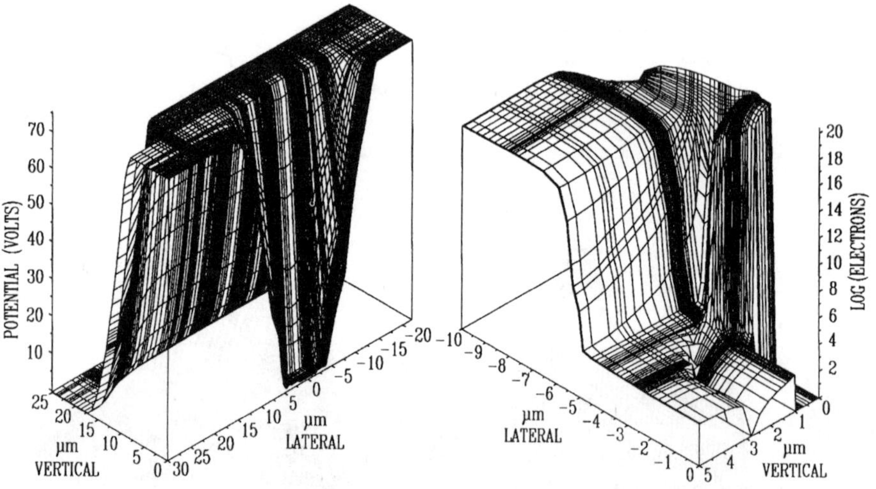

Figure 11: Electric potential (left) and electron density (right) distributions.

For the calculation of distributions of internal physical quantities the bias conditions $U_{DS} = 75\text{V}$, $U_{GS} = 20\text{V}$, and $U_{BS} = 0\text{V}$ were used. Figure 11 shows the electric potential distribution (left part) and the electron density in a logarithmic scale (right part). Note that the viewing directions are not the same: while in the left drawing the front point (with the coordinates 0 lateral/30 vertical) corresponds to the right edge at the surface in Figure 9 (drain contact) the front point of the right drawing is located 5μm below the center of the surface. The right axis ("lateral") of the potential distribution represents, therefore, the surface where the source region with zero voltage can easily be located. The gate-source voltage of

20V is not visible at first glance. It is indicated only by a slight kink in the contour of the potential distribution near the source area. Following the potential from the source in vertical direction it reaches the applied drain voltage (75V) and decreases again to the value of the substrate contact (0V).

As mentioned before the view of the electron density distribution (right drawing of Figure 11) is different: the left axis ("lateral") is parallel to the surface but in a depth of $5\mu m$ whereas the right axis ("vertical") matches the origin. As it can be seen from the simulation geometry Figure 9 the origin is located outside the device, thus, in that region the electron concentration is vanishing. (It should be mentioned that the logarithmic representation is dropped at zero value.) Starting the path of the electrons at the source one can nicely observe the high concentration in the channel and the bending of the electron flow into the depth.

REFERENCES

[1] M. ALI-OMAR AND L. REGGIANI, *Drift and Diffusions of Charge Carriers in Silicon and Their Empirical Relation to the Electric Field*, Solid-State Electron, 30, 7 (1987), pp. 693–697.

[2] M. AOKI, K. YANO, T. MASUHARA, S. IKEDA, AND S. MEGURO, *Optimum Crystallographic Orientation of Submicrometer CMOS Devices Operated at Low Temperatures*, IEEE Trans. Electron Devices ED-34, 1 (1987), pp. 52–57.

[3] N. ARORA, J. HAUSER, AND D. ROULSTON, *Electron and Hole Mobilities in Silicon as a Function of Concentration and Temperature*, IEEE Trans. Electron Devices ED-29 (1982), pp. 292–295.

[4] G. BACCARANI, M. RUDAN, R. GUERRIERI, AND P. CIAMPOLINI, *Physical Models for Numerical Device Simulation*, In European School on Device Modeling (University of Bologna, Italy, March 1991), pp. 1–70.

[5] G. BACCARANI AND M. WORDEMAN, *An Investigation of Steady-State Velocity Overshoot in Silicon*, Solid-State Electron, 28, 4 (1985), pp. 407–416.

[6] K. BLOTEKJAER, *Transport Equations for Electrons in Two-Valley Semiconductors*, IEEE Trans. Electron Devices ED-17, 1 (1970), pp. 38–47.

[7] E. BUTURLA, P. COTTRELL, B. GROSSMAN, K. SALSBURG, M. LAWLOR, AND C. MCMULLEN, *Three-Dimensional Finite Element Simulation of Semiconductor Devices*, In ISSC (1980), pp. 76–77.

[8] C. CANALI, G. MAJNI, R. MINDER AND G. OTTAVIANI, *Electron and Hole Drift Velocity Measurements in Silicon and Their Empirical Relation to Electric Field and Temperature*, IEEE Trans. Electron Devices ED-22 (1975), pp. 1045–1047.

[9] C. CANALI AND G. OTTAVIANI, *Saturation Values of the Electron Drift Velocity in Silicon between 300K and 4.2K*, Physics Lett. 32A, 3 (1970), pp. 147–148.

[10] D. CAUGHEY AND R. THOMAS, *Carrier Mobilities in Silicon Empirically Related to Doping and Field*, Proc. IEEE 52 (1967), pp. 2192–2193.

[11] S. CHAMBERLAIN AND A. HUSAIN, *Three-Dimensional Simulation of VLSI MOSFET's*, In Proc. Int. Electron Devices Meeting (1981), pp. 592–595.

[12] A. CHYNOWETH, *Ionization Rates for Electrons and Holes in Silicon*, Physical Review 109 (1958), pp. 1537–1540.

[13] P. CIAMPOLINI, A. PIERANTONI, M. MELANOTTE, C. CECCHETTI, C. LOMBARDI, AND G. BACCARANI, *Realistic Device Simulation in Three Dimensions*, In Proc. Int. Electron Devices Meeting (1989), pp. 131–134.

[14] S. COLAK, B. SINGER AND E. STUPP, *Lateral DMOS Power Transistor Design*, IEEE Electron Device Lett. EDL-1 (1980), pp. 51–53.

[15] E. CONWELL, *High Field Transport in Semiconductors*, Academic Press, 1967.

[16] C. GROWELL AND S. SZE, *Temperature Dependence of Avalanche Multiplication in Semiconductors*, Appl. Phys. Lett. 9 (1966), pp. 242–244.

[17] D. DECKER AND C. DUNN, *Temperature Dependence of Carrier Ionization Rates and Saturated Velocities in Silicon*, J. Electronic Mat. 4,3 (1975), pp. 527–547.

[18] P. DHANASEKARAN AND B. GOPALAM, *The Physical Behaviour of an $n+p$ Silicon Solar Cell in Concentrated Sunlight*, Solid-State Electron 25, 8 (1982), pp. 719–722.

[19] P. DICKINGER, *New models of high voltage DMOS devices for circuit simulation*, Electrosoft 1, 4 (December 1990), pp. 298–308.

[20] J. DORKEL AND P. LETURCQ, *Carrier Mobilities in Silicon Semi-Empirically Related to Temperature, Doping and Injection Level*, Solid-State Electron. 24 (1981), pp. 821–825.

[21] J. DZIEWIOR AND W. SCHMID, *Auger Coefficients for Highly Doped and Highly Excited Silicon*, Appl. Phys. Lett. 31 (1977), pp. 346–348.

[22] A. FRANZ AND G. FRANZ, *BAMBI – A Design Model for Power MOSFET's*, IEEE Trans. Computer-Aided Design CAD-4, 3 (1985), pp. 177–189.

[23] F. GAENSSLEN, R. JAEGER, AND J. WALKER, *Low-Temperature Threshold behavior of Depletion Mode Devices – Characterization and Simulation*, In Proc. Int. Electron Devices Meeting (1976), pp. 520–524.

[24] I. GAMBA AND M. SQUEFF, *Simulation of the Transient Behavior of a One-Dimensional Semiconductor Device II*, SIAM J. Numer. Anal. 26, 3 (1989), pp. 539–552.

[25] H. GUMMEL, *A Self-Consistent Iterative Scheme for One-dimensional Steady State transistor Calculations*, IEEE Trans. Electron Devices ED-11 (1964), pp. 455–465.

[26] P. HABAŠ AND S. SELBERHERR, *Impact of the Non-Degenerate Gate Effect on the Performance of Submicron MOS-Devices*, MIDEM – Electronic Components and Materials (December 1990), pp. 185–188.

[27] W. HÄNSCH, *Carrier Transport in Semiconductor Devices of Very Small Dimensions*, In Two-Dimensional Systems: Physics and Devices (Berlin, 1986), vol. 67, Springer, pp. 296–303.

[28] W. HÄNSCH AND H. JACOBS, *Enhanced Transconductance in Deep Submicrometer MOSFET*, IEEE Electron Device Lett. EDL-10, 7 (1989), pp. 285–287.

[29] W. HÄNSCH AND M. MIURA-MATTAUSCH, *The Hot-Electron Problem in Small Semiconductor Devices*, J. Appl. Phys. 60 (1986), p. 650.

[30] W. HÄNSCH, M. ORLOWSKI, AND E. WEBER, *The Hot-Electron Problem in Submicron MOSFET's*, In ESSDERC (1988), pp. 597–606.

[31] W. HÄNSCH AND S. SELBERHERR, *MINIMOS 3: A MOSFET Simulator that Includes Energy Balance*, IEEE Trans. Electron Devices ED-34, 5 (May 1987), pp. 1074–1078.

[32] W. HÄNSCH AND W. WEBER, *The Effect of Transients on Hot Carriers*, IEEE Electron Device Lett. EDL-10, 6 (1989), pp. 252–254.

[33] A. HENNING, N. CHAN, J. WATT, AND J. PLUMMER, *Substrate Current at Cryogenic Temperatures: Measurements and a Two-Dimensional Model for CMOS Technology*, IEEE Trans. Electron Devices ED-34, 1 (1987), pp. 64–74.

[34] P. HEREMANS, G. VANDENBOSCH, R. BELLENS, G. GROESENEKEN, AND H. MAES, *Temperature Dependence of the Channel Hot-Carrier Degradation of n-Channel MOSFET's*, IEEE Trans. Electron Devices ED-37, 4 (1990), pp. 980–993.

[35] K. HESS AND G. IAFRATE, *Theory and Applications of Near Ballistic Transport in Semiconductors*, Proc. IEEE 76, 5 (1988), pp. 519–532.

[36] W. HEYWANG AND H. PÖTZL, *Bandstruktur and Stromtransport*, Springer, 1976.

[37] A. HIROKI, S. ODANAKA, K. OHE, AND H. ESAKI, *A Mobility Model for Submicrometer MOSFET Device Simulations*, IEEE Electron Device Lett. EDL-8, 5 (1987), pp. 231–233.

[38] R. HORI, H. MASUDA, O. MINATO, S. NISHIMATU, K. SATO, AND M. SUBO, *Short Channel MOS-IC Based on Accurate Two-Dimensional Device Design*, Jap. J. Appl. Phys. 15 (1976), pp. 193–199.

[39] A. HUSAIN AND S. CHAMBERLAIN, *Three-Dimensional Simulation of VLSI MOSFET's: The Three-Dimensional Simulation Program WATMOS*, IEEE J. Solid-State Circuits SC-17, 2 (1982), pp. 261–268.

[40] C. JACOBONI, *Monte Carlo Techniques*, In European School on Device Modeling (University of Bologna, Italy, March 1991), pp. 101–124.

[41] C. JACOBONI AND L. REGGIANI, *The Monte Carlo Method for the Solution of Charge Transport in Semiconductors with Applications to Covalent Materials*, Review of Modern Physics 55, 3 (1983), pp. 645–705.

[42] D. KAHNG AND M. ATALLA, *Silicon-Silicondioxide Field Induced Surface Devices*, In IRE-AIEE Solid-State Device Res. Conf. (1960).

[43] W. KAUSEL, H. PÖTZL, G. NANZ, AND S. SELBERHERR, *Two-Dimensional Transient Simulation of the Turn-Off Behavior of a Planar MOS-Transistor*, Solid-State Electron 32, 9 (1989), pp. 685–709.

[44] M. KINUGAWA, M. KAKUMU, T. USAMI, AND J. MATSUNAGA, *Effects of Silicon Surface Orientation on Submicron CMOS Devices*, In Proc. Int. Electron Devices Meeting (1985), pp. 581–584.

[45] H. KOSINA AND S. SELBERHERR, *Coupling of Monte Carlo and Drift Diffusion Method with Applications to Metal Oxide Semiconductor Field Effect Transistors*, Jap. J. Appl. Phys. 29, 12 (December 1990), pp. 2282–2285.

[46] N. KOTANI AND S. KAWAZU, *A Numerical Analysis of Avalanche Breakdown in Short-Channel MOSFET's*, Solid-State Electron 24 (1981), pp. 681–687.

[47] S. LAUX AND M. FISCHETTI, *Monte Carlo Simulation of Submicrometer Si n-MOSFET's at 77 and 300K*, IEEE Electron Device Lett. EDL-9 (1988), pp. 467–469.

[48] S. LI AND W. THURBER, *The Dopant Density and Temperature Dependence of Electron Mobility and Resistivity in n-Type Silicon*, Solid-State Electron 20 (1977), pp. 609–616.

[49] T. LINTON AND P. BLAKEY, *A Fast, General Three-Dimensional Device Simulator and Its Application in a Submicron EPROM Design Study*, IEEE Trans. Computer-Aided Design CAD-8, 5 (1989), pp. 508–515.

[50] H. LOEB, R. ANDREW, AND W. LOVE, *Application of 2-Dimensional Solutions of the Shockley-Poisson Equation to Inversion-Layer M. O. S. T. Devices*, Electron Lett. 4 (1968), pp. 352–354.

[51] C. LU, J. SUNG, H. KIRSCH, S. HILLENIUS, T. SMITH, AND L. MANCHANDA, *Anomalous C-V Characteristics of Implanted Poly MOS Structures in n+/p+ Dual-Gate CMOS Technology*, IEEE Electron Device Lett. EDL-10, 5 (1989), pp. 192–194.

[52] M. MOCK, *A Two-Dimensional Mathematical Model of the Insulated-Gate Field-Effect Transistor*, Solid-State Electron 16 (1973), pp. 601–609.

[53] M. MOCK, *A Time-Dependent Numerical Model of the Insulated-Gate Field-Effect Transistor*, Solid-State Electron 24 (1981), pp. 959–966.

[54] C. MOGLESTUE, *A Monte Carlo Particle Model Study of the Influence of the Doping Profiles on the Characteristics of Field-Effect Transistors*, In NASECODE II Conf. (1981), pp. 244–249.

[55] D. NAVON AND C. WANG, *Numerical Modeling of Power MOSFET's*, Solid-State Electron 26, 4 (1983), pp. 287–290.

[56] T. NISHIDA AND C. SAH, *A Physically Based Mobility Model for MOSFET Numerical Simulation*, IEEE Trans. Electron Devices ED-34, 2 (1987), pp. 310–320.

[57] S. OH, D. WARD, AND R. DUTTON, *Transient Analysis of MOS Transistors*, IEEE Trans. Electron Devices ED-27 (1980), pp. 1571–1578.

[58] H. OKA, K. NISHIUCHI, T. NAKAMURA, AND H. ISHIKAWA, *Two-Dimensional Numerical Analysis of Normally-Off Type Buried Channel MOSFET's*, In Proc. Int. Electron Devices Meeting (1979), pp. 30–33.

[59] Y. OKUTO AND C. CROWELL, *Ionization Coefficients in Semiconductors: A Nonlocalized Property*, Physical Review B10 (1974), pp. 4284–4296.

[60] T. ONG, P. KO, AND C. HU, *50-A Gate-Oxide MOSFET's at 77K*, IEEE Trans. Electron Devices ED-34, 10 (1987), pp. 2129–2135.

[61] M. ORLOWSKI AND C. WERNER, *Model for the Electric Fields in LDD MOSFET's – Part II: Field Distribution on the Drain Side*, IEEE Trans. Electron Devices ED-36, 2 (1989), pp. 382–391.

[62] M. ORLOWSKI, C. WERNER, AND J. KLINK, *Model for the Electric Fields in LDD MOSFET's – Part I: Field Peaks on the Source Side*, IEEE Trans. Electron Devices ED-36, 2 (1989), pp. 375–381.

[63] Y. PARK, D. NAVON, AND T. TANG, *Monte Carlo Simulation of Bipolar Transistors*, IEEE Trans. Electron Devices ED-31, 12 (1984), pp. 1724–1730.

[64] P. ROBERTSON AND D. DUMIN, *Ballistic Transport and Properties of Submicrometer Silicon MOSFET's from 300 to 4.2K*, IEEE Trans. Electron Devices ED-33, 4 (1986), pp. 494–498.

[65] M. RUDAN AND A. GNUDI, *The Hydrodynamic Model of Current Transport in Semiconductors*, In European School on Device Modeling (University of Bologna, Italy, March 1991), pp. 125–160.

[66] G. SAI-HALASZ, *Processing and Characterization of Ultra Small Silicon Devices*, In ESSDERC (1987), pp. 71–80.

[67] E. SANGIORGI, B. RICCO, AND F. VENTURI, *MOS2: An Efficient Monte Carlo Simulator for MOS Devices*, IEEE Trans. Computer-Aided Design CAD-7, 2 (1988), pp. 259–271.

[68] J. SCHROEDER AND R. MULLER, *IGFET Analysis Through Numerical Solution of Poisson's Equation*, IEEE Trans. Electron Devices ED-15, 12 (1968), pp. 954–961.

[69] A. SCHÜTZ, S. SELBERHERR, AND H. PÖTZL, *Numerical Analysis of Breakdown Phenomena in MOSFET's*, In NASECODE II Conf. (1981), pp. 270–274.

[70] M. SEAVEY, Private Communication, 1987.

[71] S. SELBERHERR, *Analysis and Simulation of Semiconductor Devices*, Springer, 1984.

[72] S. SELBERHERR, *MOS Device Modeling at 77K*, IEEE Trans. Electron Devices ED-36, 8 (1989), pp. 1464–1474.

[73] S. SELBERHERR, W. FICHTNER, AND H. PÖTZL, *MINIMOS – a Program Package to Facilitate MOS Device Design and Analysis*, In NASECODE I Conf. (1979), pp. 275–279.

[74] S. SELBERHERR AND E. LANGER, *Low Temperature MOS Device Modeling*, In Workshop on Low Temperature Semiconductor Electronics (Burlington, Vermont, 1989), pp. 68–72.

[75] S. SELBERHERR AND E. LANGER, *Numerical Simulation of Semiconductor Devices*, In European Simulation Multiconf. (Rome, Italy, 1989), pp. 291–296.

[76] S. SELBERHERR AND E. LANGER, *Three-Dimensional Process and Device Modeling*, Microelectronics Journal 20, 1-2 (1989), pp. 113–127.

[77] S. SELBERHERR, A. SCHÜTZ, AND H. PÖTZL, *MINIMOS – A Two-Dimensional MOS Transistor Analyzer*, IEEE Trans. Electron Devices ED-27 (1980), pp. 1540–1550.

[78] M. SEVER, *Analysis of a Discretization Algorithm for Time-Dependent Semiconductor Models*, Compel 6, 3 (1987), pp. 171–189.

[79] G. SHAHIDI, D. ANTONIADIS, AND H. SMITH, *Electron Velocity Overshoot at Room and Liquid Nitrogen Temperatures in Silicon Inversion Layers*, IEEE Electron Device Lett. EDL-9, 2 (1988), pp. 94–96.

[80] N. SHIGYO, M. KONAKA, AND R. DANG, *Three-Dimensional Simulation of Inverse Narrow-Channel Effect*, Electron Lett. 18, 6 (1982), pp. 274–275.

[81] J. SLOTBOOM AND G. STREUTKER, *The Mobility Model in MINIMOS*, In ESSDERC (1989), pp. 87–91.

[82] Y.-C. SUN, Y. TAUR, R. DENNARD, AND S. KLEPNER, *Submicrometer-Channel CMOS for Low-Temperature Operation*, IEEE Trans. Electron Devices ED-34, 1 (1987), pp. 19–27.

[83] S. SVENSSON, *Theoretical Analysis of the Layer Design of Inverted Single-Channel Heterostructure Transistors*, IEEE Trans. Electron Devices ED-34, 5 (1987), pp. 992–1000.

[84] S. SZE, *Physics of Semiconductor Devices*, Wiley, 1969.

[85] A. TAMER, K. RAUCH, AND J. MOLL, *Numerical Comparison of DMOS, VMOS and UMOS Power Transistors*, IEEE Trans. Electron Devices ED-30, 1 (1983), pp. 73–76.

[86] M. THURNER AND S. SELBERHERR, *The Extension of MINIMOS to a Three-Dimensional Simulation Program*, In NASECODE V Conf. (Dublin, 1987), Boole Press, pp. 327–332.

[87] T. TOYABE, K. YAMAGUCHI, S. ASAI, AND M. MOCK, *A Numerical Model of Avalanche Breakdown in MOSFET's*, IEEE Trans. Electron Devices ED-25 (1978), pp. 825–832.

[88] T. TOYABE, K. YAMAGUCHI, S. ASAI, AND M. MOCK, *A Two-Dimensional Avalanche Breakdown Model of Submicron MOSFET's*, In Proc. Int. Electron Devices Meeting (1980), pp. 432–435.

[89] C. TURCHETTI, P. PRIORETTI, G. MASETTI, E. PROFUMO, AND M. VANZI, *A Meyer-Like Approach for the Transient Analysis of Digital MOS IC's*, IEEE Trans. Computer-Aided Design 5, 4 (1986), pp. 499–507.

[90] D. VANDORPE, J. BOREL, G. MERCKEL, AND P. SAINTOT, *An Accurate Two-Dimensional Numerical Analysis of the MOS Transistor*, Solid-State Electron 15 (1972), pp. 547–557.

[91] A. WALKER AND P. WOERLEE, *A Mobility Model for MOSFET Device Simulation*, In ESSDERC (1988), pp. 265–269.

[92] C. WILSON AND J. BLUE, *Two-Dimensional Finite Element Charge-Sheet Model of a Short Channel MOS Transistor*, Solid-State Electron 25, 6 (1982), pp. 461–477.

[93] J. WOO AND J. PLUMMER, *Short Channel Effects in MOSFET's at Liquid-Nitrogen Temperature*, IEEE Trans. Electron Devices ED-33, 7 (1986), pp. 1012–1019.

[94] K. YAMAGUCHI, *A Time Dependent and Two-Dimensional Numerical Model for MOSFET Device Operation*, Solid-State Electron 26, 9 (1983), pp. 907–916.

SCATTERING THEORY OF HIGH FREQUENCY QUANTUM TRANSPORT

H. C. LIU[*]

Abstract. The well-known scattering approach to quantum transport is extended to the time-dependent case where both dc and ac voltages are applied to a device. Such a formalism is useful for obtaining physical insight into the mechanism governing device operation, e.g., the high frequency characteristics. Scattering theory is most conveniently used in treating the ballistic transport regime, and is therefore suited to studies of new devices such as resonant tunneling diodes and ballistic quantum constrictions.

Key words: quantum transport, ballistic, resonant tunneling, constriction

The quest for faster and faster devices has pushed technology to make smaller and smaller structures. The current nanometer scale devices are made by (a) epitaxial growths with atomic layer accuracy, such as molecular beam epitaxy (MBE) and (b) nano-scale lithography by, for example, electron beams and focused ion beams. As the dimension of the device active region becomes small (shorter than the carrier mean free path and on the order of the de Broglie wavelength), quantum effects naturally come into play and ballistic transport properties start to govern device operation. The scattering approach gives a clear physical picture and is well suited to treat quantum transport in nano-scale devices. The idea of treating a device as a scattering "target" connected with carrier reservoirs (see Fig. 1) was originally proposed by Landauer.[1, 2] This approach has been used extensively in treating tunneling in quantum wells and superlattices with satisfactory results after Tsu and Esaki,[3, 4] and has been extended to the multi-channel case by Büttiker.[5] This extension has proved to be extremely useful in studying transport phenomena in confined quantum channels made from high mobility two-dimensional electron gases.[6] We extend the scattering approach to the time-dependent case, where both dc and ac voltages are applied to the device. Since this research is on-going, I will provide only an extended abstract here, i.e., I will give a introduction to the basic approach and examples of devices treated, cite the published papers for details, and point to what is being investigated.

It is important to be aware of the basic assumptions made in modeling devices. First, almost all the work in device modeling uses the effective mass approach, and, especially when dealing with the conduction band, single band effective mass approximation is used. Strictly speaking, when treating heterostructures, this is valid only for systems composed of very similar materials, such as GaAs-Al$_x$Ga$_{1-x}$As and Si-Si$_{1-x}$Ge$_x$ with relatively small alloy fraction x.[7] Although the effective mass approximation is used commonly for structures with very dissimilar materials, e.g., GaAs-AlAs, it must be borne in mind that such a treatment is only an *approximation* using empirical parameters consistent with experiments. In addition, the effective

* Institute for Microstructural Sciences, National Research Council, Ottawa, Ontario K1A 0R6, Canada

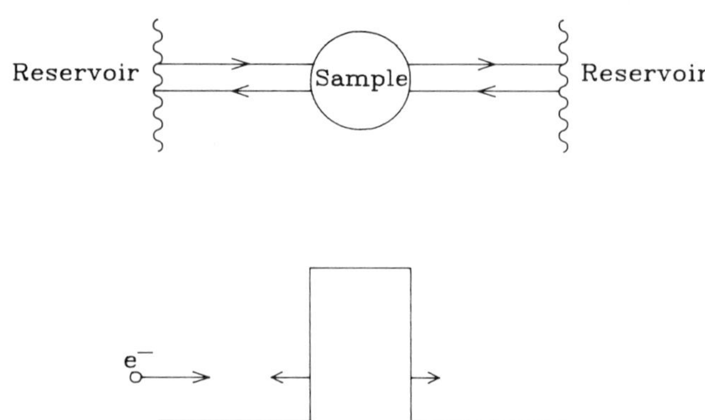

FIG. 1. *The scattering model (above) and an example of a single barrier device (below). The device active part (the sample) is treated as the scattering "target". Reservoirs "fire" carriers (normally electrons) into the sample and collect carriers scattered off the sample similar to a blackbody source. The sample can have any potential profile.*

mass approximation is sometimes used in structures made of very thin (a few mono-layers) layers. Again, strictly speaking, it is no longer valid to employ the effective mass approximation which assumes that the envelope wavefunction is slowly varying. Furthermore, the conceptual separation of a device into the active region, where transport is ballistic or quasi-ballistic, and the carrier reservoirs is artificial. In some cases, the assumption of an ideal reservoir which "fires" carriers with well defined energies into the active region may not be valid.[8] Lastly, higher order effects such as those related to many particles, image force, band non-parabolicity, and multiple bands sometimes produce measurable results and may have to be included.

The basic approach for the time-independent (dc) transport case has been well documented.[1, 2, 3, 4, 5] The approach has been extended to the time-dependent (ac) case in a series of papers by Coon and the present author.[9, 10, 11, 12, 13] These papers all dealt with the one-dimensional case suitable for treating transport in quantum wells and finite superlattices, and special attention was paid to the high frequency behavior of the double barrier resonant tunneling diode.[14, 15] The early papers[9, 10] discussed the general approach, including symmetry relations of the scattering matrix, while the later papers[11, 12, 13] concentrated on the double barrier structures; the most recent[13] provided a review and the analytical formalism. The well-known double barrier resonant tunneling structure is shown schematically in Fig. 2. The current interest in conduction through constrictions or quantum point contacts has been stimulated by recent experiments.[16, 17] The understanding of quantum transport through reduced dimensional regions is of great importance to future electronic devices. An example of an ideal constriction is shown in Fig. 3. Striking step features (in integer times $2e^2/h$) in conductance through a channel have been observed.[16, 17] Calculations have been made to explain the observed conductance steps.[18, 19, 20] These experiments and calculations dealt with dc conductance, while for device applications high frequency behavior is of key importance. I have started the theoretical work on conductance at high frequency.[21]

Here I will give a brief introduction of the basic model in one dimension and emphasize the physical idea. The basic problem consists of (a) solving for the scattering

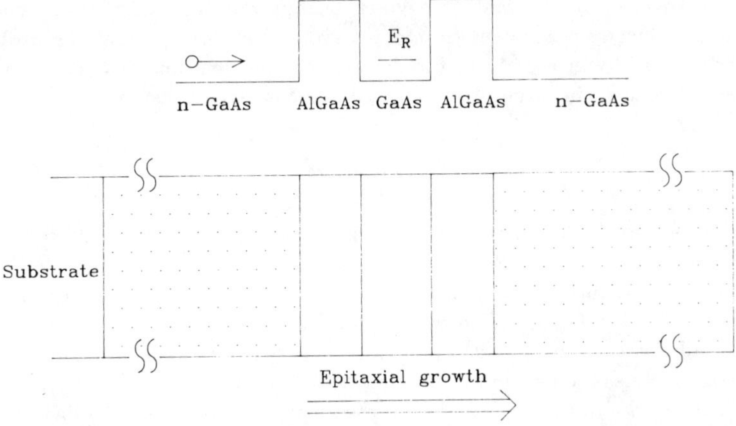

FIG. 2. *The double barrier resonant tunneling structure made from the GaAs-AlGaAs heterosystem. The upper part shows the conduction bandedge profile, and the lower part is the layer sequence normally grown by molecular beam epitaxy. When the incident electron energy coincides with the quantized resonant state in the well (E_R), an enhancement in tunneling probability occurs (resonant tunneling).*

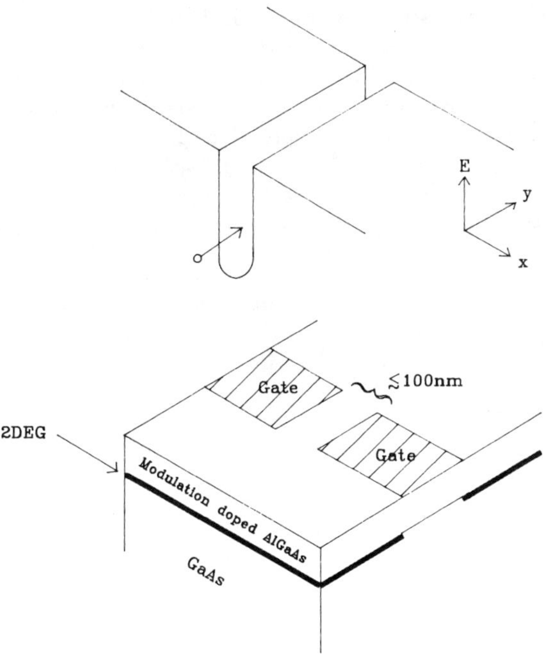

FIG. 3. *An ideal quantum channel made by gating a high mobility two-dimensional electron gas (2DEG). The upper part shows the channel potential profile and the lower part is the split gate structure on a 2DEG made from a modulation doped high mobility GaAs-AlGaAs heterostructure. Applying a suitable voltage depletes electrons under the gates and forms the quantum channel. Ohmic contacts (not shown) are made to the 2DEG on either side of the channel, and the conductance through the channel is measured.*

matrix for an electron incident at a given energy and (b) summing up contributions from all electrons incident from a reservoir. The first part of the problem is shown schematically in Fig. 4. In the dc case, the incident and scattered waves are planewaves because the asymptotic contact regions have constant potentials (zero electric fields):

$$\psi = e^{ikz - iEt/\hbar}, \tag{1}$$

where the wavevector k and the energy E are related by $E - V_{dc} = \hbar^2 k^2/2m$, m is the effective mass, and the transport direction is chosen as z-coordinate. Solving the time-independent Schrödinger equation, one finds the scattering (reflection and transmission) amplitudes. One method to solve for the reflection (r) and the transmission (t) amplitudes is the transfer matrix technique[3] which is easily implemented even for an arbitrarily shaped potential profile.[22] In the dc plus ac case, we formulated the problem in a similar way to the dc case using planewave-like solutions to the time-dependent Schrödinger equation with potential $V_{dc} + V_{ac}\cos\omega t$, where V_{dc} and V_{ac} are constants in the contact regions.[9, 13] The planewave-like solutions are:

$$\psi = e^{ikz - iEt/\hbar} e^{-(iV_{ac}/\hbar\omega)\sin(\omega t)}. \tag{2}$$

This solution is shown to satisfy the time-dependent Schrödinger equation by simple substitution. Again, one can use the transfer matrix technique to solve for all the reflection and transmission amplitudes, including side bands caused by the ac potential.[9, 11, 13] The second part of the problem, namely, summing up contributions from all electrons incident from a reservoir, is normally easy to do.[3] This takes the other two dimensions (assumed to have translation invariance) into account in the above one-dimensional case. The approach outlined here has been applied to the double barrier resonant tunneling diode.[11, 12, 13] High frequency device characteristics have been studied in detail in Ref. [13]. Initial work has been undertaken in studying a quantum constriction which is in fact a two-dimensional problem.[21]

In conclusion, I have given a brief introduction to the scattering approach to the modeling of quantum ballistic transport devices. This approach has been used in studying the tunneling devices made from quantum well structures. Further investigation is underway to understand the high frequency behavior of the quantum ballistic constrictions and quantum point contacts.

Acknowledgements – The initial part of the work was done in collaboration with Prof. D. D. Coon of the University of Pittsburgh. The author benefited from many discussions with Dr. G. C. Aers.

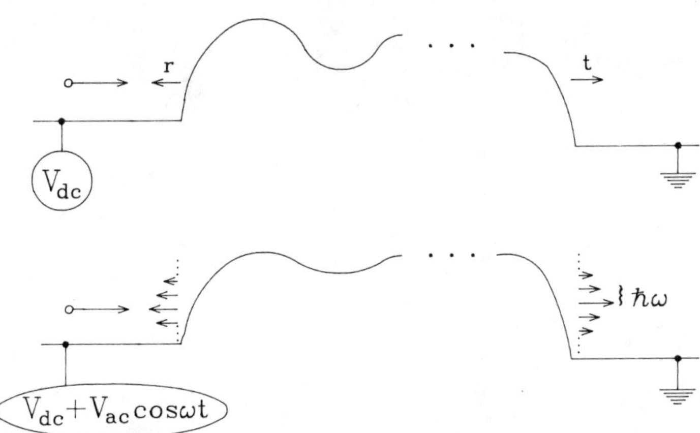

FIG. 4. *An arbitrarily shaped one-dimensional potential profile is biased with a dc voltage (above) and both dc and ac voltages (below). The asymptotic regions are connected to the voltage leads and are flat (zero electric field) in potential because these regions are normally heavily doped (metallic). In the dc case, the scattering matrix consists of only the reflection (r) and the transmission (t) amplitudes, whereas in the dc plus ac case, many side bands occur and the scattering matrix consists of reflection and transmission amplitudes with out-going energies $E + n\hbar\omega$, where E is the incident energy and n is a integer.*

REFERENCES

[1] R. Landauer, Philos. Mag. **21**, 863 (1970).

[2] R. Landauer, Z. Phys. B. – Condensed Matter **68**, 217 (1987).

[3] R. Tsu and L. Esaki, Appl. Phys. Lett. **22**, 562 (1973).

[4] D. D. Coon and H. C. Liu, Appl. Phys. Lett. **47**, 172 (1985).

[5] M. Büttiker, Phys. Rev. Lett. **57**, 1761 (1986).

[6] A. Yacoby, U. Sivan, C. P. Umbach, and J. M. Hong, Phys. Rev. Lett. **66**, 1938 (1991).

[7] H. C. Liu, Superlattices and Microstructures **3**, 413 (1987).

[8] H. C. Liu, M. Buchanan, G. C. Aers, Z. R. Wasilewski, T. W. Moore, R. L. S. Devine, and D. Landheer, Phys. Rev. B **43**, 7086 (1991).

[9] D. D. Coon and H. C. Liu, J. Appl. Phys. **58**, 2230 (1985).

[10] D. D. Coon and H. C. Liu, Solid State Commun. **55**, 339 (1985).

[11] H. C. Liu, Appl. Phys. Lett. **52**, 483 (1988).

[12] H. C. Liu, J. Appl. Phys. **69**, 2705 (1991).

[13] H. C. Liu, Phys. Rev. B. **43**, 12538 (1991).

[14] T. C. L. G. Sollner, W. D. Goodhue, P. E. Tannenwald, C. D. Parker, and D. D. Peck, Appl. Phys. Lett. **43**, 588 (1983).

[15] E. R. Brown, T. C. L. G. Sollner, C. D. Parker, W. D. Goodhue, and C. L. Chen, Appl. Phys. Lett. **55**, 1777 (1989).

[16] B. J. van Wees, H. van Houten, C. W. J. Beenakker, J. G. Williamson, L. P. Kouwenhoven, D. van der Marel, and C. T. Foxon, Phys. Rev. Lett. **60**, 848 (1988).

[17] D. A. Wharam, T. J. Thornton, R. Newbury, M. Pepper, H. Ahmed, J. E. F. Frost, D. G. Hako, D. C. Peacock, D. A. Ritchie, and G. A. C. Jones, J. Phys. C. **21**, L209 (1988).

[18] Aaron Szafer and A. Douglas Stone, Phys. Rev. Lett. **62**, 300 (1989).

[19] G. Kirczenow, Phys. Rev. B. **39**, 10452 (1989).

[20] E. Tekman and S. Ciraci, Phys. Rev. B. **43**, 7145 (1991).

[21] H. C. Liu, unpublished.

[22] Y. Ando and T. Itoh, J. Appl. Phys. **61**, 1497 (1987).

ACCELERATING DYNAMIC ITERATION METHODS WITH APPLICATION TO SEMICONDUCTOR DEVICE SIMULATION*

ANDREW LUMSDAINE[†] AND JACOB K. WHITE[‡]

Abstract. In this paper, we apply a Galerkin method to solving the system of second-kind Volterra integral equations which characterize the classical dynamic iteration methods for the linear time-varying initial value problem. It is shown that the Galerkin approximations can be computed iteratively using conjugate-direction algorithms. The resulting iterative methods are combined with an operator Newton method and applied to solving the differential-algebraic system generated by spatial discretization of the time-dependent semiconductor device equations. Experimental results are included which demonstrate the conjugate-direction methods are significantly faster than classical dynamic iteration methods.

Key Words. Conjugate-direction methods, dynamic iteration, Galerkin method, waveform relaxation

AMS(MOS) subject classification. 65L60, 65L05, 65R20, 65J10

1. Introduction. The enormous computational expense and the growing importance of mixed circuit/device simulation, as well as the increasing availability of parallel computers, suggest that specialized, easily parallelized, algorithms be developed for transient simulation of MOS devices [7]. Recently, the easily parallelized waveform relaxation (WR) algorithm was shown to be a computationally efficient approach to device transient simulation [19]. However, the WR algorithm typically requires hundreds of iterations to achieve an accurate solution, which suggests that significant performance gains can still be realized by the application of methods for accelerating the convergence of the WR algorithm.

For linear algebra problems, conjugate-direction algorithms have enjoyed huge success as techniques for accelerating classical relaxation methods. Since the WR algorithm can in some sense be considered a function-space generalization of a linear algebra relaxation method, it seems only natural that conjugate-direction methods can be similarly generalized. Such a generalization can be rigorously analyzed, at least with respect to asymptotic behavior, by formulating the conjugate-direction method as a Galerkin method.

In the next section, we begin by developing a Galerkin method for solving an operator equation formulation of the linear time-varying initial-value problem. It is then shown that certain conjugate-direction methods iteratively generate the Galerkin approximations. The resulting methods are then combined with an operator Newton method in a hybrid scheme for solving the nonlinear initial-value problem. The semiconductor device transient simulation problem is described in Section 3. In Section 4,

* This work was supported by a grant from IBM, the Defense Advanced Research Projects Agency contract N00014-91-J-1698, and the National Science Foundation.

† Research Laboratory of Electronics, Dept. of Electrical Engineering and Computer Science, Massachusetts Institute of Technology, Cambridge, MA 02139. (`lums@rle-vlsi.mit.edu`)

‡ Research Laboratory of Electronics, Dept. of Electrical Engineering and Computer Science, Massachusetts Institute of Technology, Cambridge, MA 02139. (`white@rle-vlsi.mit.edu`)

experimental results are presented which demonstrate that the conjugate-direction acceleration significantly reduces the computation time for device transient simulation. Finally, our conclusions and suggestions for further work are contained in Section 5.

2. Description of the Method. We begin by considering the linear time-varying initial value problem (IVP),

$$(1) \qquad (\tfrac{d}{dt} + \boldsymbol{A}(t))\boldsymbol{x}(t) \;=\; \boldsymbol{b}(t)$$
$$\boldsymbol{x}(0) \;=\; \boldsymbol{x}_0,$$

where $\boldsymbol{A}(t) \in \mathbf{R}^{N \times N}$, $\boldsymbol{b}(t) \in \mathbf{R}^N$ is a given right-hand side, and $\boldsymbol{x}(t) \in \mathbf{R}^N$ is the unknown vector to be computed over the simulation interval $t \in [0, T]$. There are several approaches to solving the IVP. The traditional numerical approach is to begin by discretizing (1) in time with an implicit integration rule (since large dynamical systems are typically stiff) and then solving the resulting matrix problem at each time step. This approach can be disadvantageous for a parallel implementation, especially for parallel computers having a high communication latency, since the processors will have to synchronize repeatedly for each timestep.

A more effective approach to solving the IVP with a parallel computer is to decompose the problem at the ODE level. That is, the large system is decomposed into smaller subsystems, each of which is assigned to a single processor. The IVP is solved iteratively by solving the smaller IVP's for each subsystem, using fixed values from previous iterations for the variables from other subsystems. This dynamic iteration process is known as waveform relaxation (WR) or sometimes as the Picard-Lindelöf iteration [15].

In this section, we consider conjugate-direction methods for accelerating the classical dynamic iteration methods. Our approach is to first convert the IVP to a system of second-kind Volterra integral equations by using a "dynamic preconditioner." Next, we show that the classical dynamic iteration methods are obtained by applying the Richardson iteration to the integral equation system. Finally, we develop conjugate-direction methods for accelerating the classical dynamic iteration methods. This development is approached by considering conjugate-direction methods as Galerkin methods.

2.1. Operator Equation Formulation. Let $\boldsymbol{A}(t) = \boldsymbol{M}(t) - \boldsymbol{N}(t)$ and consider the system of second kind Volterra integral equations given by

$$(2) \qquad \boldsymbol{x}(t) - \boldsymbol{\Phi}_M(t, 0)\boldsymbol{x}(0) - \int_0^t \boldsymbol{\Phi}_M(t, s)\boldsymbol{N}(s)\boldsymbol{x}(s)ds = \int_0^t \boldsymbol{\Phi}_M(t, s)\boldsymbol{b}(s)ds,$$

where $\boldsymbol{\Phi}_M$ is the state transition matrix [3] for the equation

$$(3) \qquad \tfrac{d}{dt}\boldsymbol{x}(t) = \boldsymbol{M}(t)\boldsymbol{x}(t).$$

We assume throughout that $\boldsymbol{A}$, $\boldsymbol{M}$, and $\boldsymbol{N}$ are such that (1) and (2) each have a unique solution. A sufficient condition for this assumption is that $\boldsymbol{A}$, $\boldsymbol{M}$, and $\boldsymbol{N}$ be piecewise continuous; a weaker sufficient condition is that $\boldsymbol{A}$, $\boldsymbol{M}$, and $\boldsymbol{N}$ be measurable. Note that the solution $\boldsymbol{x}$ to (2) also satisfies (1). In some sense, (2) is obtained from (1) by the application of a "dynamic preconditioner," to both sides of (1). More precisely, this preconditioner, denoted $\mathcal{M}^{-1}$, is defined by:

$$(\mathcal{M}^{-1}\boldsymbol{x})(t) = \int_0^t \boldsymbol{\Phi}_M(t,s)\boldsymbol{x}(s)ds.$$

Intuitively, one can think of $\mathcal{M}^{-1}$ as roughly being $(\frac{d}{dt} + \boldsymbol{M}(t))^{-1}$.

Equation (2) can be expressed as an operator equation over a space $\mathbf{H}$ as

(4) $$(\boldsymbol{I} - \mathcal{K})\boldsymbol{x} = \boldsymbol{\psi},$$

where $\mathbf{H} = \mathbf{L}_2([0,T], \mathbf{R}^N)$, $\boldsymbol{I} : \mathbf{H} \to \mathbf{H}$ is the identity operator, $\mathcal{K} : \mathbf{H} \to \mathbf{H}$ is defined by

$$\begin{aligned}(\mathcal{K}\boldsymbol{x})(t) &= \int_0^t \boldsymbol{K}_M(t,s)\boldsymbol{x}(s)ds \\ &= \int_0^t \boldsymbol{\Phi}_M(t,s)\boldsymbol{N}(s)\boldsymbol{x}(s)ds\end{aligned}$$

and $\boldsymbol{\psi} \in \mathbf{H}$ is given by

$$\boldsymbol{\psi}(t) = \boldsymbol{\Phi}_M(t,0)\boldsymbol{x}(0) + \int_0^t \boldsymbol{\Phi}_M(t,s)\boldsymbol{b}(s)ds.$$

The following are standard results (see, e.g., [5, 9]) which will be used in subsequent discussions of (4).

LEMMA 2.1. *If $\boldsymbol{M}$ and $\boldsymbol{N}$ are piecewise continuous (or measurable) then $\boldsymbol{\Phi}_M$ is measurable and hence $\boldsymbol{K}_M \in \mathbf{L}_2([0,T] \times [0,T], \mathbf{R}^{N \times N})$.*

LEMMA 2.2. *If $\boldsymbol{K}_M \in \mathbf{L}_2([0,T] \times [0,T], \mathbf{R}^{N \times N})$, then the operator $\mathcal{K}$ has spectral radius zero.*

LEMMA 2.3. *If $\boldsymbol{K}_M \in \mathbf{L}_2([0,T] \times [0,T], \mathbf{R}^{N \times N})$, then the operator $\mathcal{K}$ is compact.*

LEMMA 2.4. *If $\boldsymbol{K}_M \in \mathbf{L}_2([0,T] \times [0,T], \mathbf{R}^{N \times N})$, then $\mathcal{K}^*$, the adjoint operator for $\mathcal{K}$, is given by*

$$\begin{aligned}(\mathcal{K}^*\boldsymbol{x})(t) &= \int_t^T [\boldsymbol{K}_M(s,t)]^\dagger \boldsymbol{x}(s)ds \\ &= \int_t^T [\boldsymbol{\Phi}_M(s,t)\boldsymbol{N}(t)]^\dagger \boldsymbol{x}(s)ds,\end{aligned}$$

where superscript † denotes algebraic transposition.

Remark. It should be apparent from Lemma 2.4 that, in general, $\mathcal{K}$ is not self adjoint. We therefore restrict our attention to those conjugate-direction methods which are appropriate for non-self-adjoint operators.

2.2. Classical Dynamic Iteration Methods. The classical dynamic iteration is obtained by applying the Richardson iteration to the "preconditioned" problem (4):

(5) $$\boldsymbol{x}^{k+1} = \mathcal{K}\boldsymbol{x}^k + \boldsymbol{\psi}.$$

This approach is known as the method of successive approximations, waveform relaxation, or the Picard-Lindelöf iteration [1, 9, 11, 15, 26].

290

Example. Let $M(t) = 0$. Then $\Phi_M = I$ so that (2) becomes

$$x(t) - x(0) + \int_0^t A(s)x(s)ds = \int_0^t b(s)ds.$$

The corresponding dynamic iteration is

$$x^{k+1}(t) = x(0) + \int_0^t \left(b(s) - A(s)x^k(s) \right) ds$$

which is the familiar Picard iteration.

Example. Let $M(t)$ be the diagonal part of $A(t)$. Then (5) becomes the Jacobi WR algorithm in which we solve the following IVP at each iteration k for each $x_i^{k+1}(t)$:

$$\left(\tfrac{d}{dt} + a_{ii}(t) \right) x_i^{k+1}(t) + \sum_{j \neq i} a_{ij}(t)x_j^k(t) = b_i(t)$$
$$x_i(0) = x_{0i}.$$

As a direct consequence of Lemma 2.2, we have:

THEOREM 2.5. *Under the assumptions of Lemma 2.1, the method of successive approximations, defined in (5), converges.*

Remark. Theorem 2.5 only provides a description of the asymptotic behavior of the dynamic iteration. In [15], Miekkala and Nevanlinna examine the dynamic iteration on the infinite interval. In that case, more intuition about the convergence rate of the dynamic iteration can be obtained, although the extent to which that intuition applies for a given dynamic iteration on a finite interval depends on the stiffness of the problem.

2.3. Accelerating Dynamic Iteration Methods. Another approach to solving (4) is to apply a Galerkin method to solving a variational formulation of the problem. This approach leads directly to the conjugate-direction methods. Galerkin methods have been well studied for second-kind Fredholm integral equations [1, 9], of which second-kind Volterra equations are a special case, but infrequently studied for second-kind Volterra equations in particular (see, however, [12]). With the conjugate-direction approach, instead of applying the Galerkin method over a space of polynomials or splines, as is typical, one applies the Galerkin method over a Krylov space generated by $(I - K)$. The use of a Galerkin method over a Krylov space generated by $(I - K)$ is discussed in [16] and [18] where the approach is called the method of moments (see also [25]).

2.3.1. The Galerkin Method. Let $\mathbf{X}$ and $\mathbf{Y}$ be Hilbert spaces and consider the operator equation

$$(6) \qquad \qquad \mathcal{A}x = b$$

where $x \in \mathbf{X}, b \in \mathbf{Y}$ and $\mathcal{A} : \mathbf{X} \to \mathbf{Y}$ is a bounded injective operator.

By a Galerkin method, we mean any scheme by which the solution x (6) is computed by solving the problem in a sequence of finite-dimensional subspaces via the use of orthogonal projections. That is, we take the subspaces $\mathbf{X}_n \subset \mathbf{X}$ and $\mathbf{Y}_n \subset \mathbf{Y}$ with $\dim \mathbf{X}_n = \dim \mathbf{Y}_n = n$ and require the Galerkin approximation $\boldsymbol{x}_n$ to satisfy

$$(7) \qquad \langle \boldsymbol{b} - \mathcal{A}\boldsymbol{x}_n, \boldsymbol{y} \rangle = 0 \quad \forall \boldsymbol{y} \in \mathbf{Y}_n.$$

In general, it is sufficient to satisfy (7) over some basis of $\mathbf{Y}_n$. That is, we define $\mathbf{X}_n = \mathrm{span}\{\boldsymbol{u}_0, \boldsymbol{u}_1, \ldots, \boldsymbol{u}_{n-1}\}$ and $\mathbf{Y}_n = \mathrm{span}\{\boldsymbol{v}_0, \boldsymbol{v}_1, \ldots, \boldsymbol{v}_{n-1}\}$, so that the solution $\boldsymbol{x}_n$ must satisfy

$$(8) \qquad \langle \boldsymbol{b} - \mathcal{A}\boldsymbol{x}_n, \boldsymbol{v}_j \rangle = 0 \quad j = 0, 1, \ldots, n-1.$$

If we take $\boldsymbol{x}_n$ to be

$$\boldsymbol{x}_n = \sum_{i=0}^{n-1} \gamma_i \boldsymbol{u}_i$$

then (8) generates a linear system of equations for $\{\gamma_i\}$:

$$\langle \mathcal{A} \sum_{i=0}^{n-1} \gamma_i \boldsymbol{u}_i, \boldsymbol{v}_j \rangle = \langle \boldsymbol{b}, \boldsymbol{v}_j \rangle.$$

The crucial question, of course, is whether or not a particular Galerkin method converges. To answer this, we use the following notion of convergence (which is standard for the Galerkin method [9, 27]).

DEFINITION 2.6. The Galerkin method is said to be *convergent* for the operator $\mathcal{A}$ if the following hold for every $\boldsymbol{b} \in \mathbf{Y}$,

1. The solution $\boldsymbol{x} \in \mathbf{X}$ to the original equation (6) exists and is unique.
2. Either:
    (a) There exists an index M such that for every $n \geq M$, the Galerkin equation (9) has a unique solution $\boldsymbol{x}_n$.
    (b) The approximate solutions $\boldsymbol{x}_n$ converge, i.e., $\boldsymbol{x}_n \to \boldsymbol{x}$ as $n \to \infty$.
    or
    (a) There exist indices M and N such that for every n, $M \leq n \leq N$, the Galerkin equation (9) has a unique solution $\boldsymbol{x}_n$.
    (b) The solution $\boldsymbol{x}_N = \boldsymbol{x}$.

The particular Galerkin method in which $\mathbf{Y} = \mathbf{X}$ and $\mathbf{Y}_n = \mathbf{X}_n$ is often called the Bubnov-Galerkin method. If $\mathcal{A}$ is positive definite in addition to being bounded and injective, it is well known that the Bubnov-Galerkin method is convergent for (6) [17]. Furthermore, if $\mathcal{A}$ is self-adjoint, the Galerkin approximations can be computed iteratively with the conjugate-gradient method (appropriately extended from $\mathbf{R}^N$ to $\mathbf{X}$, of course) [9].

For our particular problem, the operator $(\boldsymbol{I} - \mathcal{K})$ is not self-adjoint, yet we still seek a conjugate-direction method appropriate for solving (4). Such methods can be derived by considering the Galerkin method where $\mathbf{Y} = \mathcal{A}(\mathbf{X})$ and $\mathbf{Y}_n = \mathcal{A}(\mathbf{X}_n)$. That is, we require $\boldsymbol{x}_n$ to satisfy

$$(9) \qquad \langle \boldsymbol{b} - \mathcal{A}\boldsymbol{x}_n, \mathcal{A}\boldsymbol{u}_j \rangle = 0 \quad j = 0, 1, \ldots, n-1.$$

Note that this method can also be derived from a variational formulation of the problem. That is, instead of solving (6) directly, we minimize the following functional:

$$\phi(\boldsymbol{x}) = \|\boldsymbol{b} - \mathcal{A}\boldsymbol{x}\|^2,$$

where $\|\boldsymbol{x}\|^2 = \langle \boldsymbol{x}, \boldsymbol{x} \rangle$. Clearly, if (6) has a solution, the $\boldsymbol{x}$ which minimizes ϕ will also satisfy (6).

To be a minimizer of $\phi(\boldsymbol{x})$ over $\boldsymbol{x} \in \mathbf{X}_n$, the projection of the gradient of ϕ onto $\mathbf{X}_n$ must be zero, that is,

$$(10) \qquad \langle \nabla\phi(\boldsymbol{x}_n), \boldsymbol{u}_j \rangle = 0 \quad j = 0, 1, \ldots, n-1.$$

An expression for the gradient, $\nabla\phi(\boldsymbol{x})$, can be obtained by straightforward calculation and is given in the following claim.

CLAIM 2.7.

$$\langle \nabla\phi(\boldsymbol{x}), \boldsymbol{y} \rangle = \langle \boldsymbol{b} - \mathcal{A}\boldsymbol{x}, \mathcal{A}\boldsymbol{y} \rangle$$

It should be obvious then that (10) is equivalent to (9). This particular method is also known as the method of least squares [9, 17].

Now we state our main result.

THEOREM 2.8. *Let $\mathbf{X}$ be a Hilbert space and let $\mathcal{A} : \mathbf{X} \to \mathbf{X}$ be a bounded bijective linear operator. Let $\mathbf{X}_n \subset \mathbf{X}$ be a finite-dimensional subspace with $\mathbf{X}_n \subset \mathbf{X}_{n+1}$ for all $n \in \mathbf{N}$. If $\boldsymbol{x}$ is in the closure of $\mathbf{S} = \cup_{n=1}^{\infty} \mathbf{X}_n$, then the Galerkin method for (6) is convergent. Moreover, there exists the estimate*

$$(11) \qquad \|\boldsymbol{x} - \boldsymbol{x}_n\| \leq C\|\boldsymbol{b} - \mathcal{A}\boldsymbol{x}_n\|$$

for some constant C depending only on $\mathcal{A}$.

Proof. We verify the conditions for the method to be convergent. Let $\mathbf{X}_n = \text{span}\{\boldsymbol{u}_0, \boldsymbol{u}_1, \ldots \boldsymbol{u}_{n-1}\}$.

Since $\mathcal{A}$ is bijective, the solution to (6) exists and is unique. Furthermore, there exists a constant $C = \|\mathcal{A}^{-1}\|$ such that

$$(12) \qquad \|\boldsymbol{x}\| \leq C\|\mathcal{A}\boldsymbol{x}\|.$$

<u>Case 1:</u> We first assume that $\dim \mathbf{X}_n = n$ for all $n \in \mathbf{N}$. If the Galerkin equations (9) are not uniquely solvable for some n, then the homogeneous portion of (9) has a non-trivial solution, i.e., there exists a set of coefficients $\{\gamma_k\}$, not all zero, such that

$$\langle \sum_{k=0}^{n-1} \gamma_k \mathcal{A}\boldsymbol{u}_k, \mathcal{A}\boldsymbol{u}_j \rangle = 0 \quad j = 0, 1, \ldots, n-1.$$

We take a linear combination of $\boldsymbol{u}_j$'s to obtain

$$\langle \sum_{k=0}^{n-1} \gamma_k \mathcal{A}\boldsymbol{u}_k, \sum_{j=0}^{n-1} \gamma_j \mathcal{A}\boldsymbol{u}_j \rangle = 0,$$

which contradicts the assumptions that $\mathcal{A}$ is injective and $\dim \mathbf{X}_n = n$. Therefore, (9) is solvable for all $n \geq 1$.

If $\boldsymbol{x} \in \mathrm{cl}\,\mathbf{S}$, then for any $\epsilon > 0$, there exists a $\boldsymbol{y} \in \mathbf{S}$ such that

$$\|\mathcal{A}(\boldsymbol{x} - \boldsymbol{y})\| \leq \frac{\epsilon}{C}.$$

Since $\boldsymbol{y} \in \mathbf{S}$, there must be an integer N such that $\boldsymbol{y} \in \mathbf{X}_N$. But in that case, we must also have for $n \geq N$

$$\|\mathcal{A}(\boldsymbol{x} - \boldsymbol{x}_n)\| \leq \|\mathcal{A}(\boldsymbol{x} - \boldsymbol{y})\| \leq \frac{\epsilon}{C},$$

because, from (10), $\boldsymbol{x}_N$ minimizes $\|\mathcal{A}(\boldsymbol{x} - \boldsymbol{y})\|$ for all $\boldsymbol{y} \in \mathbf{X}_N$ and, as $\mathbf{X}_n \supset \mathbf{X}_N$ for all $n \geq N$, the minimum does not increase for $n \geq N$. Therefore, by (12), $\|\boldsymbol{x} - \boldsymbol{x}_n\| \leq \epsilon$ for $n \geq N$. We conclude that $\boldsymbol{x}_n \to \boldsymbol{x}$ as $n \to \infty$.

<u>Case 2:</u> Next, assume that $\dim \mathbf{S} = N$. Without loss of generality, we can take $\mathbf{S} = \mathrm{span}\{\boldsymbol{u}_0, \boldsymbol{u}_1, \ldots, \boldsymbol{u}_{N-1}\}$ and form $\mathbf{X}_n = \mathrm{span}\{\boldsymbol{u}_0, \boldsymbol{u}_1, \ldots, \boldsymbol{u}_{n-1}\}$. The Galerkin equations are then uniquely solvable for $n = 1, 2, \ldots, N$. If $\boldsymbol{x} \in \mathrm{cl}\,\mathbf{S}$ we can express $\boldsymbol{x}$ as

$$\boldsymbol{x} = \sum_{i=0}^{N-1} \alpha_i u_i.$$

Since $\boldsymbol{b} = \mathcal{A}\boldsymbol{x}$, (9) is equivalent to

$$\langle \mathcal{A}(\boldsymbol{x} - \boldsymbol{x}_N), \mathcal{A}\boldsymbol{u}_j \rangle = \langle \mathcal{A} \sum_{i=0}^{N-1} (\alpha_i - \gamma_i)\boldsymbol{u}_i, \mathcal{A}\boldsymbol{u}_j \rangle = 0 \quad j = 0, 1, \ldots, N-1.$$

In particular,

$$\langle \mathcal{A} \sum_{i=0}^{N-1} (\alpha_i - \gamma_i)\boldsymbol{u}_i, \mathcal{A} \sum_{i=0}^{N-1} (\alpha_i - \gamma_i)\boldsymbol{u}_i, \rangle = 0$$

implying that $\alpha_i = \gamma_i$, $i = 0, \ldots, N-1$, i.e., that $\boldsymbol{x}_N = \boldsymbol{x}$.

The estimate (11) follows from (12):

$$\|\boldsymbol{x} - \boldsymbol{x}_n\| \leq C\|\mathcal{A}(\boldsymbol{x} - \boldsymbol{x}_n)\| = C\|\mathcal{A}\boldsymbol{x} - \mathcal{A}\boldsymbol{x}_n\|$$

so that

$$\|\boldsymbol{x} - \boldsymbol{x}_n\| \leq C\|\boldsymbol{b} - \mathcal{A}\boldsymbol{x}_n\|.$$

$\square$

Remark. The case for which $\dim \mathbf{S} = \infty$ but $\dim \mathbf{X}_n \neq n$ for all n is handled by redefining the $\mathbf{X}_n$'s so that the assumption $\dim \mathbf{X}_n = n$ holds (see [17], p. 93).

COROLLARY 2.9. *Given any $\boldsymbol{b} \in \mathbf{H}$, the Galerkin method described in Theorem 2.8 is convergent for (4) with $\mathbf{H}_n = \{\boldsymbol{b}, \mathcal{K}\boldsymbol{b}, \ldots \mathcal{K}^{n-1}\boldsymbol{b}\}$ for all $n \in \mathbf{N}$.*

Proof. By the Riesz theory for compact operators, $(\boldsymbol{I} - \mathcal{K})$ is bounded and bijective since $\mathcal{K}$ is compact. Let $\mathbf{S} = \cup_{n=1}^{\infty} \mathbf{H}_n$. The recursive definition of $\mathbf{H}_n$ guarantees that

---

ALGORITHM 1 (WGMRES).

Set $r^0 = b - \mathcal{A}x^\circ$, $\beta = \|r^0\|$, and $v^0 = r^0/\beta$
For $k = 1, 2, \ldots$ until $\langle r^k, r^k \rangle < \epsilon$,
$\quad h_{j,k} = \langle \mathcal{A}v^j, v^k \rangle$, $i = 1, 2, \ldots, k$
$\quad \hat{v}^{k+1} = \mathcal{A}v^k - \sum_{j=1}^{k} h_{j,k}v^j$
$\quad h_{k+1,k} = \|\hat{v}^{k+1}\|$
$\quad v^{k+1} = \hat{v}^{k+1}/h_{k+1,k}$
Set $x^k = x^0 + V^k y^k$
$\quad$ Here, y^k minimizes $\|\beta e_1 - \bar{H}^k y^k\|$ where
$\quad\quad \bar{H}^k$ is the $(k+1) \times k$ matrix with nonzero entries $h_{i,j}$,
$\quad\quad V^k = [v_1, \ldots, v_k]$, and
$\quad\quad e_1 = [1, 0, \ldots, 0]^T$

---

1. If $\dim \mathbf{S} = \infty$, then $\dim \mathbf{H}_n = n$ for all $n \in \mathbf{N}$, in which case the Galerkin equations are uniquely solvable for all $n \in \mathbf{N}$.
2. If $\dim \mathbf{S} = N$, then $\dim \mathbf{H}_n = n$ for all $1 \leq n \leq N$ and $\dim \mathbf{H}_n = N$ for all $n > N$.

To show that $x \in \operatorname{cl} \mathbf{S}$, note that

$$x = (I - \mathcal{K})^{-1}b = \sum_{j=0}^{\infty} \mathcal{K}^j b$$

where the Neumann series for $(I - \mathcal{K})^{-1}$ converges since the spectral radius of $\mathcal{K}$ is zero. Since the sum $\sum_{j=0}^{\infty} \mathcal{K}^j b$ is clearly contained in $\operatorname{cl} \operatorname{span}\{b, \mathcal{K}b, \mathcal{K}^2 b, \ldots\}$, we have $x \in \operatorname{cl} \operatorname{span}\{u_0, u_1, \ldots\}$. $\quad\square$

Remark. Corollary 2.9 can be applied to the case where $\mathcal{A} = \mathcal{J} - \mathcal{K}$, provided $\mathcal{J}$ has a bounded inverse and the spectral radius of $\mathcal{J}^{-1}\mathcal{K}$ is strictly less than unity. In that case, we can apply $\mathcal{J}^{-1}$ to the system to obtain

$$(I - \mathcal{J}^{-1}\mathcal{K})x = \mathcal{J}^{-1}\psi.$$

We then apply Corollary 2.9, since $\mathcal{J}^{-1}\mathcal{K}$ is compact.

2.3.2. Iterative Algorithms. Various iterative algorithms exist which can be used to implement the Galerkin method described in Corollary 2.9. The foremost of these in the linear algebra setting are the generalized minimum residual algorithm (GMRES) [20] and the generalized conjugate residual algorithm (GCR) [6]. The GMRES and GCR algorithms can be adapted quite readily to the space $\mathbf{H}$ instead of $\mathbf{R}^N$ and are shown in Algorithm 1 and Algorithm 2, respectively. We refer to the function-space versions of these algorithms as WGMRES and WGCR (waveform GMRES and waveform GCR) to distinguish them from their linear algebra counterparts.

The two fundamental operations in Algorithm 1 and Algorithm 2 are the operator-function product $\mathcal{A}p$ and the inner product $\langle \cdot, \cdot \rangle$. When solving (4) in the space $\mathbf{H}$, these operations are as follows:

> **ALGORITHM 2 (WGCR).**
>
> Set $p^0 = r^0 = b - \mathcal{A}x^\circ$
> For $k = 0, 1, \ldots$ until $\langle r^k, r^k \rangle < \epsilon$
> $\quad \alpha = \frac{\langle \mathcal{A}p^k, r^k \rangle}{\langle \mathcal{A}p^k, \mathcal{A}p^k \rangle}$
> $\quad x^{k+1} = x^k + \alpha p^k$
> $\quad r^{k+1} = r^k - \alpha \mathcal{A}p^k$
> $\quad p^{k+1} = r^{k+1} + \sum_{j=0}^{k} \beta_j^{(k)} p^j$
> $\qquad$ where $\{\beta_j^{(k)}\}$ are chosen so that
> $\qquad \langle \mathcal{A}p^{k+1}, \mathcal{A}p^j \rangle = 0$ for $0 \leq j \leq k$

Operator-Function Product: To calculate $w \equiv (I - \mathcal{K})p$:
$\qquad$ 1. Solve the IVP

$$
\begin{aligned}
(\tfrac{d}{dt} + M(t))y(t) &= N(t)p(t) \\
y(0) &= p_0 = 0
\end{aligned}
$$

$\qquad$ for $y(t)$, $t \in [0, T]$; this gives us $y = \mathcal{K}p$.
$\qquad$ 2. Set $w = p - y$

Inner Product: The inner product $\langle x, y \rangle$ is given by

$$
\langle x, y \rangle = \sum_{i=1}^{N} \int_0^T x_i(t) y_i(t) dt.
$$

Remark. Step 1 of the operator-function product is equivalent to one step of the classical dynamic iteration. Hence, we can speak of the WGMRES and WGCR algorithms as being accelerations of the classical dynamic iterations.

The conjugate-direction methods are just as amenable to parallel implementation as the classical dynamic iteration methods. As already discussed, one step of the dynamic iteration is embedded within the operator-function product. Moreover, the inner product is accomplished by N separate integrations of the pointwise product $x_i(t)y_i(t)$, which can be performed in parallel, followed by a global sum of the results.

2.4. Hybrid Methods for Nonlinear Systems. Many interesting applications are not necessarily described by a linear system of ODE's, but rather by a nonlinear system of ODE's:

$$
\begin{aligned}
\tfrac{d}{dt}x(t) + F(x(t), t) &= 0 \tag{13} \\
x(0) &= x_0.
\end{aligned}
$$

To solve (13), we apply Newton's method directly to the nonlinear ODE system (in a process sometimes referred to as the waveform Newton method (WN) [21]) to obtain the following iteration:

$$
\begin{aligned}
\left(\tfrac{d}{dt} + J_F(x^m)\right) x^{m+1} &= J_F(x^m)x^m - F(x^m) \tag{14} \\
x^{m+1}(0) &= x_0.
\end{aligned}
$$

296

ALGORITHM 3 (NONLINEAR GMRES/GCR).

Pick $\boldsymbol{x}^0$, ϵ^0, $\nu < 1$
For $m = 0, 1, \ldots$ until $\langle \boldsymbol{r}^m, \boldsymbol{r}^m \rangle < \phi$,
    Linearize (13) to form (14)
    Solve (14) with Algorithm 1 or Algorithm 2 using ϵ^m
    Update $\boldsymbol{x}^{m+1}$ and $\boldsymbol{r}^{m+1}$
    Set $\epsilon^{m+1} = \epsilon^m \cdot \nu$

Here, $\boldsymbol{J}_F$ is the Jacobian of $\boldsymbol{F}$. We note that (14) is a linear time-varying IVP to be solved for $\boldsymbol{x}^{m+1}$, which can be accomplished with a waveform conjugate direction method. The resulting operator Newton/conjugate-direction algorithm is shown in Algorithm 3, and we note the method is in the class of hybrid Krylov methods [4].

3. Device Transient Simulation. A device is assumed to be governed by the Poisson equation, and the electron and hole continuity equations:

$$\frac{\epsilon kT}{q} \nabla^2 u + q\,(p - n + N_D - N_A) = 0$$

$$\nabla \cdot \boldsymbol{J}_n - q\left(\frac{\partial n}{\partial t} + R\right) = 0$$

$$\nabla \cdot \boldsymbol{J}_p + q\left(\frac{\partial p}{\partial t} + R\right) = 0$$

where u is the normalized electrostatic potential, n and p are the electron and hole concentrations, $\boldsymbol{J}_n$ and $\boldsymbol{J}_p$ are the electron and hole current densities, N_D and N_A are the donor and acceptor concentrations, R is the net generation and recombination rate, q is the magnitude of electronic charge, and ϵ is the dielectric permittivity [2, 23].

The current densities $\boldsymbol{J}_n$ and $\boldsymbol{J}_p$ are given by the drift-diffusion approximations:

$$\boldsymbol{J}_n = -qD_n\,(n\,\nabla u - \nabla n)$$
$$\boldsymbol{J}_p = -qD_p\,(p\,\nabla u + \nabla p)$$

where D_n and D_p are the diffusion coefficients, which are assumed here to be related to the electron and hole mobilities by the Einstein relations, that is $D = \frac{kT}{q}\mu$. $\boldsymbol{J}_n$ and $\boldsymbol{J}_p$ are typically eliminated from the continuity equations using the drift-diffusion approximations, leaving a differential-algebraic system of three equations in three unknowns, u, n, and p.

Given a rectangular mesh that covers a two-dimensional slice of a MOSFET, a common approach to spatially discretizing the device equations is to use a finite-difference formula to discretize the Poisson equation, and an exponentially-fit finite-difference formula to discretize the continuity equations (the Scharfetter-Gummel method) [22]. On an N-node rectangular mesh, the spatial discretization yields a differential-algebraic system of $3N$ equations in $3N$ unknowns denoted by

$$\begin{aligned}
(15) \qquad \boldsymbol{f}_1(\boldsymbol{u}(t), \boldsymbol{n}(t), \boldsymbol{p}(t)) &= 0 \\
(16) \qquad \boldsymbol{f}_2(\boldsymbol{u}(t), \boldsymbol{n}(t), \boldsymbol{p}(t)) &= \tfrac{d}{dt}\boldsymbol{n}(t) \\
(17) \qquad \boldsymbol{f}_3(\boldsymbol{u}(t), \boldsymbol{n}(t), \boldsymbol{p}(t)) &= \tfrac{d}{dt}\boldsymbol{p}(t)
\end{aligned}$$

297

ALGORITHM 4 (WR FOR DEVICE SIMULATION).

```
guess u⁰, n⁰, p⁰ waveforms at all nodes
for k = 0, 1, ... until converged
    for each node i
        solve for uᵢᵏ⁺¹, nᵢᵏ⁺¹, pᵢᵏ⁺¹ waveforms:
```

$$f_{1_i}(u_i^{k+1}, n_i^{k+1}, p_i^{k+1}, u_j^k) = 0$$

$$f_{2_i}(u_i^{k+1}, n_i^{k+1}, u_j^k, n_j^k) = \frac{d}{dt} n_i^{k+1}$$

$$f_{3_i}(u_i^{k+1}, p_i^{k+1}, u_j^k, p_j^k) = \frac{d}{dt} p_i^{k+1}$$

---

where $t \in [0, T]$, and $\boldsymbol{u}(t), \boldsymbol{n}(t), \boldsymbol{p}(t) \in \mathbf{R}^N$ are vectors of normalized potential, electron concentration, and hole concentration, respectively. Here, $\boldsymbol{f}_1, \boldsymbol{f}_2, \boldsymbol{f}_3 : \mathbf{R}^{3N} \rightarrow \mathbf{R}^N$ are specified component-wise as

$$f_{1_i}(u_i, n_i, p_i, u_j) = \frac{\epsilon kT}{q} \sum_j \frac{d_{ij}}{L_{ij}}(u_i - u_j) - qA_i(p_i - n_i + N_D - N_A)$$

$$f_{2_i}(u_i, n_i, u_j, n_j) = \frac{D_n}{A_i} \sum_j \frac{d_{ij}}{L_{ij}}\Big[n_j B(u_j - u_i) - n_i B(u_i - u_j)\Big] - R_i$$

$$f_{3_i}(u_i, p_i, u_j, p_j) = \frac{D_p}{A_i} \sum_j \frac{d_{ij}}{L_{ij}}\Big[p_j B(u_i - u_j) - p_i B(u_j - u_i)\Big] - R_i.$$

The sums above are taken over the four nodes adjacent to node i (north, south, east, and west), L_{ij} is the distance from node i to node j, d_{ij} is the length of the side of the Voronoi box that encloses node i and bisects the edge between nodes i and j, and $B(v) = v/(e^v - 1)$ is the Bernoulli function, used to exponentially fit potential variation to electron concentration variation.

On the order of a thousand mesh nodes are typically needed to accurately represent a 2-D slice of a MOS transistor, so that simulating a circuit where even a few transistors are treated by numerically solving the device equations leads to an enormous coupled system of algebraic and differential equations.

The standard approach used to solve the differential-algebraic system generated by spatial discretization of the device equations is to discretize the d/dt terms with a low-order implicit integration method such as the second-order backward difference formula. The result is a sequence of nonlinear algebraic systems in $3N$ unknowns, each of which can be solved with some variant of Newton's method and/or relaxation [13]. Another approach is to apply relaxation directly to the differential-algebraic equation system with a WR algorithm [10, 19], as given in Algorithm 4.

In our approach, we apply the hybrid Krylov method described in Section 2.4 to (15)–(17) and obtain the following IVP at each Newton iteration m:

$$\begin{bmatrix} 0 \\ \frac{d}{dt}\boldsymbol{n}^{m+1} \\ \frac{d}{dt}\boldsymbol{p}^{m+1} \end{bmatrix} + \begin{bmatrix} \boldsymbol{J}_{f_{11}} & \boldsymbol{J}_{f_{12}} & \boldsymbol{J}_{f_{13}} \\ \boldsymbol{J}_{f_{21}} & \boldsymbol{J}_{f_{22}} & \boldsymbol{J}_{f_{23}} \\ \boldsymbol{J}_{f_{31}} & \boldsymbol{J}_{f_{32}} & \boldsymbol{J}_{f_{33}} \end{bmatrix} \begin{bmatrix} \boldsymbol{u}^{m+1} \\ \boldsymbol{n}^{m+1} \\ \boldsymbol{p}^{m+1} \end{bmatrix}$$

298

$$
= \begin{bmatrix} J_{f_{11}} & J_{f_{12}} & J_{f_{13}} \\ J_{f_{21}} & J_{f_{22}} & J_{f_{23}} \\ J_{f_{31}} & J_{f_{32}} & J_{f_{33}} \end{bmatrix} \begin{bmatrix} u^m \\ n^m \\ p^m \end{bmatrix} - \begin{bmatrix} f_1(u^m, n^m, p^m) \\ f_2(u^m, n^m, p^m) \\ f_3(u^m, n^m, p^m) \end{bmatrix}
$$

$$
\begin{bmatrix} u^{m+1}(0) \\ n^{m+1}(0) \\ p^{m+1}(0) \end{bmatrix} = \begin{bmatrix} u_0 \\ n_0 \\ p_0 \end{bmatrix}.
$$

4. Experimental Results. The nonlinear WGMRES and WGCR algorithms described in Section 2 were implemented in the WR-based device transient simulation program *WORDS* [19]. In addition, a waveform-relaxation-Newton algorithm (WRN) and a waveform conjugate-gradient-squared algorithm (WCGS) were implemented [24, 26]. The *WORDS* program uses a red/black block Gauss-Seidel scheme for WR and WRN, where the blocks correspond to vertical mesh lines; the corresponding Gauss-Seidel preconditioner is used for the implementations of the conjugate-direction methods. For all methods, the equations governing nodes in the same block are solved simultaneously using the second-order backward-difference formula. The implicit algebraic systems generated by the backward difference formula are solved with Newton's method and the linear equation systems generated by Newton's method are solved with sparse Gaussian elimination. Note that for the conjugate-direction algorithms, the linear algebraic equations at each timestep require only one algebraic Newton iteration.

Four N-channel MOSFET examples were used to compare the performance of the relaxation and conjugate-direction waveform methods:

kG: $2.2\,\mu$m channel-length; 50 psec, 0-5V ramp on the gate with the drain at 5V.
kD: $2.2\,\mu$m channel-length; 50 psec, 0-5V ramp on the drain with the gate at 5V.
jG: $0.17\,\mu$m channel-length; 5 psec, 0-1V ramp on the gate with the drain at 1V.
jD: $0.17\,\mu$m channel-length; 5 psec, 0-1V ramp on the drain with the gate at 1V.

The parameters used with the conjugate-direction methods were: $\epsilon^0 = 0.1$, $\nu = \sqrt{0.1}$, and $\phi = 1 \times 10^{-18}$. To simplify comparisons, 32 equally-spaced timesteps were used in all experiments.

Table 1 shows the number of function evaluations and the CPU time required for each of the waveform methods to reduce the max-norm of the drain terminal current error below 0.01% of the max-norm of the drain terminal current. As Table 1 indicates, conjugate-direction methods significantly reduced the number of function evaluations and CPU time over WR and WRN. In fact, in the **jG** example, WCGS is 22 times faster than ordinary WR. As is common in the algebraic case, WGMRES and WGCR perform similarly in terms of function evaluations, but WGMRES is computationally more efficient because it avoids several waveform inner products on each iteration. As is also common in the algebraic case, WCGS performs very well on most problems, but can also exhibit convergence difficulty on others.

We note that the speedup in CPU time is not as impressive as the speedup in terms of function evaluations. This is partly due to the extra work required at each iteration of the conjugate-direction methods — however, careful profiling and hand optimization of the conjugate-direction code will improve the CPU time comparison. The difference is especially apparent with WGMRES and WGCR, because the amount

TABLE 1

Comparison of WR, WRN, WGCR, WGMRES, and WCGS. CPU times shown are for an IBM RS/6000 model 540.

Example	Method	FEvals	CPU sec
jD	WR	8.43×10^6	14469
	WRN	3.77×10^6	7088
	WGCR	2.21×10^5	1138
	WGMRES	2.21×10^5	991
	WCGS	2.77×10^5	820
jG	WR	7.48×10^6	12615
	WRN	3.41×10^6	6214
	WGCR	1.97×10^5	1011
	WGMRES	1.97×10^5	877
	WCGS	1.97×10^5	568
kD	WR	1.22×10^6	1526
	WRN	3.94×10^5	559
	WGCR	9.03×10^4	315
	WGMRES	9.03×10^4	280
	WCGS	9.92×10^4	214
kG	WR	1.43×10^6	1756
	WRN	4.09×10^5	578
	WGCR	1.03×10^5	353
	WGMRES	1.03×10^5	316
	WCGS	Newton Non-Convergence	

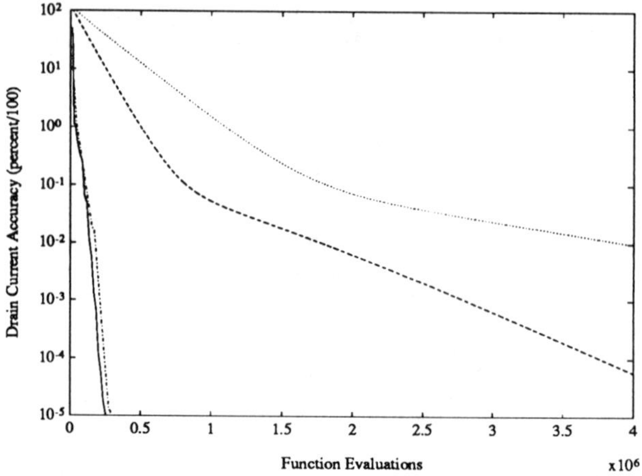

FIG. 1. *Convergence comparison between WR (dotted), WRN (dashed), WGCR/WGMRES (solid), and WCGS (dash-dotted) for* **jD** *example. The max-norm of the relative drain terminal current error is plotted against the number of function evaluations.*

of extra work per iteration grows linearly with the number of iterations (the number of *linear* iterations per Newton iteration in the nonlinear versions). On the other hand, WCGS maintains the same amount of work per iteration but can suffer from occasional stability problems. Recently, Freund and Nachtigal have developed the QMR algorithm which is characterized by a minimization property over a Krylov space but which does not have a linearly increasing amount of work per iteration [8]. A function-space generalization of the QMR algorithm is a topic which we are currently investigating.

The graphs in Figures 1 and 2 compare the convergence of WR, WRN, WGCR, WGMRES, and WCGS for the **jD** and **kD** examples, respectively. In the graphs, the terminal current error versus number of function evaluations is plotted and clearly demonstrates the rapid convergence of the conjugate-direction methods.

5. Conclusion. In this paper we presented some new dynamic iterative methods to accelerate the convergence of WR algorithm. The methods are based on the application of the Galerkin method to an operator equation formulation of the linear time-varying initial-value problem. Experimental results demonstrated that this acceleration significantly reduces the computation time for device transient simulation.

Future work is primarily focused on developing theoretical results about the convergence of linear and nonlinear conjugate-direction methods for differential-algebraic systems of equations. Our hope is to be able to provide intuitive convergence rates as in [14] and [15]. Other open questions remain regarding the behavior of the conjugate direction methods in the presence of multirate integration methods as well as in the presence of errors due to numerical quadrature. Finally, we are studying the function-space generalization of the QMR algorithm.

Acknowledgments. The authors would like to thank Ibrahim Elfadel at MIT and F. Odeh at the IBM T. J. Watson research center for many valuable discussions.

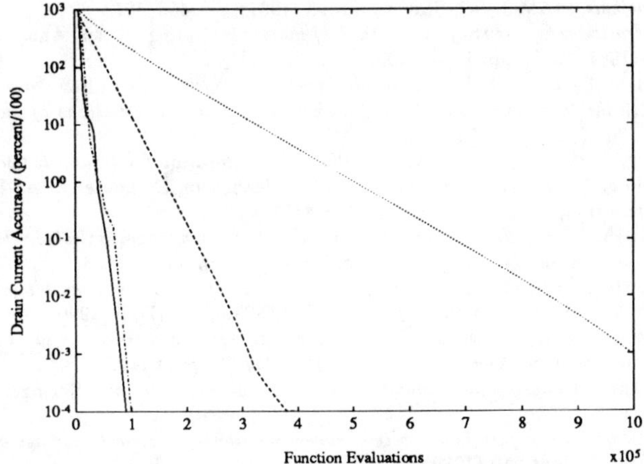

FIG. 2. *Convergence comparison between WR (dotted), WRN (dashed), WGCR/WGMRES (solid), and WCGS (dash-dotted) for* **kD** *example. The max-norm of the relative drain terminal current error is plotted against the number of function evaluations.*

The authors also express their deep appreciation to Mark Reichelt for his implementation of, and help with, his *WORDS* program.

REFERENCES

[1] K. E. ATKINSON, *A Survey of Numerical Methods for the Solution of Fredholm Integral Equations of the Second Kind*, SIAM, Philadelphia, 1976.

[2] R. BANK, W. COUGHRAN, JR., W. FICHTNER, E. GROSSE, D. ROSE, AND R. SMITH, *Transient simulation of silicon devices and circuits*, IEEE Trans. CAD, 4 (1985), pp. 436–451.

[3] R. W. BROCKETT, *Finite Dimensional Linear Systems*, Wiley, New York, 1970.

[4] P. BROWN AND Y. SAAD, *Hybrid Krylov methods for nonlinear systems of equations*, SIAM J. Sci. Statist. Comput., 11 (1990), pp. 450–481.

[5] J. B. CONWAY, *A Course in Functional Analysis, Second Edition*, Springer-Verlag, New York, 1990.

[6] H. C. ELMAN, *Iterative Methods for Large Sparse Nonsymmetric Systems of Linear Equations*, PhD thesis, Computer Science Dept., Yale University, New Haven, CT, 1982.

[7] W. ENGL, R. LAUR, AND H. DIRKS, *MEDUSA – A simulator for modular circuits*, IEEE Trans. CAD, 1 (1982), pp. 85–93.

[8] R. W. FREUND AND N. M. NACHTIGAL, *QMR: A quasi-minimal residual method for non-Hermitian linear systems*, Tech. Rep. 90.51, RIACS, NASA Ames Research Center, December 1990.

[9] R. KRESS, *Linear Integral Equations*, Springer-Verlag, New York, 1989.

[10] E. LELARASMEE, A. E. RUEHLI, AND A. L. SANGIOVANNI-VINCENTELLI, *The waveform relaxation method for time domain analysis of large scale integrated circuits*, IEEE Transactions on Computer-Aided Design of Integrated Circuits and Systems, 1 (1982), pp. 131–145.

[11] P. LINZ, *Analytical and Numerical Methods for Volterra Equations*, SIAM, Philadelphia, 1985.

[12] R. C. MACCAMY AND P. WEISS, *Numerical solution of Volterra integral equations*, Nonlinear Anal., 3 (1979), pp. 677–695.

[13] K. MAYARAM AND D. PEDERSON, *CODECS: A mixed-level device and circuit simulator*, in International Conference on Computer Aided-Design, Santa Clara, California, November 1988, pp. 112–115.

[14] U. MIEKKALA, *Dynamic iteration methods applied to linear DAE systems*, J. Comput. Appl. Math., 25 (1989), pp. 133–151.

[15] U. MIEKKALA AND O. NEVANLINNA, *Convergence of dynamic iteration methods for initial*

    value problems, SIAM J. Sci. Stat. Comp., 8 (1987), pp. 459–467.

[16] G. MIEL, *Iterative refinement of the method of moments*, Numer. Funct. Anal. and Optimiz., 9(11-12) (1987–1988), pp. 1193–1200.

[17] S. G. MIKHLIN, *Variational Methods in Mathematical Physics*, Macmillan, New York, 1964.

[18] P. OMARI, *On the fast convergence of a Galerkin-like method for equations of the second kind*, Math. Z., 201 (1989), pp. 529–539.

[19] M. REICHELT, J. WHITE, AND J. ALLEN, *Waveform relaxation for transient two-dimensional simulation of MOS devices*, in International Conference on Computer Aided-Design, Santa Clara, California, November 1989, pp. 412–415.

[20] Y. SAAD AND M. SCHULTZ, *GMRES: A generalized minimum residual algorithm for solving nonsymmetric linear systems*, SIAM J. Sci. Statist. Comput., 7 (1986), pp. 856–869.

[21] R. SALEH AND J. WHITE, *Accelerating relaxation algorithms for circuit simulation using waveform-newton and step-size refinement*, IEEE Trans. CAD, 9 (1990), pp. 951–958.

[22] D. SCHARFETTER AND H. GUMMEL, *Large-signal analysis of a silicon read diode oscillator*, IEEE Transactions on Electron Devices, ED-16 (1969), pp. 64–77.

[23] S. SELBERHERR, *Analysis and Simulation of Semiconductor Devices*, Springer-Verlag, New York, 1984.

[24] P. SONNEVELD, *CGS, a fast Lanczos-type solver for nonsymmetric linear systems*, SIAM J. Sci. Statist. Comput., 10 (1989), pp. 36–52.

[25] Y. V. VOROBYEV, *Method of Moments in Applied Mathematics*, Gordon and Breach, New York, 1965.

[26] J. K. WHITE AND A. SANGIOVANNI-VINCENTELLI, *Relaxation Techniques for the Simulation of VLSI Circuits*, Engineering and Computer Science Series, Kluwer Academic Publishers, Norwell, Massachusetts, 1986.

[27] E. ZEIDLER, *Nonlinear Functional Analysis and its Applications*, Springer-Verlag, New York, 1985.

BOUNDARY VALUE PROBLEMS IN SEMICONDUCTORS
FOR THE STATIONARY
VLASOV-MAXWELL-BOLTZMANN EQUATIONS

F. POUPAUD*

Abstract. In submicron devices the transport of carriers are govern by the Boltzmann transport equation coupled with electromagnetics. The aim of this paper is to prove the existence of stationary solutions for the corresponding boundary value problems with arbitrary large data. The techniques are those developed by the author for the mathematical analysis of the stationary Vlasov Maxwell system. They are based on the use of upper solutions of the Boltzmann transport equation that give enough a-priori estimates to apply the Schauder fixed-point theorem.

1. Introduction. For a long time, the drift-diffusion equations [19] have been the basic equations to model the transport of carriers in semiconductor devices. They have been widely studied and we refer for some mathematical results to [10,11,18] and to the enclosed references.

However, transport phenomena that occur is submicron devices are due to hot and balistic particles and they are not taken into account by drift-diffusion models [17]. In these conditions the kinetic models give the most accurate description of the physics. Recently, the works of R.J. Diperna and P.L. Lions [5,6] have allowed significant progress in the mathematical analysis of kinetic equations. In particular, they have proved existence of solutions of the Vlasov-Maxwell system for the Cauchy problem in a free space [6]. This result is based on mean compactness results [6,8] that appear to be very useful in this field. For semiconductor kinetic equations, some results can be found in [3,12,13,16].

However there are still few mathematical results on boundary value problem. One pioneer work is due to C. Greengard and P.A. Raviart [9]. They obtain existence of stationary solutions of the Vlasov Poisson system for a one-dimensional boundary value problem. The proof relies on the use of explicit solutions of the Vlasov problem. Therefore it seems not to be possible to use the same technique for more complicated geometries. On the contrary, the new strategy of proof introduced by the author in [14,15] is quite general. It has allowed to obtain stationary solutions of Vlasov-Poisson and Vlasov-Maxwell system for boundary value problems in any kind of geometries and with arbitrary large data. The aim of this paper is to use the technique developed in [14,15] for kinetic models for semiconductors.

2. Basic equations and the result of existence. In this paper, we consider electrons and holes with position x in a bounded set Ω (the device geometry) and

---

*CNRS – Département de Mathématiques – Université de Nice – Parc Valrose – F.06108 Nice Cédex 2, BP71.

with velocity v belonging to $\mathbf{R}^3$. The distribution functions $f_n(x,v)$ and $f_p(x,v)$ of electrons and holes satisfy the following Vlasov-Boltzmann equations

$$v \cdot \nabla_x f_n - \frac{q}{m_n^*}(-\nabla_x \phi(x) + v \wedge B(x)) \cdot \nabla_v f_n - Q_n(f_n) = 0,$$

$$(1) \qquad v \cdot \nabla_x f_p + \frac{q}{m_p^*}(-\nabla_x \phi(x) + v \wedge B(x)) \cdot \nabla_v f_p - Q_p(f_p) = 0,$$

$$\text{for } x \in \Omega \subset \mathbf{R}^3, \quad v \in \mathbf{R}^3.$$

q, m_n^* and m_p^* are respectively the elementary charge, the effective mass of electrons and holes. The scattering operators Q_n and Q_p are intended to model the collisions of the particles with the crystal. ϕ is the electrostatic field and B is the magnetic field given by the stationary Maxwell equations.

$$-\Delta_x \phi(x) = \frac{q}{\varepsilon}(N_d(x) - N_a(x) + n_p(x) - n_n(x)),$$

$$(2) \qquad \text{for } x \in \Omega, \quad n_p(x) = \int_{\mathbf{R}^3} f_p(x,v)dv, \quad n_n(x) = \int_{\mathbf{R}^3} f_n(x,v)dv,$$

$$\nabla_x \times B(x) = \mu q(j_p(x) - j_n(x)), \quad \nabla_x \cdot B(x) = 0,$$

$$(3) \qquad \text{for } x \in \Omega, \quad j_p(x) = \int v f_p(x,v)dv, \quad j_n(x) = \int v f_n(x,v)dv.$$

The known bounded functions N_d and N_a are the doping profiles. The constants ε and μ are the permittivity and the permeability of the semiconductor. The scattering operators Q_n and Q_p are given by

$$(4) \qquad Q_{n,p}(f)(x,v) = \int_{\mathbf{R}^3} [s_{n,p}(w,v)f(x,w) - s_{n,p}(v,w)f(x,v)]\,dw.$$

The transition rates s_n and s_p are known. To give an example, for the polar optical scattering the transition rate reads

$$(5)$$

$$s_n(v,w) = \frac{S_{\text{opt}}}{|v-w|^2}\left[(N_0 + 1)\delta\left(w^2 - v^2 + \frac{2}{m_n^*}\varepsilon_0\right) + N_0\delta\left(w^2 - v^2 - \frac{2}{m_n^*}\varepsilon_0\right)\right]$$

where S_{opt}, N_0 and ε_0 are constants and δ is the Dirac distribution. Let us notice that the polar occupation number N_0 is given by the Bose-Einstein statistics

$$(6) \qquad N_0 = \left[\exp\left(\frac{\varepsilon_0}{k_B T}\right) - 1\right]^{-1},$$

where k_B and T are the Boltzmann constant and the temperature of the semiconductor. So, in general, the transition rates are not functions but measures. However for all scatterings the transition rates satisfy

$$(7) \qquad s_{n,p} \in L^\infty\left[\mathbf{R}^3; m(\mathbf{R}^3)\right], \quad s_{n,p} \geq 0,$$

where $m(\mathbf{R}^3)$ denotes the space of bounded measures. Moreover the principle of detailed balance [2] yields

$$
\begin{aligned}
s_n(v,w)M_n(v) &= s_n(w,v)M_n(w), \\
s_p(v,w)M_p(v) &= s_p(w,v)M_p(w)
\end{aligned}
\tag{8}
$$

with the maxwellian distributions M_n and M_p given by

$$
\begin{aligned}
M_n(v) &= (2\pi k_B T/m_n^*)^{-3/2} \exp(-m_n^* v^2/2k_B T), \\
M_p(v) &= (2\pi k_B T/m_p^*)^{-3/2} \exp(-m_p^* v^2/2k_B T).
\end{aligned}
\tag{9}
$$

As an exercise, we let the reader verify that the transition rate given by (5) satisfies (7) and that the relationship (6) leads to (8).

Finally the system of equations (1),(2),(3) is completed by boundary conditions. We assume that the boundary $\partial\Omega$ of Ω can be decomposed into ohmic contacts $\partial\Omega_o$ and insulating boundaries $\partial\Omega_i$. We impose the entering part of the distributions f_n and f_p on $\partial\Omega_o$ and a reflection condition on $\partial\Omega_i$

$$
\begin{aligned}
f_n(x,v) &= g_n(x,v), \quad f_p(x,v) = g_p(x,v), \ \text{for } x \in \partial\Omega_o \\
&\text{and } v \in \mathbf{R}^3, \quad v \cdot \nu(x) < 0, \\
f_n(x,v) &= f_n(x, v - 2v \cdot \nu(x)\nu(x)), \quad f_p(x,v) = f_p(x, v - 2v \cdot \nu(x)\nu(x)) \\
&\text{for } x \in \partial\Omega_i \text{ and } v \in \mathbf{R}^3,
\end{aligned}
\tag{10}
$$

where $\nu(x)$ is the unit normal at x pointing outwards. We assume mixed Neumann-Dirichlet conditions for the potential ϕ

$$
\phi(x) = \phi_o(x), \quad \text{for } x \in \partial\Omega_o, \quad \frac{\partial\phi}{\partial\nu}(x) = 0 \text{ for } x \in \partial\Omega_i.
\tag{11}
$$

The normal component of the magnetic field is prescribed

$$
B(x) \cdot \nu(x) = b_o(x), \quad \text{for } x \in \partial\Omega.
\tag{12}
$$

We denote by (VMB) the Vlasov-Maxwell-Boltzmann system of equations (1),(2) and (3) completed with the boundary condition (10),(11) and (12). The main result of this paper is

THEOREM 1. *Let us assume that the data satisfy*

$$
0 \le g_n(x,v) \le c_n M_n(v), \quad 0 \le g_p(x,v) \le c_p M_p(v),
\tag{13}
$$

$$
\phi_0 \in H^{1/2}(\partial\Omega) \cap L^\infty(\partial\Omega),
\tag{14}
$$

$$
b_0 \in H^{-1/2}(\partial\Omega), \quad \int_{\partial\Omega} b_0(x)d\gamma(x) = 0,
\tag{15}
$$

for some constants c_n and c_p. Then the problem (VMB) has at least a weak solution (f_n, f_p, ϕ, B) that verifies

$$
0 \le f_n(x,v) \le C_n' M_n(v), \quad 0 \le f_p(x,v) \le C_p' M_p(v),
\tag{16}
$$

$$
\text{for } x \in \Omega, v \in \mathbf{R}^3,
$$

$$
\phi \in H^1(\Omega) \cap L^\infty(\Omega),
\tag{17}
$$

$$
B \in H(\mathrm{div}, \mathrm{curl}, \Omega),
\tag{18}
$$

for some constants C_n' and C_p'.

Remark 1. In the above model we have used the assumption of parabolic band diagrams for electrons and holes. Rigorously [2,13] the velocity v and the energy E of particles depends on their wave vectors k that belong to the reciprocal lattice of the semiconductor through the relationships

$$(19) \qquad E = E(k), \quad v(k) = \frac{1}{\hbar} \nabla_k E(k)$$

where $\hbar$ is the reduced Planck constant. But without any change the analysis of this paper can be applied to models with generalized band diagram.

Remark 2. The operators Q_n and Q_p given by (4) are linear. But the expression (4) is only valid for non-degenerate semiconductors, that is to say for not too high concentrations of particles. Otherwise the operators become quadratic. For instance, the collision operator for electrons reads

$$(20) \qquad Q_n(f)(x,v,w) = \int_{\mathbf{R}^3} s_n(x,w,v)f(x,w)[1 - \eta f(x,v)] -$$
$$- s_n(x,v,w)f(x,v)[1 - \eta f(x,w)]dw$$

where η is a known constant. In this case the techniques of [14,15] cannot be straightforwardly applied. However, as in [16] it is possible to prove that Fermi Dirac distributions

$$(21)$$
$$F_n(v) = \frac{1}{\eta + \exp\left(\frac{m_n^* v^2}{2k_B T} + \varphi\right)}, \quad F_p(v) = \frac{1}{\eta + \exp\left(\frac{m_p^*}{2k_B T}v^2 + \varphi\right)}, \quad \text{for } \varphi \in \mathbf{R},$$

are upper solutions of the Vlasov Boltzmann equations. Then, by using mean compactness results [6,8] to control the nonlinearity of (20) we hope to obtain an existence result for this degenerate semiconductor model.

The remainder of this paper is devoted to the proof of theorem 1. In order to motivate the following sections let us sketch this proof. In nonlinear analysis, key points are to obtain a priori estimates. We first detail this point. It is well known that the functions

$$(22)$$
$$G_n(x,v) = K_n \exp\left(\frac{q\phi(x)}{k_B T}\right) M_n(v),$$
$$G_p(x,v) = K_p \exp\left(-\frac{q\phi(x)}{k_B T}\right) M_p(v), \quad \text{for } K_n \geq 0, \quad K_p \geq 0,$$

are solutions of the Vlasov-Boltzmann equations (1). Indeed these are functions of the total energy of particles then they are solutions of the Vlasov equation and moreover they are maxwellian distributions with respect to v then they cancel the collisions terms. The last point is a consequence of (8). Now we choose the constants K_n and K_p such that

$$(23) \quad g_n(x,v) \leq G_n(x,v), \quad g_p(x,v) \leq G_p(x,v), \quad \text{for } x \in \partial\Omega_0, \quad v \cdot \nu(x) < 0.$$

For instance we may choose

$$(24) \quad \begin{aligned} K_n &= c_n \exp\left(-\frac{q \inf \text{ess } \phi_0}{k_B T}\right), \\ K_p &= c_p \exp\left(\frac{q \sup \text{ess } \phi_0}{k_B T}\right). \end{aligned}$$

Since the function G_n and G_p are upper solutions of the Vlasov-Boltzmann problem (1),(10) (they are even functions with respect to v then they satisfy the reflection conditions) we get

$$(25) \quad 0 \leq f_n(x,v) \leq G_n(x,v), \quad 0 \leq f_p(x,v) \leq G_p(x,v), \quad \text{for } x \in \Omega, v \in \mathbf{R}^3.$$

This is not an a-priori estimate because G_n and G_p depend on the unknown potential ϕ. But, by using (25) and (2) we get

$$(26) \quad \begin{aligned} \frac{q}{\varepsilon}\left[N_d(x) - N_a(x) - K_n \exp\left(\frac{q\phi(x)}{k_B T}\right)\right] &\leq -\Delta\phi(x) \leq \frac{q}{\varepsilon}\Big[N_d(x) - N_a(x)+ \\ + K_p \exp\left(-\frac{q\phi(x)}{k_B T}\right)\Big], & \text{for } x \in \Omega, \end{aligned}$$

$$\phi(x) = \phi_0(x), \text{ for } x \in \partial\Omega_0, \quad \frac{\partial\phi}{\partial\nu}(x) = 0, \quad \text{for } x \in \partial\Omega_i.$$

Let us introduce the solutions ϕ_1 and ϕ_2 of the following nonlinear elliptic problems

$$(27) \quad -\Delta\phi_1(x) + \frac{q}{\varepsilon}K_n \exp\left(\frac{q\phi_1(x)}{k_B T}\right) = \frac{q}{\varepsilon}(N_d(x) - N_a(x)),$$
$$\text{for } x \in \Omega,$$
$$\phi_1(x) = \phi_0(x) \text{ for } x \in \partial\Omega_0, \quad \frac{\partial\phi_1}{\partial\nu}(x) = 0, \quad \text{for } x \in \partial\Omega_i,$$

$$(28) \quad -\Delta\phi_2(x) - \frac{q}{\varepsilon}K_p \exp\left(-\frac{q\phi_2(x)}{k_B T}\right) = \frac{q}{\varepsilon}(N_d(x) - N_a(x)),$$
$$\text{for } x \in \Omega,$$
$$\phi_2(x) = \phi_0(x) \text{ for } x \in \partial\Omega_0, \quad \frac{\partial\phi_2}{\partial\nu}(x) = 0, \quad \text{for } x \in \partial\Omega_i.$$

Since the functions $\exp\left(\frac{q\phi}{k_B T}\right)$ and $-\exp\left(-\frac{q\phi}{k_B T}\right)$ are increasing with respect to ϕ, standard results on elliptic problems assure the existence and the uniqueness of the solutions of (27) and (28). Moreover we get

PROPOSITION 1. *Let ϕ be a function that satisfies (26), then we get*

$$(29) \qquad \phi_1(x) \le \phi(x) \le \phi_2(x), \quad \text{for } x \in \Omega.$$

This proposition is a consequence of the maximum principle. For more details we refer to [15]. Let us remark that the functions $\frac{q}{\varepsilon}(N_d - N_a)$ and ϕ_0 are supposed to be bounded. It follows that ϕ_1 and ϕ_2 belong to $L^\infty(\Omega)$. Then (29) provides an a-priori L^∞-estimate on the potential. So, we deduce from (22), (25) that there are two constants C'_n and C'_p such that (16) holds true. Let us point out that we take advantage of the fact that the self consistent electric field is repulsive to obtain the above estimates. Indeed, if we consider attractive forces, instead of (26) we get the opposite inequalities and then we cannot conclude. Now the strategy of proof is clear. Starting from a given electromagnetic field we compute the solution of the linear Vlasov-Boltzmann problem (1),(10). From the related concentrations and fluxes, we can solve the stationary Maxwell problem (2),(3),(11),(12). This defines a map on electromagnetic fields. Then, we hope to apply the fixed point Schauder theorem thanks to the previous a priori estimates and the regularizing effects of the elliptic equations (2),(3). There are two main difficulties to follow this program. First, in order to solve the Vlasov-Boltzmann problem we have to apply classic results on first order hyperbolic equations that require a very smooth convective field. At least we need that ϕ belongs to $W^{2,1}(\Omega)$ and B to $W^{1,1}(\Omega)$ or more classically that ϕ belongs to $C^2(\Omega)$ and B to $C^1(\Omega)$, cf. [1,7]. Secondly we may have no uniqueness for this linear Vlasov-Boltzmann problem. Indeed let us consider the case where the collision terms vanish. Then we can impose any value of the distributions functions f_n and f_p on the closed characteristics. Some such counterexamples are given in [15]. To overcome these two difficulties we have to introduce a perturbed Vlasov-Maxwell-Boltzmann problem (VMB)$_\lambda$, for $\lambda > 0$, where we have regularized the electromagnetic field and introduced an absorption term in the Vlasov-Boltzmann equation to recover a result of existence and uniqueness for (1),(10). For this modified problem we can follow the previous strategy of proof. It remains to pass to the limit as λ vanishes, to obtain solutions to the original problem. Thus the first step is to give a precise mathematical analysis of the linear Vlasov-Boltzmann problem.

3. The linear Vlasov-Boltzmann problem. In this section we impose that the electromagnetic field satisfies the very strong assumption

$$(30) \qquad \phi \in C^2(\Omega), \quad B \in C^1(\Omega)$$

for the sake of compactness, we will only consider electrons. Of course the same results hold true for holes. First we recall the following classical theorem.

THEOREM 2. *Let $\Sigma_n = \Sigma_n(x,v)$ be a function that satisfies*

$$(31) \qquad \Sigma_1 \ge \Sigma_n(x,v) \ge \Sigma_2 > 0, \quad \text{for } x \in \Omega, \quad v \in \mathbf{R}^3,$$

for some constants Σ_1 and Σ_2. Then for any data g_n and S_n that satisfy

(32)
$$\int_{\Gamma_o^-} |v \cdot \nu(x) g_n(x,v)| d\gamma(x) dv < \infty,$$
$$\int_{\Omega \times \mathbf{R}^3} |S_n(x,v)| dx\; dv < \infty,$$

with $\Gamma_o^- = \{(x,v), x \in \partial\Omega_0, v \cdot \nu(x) < 0\}$ and with $d\gamma$ being the superficial measure of $\partial\Omega$, the linear Vlasov equation

(33) $$\Sigma_n(x,v) f_n + v \cdot \nabla_x f_n - \frac{q}{m_n^*}(-\nabla_x \phi(x) + v \wedge B(x)) \cdot \nabla_v f_n = S_n$$

completed with the boundary conditions (10) has a unique solution f_n in $L^1(\Omega \times \mathbf{R}^3)$. This solution verifies

(34)
$$\int_{\Omega \times \mathbf{R}^3} |f_n(x,v)| \Sigma(x,v) dx dv \leq \int_{\Omega \times \mathbf{R}^3} |S_n(x,v)| dx dv +$$
$$+ \int_{\Gamma_\sigma^-} |v \cdot \nu(x) g_n(x,v)| d\gamma(x) dv,$$

Moreover we get the following weak maximum principle. If the data are nonnegative then the solution is nonnegative.

For a proof of the previous theorem we refer to [1,15]. Now we are ready to prove the following result on the linear Vlasov-Boltzmann problem.

THEOREM 3. *Let $\lambda > 0$, and g_n be a data that satisfies (13) then the equation*

(35) $$\lambda f_n + v \nabla_x f_n - \frac{q}{m_n^*}(-\nabla_x \phi(x) + v \wedge B(x)) \cdot \nabla_v f_n - Q_n(f_n) = 0$$

completed with the boundary conditions (10) has a unique solution in $L^1(\Omega \times \mathbf{R}^3)$. Moreover, independently of λ, this solution verifies (25) if (23) is satisfied.

In order to prove theorem 3, we state the following

LEMMA 1. *The operator Q_n is bounded on $L^1(\Omega \times \mathbf{R}^3)$. Let σ_n be the scattering rate defined by*

(36) $$\sigma_n(v) = \int_{\mathbf{R}^3} s_n(v,w) dw,$$

then there is a positive constant σ_1 such that

(37) $$0 \leq \sigma_n(v) \leq \sigma_1, \quad \text{for } v \in \mathbf{R}^3.$$

The operator Q_n reads

$$(38) \qquad Q_n(f) = Q_n^+(f) - \sigma_n f$$

with

$$(39) \qquad Q_n^+(f)(x,v) = \int_{\mathbf{R}^3} s_n(v,v) f(x,w) dw$$

and we have the relationships

$$(40) \qquad \int_{\mathbf{R}^3} Q_n^+(f)(x,v) dv = \int_{\mathbf{R}^3} \sigma_n(v) f(x,v) dv, \text{ for } x \in \Omega, \text{ for}$$

$$\text{any } f \text{ in } L^1(\Omega \times \mathbf{R}^3),$$

$$(41) \qquad Q_n^+(M_n)(v) = \sigma_n(v) M_n(v), \text{ for } v \in \mathbf{R}^3.$$

Proof of Lemma 1. The estimate (37) is a consequence of (7). For smooth functions f, we immediately obtain (40). Then we get

$$\int_{\Omega \times \mathbf{R}^3} |Q_n^+(f)(x,v)| dx dv \leq \int_{\Omega \times \mathbf{R}^3} Q_n^+(|f|)(x,v) dx dv = \int_{\Omega \times \mathbf{R}^3} \sigma_n(v) |f_n(x,v)| dx dv.$$

It follows that Q_n^+ can be prolonged as a bounded operator on $L^1(\Omega \times \mathbf{R}^3)$. The same is true for Q_n thanks to (38). Finally, (8) yields (41). $\square$

Proof of Theorem 3. Let us define a map Λ on $L^1(\Omega \times \mathbf{R}^3)$ by $\Lambda(f) = h$ where h is the unique solution given by theorem 2 of

$$(\lambda + \sigma_n(v))h + v\nabla_x h - \frac{q}{m_n^*}(-\nabla_x \phi(x) + v \wedge B(x))\nabla_v h = Q_n^+(f)$$

with the boundary conditions (10). Since a solution of the Vlasov-Boltzmann problem is a fixed point for Λ, we have only to prove that Λ is a contraction to obtain existence and uniqueness. Let us introduce the following norm on $L^1(\Omega \times \mathbf{R}^3)$

$$|||f||| = \int_{\Omega \times \mathbf{R}^3} (\lambda + \sigma_n(v)) |f(x,v)| dx dv.$$

From (37) it follows that

$$\lambda \|f\|_{L^1(\Omega \times \mathbf{R}^3)} \leq |||f||| \leq (\lambda + \sigma_1) \|f\|_{L^1(\Omega \times \mathbf{R}^3)}.$$

Then the norm $||| \cdot |||$ is equivalent to the usual norm of $L^1(\Omega \times \mathbf{R}^3)$. Applying the estimate (34) for the function $\Lambda(f) - \Lambda(g)$ we obtain

$$|||\Lambda(f) - \Lambda(g)||| \leq \int_{\Omega \times \mathbf{R}^3} |Q^+(f-g)| dx dv.$$

Using (40) it follows

$$|||\Lambda(f) - \Lambda(g)||| \leq \int\limits_{\Omega \times \mathbf{R}^3} \sigma_n(v)|f - g|(x,v)dx\,dv$$

$$\leq \sup \text{ess} \left(\frac{\sigma_n(v)}{\lambda + \sigma_n(v)} \right) |||f - g|||$$

$$\leq \frac{\sigma_1}{\lambda + \sigma_1}|||f - g|||.$$

Therefore Λ is a contraction for the norm $||| \cdot |||$. It remains to prove (25). Let us consider the following sequence of functions

$$f_n^{(0)}(x,v) = 0, \quad \text{for } x \in \Omega, v \in \mathbf{R}^3,$$
$$f_n^{(i+1)} = \Lambda(f_n^{(i)}).$$

Since Λ is a contraction, this sequence converges towards f_n. Then it is enough to prove that (25) holds true for the function $f_n^{(i)}$. Using that the function G_n is a solution of (1), we obtain that the functions $f^{(i)} = G_n - f_n^{(i)}$ satisfies

$$(\lambda + \sigma_n(v))f^{(i+1)} + v \cdot \nabla_x f^{(i+1)} - \frac{q}{m_n^*}(-\nabla_x \phi(x) + v \wedge B(x)) \cdot \nabla_v f^{(i+1)} =$$

$$= \lambda G_n + Q^+(f^{(i)}), \quad \text{for } x \in \Omega, v \in \mathbf{R}^3.$$
$$f^{(i+1)}(x,v) = (G_n - g_n)(x,v), \quad \text{for } x \in \partial\Omega_0, v \cdot \nu(x) < 0$$
$$f^{(i+1)}(x,v) = f^{(i+1)}(x, v - 2v \cdot \nu(x)\nu(x)), \quad \text{for } x \in \partial\Omega_i, v \in \mathbf{R}^3.$$

Thanks to the weak maximum principle, it is easy by induction to prove that $f^{(i)} \geq 0$. It concludes. $\square$

We end this section with some reflections about weak solutions of Vlasov-Boltzmann problems. Let us introduce the space of trial functions V,

$$(42) \quad \begin{aligned} V = \big\{ \theta \in C_0^1(\mathbf{R}^6); \quad & \theta(x,v) = 0, \quad \text{for } x \in \partial\Omega_0, \\ v \cdot \nu(x) > 0, \quad & \theta(x, v - 2v \cdot \nu(x)\nu(x)) = \theta(x,v), \text{ for} \\ & x \in \partial\Omega_i, v \in \mathbf{R}^3 \big\}. \end{aligned}$$

We have

DEFINITION 1. *The function f_n is a weak solution of*

$$(43) \quad \begin{aligned} v \cdot \nabla_x f_n - \frac{q}{m_n^*}(-\nabla_x \phi + v \wedge B) \cdot \nabla_v f_n - Q_n(f_n) = S, \quad & \text{for } x \in \Omega, v \in \mathbf{R}^3, \\ f_n(x,v) = g_n(x,v), \quad \text{for } x \in \partial\Omega_0, \quad v \cdot \nu(x) < 0 \\ f_n(x,v) = f_n(x, v - 2v \cdot \nu(x)v), \quad \text{for } x \in \partial\Omega_i, v \in \mathbf{R}^3, \end{aligned}$$

if and only if for any function θ in V we get

$$(44) \quad \begin{aligned} \int\limits_{\Omega \times \mathbf{R}^3} f_n \left(-v\nabla_x \theta + \frac{q}{m^*}(-\nabla_x \phi + v \wedge B) \cdot \nabla_v \theta \right) - Q(f_n)\theta dx\,dv = \\ = \int\limits_{\Omega \times \mathbf{R}^3} s\theta dx\,dv - \int\limits_{\Gamma_0^-} v \cdot \nu(x)g_n\theta d\gamma(x)dv. \end{aligned}$$

Remark 3. To give a meaning to (44) we only need the very coarse regularity assumptions,

$$\nabla_x \phi \text{ and } B \in L^1(\Omega), \quad S \in L^1_{\text{loc}}(\Omega \times \mathbf{R}^3), \quad v \cdot \nu(x) g_n \in L^1_{\text{loc}}(\Gamma_0^-).$$

Thus, starting with a sequence of smooth electromagnetic fields we easily obtain solutions for non smooth fields. But we have no more results of uniqueness. This difficult question is related to the existence of characteristics (see [7]).

The following proposition will be useful in the sequel.

PROPOSITION 2. *Let* $f_n^{(i)}$, $\phi^{(i)}$, $B^{(i)}$, $S_n^{(i)}$, $g_n^{(i)}$ *be sequences such that* $f_n^{(i)}$ *is a weak solution of (43) with the electromagnetic field* $\phi^{(i)}$, $B^{(i)}$, *the source term* $S^{(i)}$ *and the entering data* $g_n^{(i)}$. *Then if we have*

$$\phi^{(i)} \to \phi \text{ in } H^1(\Omega), \quad B^{(i)} \to B \text{ in } L^2(\Omega),$$
$$S_n^{(i)} \to S_n \text{ in } L^1_{\text{loc}}(\Omega \times \mathbf{R}^3) \text{ weak }, g_n^{(i)} \to g_n \text{ in } L^1_{\text{loc}}(\Gamma_0^-) \text{ weak}$$
$$f_n^{(i)} \to f_n \text{ in } L^\infty(\Omega \times \mathbf{R}^3) \text{ weak } *, \|f_n^{(i)}\|_{L^1(\Omega \times \mathbf{R}^3)} \leq C,$$

for some constant C, *the function* f_n *is a weak solution of (43) with the data* ϕ, B, g_n *and* S.

In view of (44) this proposition is obvious. We refer for more details to [15].

4. The modified Vlasov Maxwell Boltzmann system. In order to avoid singularities of solution between the ohmic contact and the insulating boundary, in this section we assume that there is a part ω of the boundary such that

$$(45) \qquad \omega \Subset \partial\Omega_0, \quad \text{supp}(g_n) \subset \omega \times \mathbf{R}^3, \quad \text{supp}(g_p) \subset \omega \times \mathbf{R}^3.$$

We will get rid of this restriction at the end of this section. Now for any potential ϕ that satisfies (11) and any magnetic field B that satisfies (12), we want to define a regularized functions ϕ_λ, B_λ, $\lambda > 0$ such that $\phi_\lambda \to \phi$, $B_\lambda \to B$ for convenient topologies. Moreover, in order to keep a uniform bound for the constants K_n, K_p that have to satisfy (23), we need to control uniformly ϕ_λ on the part ω of the boundary. This explains the pretty complicated expression of ϕ_λ. Let ζ_λ be a regularizing sequence

$$(46) \qquad \zeta_\lambda(x) = \frac{1}{\lambda^3} \zeta\left(\frac{x}{\lambda}\right), \text{ for } x \in \mathbf{R}^3, \quad \zeta \in C_0^\infty(\mathbf{R}^3), \quad \int_{\mathbf{R}^3} \zeta(x)dx = 1,$$
$$\text{supp}\, \zeta \subset \{x; |x| < 1\}, \quad \zeta \geq 0.$$

Then we put

$$(47) \qquad B_\lambda(x) = B * \zeta_\lambda(x), \quad \text{for } x \in \Omega,$$

where we have prolonged B by 0 outside Ω. To define ϕ_λ we need to introduce functions η and ψ such that

$$(48) \qquad \eta \in C_0^\infty(\mathbf{R}^3), \quad 0 \le \eta \le 1, \quad \eta(x) = 1 \text{ for } d(x,\omega) \le \lambda_0,$$
$$\eta(x) = 0 \text{ for } x \in \partial\Omega_i,$$

$$(49) \qquad \psi \in H^1(\Omega), \quad -\Delta\psi(x) = \frac{q}{\varepsilon}(N_d(x) - N_a(x)) \text{ for } x \in \Omega,$$
$$\psi(x) = \phi_0(x) \text{ for } x \in \partial\Omega_0, \quad \frac{\partial\psi}{\partial\nu}(x) = 0 \text{ for } x \in \partial\Omega_i.$$

Then ϕ_λ is defined on Ω by

$$(50) \qquad \phi_\lambda = J_\lambda^2[(\phi - \psi)\eta] + \zeta_\lambda * [(1 - \eta)P\phi + \eta P\psi]$$

where P is an extension operator that have the following properties

$$(51) \qquad \begin{aligned} &P : H^1(\Omega) \to H^1(\mathbf{R}^3); P\phi(x) = \phi(x), \text{ for } x \in \Omega; \\ &\text{if } \phi \ge 0 \text{ then } P\phi \ge 0; \text{ there exists a constant } c \\ &\text{such that for any } \phi \text{ in } H^1(\Omega) : \|P\phi\|_{H^1(\mathbf{R}^3)} \le c\|\phi\|_{H^1(\Omega)} \\ &\text{and if } \phi \text{ belongs to } L^\infty(\Omega), \|P\phi\|_{L^\infty(\mathbf{R}^3)} \le c\|\phi\|_{L^\infty(\Omega)}. \end{aligned}$$

J_λ is the resolvent of the laplacian $-\Delta$ consider as an unbounded operator on $L^2(\Omega)$ whose domain is $H^2 \cap H_0^1(\Omega)$

$$(52) \qquad J_\lambda : L^2(\Omega) \to H^2 \cap H_0^1(\Omega), \quad J_\lambda = (1 - \lambda\Delta)^{-1}.$$

thus, J_λ^2 maps $L^2(\Omega)$ into $H^4(\Omega)$ and then into $C^2(\Omega)$. The properties of these regularizations are summarized below.

PROPOSITION 3. *The map* $(\phi, B) \to (\phi_\lambda, B_\lambda)$ *is continuous from* $H^1(\Omega) \times [L^2(\Omega)]^3$ *into* $C^2(\Omega) \times [C^1(\Omega)]^3$. *There exists a constant* κ *such that for* $\lambda < \lambda_0$, *uniformly with respect to* λ *and* ϕ, *we obtain*

$$(53) \qquad |\phi_\lambda(x)| \le \kappa, \text{ for } x \in \omega.$$

The map $\phi \to \phi_\lambda$ *is non decreasing with respect to* ϕ,

$$(54) \qquad \phi \ge \psi \text{ implies } \phi_\lambda \ge \psi_\lambda.$$

If a sequence $(\lambda^{(i)}, \phi^{(i)}, B^{(i)})$ *satisfies*

$$(55) \qquad \begin{aligned} &\lambda^{(i)} \to 0, \quad B^{(i)} \to B \text{ in } [L^2(\Omega)]^3, \\ &\phi^{(i)} \to \phi \text{ in } H^1(\Omega), \quad \phi^{(i)}(x) = \phi_0(x) \text{ for } x \in \partial\Omega_\sigma, \\ &\|\Delta\phi^{(i)}\|_{L^2(\Omega)} \le c, \end{aligned}$$

for some constant c, *then the regularized sequence* $(\phi_{\lambda^{(i)}}^{(i)}, B_{\lambda^{(i)}}^{(i)})$ *verifies*

$$(56) \qquad \phi_{\lambda^{(i)}}^{(i)} \to \phi \text{ in } H^1(\Omega), \quad B_{\lambda^{(i)}}^{(i)} \to B \text{ in } [L^2(\Omega)]^3.$$

Proof of Proposition 3. The first point is obvious. To obtain the second one we first remark that

$$J_\lambda^2[(\phi - \psi)\eta] \in H_0^1(\Omega), \quad (1 - \eta)P\phi(x) = 0 \text{ for } d(x,\omega) \leq \lambda_0.$$

Therefore for $\lambda < \lambda_0$, thanks to (46), we get

$$\phi_\lambda(x) = \zeta_\lambda * [\eta P\psi](x), \text{ for } x \in \omega.$$

The last equality leads to (53) with $\kappa = \|\eta P\psi\|_{L^\infty(\mathbf{R}^3)}$. The monotonicity of the regularization is a consequence of the maximum principle and of the fact that we have chosen a non negative regularizing sequence. The consistency of the regularization is obvious for the magnetic field. To obtain it for the potential we have to remark that the function $(\phi^{(i)} - \psi)\eta$ belongs to a bounded set of $H^2(\Omega) \cap H_0^1(\Omega)$. It follows (see [15] for more details)

$$J_{\lambda^{(i)}}[(\phi^{(i)} - \psi)\eta] \to (\phi - \psi)\eta \text{ in } H^2(\Omega) \cap H_0^1(\Omega).$$

It allows to conclude. □

From now on, using theorem 3, we know that for any (ϕ, B) in $H^1(\Omega) \times [L^2(\Omega)]^3$ the modified Vlasov Boltzmann problem

(57)
$$\lambda f_n + v \cdot \nabla_x f_n - \frac{q}{m_n^*}(-\nabla_x \phi_\lambda(x) + v \wedge B_\lambda(x))\nabla_v f_n - Q_n(f_n) = 0,$$
$$\lambda f_p + v \cdot \nabla_x f_p + \frac{q}{m_p^*}(-\nabla_x \phi_\lambda(x) + v \wedge B_\lambda(x))\nabla_v f_p - Q_p(f_p) = 0, \quad \text{for } x \in \Omega, v \in \mathbf{R}^3,$$

with the boundary conditions (10), has a unique solution (f_n, f_p). Furthermore this solution satisfies

(58)
$$0 \leq f_n \leq G_{n,\lambda}, \quad 0 \leq f_p \leq G_{p,\lambda},$$

where the functions $G_{n,\lambda}$, $G_{p,\lambda}$ are defined by

(59)
$$G_{n,\lambda}(x, v) = K_n \exp\left(\frac{q\phi_\lambda(x)}{k_B T}\right) M_n(v),$$
$$G_{p,\lambda}(x, v) = K_p \exp\left(-\frac{q\phi_\lambda(x)}{k_B T}\right) M_p(v).$$

The constants K_n and K_p do not depend neither on λ nor on ϕ. They are given by

(60)
$$K_n = C_n \exp\left(\frac{q\kappa}{k_B T}\right), \quad K_p = C_p \exp\left(\frac{q\kappa}{k_B T}\right)$$

where κ is the constant that appears in (53). The condition (23) is well satisfied because of the assumption (45) and of (53). It follows that we have the following estimates on the concentration and fluxes related to the distribution f_n and f_p

$$
0 \le n_n(x) \le K_n \exp\left(\frac{q\phi_\lambda(x)}{k_B T}\right),
$$

(61)
$$
0 \le n_p(x) \le K_p \exp\left(-\frac{q\phi_\lambda(x)}{k_B T}\right),
$$

$$
|j_n(x)| \le J_n \exp\left(\frac{q\phi_\lambda(x)}{k_B T}\right), \quad |j_p(x)| \le J_p \exp\left(-\frac{q\phi_\lambda(x)}{k_B T}\right),
$$

$$
\text{for } x \in \Omega,
$$

where the constants J_n and J_p do not depend neither on λ nor on ϕ. Now if the concentration and fluxes n_n, n_p, j_n, j_p are known functions of $L^2(\Omega)$, let us introduce the following modified stationary Maxwell problem

(62)
$$
-\Delta\phi(x) = \frac{q}{\varepsilon}(N_d(x) - N_a(x) + n_p(x) - n_n(x)), \ x \in \Omega,
$$

$$
\phi(x) = \phi_0(x), \text{ for } x \in \partial\Omega_0, \quad \frac{\partial\phi}{\partial\nu}(x) = 0, \text{ for } x \in \partial\Omega_i,
$$

(63)
$$
\nabla \cdot B(x) = 0, \quad \nabla_x \times B = \mu q(j_p(x) - j_n(x)) + \lambda\nabla_x\varphi(x), \text{ for } x \in \Omega,
$$

$$
B(x) \cdot \nu(x) = b_0(x), \quad \text{for } x \in \partial\Omega,
$$

where the correction potential φ is the solution of

(64)
$$
\varphi \in H_0^1(\Omega), \quad -\Delta\varphi(x) = \mu q(n_p - n_n).
$$

Let us remark that if the concentration and fluxes are related to distribution that solve (57) we have

(65)
$$
\lambda n_n(x) + \nabla_x \cdot j_n(x) = 0, \quad \lambda n_p(x) + \nabla_x \cdot j_p(x) = 0, \quad \text{for } x \in \Omega.
$$

Therefore the correction potential φ appears to be necessary to recover a divergence free flux,

$$
\nabla_x \cdot [\mu q(j_p - j_n) + \lambda\nabla_x\varphi] = 0.
$$

It is a necessary and sufficient condition to solve (63) (see [4,15] for details).

For the previous Maxwell problem we get

PROPOSITION 4. *Let (n_n, n_p, j_n, j_p) be the functions that belong to a bounded set of $L^2(\Omega)$ and that satisfy (65). Then for any $\lambda > 0$ the problem (62), (63), (64) has a unique solution (ϕ, B). Furthermore ϕ belongs to a compact set of $H^1(\Omega)$ and B belongs to a compact set of $[L^2(\Omega)]^3$, independently of λ.*

For a proof of this proposition we refer to [4,15]. We only point out that a regularity result for the potential ϕ or the function $(\phi - \psi)$ is not known because of the mixed boundary conditions for some geometries in dimension 3. However it is easy to have a compactness result. Indeed let us consider a sequence $\phi^{(i)}$ such that

$$\phi^{(i)} \to \phi \text{ in } H^1(\Omega) \text{ weak} , \quad \Delta\phi^{(i)} \to \Delta\phi \text{ in } L^2(\Omega) \text{ weak}$$

$$\phi^{(i)}(x) = \phi_0(x), \text{ for } x \in \partial\Omega_0, \quad \frac{\partial\phi^{(i)}}{\partial\nu}(x) = 0, \text{ for } x \in \partial\Omega_i.$$

Then we get

$$\int_\Omega |\nabla\phi^{(i)}|^2(x)dx = \int_\Omega -\Delta\phi^{(i)}(x)\phi^{(i)}(x)dx + \int_{\partial\Omega_0} \phi_0(x) \cdot \frac{\partial\phi^{(i)}}{\partial\nu}(x)d\gamma(x).$$

Since we have

$$\phi^{(i)} \to \phi \text{ in } L^2(\Omega) \text{ strong} , \frac{\partial\phi^{(i)}}{\partial\nu} \to \frac{\partial\phi}{\partial\nu} \text{ in } H^{-1/2}(\partial\Omega) \text{ weak} ,$$

we can pass to the limit in the second member that shows

$$\int_\Omega |\nabla\phi^{(i)}(x)|^2 dx \to \int_\Omega |\nabla\phi|^2 dx.$$

Therefore $\phi^{(i)}$ converges strongly in $H^1(\Omega)$.

Now the modified Vlasov Maxwell Boltzmann problem $(VMB)_\lambda$ is defined by coupling (57), (10) with (62), (63), (64).

PROPOSITION 5. *For any* λ, $0 < \lambda < \lambda_0$, *the problem* $(VMB)_\lambda$ *has at least a solution* (f_n, f_p, ϕ, B) *that satisfies uniformly with respect to* λ

(66) $$0 \leq f_n(x,v) \leq C'_n M_n(v), \quad 0 \leq f_p(x,v) \leq C'_p M_n(v),$$

for some constants C'_n *and* C'_p.

The following and last section is devoted to the proof of this proposition 5. Now we are able to prove theorem 1.

Proof of Theorem 1. In view of (66), the assumptions of proposition 4 are satisfied. Therefore there is a sequence of solution $(\lambda^{(i)}, f_n^{(i)}, f_p^{(i)}, \phi^{(i)}, B^{(i)})$ of the problem $(VMB)_{\lambda(i)}$ such that

$$\lambda^{(i)} \to 0, \quad f_n^{(i)} \to f_n \text{ and } f_p^{(i)} \to f_p \text{ in } L^\infty(\Omega \times \mathbf{R}^3) \text{ weak } *,$$
$$\phi^{(i)} \to \phi \text{ in } H^1(\Omega), \quad B^{(i)} \to B \text{ in } [L^2(\Omega)]^3.$$

Since the function φ lies in a bounded set of H^1, there is no difficulty to pass to the limit in (62), (63). We obtain (2), (3), (11), (12) with the concentration and fluxes of the distributions f_n and f_p. We apply proposition 3 to get

$$\phi_{\lambda(i)}^{(i)} \to \phi \text{ in } H^1(\Omega), \quad B_{\lambda(i)}^{(i)} \to B \quad \text{in } L^2(\Omega).$$

Then proposition 2 allows to conclude that f_n and f_p are weak solutions of (1), (10). Thus (f_n, f_p, ϕ, B) is a solution of (VMB). In order to end the proof, it remains to get rid of (45). For that we consider a sequence $g_n^{(i)}$, $g_p^{(i)}$ such that (45) is satisfied for a sequence $\omega^{(i)}$ and

$$g_n^{(i)}(x, v) \leq c_n M_n(v) \text{ for } x \in \partial\Omega_0, v \cdot \nu(x) < 0,$$
$$g_p^{(i)}(x, v) \leq c_p M_p(v) \text{ for } x \in \partial\Omega_0, v \cdot \nu(x) < 0,$$
$$g_n^{(i)} \to g_n, \quad g_p^{(i)} \to g_p \text{ in } L^1_{\text{loc}}(\Gamma_0^-) \text{ weak },$$

where c_n, c_p does not depend on (i). The same arguments as above show that the corresponding sequence of solutions of (VMB) converges to the desired solution. $\square$

5. Existence for the modified Vlasov Maxwell system. First let us define a map T on $H^1(\Omega) \times [(L^2(\Omega)]^3$ by the following way. Starting from (ϕ, B) in $H^1(\Omega) \times [(L^2(\Omega)]^3$, we compute the solution (f_n, f_p) of (57), (10). The corresponding concentrations and fluxes (n_n, n_p, j_n, j_p) satisfies (65) and, thanks to (61), they are bounded because ϕ_λ belongs to $C(\Omega)$. Therefore we may solve (63), (64). It gives B' in $[(L^2(\Omega)]^3$. Let ϕ' be the solution of the nonlinear elliptic equation

$$-\Delta\phi'(x) = \frac{q}{\epsilon}[N_d(x) - N_a(x) - \tilde{n}_n(\phi', x) + \tilde{n}_p(\phi', x)], \quad \text{for } x \in \Omega,$$

$$\phi'(x) = \phi_o(x), \quad \text{for } x \in \partial\Omega_o, \quad \frac{\partial\phi'(x)}{\partial\nu} = 0, \quad \text{for } x \in \partial\Omega_i,$$

(67)
$$\text{with } \tilde{n}_n(\phi', x) = \inf\left\{n_n(x), K_n \exp\left(\frac{q\phi'_\lambda(x)}{k_B T}\right)\right\},$$

$$\tilde{n}_p(\phi', x) = \inf\left\{n_p(x), K_p \exp\left(-\frac{q\phi'_\lambda(x)}{k_B T}\right)\right\}.$$

The map T is defined by $T(\phi, B) = (\phi', B')$. Our goal is to apply the Schauder fixed point theorem to the map T. Indeed a fixed point (ϕ, B) for T is a solution of (VMB)$_\lambda$ since in view of (61) we obtain $\tilde{n}_n(\phi, x) = n_n(x)$, $\tilde{n}_p(\phi, x) = n_p(x)$. Let us introduce the two solutions of the following problems

(68)
$$-\Delta\psi_1(x) + \frac{q}{\epsilon}K_n \exp\left(\frac{q\psi_{1,\lambda}(x)}{k_B T}\right) = \frac{q}{\epsilon}(N_d(x) - N_a(x)),$$

$$\psi_1'(x) = \phi_o(x), \quad \text{for } x \in \partial\Omega_o, \quad \frac{\partial\psi_1}{\partial\nu}(x) = 0, \quad \text{for } x \in \partial\Omega_i,$$

(69)
$$-\Delta\psi_2(x) - \frac{q}{\epsilon}K_p \exp\left(-\frac{q\psi_{2,\lambda}(x)}{k_B T}\right) = \frac{q}{\epsilon}(N_d(x) - N_a(x)),$$

$$\psi_2'(x) = \phi_o(x), \quad \text{for } x \in \partial\Omega_o, \quad \frac{\partial\psi_2}{\partial\nu}(x) = 0, \quad \text{for } x \in \partial\Omega_i.$$

Then we have the following

LEMMA 2. *The problems (67), (68), (69) have unique solutions* ϕ', ψ_1, ψ_2, *that satisfy*

(70)
$$\|\psi_1\|_{L^\infty(\Omega)} \leq c_1, \|\psi_2\|_{L^\infty(\Omega)} \leq c_2,$$

$$(71) \qquad\qquad \psi_1 \leq \phi' \leq \psi_2,$$

for some constants c_1, c_2 that do not depend on λ.

Proof of lemma 2. The following maps

$$\psi \to \frac{q}{\epsilon} K_n \exp\left(\frac{q\psi_\lambda}{k_B T}\right), \quad \psi \to -\frac{q}{\epsilon} K_p \exp\left(-\frac{q\psi_\lambda}{k_B T}\right), \quad \psi \to \tilde{n}_n(\psi_\lambda, .) - \tilde{n}_p(\psi_\lambda, .)$$

are non decreasing with respect to ψ because (54). Then existence and uniqueness of solutions of (67), (68) and (69) are obtained by standard techniques provided that we obtain L^∞-a-priori estimates. That is the reason why, we only detail how we obtain (70) and (71). It results from the maximum principle that we have $\psi_1 \leq \psi$ where ψ satisfied (49). Using (54) yields $\psi_{1,\lambda} \leq \psi_\lambda$ and then

$$\frac{q}{\epsilon}\left(N_d - N_a - K_n \exp\left(\frac{q\psi_\lambda(x)}{k_B T}\right)\right) \leq -\Delta \psi_1 \leq \frac{q}{\epsilon}\left(N_d - N_a\right).$$

But ψ and ψ_λ are bounded in $L^\infty(\Omega)$, uniformly with respect to λ. Therefore ψ_1 is uniformly bounded. The same analysis for ψ_2 leads to (70). Finally, in view of (67), we point out that ϕ' is an upper solution of (68) and an under solution of (69). Since the maps

$$\psi \to \frac{q}{\epsilon} K_n \exp\left(\frac{q\psi_\lambda}{k_B T}\right), \quad \psi \to -\frac{q}{\epsilon} K_p \exp\left(-\frac{q\psi_\lambda}{k_B T}\right),$$

are non decreasing, we get a maximum principle for (68) and (69). That yields (72). $\square$

Let Ξ be the convex closed subset of $H^1(\Omega) \times (L^2(\Omega))^3$ defined by

$$(72) \qquad \Xi = \{(\phi, B) \in H^1(\Omega) \times (L^2(\Omega))^3, \quad \psi_1 \leq \phi \leq \psi_2\}.$$

We have

PROPOSITION 6. *T maps Ξ into itself. T is continuous and compact for the topology of $H^1(\Omega) \times [L^2(\Omega)]^3$.*

Proof of Proposition 6. The first point is a consequence of the previous lemma. Moreover for any (ϕ, B) in Ξ the solutions f_n, f_p of (57), (10) satisfy (66) with

$$C'_n = K_n \exp\left(\frac{qc_1}{k_B T}\right), \quad C'_p = K_p \exp\left(\frac{qc_2}{k_B T}\right)$$

that do not depend on (ϕ, B) and λ. As in the proof of theorem 1, it follows that (ϕ', B') belongs to a compact set of $H^1(\Omega) \times (L^2(\Omega))^3$. It proves that T is compact. To prove the continuity let us consider a sequence $(\phi^{(i)}, B^{(i)})$ in Ξ such that

$$\phi^{(i)} \to \phi \text{ in } H^1(\Omega), \quad B^{(i)} \to B \text{ in } [L^2(\Omega)]^3.$$

319

The corresponding solutions $f_n^{(i)}$, $f_p^{(i)}$ satisfy (66). So, thanks to Proposition 2, we have

$$f_n^{(i)} \to f_n, \quad f_p^{(i)} \to f_p \quad \text{in } L^\infty(\Omega \times \mathbf{R}^3) \text{ weakly },$$

with the same results for their concentration and fluxes. It follows that

$$\phi^{(i)'} \to \phi' \text{ in } H^1(\Omega) \text{ weak }, \quad B^{(i)'} \to B' \text{ in } L^2(\Omega) \text{ weak }.$$

But T is compact, so the weak convergence of $T(\phi^{(i)}, B^{(i)})$ is in fact a strong convergence. It concludes. $\square$

Now, proposition 5 is just a consequence of proposition 4 and of the Schauder fixed point theorem.

REFERENCES

[1] C. BARDOS, *Problèmes aux limites pour les équations aux dérivées partielles du premier ordre à coefficients réels; théorèmes d'approximations; application à l'équation de transport*, Ann. Sci de l'Ec. Norm. Sup., 4ᵉ série, 3 (1970), pp. 185–233.

[2] J.S. BLAKEMORE, *Semiconductor statistics*, Dover, New York, 1987.

[3] P. DEGOND, F. POUPAUD, B. NICLOT AND F. GUYOT, *Semiconductor modelling via the Boltzmann equation*, Lectures in Appl. Math., 25, AMS, Providence, Rhode Island, (1990), pp. 51–73.

[4] P. DEGOND AND P.-A. RAVIART, *An asymptotic analysis of the Darwin model of approximation to Maxwell's equations*, Forum Math. (to appear).

[5] R.J. DIPERNA AND P.L. LIONS, *On the Cauchy problem for the Boltzmann equation: global existence and weak stability*, Ann. of Math., 130 (1989), pp. 321–366.

[6] ——————————, *Global weak solutions of Vlasov-Maxwell systems*, Com. on Pure and Appl. Math., XLII (1989), pp. 729–757.

[7] ——————————, *Ordinary differential equations, transport theory and Sobolev spaces*, Invent. Math., 98 (1989), pp. 511–547.

[8] F. GOLSE, P.L. LIONS, B. PERTHAME AND R. SENTIS, *Regularity of the moments of the solution of a transport equation*, J. Funct. Anal., 88 (1988), pp. 110–125.

[9] C. GREENGARD AND P.A. RAVIART, *A boundary value problem for the stationary Vlasov-Poisson system: the plane diode*, Com. on Pure and Appl. Math., XLII (1990), pp. 473–507.

[10] P.A. MARKOWICH, C. RINGHOFER AND C. SCHMEISER, *Semiconductor equations*, Springer, Vienna, 1990.

[11] M. MOCK, *An initial value problem from semiconductor device theory*, SIAM J. of Math. Anal., 5 (1974), pp. 597–612.

[12] F.J. MUSTIELES, *Global existence of solution for a system of nonlinear Boltzmann equations of semiconductor physics*, Math. Meth. in the Appl. Sci., 14 (1991), pp. 107–121.

[13] F. POUPAUD, *On a system of nonlinear Boltzmann equations of semiconductor physics*, SIAM J. on Appl. Math., 50 (1990), pp. 1593–1606.

[14] ——————————, *Solutions stationnaires des equations de Vlasov-Poisson*, C.R. Acad. Sci. Paris, série I, 311 (1990), pp. 307–312.

[15] ——————————, *Boundary value problems for the stationary Vlasov-Maxwell system*, Forum Math. (to appear).

[16] F. POUPAUD AND C. SCHMEISER, *Charge transport in semiconductors with degeneracy effects*, Math. Meth. in the Appl. Sci., 14 (1991), pp. 301–318.

[17] L. REGGIANI (ed.), *Hot-electron transport in semiconductors*, Topics in Appl. Phys., 58, Springer-Verlag, Berlin, Heidelberg, 1985.

[18] T.I. SEIDMAN, *The transient semiconductor problem with generation terms*, Lectures in Appl. Math., 25, AMS, Providence, Rhode Island (1990), pp. 75–87.

[19] S.M. SZE, *Physics of semiconductor devices*, J. Wiley and sons, New York, 1981.

ON THE TREATMENT OF THE COLLISION OPERATOR FOR HYDRODYNAMIC MODELS

LUIS G. REYNA* AND ANDRÉS SAÚL**

Abstract. In this work we study an alternative treatment for the collision operator for hydrodynamic models. We start with a trial displaced Maxwellian function for the electron distribution function and by explicit integration of the collision term, we obtain a hydrodynamic model. The model includes the Drift-Diffusion equations, in the low electric field limit. However, it does not include any heat transfer.

We present numerical results obtained from these equations for a simple $N^+ - N - N^+$ structure. We include the interaction of electrons with acoustic and optical phonons and discuss the possibility of including non-parabolic bands and intervalley scattering. The simulations seem to indicate that the lack of heat transfer results in a non-physical decrease in temperature at the entrance of the N-middle region. They also indicate that it is necessary to include the non-parabolicity of the band structure to properly predict the saturation velocity.

1. Introduction. It is currently believed that a complete description of sub-micron devices can be achieved by solving the Boltzmann Transport Equations (BTE) [5], [7]. This equation describes the carrier distribution in both physical and momentum spaces. Solutions are obtained using Monte-Carlo methods which are very inefficient numerically. However, the BTE description is very detailed and in some applications may not be avoided.

An alternative approach for submicron devices is obtained from the Hydrodynamic models [1], [9]. these models consist of equations describing the first few moments of the distribution function. The differential equations are solved using methods very similar to those traditionally used in Drift-Diffusion codes. Thus, solutions are obtained at very low computational costs. The details of the band structure, however, are almost ignored and only some lump parameters are kept.

It is very likely that an efficient simulation of a semiconductor device has to involve a Monte-Carlo solver for the high electric field areas combined with finite element methods for the remaining of the device. When trying to combine both methods one is reminded that there is no rigorous way of deriving a hydrodynamic model from the Boltzmann equations. Without such a derivation, any attempt to combine these methods has very little chance of succeeding.

In order to obtain a "rigorous" derivation of the Hydrodynamic model from the BTE, we start from a displaced Maxwellian for the distribution function, and by taking moments we obtain a set of differential equations. We only consider the acoustic phonon scattering and the optical phonon scattering processes and assume a parabolic band structure. By studying an homogeneous problem we find that the non-parabolicity of the band structure is important in the determination of the saturation velocity.

*Mathematical Sciences, IBM Thomas J. Watson Research Center, P.O. Box 218, Yorktown Heights, New York 10598, USA.

**Grupo de Investigación en Computación y Aplicaciones Avanzadas, IBM Argentina, Ing. Enrique Butty 275, 1300 Buenos Aires, Argentina.

Results from the $N^+ - N - N^+$ structure reveals a serious weakness of the model. We find that at the entrance of the channel the electrons accelerate but their temperature decreases. At high voltages, the temperature becomes very low and the model looses its physical significance. Heat transfer is a mechanism that would eliminate this problem.

2. The hydrodynamic model. Consider the Boltzmann Transport Equation for the electron distribution function

$$\frac{\partial f}{\partial t} + \mathbf{v}.\nabla_x f - \frac{q}{\hbar}\mathbf{E}.\nabla_k f = \left(\frac{\partial f}{\partial t}\right)_{coll},$$

where $\mathbf{x}$ is the spatial coordinate, $\mathbf{k}$ the wave-vector, $\mathbf{E}$ the electric field and $\mathbf{v} = \nabla_k \varepsilon(\mathbf{k})/\hbar$ the electron group velocity. The band structure is determined by the energy $\varepsilon(\mathbf{k})$. In the parabolic band structure case, we have $\varepsilon(\mathbf{k}) = \hbar^2 k^2/2m$, where $k = |\mathbf{k}|$.

By taking moments of the BTE equation and integrating with respect to the wavevector [8], we obtain

$$\frac{\partial \rho}{\partial t} + \frac{\partial}{\partial x_j}\left(\rho u_j\right) = Q,$$

$$\frac{\partial}{\partial t}\left(\rho u_i\right) + \frac{\partial}{\partial x_j}\left(\rho u_{ij}\right) + \frac{q}{m}\rho E_i = Q_i$$

and

$$\frac{\partial}{\partial t}\left(\rho u_{ij}\right) + \frac{\partial}{\partial x_k}\left(\rho u_{ijk}\right) + \frac{q}{m}\rho\left(E_i u_j + E_j u_i\right) = Q_{ij}.$$

Here ρ is the charge density, u_i the mean velocity, u_{ij} the energy matrix and u_{ijk} the energy flux tensor, defined by

$$\rho(\mathbf{x},t) = \int f(\mathbf{x},\mathbf{k},t)\, d^3 k,$$

$$u_i(\mathbf{x},t) = \frac{1}{\rho(\mathbf{x},t)} \int v_i\, f(\mathbf{x},\mathbf{k},t)\, d^3 k,$$

$$u_{ij}(\mathbf{x},t) = \frac{1}{\rho(\mathbf{x},t)} \int v_i v_j\, f(\mathbf{x},\mathbf{k},t)\, d^3 k,$$

$$u_{ijk}(\mathbf{x},t) = \frac{1}{\rho(\mathbf{x},t)} \int v_i v_j v_k\, f(\mathbf{x},\mathbf{k},t)\, d^3 k,$$

$$Q(\mathbf{x},t) = \int \left(\frac{\partial f}{\partial t}\right)_{coll} d^3 k,$$

$$Q_i(\mathbf{x},t) = \int v_i \left(\frac{\partial f}{\partial t}\right)_{coll} d^3 k,$$

$$Q_{ij}(\mathbf{x},t) = \int v_i\, v_j \left(\frac{\partial f}{\partial t}\right)_{coll} d^3 k.$$

These equations are coupled to Poisson's equation

$$\nabla \cdot \mathbf{E} = \frac{4\pi q}{\epsilon}(N - \rho)\,,\quad \nabla \psi = -\mathbf{E}$$

relating the electric field $\mathbf{E}$, the charge density ρ, the doping profile $N(\mathbf{x})$ and the electrostatic potential ψ. The system of equations is very similar to those used by other authors [1], [4], [9].

Related moments are the temperature tensor

$$T_{ij} = \frac{m}{k_B}\left(u_{ij} - u_i u_j\right),$$

the energy density

$$\omega = \frac{m}{2}\sum_{j=1}^{3} u_{jj},$$

and the heat flux vector

$$\mathbf{F}_i = \rho\frac{m}{2}\sum_{j=1}^{3}\left(u_{jji} - 2u_j u_{ij} - u_{jj}u_i + 2u_i u_j^2\right).$$

In order to obtain a complete system of equations one needs some closure assumptions. It is common to assume a diagonal temperature tensor and a semiempirical relation between the heat flux vector and the temperature gradient $\mathbf{F}_i = -\kappa\nabla T$ [1], [9]. An alternative approach is to model the distribution function with a finite number of free functions, for example the charge density ρ, the drift velocity $\mathbf{u}$, and the electron temperature T. Higher moments can then be related to the undetermined functions. This is going to be our approach.

For a displaced Maxwellian, we have

$$f(\mathbf{x},\mathbf{k},t) = \rho\left(\frac{\hbar^2}{2\pi k_B T m}\right)^{3/2} e^{-m|\mathbf{v}-\mathbf{u}|^2/(2k_B T)},$$

where

$$\mathbf{v} = \frac{\hbar\mathbf{k}}{m} \quad\text{and}\quad \mathbf{u} = \mathbf{u}(\mathbf{x},t)\ ,\ T = T(\mathbf{x},t)\ ,\ \rho = \rho(\mathbf{x},t).$$

If we now consider a one dimensional stationary problem ($\mathbf{E} = E(z)\hat{z}, \mathbf{u} = u(z)\hat{z}, T = T(z), \rho = \rho(z)$), the moments of the distribution function are

$$\text{zero order} \qquad \rho = \rho(z)$$

$$\text{first order} \qquad u_x = u_y = 0$$
$$u_z = u$$

$$\text{second order} \qquad u_{xy} = u_{xz} = u_{yz} = 0$$
$$u_{xx} = u_{yy} = \frac{k_B T}{m}$$
$$u_{zz} = \frac{k_B T}{m} + u^2$$

$$\text{third order} \qquad u_{xxz} = u_{yyz} = \frac{k_B T u}{m}$$
$$u_{zzz} = u\left(\frac{3k_B T}{m} + u^2\right)$$
$$u_{xyz} = u_{xxy} = \ \dots\ = 0.$$

The moments of the collision operator are more complicated. If we assume that the collision operator preserves the number of particles (i.e., no recombination process is allowed), and that the scattering probability is isotropic in the plane perpendicular to the displacement vector $\mathbf{u}$, the only moments of the collision operator different from zero are Q_z and Q_{zz}. Moments of the collision operator have been previously studied by Bløtekjær and Lunde [2].

The Hydrodynamic equations in the one dimensional case read

$$\frac{\partial \rho}{\partial t} + \frac{\partial}{\partial z}(\rho u) = 0$$

$$\frac{\partial}{\partial t}(\rho u) + \frac{\partial}{\partial z}\left[\rho\left(\frac{k_B T}{m} + u^2\right)\right] + \frac{q}{m}\rho E = Q_z(\rho, u, T)$$

$$\frac{\partial}{\partial t}\left[\rho\left(\frac{3k_B T}{m} + u^2\right)\right] + \frac{\partial}{\partial z}\left[\rho u\left(\frac{5k_B T}{m} + u^2\right)\right] + \frac{2q}{m}\rho E u = Q_{zz}(\rho, u, T).$$

Note that due to symmetries of the displaced Maxwellian, we have no heat flux in the model. Also, the collision operator is now rather complicated.

Most hydrodynamic models, however, use a simpler version of the collision operator referred to as the relaxation time approximation

$$Q_z = -\frac{\rho u}{\tau_p} \quad \text{and} \quad Q_{zz} = -\frac{\rho}{\tau_w}\left[u^2 + 3k_B\frac{(T - T_0)}{m}\right],$$

where τ_p and τ_w are momentum and energy relaxation times [1], [9].

We now introduce dimensionless variables in the primed notation by

$$\rho = N_0\rho' \ , \ N = N_0 N' \ , \ T = T_0 T' \ , \ \psi = \frac{k_B T_0}{q}\psi' \ , \ z = \lambda_D z' = \left(\frac{k_B T_0 \epsilon}{4\pi q^2 N_0}\right)^{1/2} z' \ ,$$

$$E = \frac{k_B T_0}{q\lambda_D}E' \ , \ u = u_0 u' = \left(\frac{2k_B T_0}{m}\right)^{1/2}u' \ \text{and} \ t = \frac{\lambda_D}{u_0}t'.$$

Here N_0 is a typical charge density and T_0 the lattice temperature. Dropping the primes the scaled equations are

$$\frac{\partial \rho}{\partial t} + \frac{\partial}{\partial z}\left(\rho u\right) = 0$$

$$\frac{\partial}{\partial t}(2\rho u) + \frac{\partial}{\partial z}\left[\rho\left(T + 2u^2\right)\right] + \rho E = \frac{2\lambda_D}{N_0 u_0^2}Q_z(\rho, u, T)$$

$$\frac{\partial}{\partial t}\left[\rho\left(\frac{3}{2}T + u^2\right)\right] + \frac{\partial}{\partial z}\left[\rho u\left(\frac{5T}{2} + u^2\right)\right] + \rho E u = \frac{\lambda_D}{N_0 u_0^3}Q_{zz}(\rho, u, T),$$

$$\frac{\partial E}{\partial z} = (N - \rho), \frac{\partial \psi}{\partial z} = -E.$$

3. Treatment of the collision operator. We will consider the interaction of the electrons with acoustic phonons as an elastic process and will consider one inelastic process that can be thought as the interaction with optical phonons or intervalley scattering as well. The collision operator is then

$$\left(\frac{\partial f}{\partial t}\right)_{coll} = \int T(\mathbf{k}',\mathbf{k})f(\mathbf{k}')d\mathbf{k}' - \int T(\mathbf{k},\mathbf{k}')f(\mathbf{k})d\mathbf{k}',$$

with $T(\mathbf{k},\mathbf{k}')$, the probability rate of the transition from $\mathbf{k}$ to $\mathbf{k}'$, defined by

$$T(\mathbf{k}',\mathbf{k}) = \sum_{i=1}^{3} C_i \delta(\varepsilon(\mathbf{k}') - \varepsilon(\mathbf{k}) + \hbar\omega_i) \ , \ T(\mathbf{k},\mathbf{k}') = \sum_{i=1}^{3} C_i \delta(\varepsilon(\mathbf{k}) - \varepsilon(\mathbf{k}') + \hbar\omega_i)$$

Table

i		ω_i	C_i
1	optical abs.	ω_{op}	C_{op}
2	optical em.	$-\omega_{\mathrm{op}}$	C'_{op}
3	acoustic	0	C_{ac}

The transition rates are given by, see [7],

$$C_{op} = \frac{1}{8\pi^2}\frac{(D_t K)^2}{\varrho\omega_{\mathrm{op}}}N_{op} \ , \ C'_{op} = \frac{1}{8\pi^2}\frac{(D_t K)^2}{\varrho\omega_{\mathrm{op}}}(N_{op}+1) = C_{op}e^{\hbar\omega_{\mathrm{op}}/k_B T_0} \ ,$$

$$C_{ac} = \frac{1}{4\pi^2}\frac{\xi^2 k_B T_0}{\varrho\hbar v_s^2}$$

where for Si:

$\hbar\omega_{\mathrm{op}} \sim .063eV$ Optical phonon energy

$N_{op} = \dfrac{1}{e^{\hbar\omega_{\mathrm{op}}/k_B T_0} - 1}$ Phonon occupation number at the lattice temperature T_0

$DtK \sim 5.5eV/cm$ Optical deformation potential constant

$v_s \sim 6.5\ 10^5 cm/s$ Sound velocity

$\xi \sim 9eV$ Acoustic deformation potential

$\varrho \sim 2.33gr/cm^3$ Density

At room temperature, $T_0 = 300K$, the rates values are

$$C_{op} = 2.634\ 10^{-10}eV\,cm^3/s \text{ and } C_{ac} = 1.311\ 10^{-12}eV\,cm^3/s.$$

We make the following change of variables

$$d\mathbf{k}' = k^2 dk\,d\Omega = g(\varepsilon)d\varepsilon d\Omega \ , \ g(\varepsilon) = \frac{\sqrt{2m^3}}{\hbar^3}\varepsilon^{1/2}$$

so that the collision operator results in

$$\left(\frac{\partial f}{\partial t}\right)_{coll} = C_{op}g(\varepsilon + \hbar\omega_{op})\left(e^{\beta_0\hbar\omega_{op}}\int d\Omega' f(\varepsilon + \hbar\omega_{op}, \Omega') - 4\pi f(\varepsilon, \Omega)\right)$$

$$+ C_{ac}g(\varepsilon)\left(\int d\Omega' f(\varepsilon, \Omega') - 4\pi f(\varepsilon, \Omega)\right)$$

$$+ C_{op}g(\varepsilon - \hbar\omega_{op})\Theta(\varepsilon - \hbar\omega_{op})\left(\int d\Omega' f(\varepsilon - \hbar\omega_{op}, \Omega') - 4\pi e^{\beta_0\hbar\omega_{op}}f(\varepsilon, \Omega)\right),$$

where Ω and Ω' are angular variables and Θ the Heaviside function. The displaced Maxwellian is now

$$f(\varepsilon, \Omega) = \rho\left(\frac{\hbar^2}{2\pi k_B T m}\right)^{3/2} e^{-\beta\varepsilon}e^{-\beta\varepsilon_c}e^{2\sqrt{\varepsilon\varepsilon_c}\beta\cos\theta},$$

where

$$\beta = \frac{1}{k_B T} \ , \ \ \beta_0 = \frac{1}{k_B T_0} \ , \ \ \varepsilon = \frac{1}{2}mv^2 \ , \ \ \varepsilon_c = \frac{1}{2}mu^2.$$

Integrating over the angular variables we obtain

$$\left(\frac{\partial f}{\partial t}\right)_{coll} = \rho\left(\frac{\beta}{\pi}\right)^{1/2} e^{-\beta\varepsilon_c}e^{-\beta\varepsilon}$$

$$\left[C_{op}\left\{\frac{e^{(\beta_0-\beta)\hbar\omega_{op}}}{\sqrt{\varepsilon_c}}\sinh(2\beta\sqrt{(\varepsilon + \hbar\omega_{op})\varepsilon_c}) - 2\beta\sqrt{\varepsilon + \hbar\omega_{op}}e^{2\sqrt{\varepsilon\varepsilon_c}\beta\cos\theta}\right\} + \right.$$

$$C_{ac}\left\{\frac{\sinh(2\beta\sqrt{\varepsilon\varepsilon_c})}{\sqrt{\varepsilon_c}} - 2\beta\sqrt{\varepsilon}e^{2\sqrt{\varepsilon\varepsilon_c}\beta\cos\theta}\right\} +$$

$$\left. C_{op}\left\{\frac{e^{\beta\hbar\omega_{op}}}{\sqrt{\varepsilon_c}}\sinh(2\beta\sqrt{(\varepsilon - \hbar\omega_{op})\varepsilon_c}) - 2\beta\sqrt{\varepsilon - \hbar\omega_{op}}e^{\beta_0\hbar\omega_{op}}e^{2\sqrt{\varepsilon\varepsilon_c}\beta\cos\theta}\right\}\right].$$

The collision operator involves six terms. The terms which do not include any angular dependence do not contribute to Q_z, while all the terms contribute to Q_{zz}. When computing moments of the collision operator, we use

$$v_z k^2 dk d\Omega = \sqrt{\frac{2\varepsilon}{m}}\cos\theta g(\varepsilon)d\varepsilon d\Omega$$

and

$$v_z^2 k^2 dk d\Omega = \frac{2\varepsilon}{m}\cos^2\theta g(\varepsilon)d\varepsilon d\Omega.$$

Using the expression for $(\partial f/\partial t)_{coll}$ given above we now evaluate Q_z and Q_{zz}, as defined in section 2. In dimensionless variables this yields

$$\frac{2\lambda_D}{N_0 u_0^2}Q_z(\rho, u, T) = \frac{\rho T e^{-x_c}}{\sqrt{x_c}}\left(C_{acz}I_4(x_c) + C_{opz}\{I_2(x_c, x_0) + I_6(x_c, x_0, x_{00})\}\right)$$

and

$$\frac{\lambda_D}{N_0 u_0^3}Q_{zz}(\rho, u, T) = \rho T^{3/2}e^{-x_c}\left(C_{aczz}I_{34}(x_c) + C_{opzz}I_{1256}(x_c, x_0, x_{00})\right),$$

where

$$I_2(x_c, x_0) = \int_0^\infty e^{-x}\sqrt{x+x_0}\sqrt{x}\left\{\cosh(2\sqrt{xx_c}) - \frac{\sinh(2\sqrt{xx_c})}{2\sqrt{xx_c}}\right\} dx,$$

$$I_4(x_c) = \int_0^\infty e^{-x}x\left\{\cosh(2\sqrt{xx_c}) - \frac{\sinh(2\sqrt{xx_c})}{2\sqrt{xx_c}}\right\} dx$$

$$= \left(x_c^{3/2} + x_c^{1/2} - \frac{1}{4x_c^{1/2}}\right)\sqrt{\pi}e^{x_c}\mathrm{erf}(\sqrt{x_c}) + x_c + \frac{1}{2},$$

$$I_6(x_c, x_0, x_{00}) = e^{x_{00}}\int_{x_0}^\infty e^{-x}\sqrt{x-x_0}\sqrt{x}\left\{\cosh(2\sqrt{xx_c}) - \frac{\sinh(2\sqrt{xx_c})}{2\sqrt{xx_c}}\right\} dx,$$

and

$$I_{34}(x_c) = \pi\left\{\left(-\frac{8}{3}x_c^{3/2} - 4x_c^{1/2} + \frac{2}{x_c^{1/2}} - \frac{1}{x_c^{3/2}}\right)\sqrt{\pi}e^{x_c}\mathrm{erf}(\sqrt{x_c}) + \left(-\frac{8}{3}x_c - \frac{8}{3} + \frac{2}{x_c}\right)\right\}.$$

(We omit the remaining integral.) Here

$$x_c = \beta\varepsilon_c = \frac{1/2\,mu^2}{k_BT} = \left\{\frac{u^2}{T}\right\}_{\mathrm{adim}},$$

$$x_{00} = \beta_0\hbar\omega_{\mathrm{op}} = \frac{\hbar\omega_{\mathrm{op}}}{k_BT_0}, \quad x_0 = \beta\hbar\omega_{\mathrm{op}} = \frac{\hbar\omega_{\mathrm{op}}}{k_BT} = \left\{\frac{x_{00}}{T}\right\}_{\mathrm{adim}}$$

$$C_{acz} = -8\sqrt{\pi}\frac{m^2\lambda_D}{\hbar^3}C_{ac}, \quad C_{opz} = -8\sqrt{\pi}\frac{m^2\lambda_D}{\hbar^3}C_{op}$$

$$C_{aczz} = -\frac{1}{8\pi}C_{acz} \text{ and } C_{opzz} = -\frac{1}{8\pi}C_{opz}.$$

The integrals I_4 and I_{34} have been evaluated analytically and arise from the acoustic phonons scattering. The remaining integrals have to be computed numerically. Series expansions for these integrals are included in [2].

In the limit of small velocities (or low temperatures), we obtain for the acoustic phonons that

$$\lim_{x_c\to 0}\frac{\rho T e^{-x_c}}{\sqrt{x_c}}C_{acz}I_4(x_c) = \frac{8}{3}C_{acz}\rho u\sqrt{T} \sim -\frac{2\rho u}{\tau_{\mathrm{eff}}}$$

where

$$\tau_{\mathrm{eff}} = -\frac{3}{8}\frac{1}{C_{acz}\sqrt{T}},$$

i.e.

$$\tau_{\mathrm{eff}} = \frac{\mu_{\mathrm{eff}}m}{q} = \frac{3\pi^{3/2}}{8\sqrt{2}}\frac{\hbar^4 v_s^2\varrho}{k_BT_0(k_BT)^{1/2}\xi^2 m^{3/2}}.$$

For the optical phonons, again in the limit of small velocities, we obtain

$$\lim_{x_c\to 0}\frac{\rho T e^{-x_c}}{\sqrt{x_c}}C_{opz}\left(I_2(x_c, x_0) + I_6(x_c, x_0, x_{00})\right) = -\frac{2\rho u}{\tau_{\mathrm{eff}}},$$

where

$$\tau_{\text{eff}} = \frac{3\pi^{3/2}}{8\sqrt{2}} \frac{\hbar^2(\hbar\omega_{\text{op}})^2\varrho}{k_B T_0(k_B T)^{1/2}(D_t K)^2 m^{3/2}} \qquad \text{for} k_B T \gg \hbar\omega_{\text{op}}$$

i.e.

$$\tau_{\text{eff}} = \sqrt{2}\pi \frac{\hbar^2\sqrt{\hbar\omega_{\text{op}}}\varrho}{N_{op}(D_t K)^2 m^{3/2}} \qquad \text{for } k_B T \ll \hbar\omega_{\text{op}} .$$

Approximate expressions can also be determined for large electron temperatures

$$Q_z \sim \frac{8}{3}\rho u \sqrt{T}\left(C_{\text{acz}} + C_{\text{opz}}(e^{x_{00}} + 1)\right)$$

$$Q_{zz} \sim \frac{8\pi}{3}\rho\sqrt{T}\left(C_{\text{aczz}}\frac{16}{5}u^2 + C_{\text{opzz}}\left\{x_{00}(1 - e^{x_{00}}) - u^2\frac{16}{5}(1 + e^{x_{00}})\right\}\right).$$

We can also analyze the behavior of the collision operator by studying the homogenous problem:

$$\frac{q}{m}\rho E = Q_z(\rho, u, T) , \quad \frac{2q}{m}\rho E u = Q_{zz}(\rho, u, T)$$

Since Q_z and Q_{zz} are both proportional to ρ. The homogeneous system gives two relations between E, u and T. In Fig. 1 and Fig. 2 we show the relation between u and T and u and E respectively. Note the slow growth of the velocity vs. the electric field. This growth is a result of the $\sqrt{T}$ growth of the collision terms. Monte Carlo results for homogeneous problem seem to indicate a linear growth in T [6].

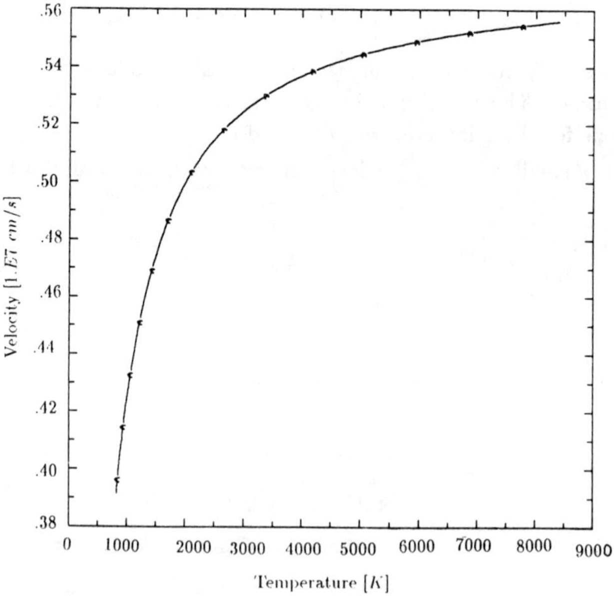

Figure 1: Drift velocity vs. temperature

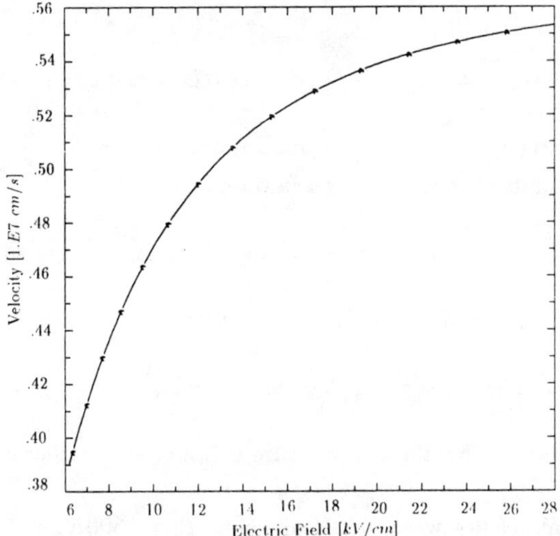

Figure 2: Drift velocity vs. electric field

4. Numerical results. For the numerical solution of the hydrodynamic equations we used a finite-difference method on a non-uniform mesh. Solutions for different voltages were obtained starting from equilibrium voltage and using a continuation method on the voltage.

In order to reduce the stiffness of the equations we used a modification of the Scharfetter-Gummel method and upwinding methods. This combination was used in order to increase the accuracy of the solutions in the highly doped region, where the electric field is nearly constant.

The equilibrium conditions were computed using a non-linear Gauss-Seidel on the non-linear Poisson equation

$$\psi_{zz} = e^{\psi} - N.$$

An iterative mesh refinement procedure was used at this stage in order to resolve the different layers of this solution. The resulting mesh was then used, without modifications, for higher voltages. A similar approach can be found in [10].

The discretization is as follows,

$$E_i - E_{i-1} + \frac{1}{2}(\Delta x_{i-1} + \Delta x_i)(\rho_i - N_i) = 0,$$

$$\rho_i v_i - \rho_{i+1} v_{i+1} = 0,$$

$$(\rho_{i+1}B(c) - \rho_i B(-c))T_i + \rho_{i+1}(T_{i+1} - T_i) + 2\rho_{i+1} v_{i+1}(v_{i+1} - v_i) = \Delta x_i \frac{\lambda_0}{N_0 u_0^3} Q_z(v_i, T_i, \rho_i)$$

and

$$\rho_i u_i \left(5/2 T_{i+1} + u_{i+1}^2 - 5/2 T_i - u_i^2\right) + \Delta x_i E_i \rho_i v_i = \Delta x_i \frac{2\lambda_0}{N_0 u_0^2} Q_{zz}(v_i, T_i, \rho_i),$$

where

$$\Delta x_i = x_{i+1} - x_i \ , \ E_i = -(\psi_{i+1} - \psi_i)/\Delta x_i \text{ and } c = \Delta x_i E_i/T_i.$$

Here $B(x) = x/(\exp(x) - 1)$ is the Bernoulli function.

Regarding boundary conditions: we impose at $x = 0$

$$\rho_0 = N_0 \ , \ \psi_0 = \log(N_0) + \Delta\psi$$

and, at $x = x_N = L$, the total length of the device,

$$\rho_N = N_N \ , \ \psi_N = \log(N_N) \text{ and } 5/2(T_N - 1) + v_N^2 = 0.$$

The boundary condition for the temperature is imposed at either end of the device, depending on the sign of the current.

The numerical results were obtained using $T_0 = 300K$, $m = .33m_e$, $N_0 = 2 \ 10^{15}cm^{-3}$, $\epsilon = 12$, $N = 2 \ 10^{15}cm^{-3}$, $N^+ = 2 \ 10^{17}cm^{-3}$. Thus, $\psi_0 = .026eV$, $\lambda_D = .093\mu m$, $u_0 = 1.66 \ 10^7cm/s$, $E_0 = 2.792KV/cm$ and $t_0 = .557ps$. The constants involved in the moments of the collision operator are: $C_{acz} = -2.135$ and $C_{opz} = -4.292 \ 10^{-2}$.

Fig. 3 shows the dimensionless temperature vs. the position for a 0.4μ N region at $V = 1.61$ Volts. Fig. 4 shows the dimensionless velocity. Note the decrease in temperature at the entrance of the low doped region. Also, the temperature rises to about 16 times the temperature of the lattice. This raise would be diminished in the presence of non-parabolic bands.

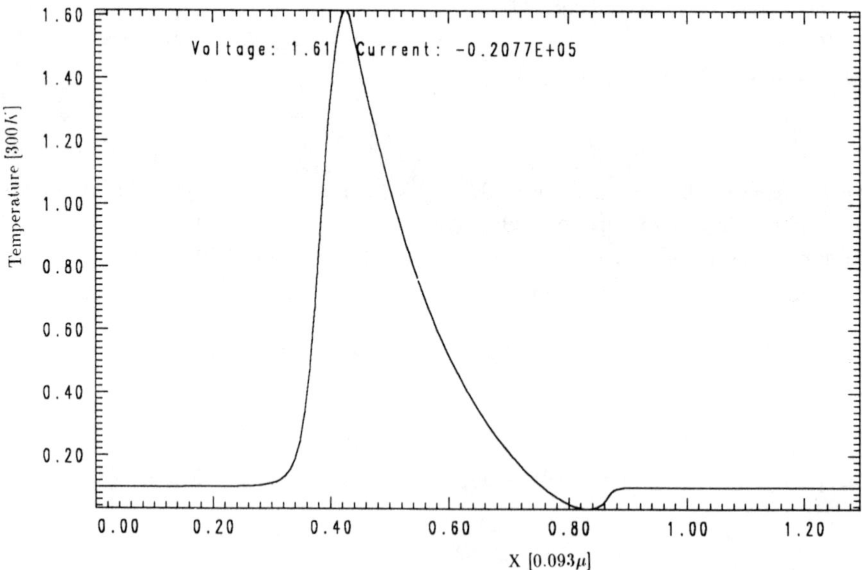

Figure 3: Dimensionless temperature along device: $L = 0.4\mu$

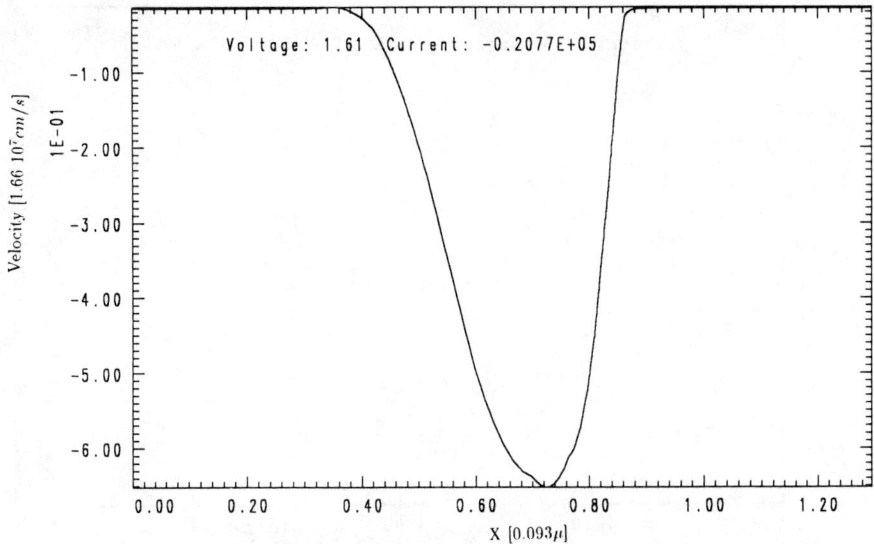

Figure 4: Dimensionless velocity along device: $L = 0.4\mu$

Fig. 5 shows the temperature vs. position for a 0.25μ device and Fig. 6 the velocity. Now the temperature decrease is just too high.

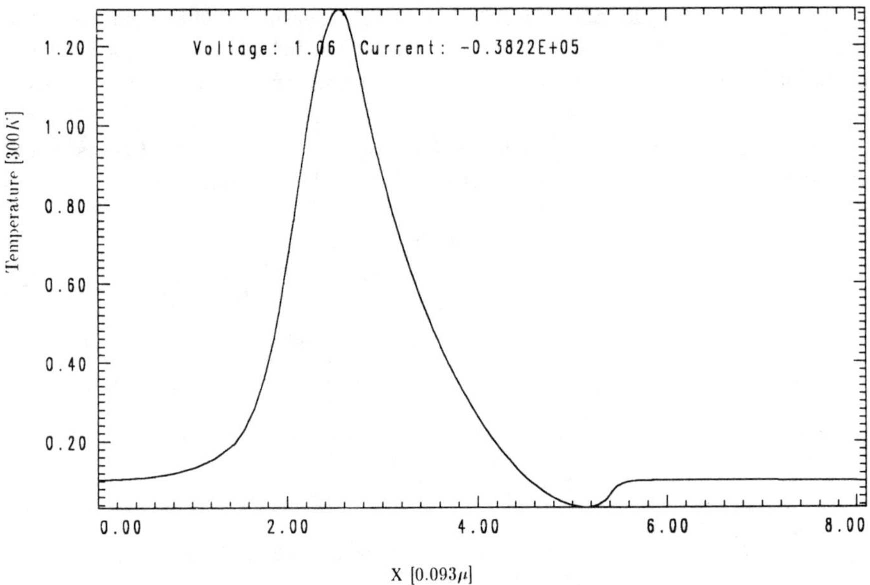

Figure 5: Dimensionless temperature along device: $L = 0.25\mu$

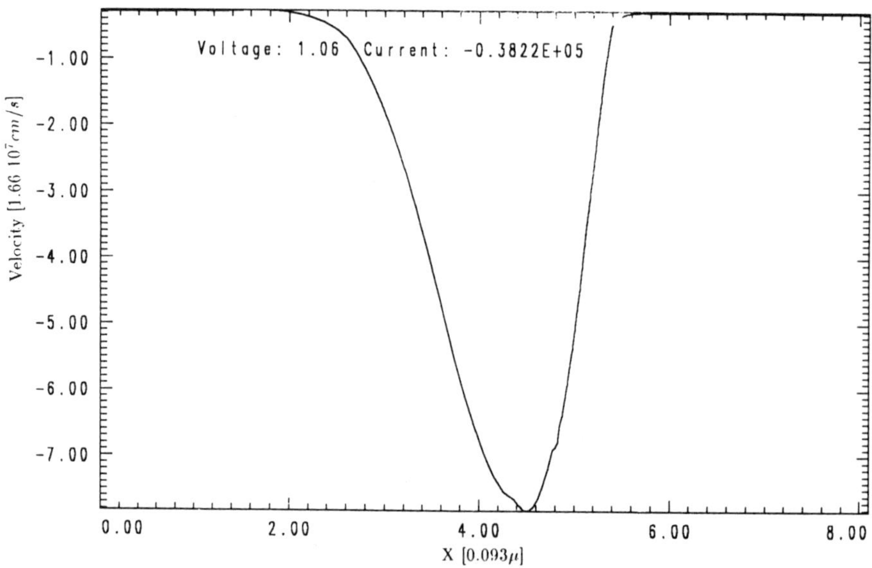

Figure 6: Dimensionless velocity along device: $L = 0.25\mu$

5. Conclusions. We have studied a possible derivation of a hydrodynamic model starting from a drift Maxwellian distribution. We find that the absence of heat conduction leads to unphysically low temperatures for the electrons at the entrance of the middle junction in a $N^+ - N - N^+$ structure. We also find that the scattering rates predicted for parabolic bands are too low. For these low rates the saturation of the electron velocity only takes place at very high electron temperatures.

We are currently studying alternative approaches that include heat transfer and are also including a more detailed description of the band structure.

REFERENCES

[1] K. BLØTEKJÆR, *Transport Equations for Electrons in Two-Valley Semiconductors*, IEEE Trans. Elec. Dev. **ED-17** (1970), pp. 38–47.

[2] K. BLØTEKJÆR, E.B. LUNDE, *Collision Integrals for Displaced Maxwellian Distributions*, Phys. Stat. Sol. **35** (1969), pp. 581–592.

[3] C. CANALI, C. JACOBONI, F. NAVA, G. OTTAVIANI, A. ALBERIGI-QUARANTA,, *Electron Drift Velocity in Silicon*, Phys. Rev. B **12** (1975), pp. 2265–2284.

[4] R.K. COOK, J. FREY, *An Efficient Technique for Two-Dimensional Simulation of Velocity Overshoot Effects in Si and GaAs Devices*, COMPEL **1** (1982), pp. 65–87.

[5] M.V. FISCHETTI, S.E. LAUX, *Monte Carlo Analysis of Electron Transport in Small Semiconductor Devices Including Band Structure and Space-Charge Effects*, Phys. Rev. B **38** (1988), pp. 9721–9745.

[6] A. GNUDI, F. ODEH, M. RUDAN, *Non-Local Effects in Small Semiconductor Devices*, Eur. Trans. Telecomm. **1** (1990), pp. 307–312.

[7] C. JACOBONI, L. REGGIANI, *The Monte Carlo Method for the Solution of Charge Transport in Semiconductor with Applications to Covalent Materials*, Rev. Modern Phys. **55** (1983), pp. 645–705.

[8] F. POUPAUD, *Derivation of a Hydrodynamic System Hierarchy for Semiconductors form the Boltzmann Equation*, Appl. Math. Lett. **4** (1991), pp. 75–79.

[9] M. RUDAN, F. ODEH, *Multi-Dimensional Discretization Scheme for the Hydrodynamic Model of Semiconductor Devices*, COMPEL **5** (1986), pp. 149–183.

[10] M. WARD, L. REYNA, F. ODEH, *Multiple Steady State Solutions in a Multi-Junction Semiconductor Device*, SIAM J. Appl. Math. **51** (1990), pp. 1099–1125.

ADAPTIVE METHODS FOR THE SOLUTION OF THE WIGNER-POISSON SYSTEM*

CHRISTIAN RINGHOFER**

Abstract. A numerical method, based on spectral collocation, for the solution of the Wigner-Poisson system is presented. A numerical example is briefly discussed.

1. Introduction. Quantum mechanical transport phenomena in small semi-conductor devices, such as c. f. resonant tunneling diodes, can be described either by the Schrödinger equation or equivalently by the Wigner-Poisson system [3], [6]

$$
\begin{aligned}
&\text{a)} \quad \partial_t w + \frac{1}{m} p \cdot \nabla_x w - \theta[V_E + V_B]w = 0 \\
(1) \\
&\text{b)} \quad -\epsilon \Delta_x V_E + q(n - D(x)) = 0, \qquad \text{c)} \quad n(x,t) = \int_{R_p^d} w(x,p,t)dp.
\end{aligned}
$$

Here x denotes position, p momentum, and t denotes time. d is the dimensionality of the problem (i.e. $d = 1$, 2 or 3 holds). $w(x,p,t)$ is the Wigner distribution funciton (WDF), $V_E(x,t)$ is the electric potential and $n(x,t)$ denotes the charge density. V_B is a time independent step function modelling the quantum barriers. $D(x)$ models the doping profile of the device. The pseudo differential operator $\theta[V]$ in (1)a), given by

$$
(2) \qquad \theta[V] = \frac{1}{i\bar{h}} \left[V\left(x + \frac{i\bar{h}}{2}\nabla_p, t\right) - V\left(x - \frac{i\bar{h}}{2}\nabla_p, t\right) \right],
$$

models the impact of the electric field $-\nabla_x V$ on the WDF. Here $\bar{h}$ denotes Planck's constant. (See [3] for alternative definitions of θ.) The system (1) reduces to the classical Boltzmann transport equation in the classical limit for $\bar{h} \to 0$. However, for very small devices with an active region of the order of $10^{-8}m$, the system (1) is quite far from the classical regime [2].

In this paper we discuss the numerical solution of the Wigner – Poisson problem. In Section 2 we review a pseudospectral discretization of the Wigner – Poisson problem which was previously presented and analyzed in [4], [5]. In Section 3 we discuss the use of adaptive meshes in the context of this method to reduce the over all computational cost. In Section 4 we present some numerical results.

*Research supported by NSF Grant No. DMS 8801153.
**Department of Mathematics, Arizona State University, Tempe, AZ 85287, USA.

2. Numerical method. Because of the definition of the pseudo differential operator θ in (1) via Fourier transforms, spectral methods using trigonometric basis functions are a natural choice for its discretization. Thus the WDF is approximated by

$$(3) \qquad u(x,p,t) = \sum_{|\mu| \le N} \hat{u}(x,\mu,t) \exp(i\alpha\mu \cdot p),$$

where α denotes a small parameter. Choosing the values of u at the collocation points $p_j = \frac{\pi}{\alpha N} j$, $j = (j_1, \ldots, j_d)$ as variables, we obtain the following discretization of θ:

$$
\begin{aligned}
\text{a)} \quad & \theta[V]u(x,p_j,t) = \sum_{|k| \le N} A_{jk}(x,t)u(x,p_k,t) \\
\text{b)} \quad & A_{jk}(x,t) = (2N)^{-d} \sum_{|\mu| \le N} \frac{i}{h} \delta V\left(x, \frac{i\bar{h}}{2}\mu, t\right) \exp[i\alpha\mu \cdot (p_j - p_k)] \\
\text{c)} \quad & \delta V(x,z,t) := V(x+z,t) - V(x-z,t).
\end{aligned}
$$

(4)

After truncation of the integral in (1)c) outside the cube $\left[-\frac{\pi}{\alpha}, \frac{\pi}{\alpha}\right]^d$ the charge density n is given by

$$(5) \qquad n(x,t) = \left(\frac{2\pi}{\alpha}\right)^d \hat{u}(x,0,t) = \left(\frac{\pi}{\alpha N}\right)^d \sum_{|k| \le N} u(x,p_k,t).$$

This yields the mixed hyperbolic-elliptic systems

$$
\begin{aligned}
\text{a)} \quad & \partial_t u_j + \frac{1}{m} p_j \cdot \nabla_x u_j - \sum_{|k| \le N} A_{jk} u_k = 0, \quad u_j(x,t) := u(x,p_j,t) \\
\text{b)} \quad & -\epsilon \Delta_x V_E + q\left[\left(\frac{\pi}{\alpha N}\right)^d \sum_{|k| \le N} u(x,p_k,t) - D(x)\right] = 0.
\end{aligned}
$$

(6)

This system can be discretized by any standard discretization method for hyperbolic systems. Multiplication with the matrix A in (6)a) can be performed in $O(N \log N)$ operations using FFT algorithms. It is important to notice that the matrix $A(x,t)$ is skew, i.e. that $z^* A z = 0$ holds for all vectors z. This implies that the Poisson equation can be solved on the previous time step without any additional step restriction [4].

To obtain convergence of the solution of (6) to the solution w of the Wigner – Poisson problem (1) it is necessary to let the diameter of the simulation domain $\left[-\frac{\pi}{\alpha}, \frac{\partial}{\alpha}\right)^d$ tend to infinity together with the number of modes. From [5] we hve the following theorem establishing a spectral order of convergence for finite but arbitrary time intervals.

THEOREM. *Let w be sufficiently smooth and decay sufficiently fast int he p-direction; i.e. let*

$$(7) \qquad \begin{aligned} &p_1^{j_1}\ldots p_d^{j_d}\,\partial_{p_1}^{k_1}\ldots\partial_{p_d}^{k_d} w(x,.,t)\in L^2(R_p) \quad \forall j_s,k_s \\ &\text{with } j_1+\cdots+j_d=m,\ \ k_1+\cdots+k_d=n \text{ hold}. \end{aligned}$$

Then, for any final time T, there exists a constant $C(T)$ such that

$$(8) \qquad \max_{t\in[0,T]}\|u-w\|_{L^2(R_x^d\times R_p^d)}\le C(T)[(\alpha N)^{-n}+\alpha^m]$$

holds.

3. Adaptive grids. We now discuss the use of adaptive grid strategies to minimize the computation cost of the solution of (6). Unfortunately the use of Fast Fourier Transform algorithms prohibits the use of a variable meshspacing in the $p-$ direction. So the basic idea of the strategy is to use only those, equally spaced gridpoints which are necessary to represent the support of the solution u. To this end, it is advantageous to employ an operator splitting method for the numerical solution of the hyperbolic system (6). In the splitting approach the free flight of particles and the interactions with the electric field can be treated comletely independently. For reasons of notatonal simplicity we will present this idea only in the one dimensional case ($d=1$). The following equatiosn are solved to carry out one time step.

$$(9) \qquad \begin{aligned} &\text{a)}\quad \partial_s y+\frac{1}{m}p\partial_p y=0,\quad t\le s\le t+\Delta t,\quad y(x,p,s=t)=u(x,p,t) \\[4pt] &\text{b)}\quad \partial_s z=\theta[V_E]z,\quad \epsilon\Delta_x V=\int z\,dp-D,\quad t\le s\le +\Delta t \\[4pt] &\qquad\quad z(x,p,s=t)=y(x,p,t+\Delta t) \\[4pt] &\text{c)}\quad u(x,p,t+\Delta t)=z(x,p,t+\Delta t). \end{aligned}$$

The advantage of this approach is that the second step (9)b) can be performed independently for each $x\in R_x$. A convergence analysis of the above splitting scheme for the Wigner – Poisson problem is currently in preparation [1]. The grid is now adjusted to the support of the discrete solution u. Thus we compute only with the variables

$$(10) \qquad \begin{aligned} &u_j(x,t):=u(x,p_j,t), \\[4pt] &j=J(x,t)-L(x,t)+1(1)J(x,t)+L(x,t),\quad p_j=\frac{j\pi}{\alpha N} \end{aligned}$$

Due to the nonlocal nature of the pseudo differential operator θ the support of u will immediately spread to cover all of R_p. We take this into account by filling up the vector z in (9)b) with zeros and then truncating after each time step at an appropriate cutoff value. Thus we perform the following steps to solve (9)b). For notational simplicity we suppress the dependence on x and t from now on.

Step 1: (Padding)

$$(11) \qquad z_j^0 = \begin{cases} y_j(t + \Delta t), & j = J - L + 1(1)J + L \\ 0 & j = J - 2L + 1(1)J - L \text{ and } j = J + L + 1(1)J + 2L \end{cases}$$

Step 2: (Solution of (9)b))

Equations (9)b) are solved as in the previous section by expanding

$$(12) \qquad z_j^0 = \sum_{\mu=-2L}^{2L} \hat{z}(\mu) \exp(i\alpha\mu p_j), \quad j = J - 2L + 1(1)J + 2L.$$

Step 3: (Truncation)

The z_j^0 are truncated by using an appropriate cutoff criterion. so, J_1 and L_1 are chosen such that

$$(13) \qquad \sum_{j=J-2L+1}^{J_1-L_1} (z_j^0)^2 + \sum_{j=J_1+L_1+1}^{J+2L} (z_j^0)^2 \leq (\delta\Delta t)^2 \sum_{j=J-2L+1}^{J+2L} (z_j^0)^2 \text{ holds .}$$

Step 4: (Projection)

The z_j^0 are projected onto the interval $\left[(J_1 - L_1)\frac{\pi}{\alpha N}, (J - 1 + L_1)\frac{\pi}{\alpha N}\right]$ using a Galerkin projector, such that the first two moments are preserved. We compute the moments

$$(14) \qquad M_k = \sum_{j=J-2L+1}^{J+2L} (p_j)^k z_j^0, \quad k = 0, 1,$$

and define

$$(15) \qquad z_j^1 = z_j^0 + A + Bp_j, \quad j = J_1 - L_1 + 1(1)J_1 + L_1$$

such that

$$(16) \qquad \sum_{j=J_1-L_1+1}^{J_1+L_1} p_j^k z_j^1 = M_k, \quad k = 0, 1 \text{ holds .}$$

This implies the solution of a 2×2 system for the parameters A and B for each x-gridpoint and each time step.

3. Numerical example. As a numerical example we present a one dimensional simulation of the tunneling of a Gaussian wave packet through a quantum barrier. (So x and p are taken to be in R^1.) The barrier is located at $x = 0$. In Figures 1-3 the wave packet is shown at different points in time. Figure 1 shows the initial solution at $t = 0$. Figures 2 and 3 show the wave packet during reflection. The incident and the reflected parts are visible as is the correlation around $p = 0$. Finally, Figure 4 shows the evolution of the mesh. The mesh points are equally spaced in $x - p$ space and only the meshpoints inside the depicted curves are used in the computation.

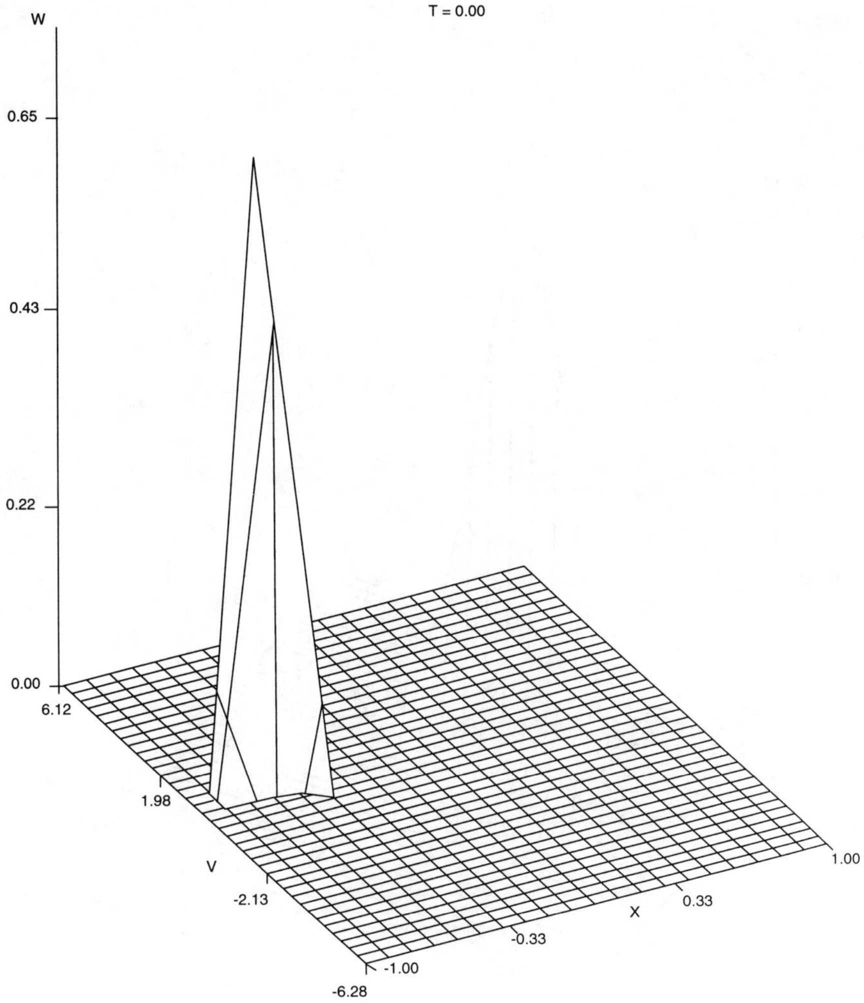

Figure 1

T=0.48

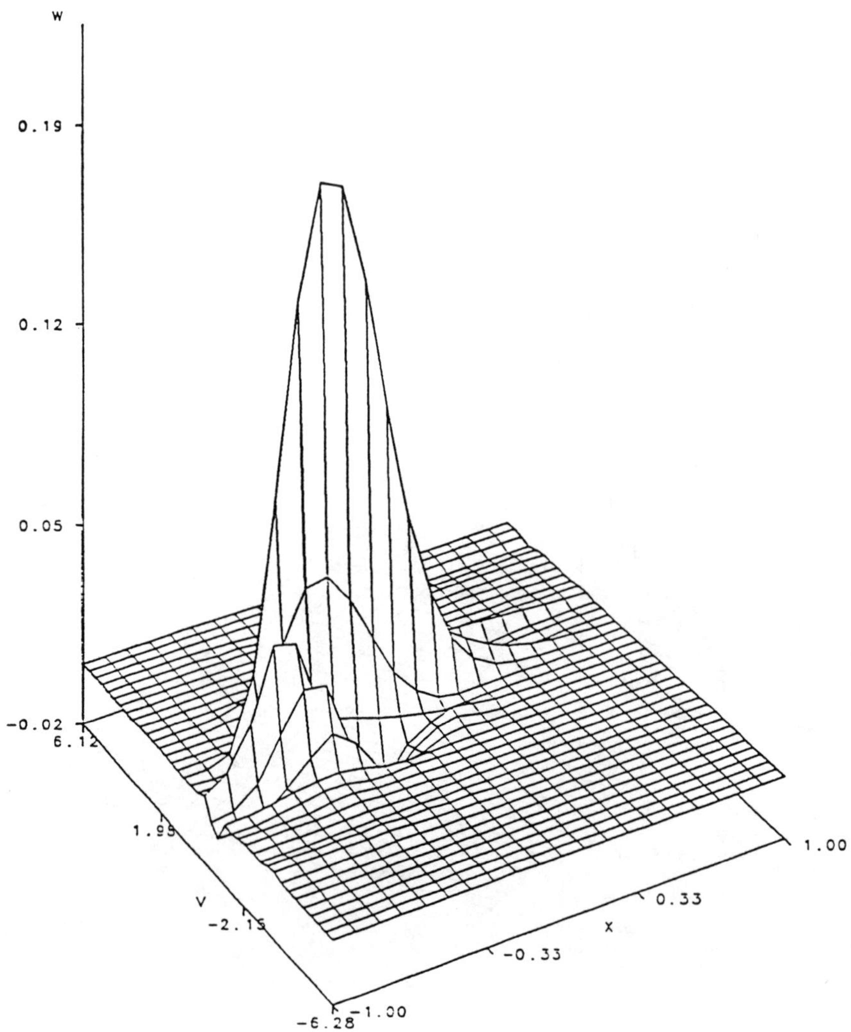

Figure 2

T=0.80

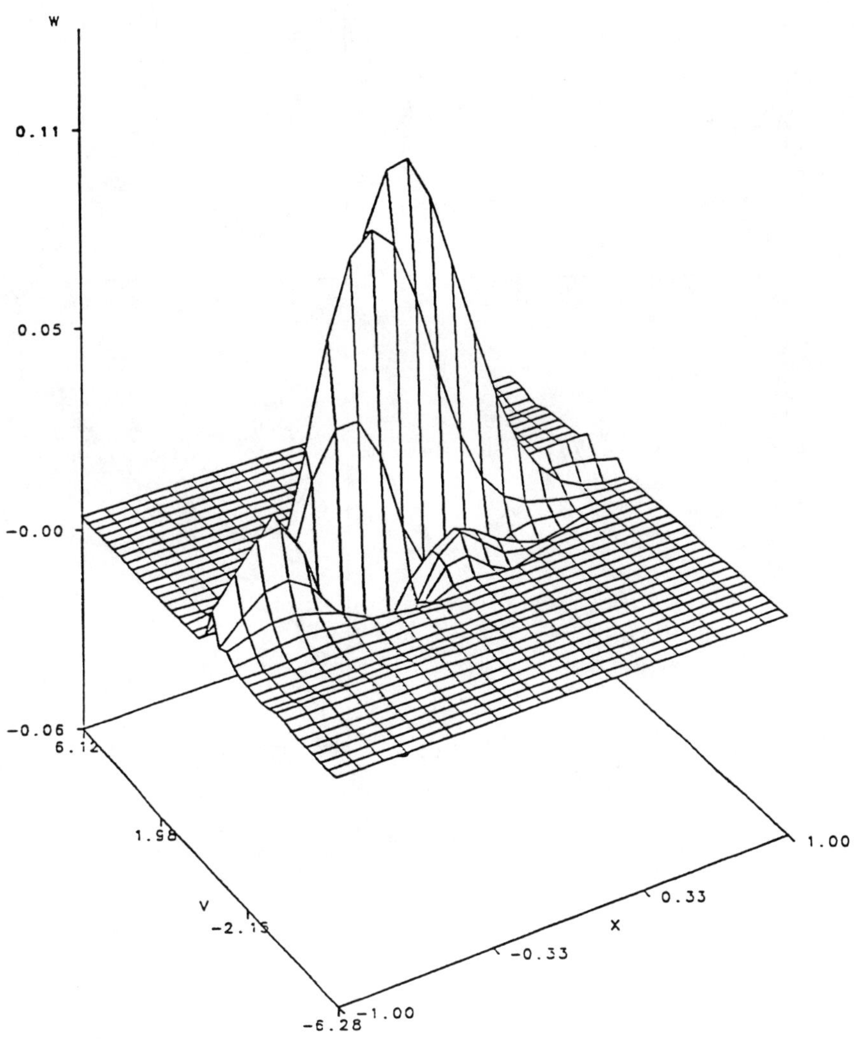

Figure 3

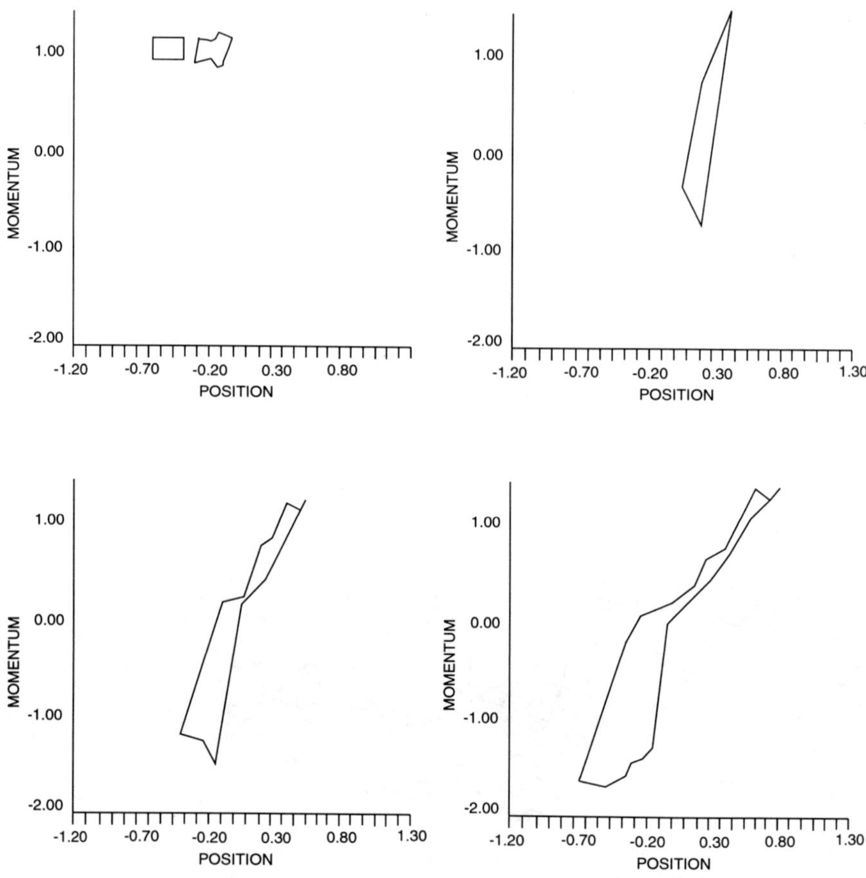

Figure 4

REFERENCES

[1] A. ARNOLD AND C. RINGHOFER, in preparation.
[2] N. KLUKSDAHL ET.AL., Phys. Rev. B, vol 39 (1989), pp. 7720–7735.
[3] P. MARKOWICH ET.AL., Semiconductor Equations, Springer Verlag (1990).
[4] C. RINGHOFER, SIAM J. Num. Anal., vol. 27 (1990), pp. 32–50.
[5] C. RINGHOFER, to appear in SIAM J. Num. Anal., (1992).
[6] V. TATARSKIJ, Soviet Phys., Uspekhi 26 (1983), pp. 311–327.

THE DERIVATION OF ANALYTIC DEVICE MODELS BY ASYMPTOTIC METHODS

CHRISTIAN SCHMEISER† AND ANDREAS UNTERREITER‡

Abstract. In circuit simulation, device models should be as simple as possible. On the other hand, physically sound models for the electrical behaviour of semiconductor devices involve nonlinear systems of partial differential equations posed on domains with complicated geometries. Therefore simplifications have to be introduced corresponding to certain idealizing assumptions. By the use of asymptotic methods the simplification procedure can be carried out in a mathematically justifiable way.

This paper presents an overview of recent results on steady state voltage-current characteristics of multi-dimensional bipolar devices [11], [21] as well as on a new approach to the modelling of the transient behaviour of pn-diodes via integral equations [23], [24]. These ideas are extended to more general bipolar devices.

1. Introduction. The starting point of our analysis is the classical drift-diffusion model for the flow of negatively charged electrons (density $n(\mathbf{x}, t)$) and positively charged holes (density $p(\mathbf{x}, t)$) in a semiconductor. In scaled form it consists of the continuity equations

$$(1.1a) \qquad \nabla \cdot \mathbf{J}_n - \frac{1}{\delta^4} \frac{\partial n}{\partial t} = R, \qquad -\nabla \cdot \mathbf{J}_p - \frac{1}{\delta^4} \frac{\partial p}{\partial t} = R,$$

the current relations

$$(1.1b) \qquad \delta^4 \mathbf{J}_n = \mu_n (\nabla n - n \nabla V), \qquad \delta^4 \mathbf{J}_p = -\mu_p (\nabla p + p \nabla V),$$

and the Poisson equation

$$(1.1c) \qquad \lambda^2 \Delta V = n - p - C$$

for the electrostatic potential $V(\mathbf{x}, t)$. The mobilities μ_n, $\mu_p > 0$ and the doping profile C are assumed to be given functions of position $\mathbf{x} \in \Omega$, where the bounded domain $\Omega \subset \mathbf{R}^k$, $k = 1, 2$ or 3, represents the semiconductor part of the device. We assume the recombination-generation rate to be of the form

$$R = Q(n, p, \mathbf{x})(np/\delta^4 - 1), \qquad Q \geq 0,$$

including the standard models for band-to-band processes and recombination-generation via traps in the forbidden band. Since this work is restricted to low injection situations, our model certainly describes the relevant physical phenomena [11].

---

†Institut für Angewandte und Numerische Mathematik, TU Wien, Austria. The work of this author has been supported by the Austrian Fonds zur Förderung der wissenschaftlichen Forschung and by the Institute for Mathematics and its Applications with funds provided by the National Science Foundation.

‡Fachbereich Mathematik, TU Berlin, Germany.

The equations (1.1) are in dimensionless form. The reference quantity for the particle densities n, p, C is the maximal doping concentration $C_{\max}$, i.e. $\max_\Omega |C| = 1$ holds. The potential has been scaled by the thermal voltage $U_T = kT/q$ where k, T and q denote the Boltzmann constant, the lattice temperature and the elementary charge, respectively. The reference length L is the diameter of the device and, thus, $\mathrm{diam}(\Omega) = 1$. The mobilities are scaled by a characteristic value $\widetilde{\mu}$ and the reference time is given by the diffusion time $L^2/(\widetilde{\mu} U_T)$. Finally, the reference value

$$(1.2) \qquad \frac{q\widetilde{\mu} U_T}{L}\,\frac{n_i^2}{C_{\max}}$$

for the electron and hole current densities $\mathbf{J}_n$ and $\mathbf{J}_p$, respectively, contains the intrinsic density n_i. In low injection situations the factor $n_i^2/C_{\max}$ is a typical minority carrier density. Thus, the value (1.2) is a characteristic value for current densities in low injection. For further details on the scaling of the drift-diffusion equations we refer to [8] and [11].

The equations (1.1) contain the dimensionless parameters

$$\lambda = \frac{1}{L}\sqrt{\frac{\varepsilon U_T}{q C_{\max}}}\,, \qquad \delta^2 = \frac{n_i}{C_{\max}}\,,$$

which can be interpreted as scaled versions of the minimal Debye length and of the intrinsic number, respectively.

Subregions of Ω where the doping profile C is positive are called n-regions because the positively charged impurity ions attract electrons. On the other hand, in p-regions the doping profile is negative. The $(k-1)$-dimensional boundaries between n-and p-regions are called pn-junctions. We assume abrupt junctions, i.e. the doping profile has jumps across these junctions and is bounded away from zero within the n- and p-regions.

We restrict our attention to bipolar devices. Therefore, the boundary $\partial\Omega$ of the device is the disjoint union of Ohmic contacts $C_1, \cdots, C_m$ and artificial or insulating boundary segments $\partial\Omega_N$. At Ohmic contacts we assume zero space charge and thermal equilibrium:

$$n - p - C = 0\,, \quad np = \delta^4\,, \qquad \text{at } C_1, \cdots, C_m\,,$$

which translates to Dirichlet boundary conditions for the charge carrier densities:

$$(1.3a)$$
$$n = \frac{1}{2}\left(C + \sqrt{C^2 + 4\delta^4}\right)\,, \quad p = \frac{1}{2}\left(-C + \sqrt{C^2 + 4\delta^4}\right)\,, \qquad \text{at } C_1, \cdots, C_m$$

For the potential, we have the boundary conditions

$$(1.3b) \qquad V = V_{bi} - U_j(t)\,, \qquad \text{at } C_j,\ j = 1, \cdots, m\,,$$

where $U_i(t) - U_j(t)$ is the external voltage between the contacts C_i and C_j, and the built-in potential is given by

$$(1.3c) \qquad V_{bi} = \ln \frac{C + \sqrt{C^2 + 4\delta^4}}{2\delta^2} \, .$$

Along the artificial and insulating boundary segments, we assume that the normal components of the electric field and of the electron and hole current densities vanish. This amounts to homogeneous Neumann conditions for V, n and p:

$$(1.3d) \qquad \frac{\partial V}{\partial \nu} = \frac{\partial n}{\partial \nu} = \frac{\partial p}{\partial \nu} = 0, \qquad \text{at } \partial\Omega_N \, ,$$

where ν denotes the unit outward normal.

The formulation of an initial-boundary value problem is completed by prescribing initial conditions for the carrier densities:

$$(1.4) \qquad n(\mathbf{x}, 0) = n_0(\mathbf{x}), \qquad p(\mathbf{x}, 0) = p_0(\mathbf{x})$$

We shall assume that the initial data n_0 and p_0 correspond to steady state solutions of (1.1), (1.3).

Important quantities are the currents through the Ohmic contacts. The current $I_j(t)$ leaving the device through the contact C_j is given by

$$I_j = \int_{C_j} \mathbf{J}_{tot} \cdot \boldsymbol{\nu} \, ds$$

in terms of the total current density

$$\mathbf{J}_{tot} = \mathbf{J}_n + \mathbf{J}_p - \frac{\lambda^2}{\delta^4} \nabla \frac{\partial V}{\partial t} \, ,$$

which is the sum of the particle current densities and the density of the displacement current.

The specific properties of a device are determined by the number and location of the n- and p-regions as well as of the Ohmic contacts. We consider devices meeting the following requirements: There is a finite number of open connected n-regions whose union is denoted by Ω_+. In the same way, the number of p-regions is finite and their union is denoted by Ω_-. Each n- or p-region has at most one contact and each contact is adjacent to only one n- or p-region. The union of the pn-junctions is denoted by $\Gamma = \overline{\Omega}_+ \cap \overline{\Omega}_-$. Note that these assumptions do not rule out so called floating regions without any contacts. In Figure 1 two-dimensional cross sections of three typical devices are depicted. The pn-diode consists of one n- and one p-region, each with a contact. The bipolar transistor has three differently doped regions with contacts. Finally, the thyristor is a $pnpn$-structure. Figure 1 shows the so called Shockley diode where the two middle layers are floating regions.

In this paper we both discuss the stationary problem corresponding to (1.1), (1.3) as well as the transient case. Our objective in both cases is to find simple representations of the steady state and transient voltage-current characteristics, i.e. the dependence of the currents through the contacts on the $m - 1$ contact voltages $U_j - U_1$, $j = 2, \cdots, m$. Obviously, the choice of U_1 does not influence the result. Note that we only need to compute $m - 1$ currents, since the total current density is divergence free, implying $I_1 + \cdots + I_m = 0$.

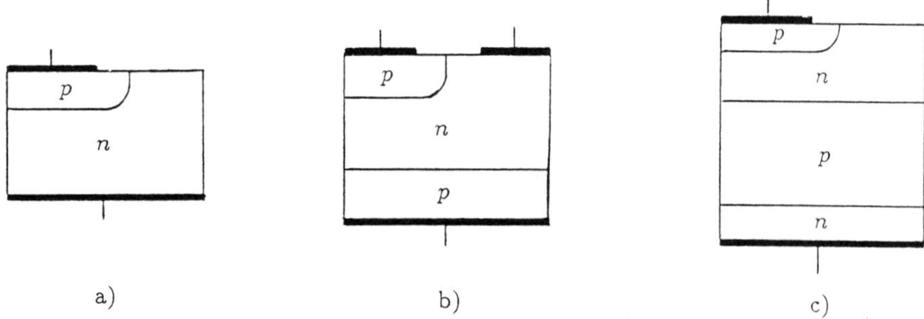

Figure 1. Cross section of a) pn-diode, b) bipolar transistor and c) thyristor.

The dimensionless parameters λ and δ^2 are small compared to 1 in practical applications. Our approach is an asymptotic analysis where we let these parameters tend to zero consecutively.

The limit $\lambda \to 0$ corresponds to the physical assumption of zero space charge. Essentially, it amounts to replacing the left hand side of the Poisson equation (1.1c) by zero. However, since the problem is singularly perturbed in terms of λ, layer behaviour has to be expected. In particular, the jump discontinuities of the doping profile cause the limiting potential and the carrier densities to have jumps across pn-junctions. Also, an initial jump has to be expected for the time dependent problem. The nature of these jumps can be analyzed by introducing slow variables which are continuous in the limit. The construction of formal asymptotic expansions — including layer corrections at the pn-junctions and initial layers in the transient case — and their use for explaining device behaviour has received a considerable amount of attention in the literature (see [4], [7], [8], [9], [10], [11], [12], [22], [27], [33] for the stationary problem and [16], [17], [30], [31] for the time dependent case).

A rigorous theory justifying the limiting procedure is incomplete. To the authors' knowledge, no results are available for the transient case. For one-dimensional steady state problems the approximations can be justified [10], [12] by general results for singularly perturbed two-point boundary value problems [25]. In higher dimensions, a justification for steady states close to thermal equilibrium, i.e. for small enough applied voltages, can be found in [7]. A weaker convergence result for arbitrary biases and a simplified problem with constant mobilities and vanishing recombination-generation rate has been derived in [4]. A generalization of this result, proven in [21], is stated in section 2 of this work.

In terms of the exponentials of the quasi Fermi levels, which are a convenient choice of variables for the steady state problem, the limiting stationary problem is a system of two nonlinearly coupled elliptic equations. The time dependent problem becomes a parabolic-algebraic system in the limit. In the language of the theory of differential-algebraic equations [3], it is an index 2 problem [1]. This means that the initial conditions have to satisfy certain compatibility relations which is guaranteed by the assumption that the initial data originate from steady state solutions. In other words this assumption implies the absence of initial jumps [16].

A further simplification of the problem is introduced by letting δ^2 tend to zero. Keeping the applied voltages fixed as the built-in potential tends to infinity (as $\delta^2 \to 0$, see (1.3b), (1.3c)) can be interpreted as a low injection condition.

For steady state problems this limit has been formally carried out in [11] and rigorously justified in [21]. However, the zero space charge and low injection assumptions have already been used for one-dimensional model problems in the early physical literature on semiconductor devices [26]. In particular, the famous Shockley equation for the voltage-current characteristic of a pn-diode and the qualitative behaviour of bipolar transistors are derived in this way (see also [29]). Thus, the results presented in section 2 can be seen as an extension of this early work. In sections 3 and 4 we demonstrate for multi-dimensional models of a pn-diode and a bipolar transistor, respectively, that the voltage-current characteristic close to thermal equilibrium can be determined explicitly in terms of the solutions of simple elliptic reference problems. In [21], the same methods have been applied for the computation of the forward and reverse bias blocking branches of the voltage-current characteristics of Shockley diodes. As in [27] for one-dimensional models, a device dependent parameter is identified whose sign determines whether a given $pnpn$-structure is a thyristor or behaves like a pin-diode.

For the transient behaviour of pn-diodes the limit $\delta^2 \to 0$ in the zero space charge problem has been carried out in [23] and [24]. In section 5 we extend this procedure to general bipolar devices. The limiting problem consists of linear parabolic equations in the n- and p-regions coupled by interface conditions at the pn-junctions. For the pn-diode and the bipolar transistor we show in sections 6 and 7, respectively, that these problems are equivalent to systems of integral equations describing the evolution of the currents in terms of the evolution of the applied voltages. Similarly to the steady state case, the kernels of the integral equations are determined by solving simple reference problems.

The question arises if the limit problems depend on the order of the limiting procedures $\lambda \to 0$ and $\delta^2 \to 0$. In [11] it has been shown that for the thermal equilibrium problem (consisting only of a nonlinear Poisson equation) the limits commute. No proof is available, however, that the limits commute also in the general case. The limit $\delta^2 \to 0$ leads to a free boundary problem [20] having practical importance in VLSI applications, since for very small devices λ can be considerably large whereas $\delta^2 \ll 1$ is always a safe assumption. It should be mentioned in this context that the distinguished limit $\lambda = \delta \to 0$ has been considered in [15]. This amounts to the assumption that the reference length in the scaling is equal to the intrinsic Debye length.

Finally we want to discuss the limitations of our approach. They originate from keeping the applied voltages fixed in the course of the limiting procedures. It has already been pointed out that the assumption of smallness of the applied voltages compared to the built-in potential means low injection. Therefore high injection effects are neglected. A second source of error is the zero space charge assumption. We neglect the depletion regions in the neighbourhoods of the pn-junctions. This is justified, because the width of these regions is of the order of the Debye length which has been assumed to be small compared to other characteristic length scales in the device. It is well known, however, that the width grows with the potential jump across the junction. Effects involving large applied biases and, therefore, widening depletion regions have to be accounted for by an asymptotic analysis of a rescaled problem (see [2], [14], [18], [19], [28] for the stationary problem and [13] for the time dependent case). Unfortunately, in this case the most appealing feature of the close-to-thermal-equilibrium results of this work is lost, namely the fact that the voltage-current characteristics can be given explicitly in terms of the solutions of reference problems independent of the biasing situation.

2. Steady state problems. The analysis of the stationary problem is greatly facilitated by introducing the exponentials u and v of the quasi Fermi levels as new variables instead of the carrier densities:

$$n = \delta^2 e^V u, \qquad p = \delta^2 e^{-V} v$$

This symmetrizing transformation for the continuity equations changes the steady state version of (1.1), (1.3) to the differential equations

$$
\begin{aligned}
\nabla \cdot \left(\mu_n \delta^2 e^V \nabla u \right) &= \delta^4 R, \\
\nabla \cdot \left(\mu_p \delta^2 e^{-V} \nabla v \right) &= \delta^4 R, \\
\lambda^2 \Delta V &= \delta^2 e^V u - \delta^2 e^{-V} v - C,
\end{aligned}
$$

(2.1)

subject to the boundary conditions

$$
\begin{aligned}
u = e^{U_j}, \quad v = e^{-U_j}, \quad V = V_{bi} - U_j, \qquad &\text{at } C_j, \; j = 1, \cdots, m, \\
\frac{\partial V}{\partial \nu} = \frac{\partial u}{\partial \nu} = \frac{\partial v}{\partial \nu} = 0, \qquad &\text{at } \partial \Omega_N.
\end{aligned}
$$

(2.2)

Now the recombination-generation rate is given by

$$R = Q(\delta^2 e^V u, \delta^2 e^{-V} v, \mathbf{x})(uv - 1).$$

Note that the differential operators in the continuity equations are formally self-adjoint. Also we expect the derivatives of u and v to be bounded uniformly with respect to λ. This makes them slow variables in the language of singular perturbation theory. Before we can state the convergence result for $\lambda \to 0$, a few regularity assumptions for the data are needed: The domain Ω is Lipschitz and the $(k-1)$-dimensional Lebesgue measure of the union of the contacts is positive. The Dirichlet boundary data for V, u and v at the contacts can be extended to Ω as functions in $H^1(\Omega)$. The nonnegative reaction rate Q in the recombination-generation term is a smooth function of the carrier densities as well as a bounded function of position. The doping profile and the mobilities satisfy

$$C \in L^\infty(\Omega) \cap W^{2,1}(\Omega_0), \qquad \mu_n, \mu_p \in L^\infty(\Omega) \cap H^1(\Omega_0),$$

where $\Omega_0 = \Omega \setminus \Gamma$ denotes the semiconductor domain without the pn-junctions. Additionally we assume the mobilities to be bounded away from zero.

THEOREM 2.1. *a) ([8]) Under the above assumptions the problem (2.1), (2.2) has a solution* $(V, u, v) \in \left(H^1(\Omega) \cap L^\infty(\Omega)\right)^3$.

b) ([21]) For every sequence $\lambda_k \to 0+$ *there exists a subsequence (again denoted by* λ_k*) such that corresponding solutions* (V_k, u_k, v_k) *of (2.1), (2.2) satisfy*

$$V_k \to V_0, \quad \text{in } L^2(\Omega),$$
$$u_k \to u_0, \quad \text{weakly in } H^1(\Omega),$$
$$v_k \to v_0, \quad \text{weakly in } H^1(\Omega),$$

where (V_0, u_0, v_0) *is a solution of (2.1), (2.2) with* λ *replaced by zero.*

Remark. Note that the limiting potential would not satisfy general Dirichlet boundary conditions since it is determined from an algebraic equation (the reduced Poisson equation). However, the assumption of zero space charge at the Ohmic contacts is compatible with the limiting problem, and therefore the limiting solution satisfies the complete set of Dirichlet conditions.

After elimination of the potential the zero space charge equations can be written as

$$(2.3) \qquad \delta^4 \mathbf{J}_n = \mu_n \frac{C + \sqrt{C^2 + 4\delta^4 uv}}{2u} \nabla u, \qquad \delta^4 \mathbf{J}_p = -\mu_p \frac{-C + \sqrt{C^2 + 4\delta^4 uv}}{2v} \nabla v,$$
$$\nabla \cdot \mathbf{J}_n = -\nabla \cdot \mathbf{J}_p = R.$$

The carrier densities are given in terms of u and v by

$$n = \frac{1}{2}\left(C + \sqrt{C^2 + 4\delta^4 uv}\right), \qquad p = \frac{1}{2}\left(-C + \sqrt{C^2 + 4\delta^4 uv}\right).$$

This shows that

$$n = C + O(\delta^4), \qquad p = O(\delta^4), \qquad \text{in } \Omega_+,$$
$$p = -C + O(\delta^4), \qquad n = O(\delta^4), \qquad \text{in } \Omega_-.$$

The statement that the density of the majority carriers (electrons in n-regions, holes in p-regions) is close to the modulus of the doping profile and the density of the minority carriers is small compared to that, is usually called a low injection condition.

Performing the formal limit $\delta^2 \to 0$ in (2.3) implies

$$(2.4a) \qquad\qquad \nabla u = 0 \quad \text{in } \Omega_+, \qquad \nabla v = 0 \quad \text{in } \Omega_-.$$

In other words, in each n- or p-region the quasi Fermi level corresponding to the majority carriers is constant. The current relations for the minority carriers are divided by δ^4 before passing to the limit. We obtain

$$(2.4b) \qquad
\begin{aligned}
\nabla \cdot \left(\frac{\mu_n}{|C|} \nabla u\right) &= Q(0, |C|)(u - v^{-1}), \qquad \text{in } \Omega_-, \\
\nabla \cdot \left(\frac{\mu_p}{C} \nabla v\right) &= Q(C, 0)(v - u^{-1}), \qquad \text{in } \Omega_+.
\end{aligned}$$

For the limiting minority carrier current densities we have

$$\mathbf{J}_n = \frac{\mu_n v}{|C|} \nabla u \quad \text{in } \Omega_-, \qquad \mathbf{J}_p = -\frac{\mu_p u}{C} \nabla v \quad \text{in } \Omega_+.$$

Concerning the limit $\delta^2 \to 0$ the following result holds:

350

THEOREM 2.2. *For every sequence $\delta_k^2 \to 0+$ there exists a subsequence (again denoted by δ_k^2) such that corresponding solutions (u_k, v_k) of (2.2), (2.3) satisfy*

$$u_k \to u_0, \quad \text{weakly in } H^1(\Omega),$$
$$v_k \to v_0, \quad \text{weakly in } H^1(\Omega),$$

where the limit (u_0, v_0) satisfies (2.2), (2.4).

Assuming the constant values of u in the n-regions and of v in the p-regions to be known, the problem has been reduced to the solution of the linear elliptic equations (2.4b). At first glance, it seems disturbing, however, that only the minority carrier current densities are determined by the limiting problem. This means that the total current density is known only at pn-junctions. On the other hand, by current continuity it is sufficient to know the currents through the pn-junctions for computing the currents through the contacts. In the following two sections we shall demonstrate that the voltage-current characteristics can be determined explicitly in terms of a number of device dependent parameters from the simplified problem (2.2), (2.4).

3. The Shockley equation for the pn-diode. We denote the n-region of a pn-diode by Ω_n, the p-region by Ω_p, the adjacent Ohmic contacts by Γ_n and Γ_p, respectively, and the pn-junction by Γ (see Figure 2). From a simple one-dimensional model problem Shockley (1949, [26]) computed the approximation

$$(3.1) \qquad\qquad I = I_s(e^U - 1)$$

for the steady state voltage-current characteristic which is now known as the Shockley equation. In (3.1), I denotes the current through the device and U the contact voltage. The reverse bias saturation current I_s has been determined by Shockley as a function of the doping levels in the n- and p-regions, the mobilities and recombination parameters. An application of the results of the preceding section will show that the Shockley equation remains valid in the multi-dimensional case with an appropriately chosen value of I_s.

In terms of the contact voltage U, the variables u and v satisfy the boundary conditions

$$u = v = 1, \quad \text{at } \Gamma_p, \qquad u = e^U, \ v = e^{-U}, \quad \text{at } \Gamma_n.$$

From (2.4a) we immediately obtain

$$u = e^U, \quad \text{in } \Omega_n, \qquad v = 1, \quad \text{in } \Omega_p.$$

For u in Ω_p and v in Ω_n we choose the representations

$$u = 1 + (e^U - 1)\varphi_p, \quad \text{in } \Omega_p, \qquad v = e^{-U} + (1 - e^{-U})\varphi_n, \quad \text{in } \Omega_n,$$

in terms of the reference functions φ_p and φ_n, respectively, solving the problem

$$(3.2a) \qquad \nabla \cdot \left(\frac{\mu_n}{|C|} \nabla \varphi_p \right) = Q(0, |C|)\varphi_p \,, \qquad \text{in } \Omega_p \,,$$

$$(3.2b) \qquad \nabla \cdot \left(\frac{\mu_p}{C} \nabla \varphi_n \right) = Q(C, 0)\varphi_n \,, \qquad \text{in } \Omega_n \,,$$

$$(3.2c)$$
$$\varphi_n = \varphi_p = 1 \,, \quad \text{at } \Gamma \,, \qquad \varphi_n = 0 \,, \quad \text{at } \Gamma_n \,, \qquad \varphi_p = 0 \,, \quad \text{at } \Gamma_p \,.$$

Important is the fact that φ_n and φ_p only depend on the device but not on the biasing situation. With the formulas for the current densities from the preceding section we obtain the Shockley equation (3.1) from an integration along the pn-junction Γ. The saturation current is given by

$$I_s = \int_\Gamma \left(\frac{\mu_n}{|C_p|} \nabla \varphi_p - \frac{\mu_p}{C_n} \nabla \varphi_n \right) \cdot \boldsymbol{\nu} \, ds \,,$$

where $\boldsymbol{\nu}$ is the unit normal vector along Γ pointing into Ω_n, and C_n (C_p) is the doping profile evaluated at the n-(p-)side of the junction. It is easy to see that both terms which sum up to the integrand are positive. If a one-dimensional model with constant mobilities and a piecewise constant doping profile is considered, the differential equations for φ_n and φ_p are linear homogeneous ODEs with constant coefficients. In this case, I_s can be computed explicitly, recovering Shockley's formulas.

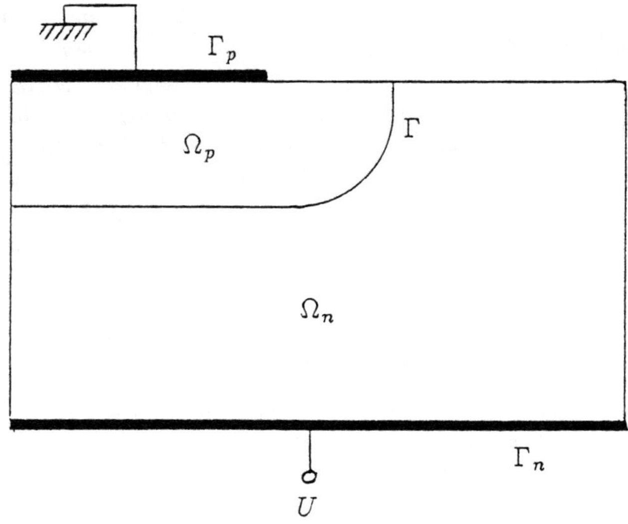

Figure 2. Cross section of a pn-diode.

4. The current gain of the bipolar transistor. A bipolar transistor consists of three differently doped regions each having a contact. Among the two possibilities of *npn-* and *pnp*-configurations we choose to consider the latter. The arguments of this section carry over to *npn*-transistors with the obvious changes.

Note that three contacts cannot be incorporated into a one-dimensional model. Therefore we have to assume $k = 2$ or 3 for the space dimension in this section. Below we shall see that multi-dimensional effects are indeed important for the performance of bipolar transistors.

The outer $(p\text{-})$regions are called emitter (Ω_E) and collector (Ω_C), the sandwiched n-region is the base (Ω_B). The corresponding contacts are denoted by Γ_E, Γ_C and Γ_B, respectively, the emitter junction by Γ_{EB} and the collector junction by Γ_{BC} (see Figure 3). Contact voltages are measured with respect to the emitter: U_{BE} is the base-emitter voltage and U_{CE} the collector-emitter voltage. The Dirichlet conditions for u and v are then given by

$$u = v = 1, \quad \text{at } \Gamma_E, \qquad u = e^{U_{CE}}, \ v = e^{-U_{CE}}, \quad \text{at } \Gamma_C,$$
$$u = e^{U_{BE}}, \ v = e^{-U_{BE}}, \quad \text{at } \Gamma_B.$$

Considering (2.4a) we have

$$v = 1, \quad \text{in } \Omega_E, \qquad u = e^{U_{BE}}, \quad \text{in } \Omega_B, \qquad v = e^{-U_{CE}}, \quad \text{in } \Omega_C.$$

For the computation of the flow of minority carriers we again use a representation in terms of reference functions:

$$u = 1 + (e^{U_{BE}} - 1)\varphi_1, \qquad \text{in } \Omega_E,$$
$$v = e^{-U_{BE}} + (1 - e^{-U_{BE}})\varphi_2 + (e^{-U_{CE}} - e^{-U_{BE}})\varphi_3, \qquad \text{in } \Omega_B,$$
$$u = e^{U_{CE}} + (e^{U_{BE}} - e^{U_{CE}})\varphi_4, \qquad \text{in } \Omega_C,$$

where φ_1 and φ_4 satisfy the differential equations (3.2a), φ_2 and φ_3 satisfy (3.2b) and the boundary conditions for $\varphi_1, \cdots, \varphi_4$ are indicated in Figure 3.

The currents I_E, entering the device through the emitter, and I_C, leaving the device through the collector, can be computed by integrations along the emitter and collector junctions. Then the base current is given by $I_B = I_E - I_C$. The bipolar transistor serves as an amplifier in the following way: A certain collector-emitter voltage is applied and the base current is used for triggering the collector current. Thus, we are interested in the dependence of I_C on I_B and U_{CE}. This is achieved by computing U_{BE} from the formula for I_B and substituting the result into the equation for I_C. Straightforward algebra gives

$$I_C = (I_B + a_3 + a_4)\frac{a_1 - a_2 e^{-U_{CE}}}{a_3 + a_4 e^{-U_{CE}}} - a_1 + a_2$$

where the parameters $a_1, \cdots, a_4$ can be given in terms of the reference functions $\varphi_1, \cdots, \varphi_4$, and, in particular,

$$a_1 = -\int_{\Gamma_{BC}} \frac{\mu_p}{C_B} \nabla\varphi_2 \cdot \boldsymbol{\nu}\, ds,$$

$$a_3 = \int_{\Gamma_{EB}} \frac{\mu_n}{|C_E|} \nabla\varphi_1 \cdot \boldsymbol{\nu}\, ds - \int_{\Gamma_B} \frac{\mu_p}{C_B} \nabla\varphi_2 \cdot \boldsymbol{\nu}\, ds,$$

holds. Here C_B denotes the doping profile evaluated at the base side of the junctions, with similar definitions for C_E and C_C.

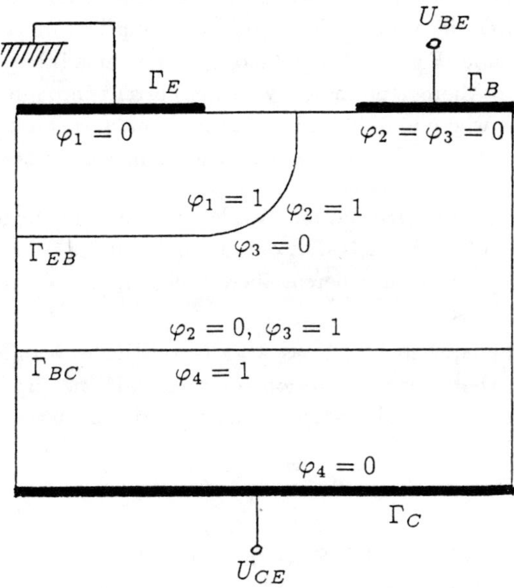

Figure 3. Cross section of a bipolar transistor.

A measure for the device performance is the common-emitter current gain

$$\beta = \frac{\partial I_C}{\partial I_B} = \frac{a_1 - a_2 e^{-U_{CE}}}{a_3 + a_4 e^{-U_{CE}}}.$$

For collector-emitter voltages significantly larger than the thermal voltage, β can be approximated by a_1/a_3 which is large iff both terms summing up to a_3 are small compared to a_1. Usually the doping in the emitter region is much higher than that in the base region implying that the ratio between a_1 and the first term in a_3 is large. However, we also require

$$(4.1) \qquad -\int_{\Gamma_B} \frac{\mu_p}{C_B} \nabla \varphi_2 \cdot \boldsymbol{\nu} \, ds \ll -\int_{\Gamma_{BC}} \frac{\mu_p}{C_B} \nabla \varphi_2 \cdot \boldsymbol{\nu} \, ds$$

which refers only to the base region. The reference function φ_2 describes a situation where the potential at the emitter junction is raised. The hole current entering through the emitter junction is split into two parts leaving through the base contact and the collector junction, respectively. The above inequality means that the current through the base contact is much smaller than that through the collector junction, i.e. essentially all the holes injected into the base reach the collector. Consider a simplified model with constant hole mobility, constant doping in the base region and neglecting recombination-generation effects. Then φ_2 solves the Laplace equation and the validity of (4.1) only depends on the geometry of the base region.

The classical analysis of bipolar transistors (see e.g. [29]) uses a one-dimensional model. As pointed out above, this means that there is no obvious way of incorporating the base contact into the model. A priori assumptions on the flow in the base region have to be made. For the classical model it is assumed that the left hand side of (6.1) vanishes, i.e. there is no minority carrier current through the base contact. However, situations where in a_3 the second term dominates can be easily imagined. Then it is necessary to use the more general theory presented here.

5. Time dependent problems.. Mathematically, the biggest difference between this section and section 2 is that only formal limiting procedures are considered for the transient problem whereas these limits have been rigorously justified for the steady state case.

The limit of the equations (1.1) as $\lambda \to 0$ constitutes a differential-algebraic system in time. If the algebraic solution component V (no time derivatives of V appear in the equations) could be determined from the algebraic equation

$$(5.1) \qquad\qquad 0 = n - p - C \,,$$

the system would be of index 1 [3]. Since this is not the case we are confronted with a system of higher index. As usual in the analysis of differential algebraic equations, we differentiate (5.1) with respect to time and obtain from (1.1a)

$$(5.2) \qquad\qquad \nabla \cdot (\mathbf{J}_n + \mathbf{J}_p) = 0 \,.$$

Since this can be interpreted as an elliptic equation for V, we obtain an index 1 system replacing (5.1) by (5.2). Therefore the original system is of index 2 in the language of the theory of differential algebraic equations, as has been pointed out by Ascher [1]. Initial data for the problem have to be compatible with both (5.1) and (5.2). It has already been observed [16], [30] that initial layers of rapid variation occur if the initial data do not satisfy (5.1), (5.2). For the steady state solutions used as initial conditions in this work, these conditions are certainly true and, thus, no initial layers are present.

For the further analysis of the transient problem it is convenient to introduce the new variables w and $\widetilde{V}$ by

$$np = \delta^4 w \,, \qquad \widetilde{V} = V - \widetilde{V}_{bi} \,,$$

where

$$\widetilde{V}_{bi} = \begin{cases} \ln(C/\delta^2) \,, & \text{in } \Omega_+; \\ -\ln(|C|/\delta^2) \,, & \text{in } \Omega_- \end{cases}$$

is an approximation for the built-in potential for small δ^2: $V_{bi} = \widetilde{V}_{bi} + O(\delta^4)$

The limiting equations for $\lambda \to 0$ in terms of w and $\widetilde{V}$ can be written as

$$\nabla \cdot \mathbf{J}_n = -\nabla \cdot \mathbf{J}_p = \frac{1}{\sqrt{C^2 + 4\delta^4 w}} \frac{\partial w}{\partial t} + Q(n,p,\mathbf{x})(w-1),$$

(5.3) $\qquad \nabla w = p\mathbf{J}_n/\mu_n - n\mathbf{J}_p/\mu_p,$

$$(n+p)\nabla\widetilde{V} = \nabla C\left(1 - \frac{n+p}{|C|}\right) - \delta^4(\mathbf{J}_n/\mu_n + \mathbf{J}_p/\mu_p)$$

with the carrier densities given by

$$(5.4) \qquad n = \frac{1}{2}\left(C + \sqrt{C^2 + 4\delta^4 w}\right), \qquad p = \frac{1}{2}\left(-C + \sqrt{C^2 + 4\delta^4 w}\right).$$

The boundary conditions at the Ohmic contacts for the new variables read

$$w = 1, \quad \widetilde{V} = -U_j(t) + O(\delta^4), \qquad \text{at } C_j, \; j = 1, \cdots, m.$$

As in the steady state case, jump conditions at the pn-junctions have to be considered. The jump conditions are the same as for the steady state problem [16]. The variables u and v introduced in section 2 — and therefore also $w = uv$ — as well as the normal components of the current densities are continuous across the junctions. From these conditions the equation

$$w = \frac{1}{4}e^{\widetilde{V}_- - \widetilde{V}_+}\left(1 + \sqrt{1 + 4\delta^4 w/C_+^2}\right)\left(1 + \sqrt{1 + 4\delta^4 w/C_-^2}\right), \qquad \text{on } \Gamma,$$

can be deduced, where the subscripts "+" and "−" refer to one-sided limits from the Ω_+- and the Ω_--sides, respectively.

Now we perform the limit $\delta^2 \to 0$. As in the steady state case we obtain from (5.4)

$$n = \max(0, C), \qquad p = \max(0, -C)$$

and, in particular, $n + p = |C|$ in the limit. Therefore the last equation in (5.3) implies that $\widetilde{V}$ is independent of position in each n- and p-region. The limiting w satisfies

$$(5.5) \qquad \frac{\partial w}{\partial t} = \begin{cases} C\left(\nabla \cdot \left(\frac{\mu_p}{C}\nabla w\right) - Q(C, 0, \mathbf{x})(w-1)\right), & \text{in } \Omega_+, \\ |C|\left(\nabla \cdot \left(\frac{\mu_n}{|C|}\nabla w\right) - Q(0, -C, \mathbf{x})(w-1)\right), & \text{in } \Omega_-. \end{cases}$$

The equation at the pn-junctions reduces to

$$(5.6) \qquad w = e^{\widetilde{V}_- - \widetilde{V}_+}, \qquad \text{on } \Gamma.$$

The initial datum $w_0(\mathbf{x})$ for w is the solution of a stationary version of (5.5), (5.6) with the initial values of $\widetilde{V}$ denoted by $\widetilde{V}(\mathbf{x}, 0) = \widetilde{V}^0(\mathbf{x})$ (constant in each n- and p-region). The new variable

$$z = w - w_0$$

solves the equations

$$
(5.7) \qquad \frac{\partial z}{\partial t} = \begin{cases} C\left(\nabla \cdot \left(\frac{\mu_p}{C}\nabla z\right) - Q(C,0,\mathbf{x})z\right), & \text{in } \Omega_+, \\[2mm] |C|\left(\nabla \cdot \left(\frac{\mu_n}{|C|}\nabla z\right) - Q(0,-C,\mathbf{x})z\right), & \text{in } \Omega_-, \end{cases}
$$

subject to the auxiliary conditions

$$
(5.8) \qquad \begin{aligned} z &= 0, \quad \text{on } C_1, \cdots, C_m, \qquad z(\mathbf{x},0) = 0, \\[2mm] z &= e^{\widetilde{V}_- - \widetilde{V}_+} - e^{\widetilde{V}^0_- - \widetilde{V}^0_+}, \quad \text{on } \Gamma, \end{aligned}
$$

and homogeneous Neumann conditions along $\partial\Omega_N$. Note that the values of z on Γ are the only inhomogeneities in (5.7), (5.8).

Similar comments as in the steady state case are also relevant here. Assuming the (spatially constant) values of $\widetilde{V}$ in each n- and p-region to be given, the flow of the minority carriers can be obtained by solving the linear problem (5.7), (5.8). The currents through the contacts are computed from the currents through pn-junctions.

6. Switching of the pn-diode. In this section we present the derivation of an equation relating the evolution of the contact voltage and the current through a pn-diode as well as the analysis of a model for a simple switching application in the form of a Volterra integral equation [23], [24].

With the notation of section 3 the last condition in (5.8) reads for the pn-diode

$$
(6.1) \qquad z = e^{U(t)} - e^{U_0}, \qquad \text{on } \Gamma,
$$

where $U(t)$ denotes the contact voltage with the initial value U_0. The current through the diode is given by $I(t) = I[w_0] + I[z](t)$ where the functional $I[\cdot]$ is defined by

$$
I[z] = \int_\Gamma \left[\left(\frac{\mu_n}{|C|}\nabla z\right)_- - \left(\frac{\mu_p}{C}\nabla z\right)_+\right] \cdot \boldsymbol{\nu}\, ds .
$$

We consider the linear operator $A: D(A) \to L^2(\Omega_0)$ with domain

$$
D(A) = \left\{ z \in H^2(\Omega_0) \,|\, z = 0 \text{ on } \Gamma_n \cup \Gamma_p, \ \frac{\partial z}{\partial \nu} = 0 \text{ on } \partial\Omega_N, \right.
$$
$$
\left. z = \text{const on } \Gamma, \ I[z] = 0 \right\},
$$

whose action on a function $z \in D(A)$ is defined by the right hand side of (5.7). It is easy to see that A is symmetric with respect to the $L^2(\Omega_0)$-inner product

$$
\langle z_1, z_2 \rangle = \int_{\Omega_+} \frac{z_1 z_2}{C}\, d\mathbf{x} + \int_{\Omega_-} \frac{z_1 z_2}{|C|}\, d\mathbf{x} .
$$

It can be shown [23] that there exists an orthonormal basis of $L^2(\Omega_0)$ consisting of eigenfunctions φ_k, $k = 1, 2, \cdots$, of a self-adjoint extension of A. The corresponding eigenvalues λ_k, $k = 1, 2, \cdots$, are negative and have $-\infty$ as their only accumulation point.

The Fourier coefficients in the representation

$$(6.2) \qquad z(\mathbf{x}, t) = \sum_{k=1}^{\infty} z_k(t)\varphi_k(\mathbf{x})$$

of the solution of (5.7), (5.8) satisfy the initial value problems

$$\dot{z}_k = \lambda_k z_k + \Phi_k I[z](t), \quad z_k(0) = 0, \qquad k = 1, 2, \cdots,$$

where Φ_k denotes the constant value of φ_k on Γ. With (6.1), evaluation of (6.2) on Γ gives

$$(6.3) \qquad e^{U(t)} - e^{U_0} = \int_0^t \widetilde{K}(t - s)I[z](s)\,ds, \qquad \widetilde{K}(t) = \sum_{k=1}^{\infty} \Phi_k^2 e^{\lambda_k t}.$$

The kernel $\widetilde{K}$ is integrable and we introduce the normalized version

$$K(t) = I_s \widetilde{K}(t), \qquad \text{with } \frac{1}{I_s} = \int_0^{\infty} \widetilde{K}(t)dt = \sum_{k=1}^{\infty} \frac{\Phi_k^2}{-\lambda_k}.$$

The choice of the symbol I_s for the normalization constant is justified. It is equal to the saturation current computed in section 3. For the initial current the Shockley equation can be used, and we have

$$I(t) = I[z](t) + I_s(e^{U_0} - 1).$$

Substitution of this relation into (6.3) gives

$$(6.4) \qquad \int_0^t K(t - s)I(s)\,ds = I_s(e^{U(t)} - 1) - I_s(e^{U_0} - 1)\int_t^{\infty} K(s)\,ds.$$

Equation (6.4) is the main result of this section. It provides the desired relation between the evolution of the current $I(t)$ and the contact voltage $U(t)$. The integral term on the left hand side accounts for the influence of the history of the evolution, i.e. in particular for charge-storage effects. The exponentially decaying kernel $K(t)$ has an integrable singularity at $t = 0$ ($K(t) = O(t^{-1/2})$ for a one-dimensional model problem [24]).

Instead of solving the eigenvalue problem for the operator $\mathcal{A}$, the kernel can also be determined from a Green's function[1] Z solving a modified version of problem (5.7), (5.8) where the last condition in (5.8) is replaced by

---

[1] We are indebted to Pierre Degond for pointing out this fact.

a) Z is independent from the position along Γ and

b) $I[Z](t) = \delta(t)$.

The kernel is then given in terms of the values of Z on Γ:

$$K(t) = I_s Z|_\Gamma$$

As in the stationary case, the voltage-current relation can be determined completely in terms of the solution of a device dependent reference problem.

As expected, (6.4) has the property that $I(t)$ converges as $t \to \infty$ if and only if $U(t)$ also converges. It is easy to see that in this case the limiting values satisfy the Shockley equation.

For given contact voltage, (6.4) is a Volterra integral equation of the first kind for $I(t)$. It is well known that this problem is mathematically ill posed. Jumps in the voltage, for example, correspond to singularities in the current. Also the purely voltage controlled problem is not very sensible from a physical point of view. The effect of a serial resistance always has to be taken into account.

The simple switching circuit depicted in Figure 4 has been considered in [23] and [24]. For $t < 0$ we assume a steady state with the contact voltage $U = U_0$ and the corresponding current $I = I_s(e^{U_0} - 1)$. At time $t = 0$ the switch S is suddenly thrown to the right. For positive t the relation

$$V = U(t) + R\,I(t)$$

holds. If this is substituted in (6.4), a nonlinear Volterra integral equation of the second kind for $I(t)$ results:

$$(6.5) \qquad \int_0^t K(t-s)I(s)\,ds = I_s(e^{V-RI(t)} - 1) - I_s(e^{U_0} - 1)\int_t^\infty K(s)\,ds\,.$$

For this equation the following result can be proven:

THEOREM 6.1. *([23]) Equation (6.5) has a unique solution $I \in C[0,\infty) \cap C^\infty(0,\infty)$ converging as $t \to \infty$ to the unique solution I_∞ of*

$$I_\infty = I_s(e^{V-RI_\infty} - 1)\,.$$

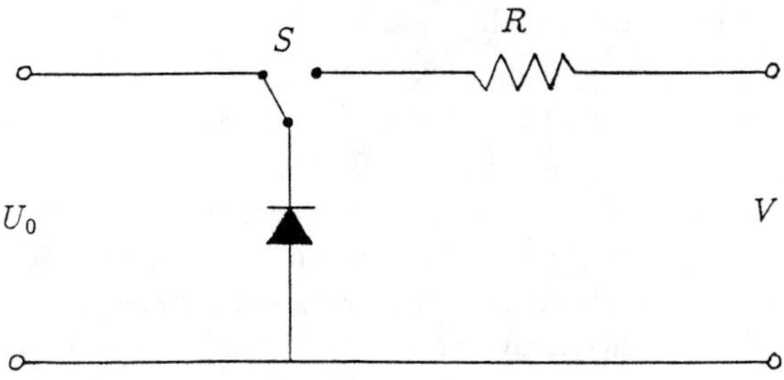

Figure 4. Basic *pn*-diode switching circuit.

Finally we wish to demonstrate the capability of the model (6.5) to describe charge-storage effects by discussing the example of switching a *pn*-diode from a forward conduction to a reverse blocking state. In this case $U_0 > 0$ and $V < 0$ holds. We rescale the current by the modulus of its initial value $I(0) = (V - U_0)/R$:

$$I = \frac{U_0 - V}{R} y$$

Equation (6.5) in terms of the new variable y reads

$$(6.6) \qquad e^{(V-U_0)(y(t)+1)} = 1 - \int_0^t K(t-s)\left(1 - e^{-U_0} - \frac{U_0 - V}{RI_s e^{U_0}} y(s)\right) ds \,.$$

We wish to discuss a situation where the involved voltages have absolute values large compared to the thermal voltage, i.e. $U_0, -V \gg 1$. The factor

$$\alpha = \frac{U_0 - V}{RI_s e^{U_0}}$$

in (6.6) is an approximation for the ratio of the initial reverse current immediately after the switching and the forward current before the switching. We consider the limits $U_0 \to \infty$ and $V \to -\infty$ keeping the value of α fixed. As a convenient small parameter we choose

$$\varepsilon = \frac{1}{U_0 - V} \,.$$

With the new parameters we write (6.6) as

$$(6.7) \qquad e^{-(y(t)+1)/\varepsilon} = 1 - \int_0^t K(t-s)[1 - \varepsilon\alpha RI_s - \alpha y(s)]ds \,.$$

The limit $\varepsilon \to 0$ in (6.7) has been carried out and justified for a simple model problem with explicitely known kernel in [24]. The limiting behaviour can be described as follows: Initially, a so called 'constant current phase' appears:

$$y(t) = -1, \qquad \text{for } 0 \le t \le t_0 \,,$$

whose length t_0 is the solution of the equation

$$(6.8) \qquad 0 = 1 - (1 + \alpha) \int_0^{t_0} K(s)\,ds\,.$$

The constant current phase is a time period where the resistivity of the diode is dominated by the serial resistance. Its occurrence is a charge-storage effect. When a sufficient amount of excess charges has been removed the current starts decaying to its steady state value. In this decay phase the exponential in (6.7) can be neglected, and the approximate solution $y(t)$ is determined from the Volterra equation of the first kind

$$0 = 1 - \int_0^t K(s)\,ds - \alpha \int_0^{t_0} K(t-s)\,ds + \alpha \int_{t_0}^t K(t-s)y(s)\,ds\,, \qquad \text{for } t > t_0\,.$$

The approximate current is continuous, but its derivative in general has a singularity at $t = t_0 +$ [24].

For the model problem treated in [24] (infinitely long one-dimensional diode) the length of the constant current phase and the solution in the decay phase have already been computed by Kingston (1954, [5]) and Lax and Neustadter (1954, [6]). As in the steady state case our analysis leads to an extension of classical results.

7. The transient behaviour of the bipolar transistor. In this section we show that the currents through a bipolar transistor and the contact voltages are related by a system of two integral equations.

We recall the transistor geometry considered in section 4 (Figure 3). Then the last condition in (5.8) can be written as

$$z = e^{U_{BE}(t)} - e^{U_{BE}^0}\,, \quad \text{on } \Gamma_{EB}\,, \qquad z = e^{U_{BC}(t)} - e^{U_{BC}^0}\,, \quad \text{on } \Gamma_{BC}\,.$$

Here U_{BE} and U_{BC} denote the base-emitter voltage and the base-collector voltage, respectively; U_{BE}^0 and U_{BC}^0 are their values at $t = 0$.

Similarly to the preceding section we define the current functionals $I_E[\cdot]$ by

$$I_E[z] = \int_{\Gamma_{EB}} \left[\left(\frac{\mu_n}{|C|}\nabla z\right)_- - \left(\frac{\mu_p}{C}\nabla z\right)_+ \right] \cdot \boldsymbol{\nu}\,ds\,,$$

and $I_C[\cdot]$ analogously by an integral over Γ_{BC}. We procede as for the pn-diode by considering the operator $\mathcal{A}$. Functions in the domain of $\mathcal{A}$ are now required to be constant along the junctions Γ_{EB} and Γ_{BC} with $I_E[z] = I_C[z] = 0$ for $z \in D(\mathcal{A})$. We again derive a Fourier series expansion of the solution z of (5.7), (5.8) in terms of eigenfunctions of $\mathcal{A}$. Evaluation of this representation on the pn-junctions leads to the system

$$e^{U_{BE}(t)} - e^{U_{BE}^0} = \int_0^t \left[K_E(t-s)(I_E(s) - I_E^0) + K_{EC}(t-s)(I_C(s) - I_C^0) \right] ds\,,$$

$$(7.1)$$

$$e^{U_{BC}(t)} - e^{U_{BC}^0} = \int_0^t \left[K_{EC}(t-s)(I_E(s) - I_E^0) + K_C(t-s)(I_C(s) - I_C^0) \right] ds\,.$$

These integral equations are the equivalent of equation (6.4) for the pn-diode. They relate the contact voltages to the emitter and collector currents $I_E(t)$ and $I_C(t)$, respectively, whose initial values are denoted by I_E^0 and I_C^0. The kernel functions are given by

$$K_E(t) = \sum_{k=1}^{\infty} \Phi_{Ek}^2 e^{\lambda_k t}, \qquad K_C(t) = \sum_{k=1}^{\infty} \Phi_{Ck}^2 e^{\lambda_k t},$$

$$K_{EC}(t) = \sum_{k=1}^{\infty} \Phi_{Ek} \Phi_{Ck} e^{\lambda_k t},$$

where Φ_{Ek} and Φ_{Ck} are the values of the k-th eigenfunction of $\mathcal{A}$ at Γ_{EB} and Γ_{BC}, respectively. An application of the Cauchy-Schwarz inequality shows that the kernel matrix of the two-dimensional system (7.1) is symmetric positive definite.

The kernel functions can again be computed by solving parabolic reference problems. We consider a function Z_E solving a version of (5.7), (5.8) with

$$I_E[Z_E](t) = \delta(t), \qquad I_C[Z_E](t) = 0,$$

as well as a function Z_C satisfying

$$I_E[Z_C](t) = 0, \qquad I_C[Z_C](t) = \delta(t).$$

Then the kernel functions are given by evaluation of these Green's functions at the pn-junctions:

$$K_E(t) = Z_E|_{\Gamma_{EB}}, \qquad K_C(t) = Z_C|_{\Gamma_{BC}}, \qquad K_{EC}(t) = Z_E|_{\Gamma_{BC}} = Z_C|_{\Gamma_{EB}}$$

8. Conclusions. Asymptotic methods have been used to reduce the computation of voltage-current characteristics of multidimensional bipolar semiconductor devices to the solution of simple, bias-point independent reference problems.

For the transient behaviour a new type of models in the form of integral equations has been presented. These models are well suited for an analysis of switching processes. Furthermore, they are simple enough to be used in circuit simulation programs as an alternative for equivalent circuit models. Preliminary numerical experiments [32] indicate that an efficient implementation is possible.

REFERENCES

[1] U. ASCHER, *On Numerical Differential Algebraic Problems with Application to Semiconductor Device Simulation*, SIAM J. Num. Anal., 26 (1989), pp. 517–538.

[2] F. BREZZI, A. CAPELO, AND L. GASTALDI, *A Singular Perturbation Analysis of Reverse Biased Semiconductor Diodes*, SIAM J. Math. Anal., 20 (1989), pp. 372–387.

[3] E. GRIEPENTROG AND R. MÄRZ, *Differential-Algebraic Equations and their Numerical Treatment*, Teubner-Texte Math. 88, Teubner, Leipzig, 1986.

[4] J. HENRY AND B. LOURO, *Singular Perturbation Theory Applied to the Electrochemistry Equations in the Case of Electroneutrality*, Nonlinear Analysis TMA, 13 (1989), pp. 787–801.

[5] R.H. KINGSTON, *Switching Time in Junction Diodes and Junction Transistors*, Proc. IRE, 42 (1954), pp. 829–834.

[6] B. LAX AND S.F. NEUSTADTER, *Transient Response of a pn-Junction*, J. Appl. Phys., 25 (1954), pp. 1148–1154.

[7] P.A. MARKOWICH, *A Singular Perturbation Analysis of the Fundamental Semiconductor Device Equations*, SIAM J. Appl. Math., 44 (1984), pp. 896–928.

[8] P.A. MARKOWICH, *The Stationary Semiconductor Device Equations*, Springer-Verlag, Wien—New York, 1986.

[9] P.A. MARKOWICH AND C. RINGHOFER, *A Singularly Perturbed Boundary Value Problem Modelling a Semiconductor Device*, SIAM J. Appl. Math., 44 (1984), pp. 231–256.

[10] P.A. MARKOWICH, C. RINGHOFER, AND C. SCHMEISER, *An Asymptotic Analysis of One-Dimensional Semiconductor Device Models*, IMA J. Appl. Math., 37 (1986), pp. 1–24.

[11] P.A. MARKOWICH, C. RINGHOFER, AND C. SCHMEISER, *Semiconductor Equations*, Springer-Verlag, Wien—New York, 1990.

[12] P.A. MARKOWICH AND C. SCHMEISER, *Uniform Asymptotic Representation of Solutions of the Basic Semiconductor Device Equations*, IMA J. Appl. Math., 36 (1986), pp. 43–57.

[13] P.A. MARKOWICH AND P. SZMOLYAN, *A System of Convection-Diffusion Equations with Small Diffusion Coefficient arising in Semiconductor Device Physics*, J. Diff. Equ., 81 (1989), pp. 234–254.

[14] R.E. O'MALLEY AND C. SCHMEISER, *The Asymptotic Solution of a Semiconductor Device Problem Involving Reverse Bias*, SIAM J. Appl. Math., 50 (1990), pp. 504–520.

[15] C.P. PLEASE, *An Analysis of Semiconductor PN-Junctions*, IMA J. Appl. Math., 28 (1982), pp. 301–318.

[16] C. RINGHOFER, *An Asymptotic Analysis of a Transient pn-Junction Model*, SIAM J. Appl. Math., 47 (1987), pp. 624–642.

[17] C. RINGHOFER, *A Singular Perturbation Analysis for the Transient Semiconductor Device Equations in One Space Dimension*, IMA J. Appl. Math., 39 (1987), pp. 17–32.

[18] C. SCHMEISER, *On Strongly Reverse Biased Semiconductor Diodes*, SIAM J. Appl. Math., 49 (1989), pp. 1734–1748.

[19] C. SCHMEISER, *A Singular Perturbation Analysis of Reverse Biased pn-Junctions*, SIAM J. Math. Anal., 21 (1990), pp. 313–326.

[20] C. SCHMEISER, *Free Boundaries in Semiconductor Devices*, Proc. Free Boundary Problems: Theory and Applications, Montreal, 1990.

[21] C. SCHMEISER, *Voltage-Current Characteristics of Multi-Dimensional Semiconductor Devices*, Quarterly of Appl. Math., 4 (1991), pp. 753–772.

[22] C. SCHMEISER AND H. STEINRÜCK, *A New Approach to the Modelling of PNPN Structures*, Solid-State Electr., 34 (1991), pp. 57–62.

[23] C. SCHMEISER AND A. UNTERREITER, *The Transient Behaviour of Multi-Dimensional PN-Diodes in Low Injection*, Math. Meth. in the Appl. Sci., 15 (1992).

[24] C. SCHMEISER, A. UNTERREITER, AND R. WEISS, *The Switching Behaviour of a One-Dimensional PN-Diode in Low Injection*, Math. Models and Meth. in Appl. Sci., 3 (1993).

[25] C. SCHMEISER AND R. WEISS, *Asymptotic Analysis of Singular Singularly Perturbed Boundary Value Problems*, SIAM J. Math. Anal., 17 (1986), pp. 560–579.

[26] W. SHOCKLEY, *The Theory of p-n Junctions in Semiconductors and p-n Junction Transistors*, Bell Syst. Tech. J., 28 (1949), p. 435.

[27] H. STEINRÜCK, *A Bifurcation Analysis of the Steady State Semiconductor Device Equations*, SIAM J. Appl. Math., 49 (1989), pp. 1102–1121.

[28] H. STEINRÜCK, *Asymptotic Analysis of the Current Voltage Curve of a PNPN Semiconductor Device*, IMA J. Appl. Math., 43 (1989), pp. 243–259.

[29] S.M. SZE, *Physics of Semiconductor Devices*, 2nd ed., John Wiley & Sons, New York, 1981.

[30] P. SZMOLYAN, *A Singular Perturbation Analysis of the Transient Semiconductor Device Equations*, SIAM J. Appl. Math., 49 (1989), pp. 1122–1135.

[31] P. SZMOLYAN, *Asymptotic Methods for the Transient Semiconductor Device Equations*, COMPEL, 8 (1989), pp. 113–122.

[32] A. UNTERREITER, *The Switching Behaviour of PN-Diodes in the Case of Low Injection*, Dissertation, TU Wien, 1991.

[33] A.B. VASIL'EVA AND V.G. STELMAKH, *Singularly Disturbed Systems of the Theory of Semiconductor Devices*, USSR Comput. Math. Phys., 17 (1977), pp. 48–58.

SYMMETRIC FORMS OF ENERGY - MOMENTUM TRANSPORT MODELS

MICHAEL SEVER*

Abstract. The energy - momentum transport models of carrier flow in a semiconductor admit entropy functions and therefore a generalized symmetric form. A simpler such form can be obtained for the reduced "mass - momentum" system, thinking of the energy equation as separately determining the carrier temperature distribution. The use of such a symmetric form leads naturally to suitable choices for boundary conditions, discretization schemes, and regularization when discontinuous solutions appear.

1. Introduction. The energy - momentum or hydrodynamic models of carrier flow in semiconductors are just the Euler equations for an ideal gas, with the addition of electric field, scattering, and heat conduction terms [4,1]. For one carrier, say electrons, we write such a model, in n space dimensions, in the form

$$(1.1) \qquad \rho_t + \sum_{i=1}^{n} (\rho u_i)_{x_i} = \rho_{scat};$$

$$(1.2) \qquad (\rho u_i)_t + \sum_{j=1}^{n} (\rho u_i u_j + \rho T \delta_{ij})_{x_j} + \rho E_i = p^i_{scat}, i = 1, \ldots, n;$$

$$(1.3) \qquad (\frac{3}{2}\rho T + \frac{1}{2}\rho |u|^2)_t + \sum_i (u_i(\frac{5}{2}\rho T + \frac{1}{2}\rho |u|^2))_{x_i} + \rho u \cdot E$$
$$= \nabla \cdot (\mu \rho T \nabla T) + e_{scat}.$$

Here the dependent variables are the carrier density ρ, velocity $u = (u_1, \cdots, u_n)$ (with $|u|$ the Euclidean norm), and the carrier temperature T. The electric field is denoted by $E = (E_1, \ldots, E_n)$; subscripts otherwise denote partial derivatives. The Wiedman - Franz model for heat condition is used [3], with μ the appropriate mobility. Factors of Boltzmann's constant, the magnitude of the election change, and the carrier effective mass have been absorbed in to the dependent variables, thus determining units in an obvious manner.

In general, the system (1.1 -1.3) is coupled with a similar system for the other carrier, Poisson's equation, and an equation for the lattice temperature. Here we shall consider only the system (1.1 - 1.3) separately.

The scattering terms $\rho_{scat}, \rho^i_{scat}, e_{scat}$ are very complicated, as they incorporate considerable information on the flow of carriers in a semiconductor, c.f. [15]. But for present purposes, these terms are remarkably docile, as they depend smoothly on the dependent variables (and the electric field, of course), but not on their space

---

*Department of Mathematics, The Hebrew University, Jerusalem, Israel

or time derivatives. Furthermore, while these terms are not expected to be small in magnitude, they are dissipative. Indeed, the momentum scattering term is typically of the form

$$(1.4) \qquad \rho^i_{scat} = -\rho u_i/\tau,$$

with a suitable, but uniformly positive and bounded, relaxation time τ. Similarly, the mobility μ in (1.3) is uniformly positive and bounded, and we shall assume it a given function of x, t.

Such energy - momentum transport models are unquestionably better than drift - diffusion models for the study of small geometry effects, carrier heating and velocity overshoot in particular, cf [2, 5]. The price of this improvement is, of course, considerable additional mathematical complexity and expense of computing solutions. In this context, the quadratic terms $(\rho u_i u_j)_{x_j}$ in (1.2) play a crucial role. If these terms are insignificant or can be added as a perturbation, then in view of (1.4), at least for stationary solutions, (1.1) and (1.2) can be combined and the drift - diffusion approximation recovered. The energy - momentum model is needed precisely when these terms are not small. Several numerical methods proposed for these models, however, do not seem to be well adapted to this situation [8, 15, 16, 17].

The heat conduction term in (1.3) may be expected to result in the temperature T remaining continuous as a function of x, t. The quadratic terms in (1.2), however, can result in discontinuities in ρ, u forming spontaneously. These "isothermal shocks" are discussed in general in [20] and in the specific context of semiconductor models in [10]. Such discontinuous stationary solutions have recently been observed in reasonable semiconductor models, at least at liquid nitrogen temperature [6, 7, 11]. Conventional wisdom suggests that they are expected whenever the flow becomes supersonic at some point.

Clearly numerical methods require some modification when discontinuities are expected. Indeed, uniqueness, even of stationary solutions, is not expected without some additional condition on the admissible discontinuities.

The system (1.1 - 1.3) is typically given in $\mathbf{R}^+ \times \Omega$, where $\Omega \subset \mathbf{R}^n$, representing the interior of a device model, is an open bounded region with piecewise smooth boundary $\partial\Omega$. In the presence of the quadratic terms in the momentum equation (1.2), appropriate boundary conditions on $\partial\Omega$ are not obvious; this question has recently been discussed in [19], and we shall emphasize this point. We also address the related questions of discretization of the system, and modification in the presence of discontinuities.

2. Symmetric form of systems of conservation laws. Our approach is based on a fairly simple change of dependent variable. A nonlinear system of conservation laws is said to be in generalized symmetric form if it is written

$$(2.1) \qquad \phi_{z,t} + \sum_{i=1}^{n} \psi^i_{z,x_i} = 0,$$

where $z = z(x,t) \in \mathbf{R}^m$ is the dependent variable, ϕ, ψ^i are smooth scalar-valued functions of z and perhaps x, t.

Much of the work of Friedrichs [9] on linear hyperbolic systems can be applied to nonlinear systems in this form; for example, if ϕ is strictly convex in z, then the system is necessarily hyperbolic.

Godunov showed that the Euler equations of gas dynamics can indeed be written in this form [12], but the result is more general [14]. A nonlinear system of conservation laws, of the form

$$(2.2) \qquad g(w)_t + \sum_{i=1}^{n}(f^i(w))_{x_i} = 0$$

with $w = w(x,t), g = g(w), f^i = f^i(w)$ vector valued in $\mathbf{R}^m$ admits an entropy function /flux U, F^i, if smooth solutions of (2.2) satisfy an additional scalar conservation law

$$(2.3) \qquad U(w)_t + \sum_{i=1}^{n}(F^i(w))_{x_i} = 0.$$

Taking derivatives and comparing terms in the obvious manner, this will be the case if there exists a (smooth) vector function $q(w)$ such that

$$(2.4) \qquad q^\dagger g_w = U_w \quad , \quad q^\dagger f_w^i = F_w^i \quad , \quad i = 1, \ldots, n$$

thus determining the scalar valued U, F^i. This is certainly true for the Euler equations, with

$$(2.5) \qquad U = \rho S \quad , \quad F^i = \rho u_i S \quad , \quad i = 1, \ldots, n$$

S the specific entropy or any smooth function thereof.

But for a system (2.2) with such an entropy function, if one simply sets

$$(2.6) \qquad \begin{aligned} z(w) = q(w) = U_g(w), \quad &\phi(z(w)) = z(w) \cdot g(w) - U(w), \\ &\psi^i(z(w)) = z(w) \cdot f^i(w) - F^i(w), i = 1, \cdots, n \end{aligned}$$

it easily follows form (2.4) that

$$(2.7) \qquad \phi_z = g, \psi_z^i = f^i, \quad i = 1, \ldots, n,$$

i.e. the systems (2.2) and (2.1) are equivalent, even for weak (discontinuous) solutions. Furthermore, from (2.4) and (2.6)

$$(2.8) \qquad \phi_{zz}(z(w)) = U_{gg}^{-1}(w)$$

so U strictly convex in g, easily obtained for the Euler equations by appropriate choice of S, implies ϕ strictly convex in z, i.e. a hyperbolic system and the transformation $g \to z$ globally invertible.

368

3. Application to the energy-momentum model. To apply this idea to the energy-momentum model, we put

$$(3.1) \qquad m = n + 2,$$

$$(3.2) \qquad g(w) = (\rho, \rho u_1, \ldots, \rho u_n, \tfrac{3}{2}\rho T + \tfrac{1}{2}|u|^2)^\dagger,$$

$$(3.3) \qquad f^j(w) = (\rho u_j, pu_1 u_j, \ldots, \rho u_n u_j, u_j(\tfrac{5}{2}\rho T + \tfrac{1}{2}\rho|u|^2))^\dagger$$
$$+ \rho T \text{ in the } j + 1 \text{ component;}$$

the vector w may be taken as convenient, provided that the mapping $w \to g$ is uniquely invertible, for example

$$(3.4) \qquad w = (\rho, u_1, \cdots, u_n, T)^\dagger.$$

The energy-momentum model (1.1-1.3) is not simply a system of conservation laws, of the form (2.2); the extra terms, corresponding to the electric field, scattering, and heat conduction will lead to a somewhat modified symmetric form

$$(3.5) \qquad \phi_{z,t} + \sum_{i=1}^{n} \psi^i_{z,x_i} + \omega(z) = \nabla \cdot (A(z)\nabla z)$$

where ω is a vector function of dimension m, and A a symmetric $m \times m$ "viscosity matrix".

The vector ω is just the transformed electric field and scattering terms, and will cause no problems. As against that, the heat conduction term is of paramount importance, for it is highly singular, and must therefore be dissipative if there is any realistic hope for a well-posed problem. In terms of the symmetric form (3.5), this means that the matrix A must be nonnegative definite. The vector $A(z)\nabla z$ has only one nonvanishing component, the m-th, equal to $\mu\rho T\nabla T$. So nonnegative definite A requires that

$$(3.6) \qquad z_m = \partial U/\partial g_m \quad \text{is an increasing function of } T \text{ alone.}$$

This has important implications on the choice of the entropy function U. For an ideal gas, it can be shown [18] that all entropy functions are of the form (2.5), but there remains the ambiguity of which function of S to use.

This is removed by the requirement (3.6), and we have

$$(3.7) \qquad S = \log \rho - \frac{3}{2}\log T;$$

obtaining U, F from (2.5), we find from (2.6), (3.2), (3.3)

$$(3.8) \qquad z = U_g = \begin{pmatrix} \log \rho - \tfrac{3}{2} \log T - |u|^2/2T + 5/2 \\ u_1/T \\ \vdots \\ u_n/T \\ -1/T \end{pmatrix},$$

$$(3.9) \qquad \phi = z \cdot g - U = \rho = (-z_m)^{-3/2} \exp\left[z_1 - \frac{5}{2} - (\sum_{j=2}^{n+1} z_j^2)/z_m \right]$$

$$(3.10) \qquad \psi^i = z \cdot f^i - F^i = \rho u_i = -\phi z_{i+1}/z_m, i = 1, \ldots, n;$$

except for z_1, the symmetric variables z and the potentials ϕ, ψ^i are of surprisingly simple form. Nonetheless, ϕ is strictly convex in z, as U is strictly convex in g, using the choice (3.7) for S. The only nonvanishing component of the viscosity matrix A is

$$(3.11) \qquad A_{mm} = \mu \rho T^3.$$

As so obtained, ϕ, ψ^i depend only on z, but ω, A may depend also on x, t.

4. Well-posedness, boundary conditions and discretization. The symmetric form (3.5) permits a classical analysis of initial boundary value problems for such systems, at least when such solutions remain sufficiently smooth. Let z be continuous is x, t and satisfy (3.5) in $\Omega \times (0, t)$, and let $\tilde{z}$ satisfy a similar system

$$(4.1) \qquad \phi_z(\tilde{z})_t + \sum_{i=1}^{n} (\psi_z^i(\tilde{z}))_{x_i} + \omega(\tilde{z}) = \nabla \cdot (A(\tilde{z})\nabla\tilde{z}) + \xi$$

in the same region, with $\tilde{z}_t, \tilde{z}_{x_i}$ uniformly bounded, but with some residual ξ, as yet unspecified.

Subtracting (4.1) from (3.5), taking the inner product with $z - \tilde{z}$, and integrating over $\Omega \times (0, t)$, we obtain a formidable looking expression

$$(4.2) \qquad I_t - I_0 + I + II_\Omega + II_\partial + III + IV_d + IV + IV_\partial + V = 0,$$

with

$$I_t = \int_\Omega [\phi(\tilde{z}(x,t)) - \phi(z,(x,t)) - (\tilde{z}(x,t) - z(x,t)) \cdot \phi_z(z(x,t))]\, dx$$

$$I_0 = \int_\Omega [\phi(\tilde{z}(x,0)) - \phi(z(x,0)) - (\tilde{z}(x,0) - z(x,0)) \cdot \phi_z(z(x,0))]\, dx$$

$$I = \int_{\Omega \times (0,t)} \left[\tilde{z}_t \cdot (\phi_z - \tilde{\phi}_z - \tilde{\phi}_{zz}(z - \tilde{z})) \right] dx\, ds$$

$$II_\Omega = \int_{\Omega \times (0,t)} \left[\sum_{i=1}^n \tilde{z}_{x_i} \cdot (\psi_z^i - \tilde{\psi}_z^i - \tilde{\psi}_{zz}^i(z - \tilde{z})) \right] dx\, ds$$

$$II_\partial = \int_{\partial\Omega \times (0,t)} \left[\sum_{i=1}^n (\nu \cdot e_i)(\tilde{\psi}^i - \psi^i - \psi_z^i \cdot (\tilde{z} - z)) \right] dx\, ds$$

$$III = \int_{\Omega \times (0,t)} (\tilde{\omega} - \omega) \cdot (\tilde{z} - z) dx\, ds$$

$$IV_d = \int_{\Omega \times (0,t)} \nabla(z - \tilde{z}) \cdot A\nabla(z - \tilde{z}) dx\, ds$$

$$IV = \int_{\Omega \times (0,t)} \nabla\tilde{z} \cdot (A - \tilde{A})\nabla(z - \tilde{z}) dx\, ds$$

$$IV_\partial = - \int_{\partial\Omega \times (0,t)} (z - \tilde{z}) \cdot \sum_{i=1}^n (\nu \cdot e_i)(Az_{x_i} - \tilde{A}\tilde{z}_{x_i}) dx\, ds$$

$$V = \int_{\Omega \times (0,t)} \xi \cdot (z - \tilde{z}) dx\, ds$$

Here ν is the outward unit normal almost everywhere on $\partial\Omega$, e_i the unit vector in the x_i direction, and we have used numerous abbreviations of the form $\tilde{\phi} = \phi(\tilde{z}(x,s))$ etc.

For $s \in [0,t]$, denote by

$$(4.3) \qquad \gamma(s) = \int_\Omega |z(x,s) - \tilde{z}(x,s)|^2\, dx;$$

then since ϕ is strictly convex in z, it follows that

$$(4.4) \qquad I_t \geq c\gamma(t)$$

where here and below, c is a positive generic constant. As ϕ is smooth in z,

$$(4.5) \qquad I_0 \leq c\gamma(0);$$

using the uniform boundedness of $\tilde{z}_t$ and $\tilde{z}_{x_i}$, we have

$$(4.6) \qquad |I| + |II_\Omega| + |III| \leq c \int_0^t \gamma(s)ds.$$

As the viscosity matrix A is smooth and nonnegative definite, IV_d is nonnegative and

$$(4.7) \qquad |IV| \leq \frac{1}{2}IV_d + c \int_0^t \gamma(s)ds$$

whereas

$$(4.8) \qquad |V| \leq c(\int_0^t \gamma(s)ds + \int_{\Omega \times (0,t)} |\xi(x,s)|^2 \, dxds)$$

Combing (4.2- 4.8), we shall have an easy proof of uniqueness and continuous dependence of (sufficiently smooth) solutions, and convergence of Galerkin discretization schemes, if only the two boundary terms in (4.2) can be controlled, e.g. made nonnegative.

The boundary $\partial\Omega$ is typically composed of insulating and Dirichlet segments, the latter corresponding to ohmic contacts. If one adopts the usual condition

$$(4.9) \qquad \nu \cdot u = \nu \cdot \nabla T = 0 \quad \text{on } \partial\Omega_N,$$

where $\partial\Omega_N$ is the insulating boundary segment, then the contributions to II_∂ and IV_∂ from this segment vanish. If the (electron) temperature T is specified on the Dirichlet segment $\partial\Omega_D$, then as $T = \tilde{T}$, IV_∂ vanishes entirely, and

$$(4.10)$$

$$II_\partial = \int_{\partial\Omega_D \times (0,t)} \left[(\tilde{\rho} - \rho)\nu \cdot (\tilde{u} - u) + (\nu \cdot u)(\tilde{\rho} - \rho - \rho\log(\tilde{\rho}/\rho)) \right.$$

$$\left. + \frac{\rho(\nu \cdot u)}{2T}|\tilde{u} - u|^2 \right] dxds.$$

In the context of semiconductor models, we shall assume that the flow is subsonic at the Dirichlet boundary,

$$(4.11) \qquad |u|^2 \quad , \quad |\tilde{u}|^2 < T \quad \text{on} \quad \partial\Omega_D$$

and that the Dirichlet boundary is the union of "inflow" and "outflow" segments,

$$(4.12) \qquad \begin{aligned} \partial\Omega_0 &= \partial\Omega_+ \cup \partial\Omega_- \\ \nu \cdot u, \nu \cdot \tilde{u} &\geq 0 \quad \text{on } \partial\Omega_+ \\ \nu \cdot u, \nu \cdot \tilde{u} &\leq 0 \quad \text{on } \partial\Omega_-. \end{aligned}$$

On the "outlfow" segment $\partial\Omega_+$, from (4.10) it clearly suffices to specify ρ as any nondecreasing function of $\nu \cdot u$; then the contribution to (4.10) from $\partial\Omega_+$ is nonnegative. As is well-known, more is required on the "inflow" segments. Typically, one could specify the tangential velocity components, e.g. vanishing, and the density ρ as a suitable decreasing function of $|\nu \cdot u|$, for example

$$(4.13) \qquad \rho(1 + 3|\nu \cdot u|^2/T) \quad \text{specified on } \partial\Omega_-.$$

Then II_∂ is nonnegative as well, and we have

$$(4.14) \qquad \gamma(t) \leq c(\gamma(0) + \int\limits_{\Omega \times (0,t)} |\xi(x,s)|^2 \, dx \, ds)$$

as a statement of uniqueness of the solution of initial-boundary value problems, and of continuous dependence on the given data . We note, however, that the constant c in (4.14) depends on t and on the uniform bounds for $\tilde{z}_{x_i} \tilde{z}_t$.

This same argument works in discrete form, and leads to an easy proof of convergence of suitable discretization schemes, e.g. Galarkin methods. Let X_h be a family of finite-dimensional subspaces of $H^1(\Omega)^m$, depending on a mesh size h. Let $Z \subset X_h$ be the set of those functions satisfying the Dirichlet boundary conditions for z, and let $Z_0 \subset X_h$ denote these functions satisfying the homogeneous form of the Dirichlet boundary conditions. Let z satisfy (3.5), and let $z_h \in Z \times [0,t]$ be determined from

$$(4.15) \qquad \phi_z(z_h)_t + \sum_{i=1}^{n} (\psi_z(z_h))_{x_i} + \omega(z_h) - \nabla \cdot (A(z_h)\nabla z_h) \perp Z_0,$$

$$(4.16) \qquad z_h(\cdot, 0) = Pz(\cdot, 0)$$

with P the L_2 projection into Z, for example. If $\tilde{z} = P_1 z$, say with P_1 the H_1 projection into Z, then $z_h - \tilde{z} \in Z_0$ at each t and from (3.5), $\tilde{z}$ satisfies (4.15) with an extra "residual" ξ added, the truncation error. Then an entirely similar analysis leads to an error estimate,

$$(4.17) \qquad \|z(\cdot,t) - z_h(\cdot,t)\|_{L_2}^2 \leq c(\|(I-P)z(\cdot,0)\|_{L_2}^2 + \int\limits_0^t \|(I-P_1)z(\cdot,s)\|_{H_1}^2 \, ds).$$

5. Weak solutions. The model (1.1 - 1.3), specifically the presence of the quadratic teams in the momentum equation (1.2), admits discontinuous, "weak" solutions, in which some of the dependent variables experience jump discontinuities, and the differential equations are satisfied only weakly, i.e. in the sense of distributions. The presence of the heat conduction term in (1.3) will keep the temperature T continuous in x, t, but jump discontinuities of two types are possible: [20]

(1) "Tangential discontinuities", where the tangential velocity components are discontinuous across some surface, but the density and normal velocity are continuous.

(2) Shocks, in which the density, normal velocity component and normal component of the temperature gradient are discontinuous across some surface, but the tangential velocity components and temperature are continuous.

In the context of semiconductor modeling, it appears highly doubtful that tangential discontinuities correspond to physically realistic phenomena, and we take the position here that they are to be avoided. In contrast, shocks are expected when the flow becomes supersonic ($|u|^2 > T$) in some region; indeed the normal velocity component must be supersonic on one side of such a shock.

But such weak solutions of (1.1 - 1.3), containing shocks, are not unique; an "entropy condition" is required to select the physically admissible weak solution. A convenient way of stating the entropy condition is by an entropy inequality [13], of the form

(5.1)
$$\int_{\Omega x(0,t)} \left[-\theta_t U(w(z)) - \sum_{i=1}^{n} \theta_{x_i} F^i(w(z)) + \theta z \cdot \omega(z) + \theta \nabla z \cdot A \nabla z \right.$$
$$\left. - \Delta\theta(z \cdot H_z - H) \right] dx\, ds \leq 0$$

where $\theta = \theta(x,s)$ is any smooth nonnegative scalar function with compact support in $\Omega \times (0,t)$, and the scalar function H is determined from $H_{zz} = A$.

Smooth solutions of (3.5) satisfy (5.1) with the equality sign; this is obtained by taking the inner product of (3.5) with θz, integrating over $\Omega \times (0,t)$, and using (2.6), (2.7).

The standard method of recovering the inequality (5.1) for weak solutions, and of avoiding tangential discontinuities, is to "mollify" the solutions by the addition of additional higher order dissipation terms. For present purposes, a viscosity term in the momentum equation will suffice.

In view of (3.8) and dimensional analysis, one would purpose the addition of a term such as

(5.2)
$$\nabla \cdot (\zeta \mu \rho T^2 \nabla(u_i/T))$$

to the right side of (1.2). In (5.2), ζ is a "switch", typically determined empirically, positive in the vicinity of shocks and zero away from them, particularly on the Dirichlet boundary segments.

The reported computations involving shocks [6, 7, 11] have used much more elaborate schemes, to obtain relatively sharp shock profiles and still avoid "overshooting" or spurious oscillations. This is clearly appropriate where the aim of the computations is to identify the presence of shocks, but in the context of realistic device modeling, in two or three space dimensions, more "smoothed out" shock profiles would appear acceptable, and a simple term such as (5.2) is easily compatible with a discretization scheme such as (4.15).

Indeed, it is questionable whether sharp shock profiles will ever be obtained for this problem. The reported computations show the apparent discontinuity immediately adjacent to a narrow transition layer, of thickness determined by the heat conductivity, in which ρ, u, T are continuous but undergo strong relative variation. The total variation of ρ, u, T over the discontinuity plus the transition layer appears to correspond, at least approximately, to what would be expected for an ideal gas shock in the absence of heat conduction. It is virtually impossible to distinguish the discontinuity from the transition layer empirically, even for the reported computations, in one space dimension and using special numerical methods.

But good resolution of the jump discontinuity in the normal component of the temperature gradient has been obtained, confirming the presence of a shock.

6. Another symmetric form. Symmetric variables such as those given in (3.8) are virtually never used, for example, in computations of flow around airframes, because of the resulting complexity of the discrete equations. While the priorities of such computations are certainly quite different from those of semiconductor modeling, this argument against the symmetric variables retains considerable force.

A procedure suggests itself, however, for simplifying the variables and thus the discrete equations considerably, while retaining many of the theoretical advantages of the symmetric form - an effective treatment of the quadratic terms of the momentum equation is particular.

We propose to decouple the system (1.1-1.3), as is common practice, identifying the energy equation (1.3) with the temperature T, and identifying the system (1.1-1.2) with ρ, u_i. If ρ, u_i are given, then (1.3) is a parabolic equation for T, readily discretized, and we shall not discuss this part further.

The system (1.1-1.2) is well-known in the context of the adiabatic approximation, with the specific entropy S constant. In the context of semiconductor modeling, however, the specific entropy is not close to constant, because of the scattering and heat conduction terms. And occurs shocks, since T is continuous the jump in S is not small, but of the same magnitude as the jump in ρ or u. Therefore we consider now the system (1.1-1.2) with T a given, continuous function of x, t.

Smooth solutions of this system satisfy something close to an entropy equation; close enough, in fact, to lead to alternative symmetric variables. With

$$(6.1) \qquad \hat{U} = \sigma(\rho, T) + \frac{1}{2}\rho|u|^2$$

$$(6.2) \qquad \hat{F}^i = u_i(\hat{U} + \rho T), i = 1, \cdots, n$$

and σ satisfying

$$(6.3) \qquad \rho\sigma_\rho = \sigma + \rho T$$

(here σ_ρ means with T constant), one readily verifies that

$$(6.4) \quad \hat{U}_t + \sum_i \hat{F}^i_{x_i} - \sigma_T(T_t + \sum_i u_i T_{x_i}) = (\sigma_\rho - \frac{1}{2}|u|^2)\rho_{scat} + \sum_i u_i(p^i_{scat} - \rho E_i).$$

375

For definiteness we take

$$(6.5) \qquad \sigma = \rho T(\log \rho - 1);$$

then $\hat{U}$ is indeed convex in

$$\hat{w} = \begin{pmatrix} \rho \\ \rho u_i \\ \vdots \\ \rho u_n \end{pmatrix}$$

and taking, analogously with section 2 above,

$$(6.6) \qquad \hat{z} = \hat{U}_{\hat{w}} = \begin{pmatrix} \sigma_\rho - \frac{1}{2}|u|^2 \\ u_i \\ \vdots \\ u_n \end{pmatrix} = \begin{pmatrix} T\log\rho - \frac{1}{2}|u|^2 \\ u_i \\ \vdots \\ u_n \end{pmatrix}$$

$$\hat{\phi} = \hat{z} \cdot \hat{\omega} - \hat{U} = \rho T$$

$$\hat{\psi}^i = \hat{z} \cdot \hat{\rho}^i - \hat{F}^i = u_i \rho T$$

($\hat{f}^i$ analogously obtained from (3.3), dropping the last component), we recover relations of the form

$$(6.7) \qquad \hat{\phi}_{\hat{z}} = \hat{w}, \hat{\psi}^i_{\hat{z}} = \hat{f}^i,$$

so the "subsystem" (1.1 - 1.2) admits another symmetric form

$$(6.8) \qquad \hat{\phi}_{\hat{z},t} + \sum_{i=1}^{n} (\hat{\psi}^i_{\hat{z}})_{x_i} = \begin{pmatrix} \rho_{scat} \\ p^i_{scat} - \rho E_i \end{pmatrix},$$

using the considerably simpler symmetric variables $\hat{z}$. Indeed, the velocity components u_i are now also symmetric variables, and $T\log\rho$ is closely related to the familiar quasi - Fermi potential.

The system (6.8) lends itself to a Galerkin discritization in the variables $\hat{z}$; the artificial viscosity term in the momentum equation would presumably now to taken

$$(6.9) \qquad \nabla \cdot (\zeta \mu \rho T \nabla u_i)$$

so as to be dissipative in the $\hat{z}$ variables.

This approach certainly leaves many unanswered questions - how to solve (1.3) and (6.8) simultaneously, how to discretize (1.3) in view of (6.6) and (6.8), and so on. As against that, successful resolution of these questions could well result in substantially improved computation schemes for this problem.

REFERENCES

1. G. BACCARANI, M. RUDAN, R. GUERRIERI AND P. CIAMPOLINI, "Physical models for numerical device simulation", in Process and Device Modeling, W. L. Engl, ed. (Elsevier, Amsterdam 1986.

2. G. BACCARANI AND M. R. WORDEMAN, "An investigation of steady-state velocity overshoot effects in Si and GaAs devices", Solid State Electronics $\underline{28}$ (1985) pp. 407-416.

3. F. J. BLATT, "Physics of electron conduction in solids", McGraw Hill, New York, 1968.

4. K. BLOTEKJAER, "Transport equations for electrons in two-valley semiconductors," IEEE Trans. Elec. Dev. $\underline{ED - 17}$ (1970), pp. 38-47.

5. R. J. COOK AND J. FREY, "Two-dimensional simulation of energy transport effects in Si and GaAs MESFETS", IEEE Trans. Elec. Dev. $\underline{ED-29}$(1982) pp. 970-977.

6. E. FATEMI, C. L. GARDNER, J. W. JEROME, S. OSHER AND D. J. ROSE, "Simulation of a steady-state electron shock wave in a submicron semiconductor device using high-order upwind methods", preprint.

7. E. FATEMI, J. W. JEROME, AND S. OSHER, "Solution of the hydrodynamic device model using high-order non-oscillatory shock capturing algorithms", IEEE Trans. on Computer Aided Design of Integrated Circuits and Systems (to appear).

8. A. FORGHIERI, R. GUERRIERI, P. CIAMPOLINI, A. GNADO, M. RUDAN AND G. BACCARANI, "A new discretization strategy of the semiconductor equations comprising momentum and energy balance", IEEE Trans. on Computer Aided Design of Integrated Circuits and Systems $\underline{7}$(1988) pp. 231-242.

9. K. O. FRIEDRICHS, "Symmetric positive linear differential equations", Comm. Pure Appl . Math. $\underline{11}$(1958) pp. 333- 418.

10. C. L. GARDNER, J. W. JEROME, AND D. J. ROSE, "Numerical methods for the hydrodynamic device model: subsonic flow", IEEE Trans. Computer Aided Design of Integrated Circuits and Systems $\underline{8}$ (1989) pp. 501-507.

11. C. L. GARDNER, "Numerical simulation of a steady-state electron shock wave in a submicron semiconductor device", preprint.

12. S. K. GODUOV, "An interesting class of quasilinear systems", Dokl. Akad. Nauk SSSR $\underline{139}$ (1961) pp. 521-523.

13. P. D. LAX, "Shock waves and entropy", in: "Contributions to Nonlinear Functional Analysis," E. H. Zarontonello, ed, Academic Press, New York, 1971.

14. M. S. MOCK, "Systems of conservation laws of mixed type", J. Diff Eq. $\underline{37}$ (1980) pp. 70-88.

15. J. W. ROBERTS AND S. G. CHAMBERLAIN, "Energy-momentum transport model suitable for small geometry silicon device simulation", Compel $\underline{9}$ (1990), pp. 1-22.

16. M. RUDAN AND F. ODEH, "Multi-dimensional discretization scheme for the hydrodynamic model of semiconductor devices", Compel $\underline{5}$ (1986), pp. 149-183.

17. M. RUDAN, F. ODEH AND J. WHITE, "Numerical solution of the hydrodynamic model for a one-dimensional semiconductor," Compel $\underline{6}$ (1987), p. 151-170.

18 S. SCHOCHET, "Examples of measure - valued solutions", Comm. in Partial Differential Equations $\underline{14}$(1989), pp. 545-575.

19. E. THOMANN AND F. ODEH, "On the well-posedness of the two-dimensional hydrodynamic model for semiconductor devices", Compel $\underline{9}$(1990), pp. 45-57.

20. YA. B. ZELDOVICH AND YU. P. RAISER, "Physics of shock waves and high temperature hydrodynamic phenomena", Academic Press New York 1967.

ANALYSIS OF THE GUNN EFFECT

H. STEINRÜCK AND P. SZMOLYAN*

Abstract. If a constant voltage above a certain threshold is applied to a piece of semiconductor material with negative differential resistance periodic current oscillations are observed in certain parameter regimes. The current peaks are due to dipole waves which are generated periodically at one contact of the device and leave at the other contact. We give a refined analysis of the classical explanation of the Gunn effect as traveling waves on an infinite domain. We show that under appropriate boundary conditions multiple steady state solutions exist and that periodic solutions are generated by a Hopf bifurcation. A singular perturbation analysis of a steady moving dipole wave on a finite domain is given.

1. Introduction. In 1963 J.B.Gunn [5] observed time periodic current oscillations in GaAs under constant voltage bias. He related these oscillations to the nonlinear dependence of the electron velocity $v(E)$ on the electric field E which is shown in Figure 1.

The electron velocity has a maximum v_{max} at the value E_T and saturates at the value v_s. The surprising fact that $v(E)$ is decreasing for $E > E_T$ is called negative differential resistance. This negative differential resistance allows the existence of moving high field pulses which give a partial explanation of the Gunn effect. The bulk negative differential resistance is due to the transferred electron effect which is outlined below.

We consider semiconductor materials, the most important of which is Gallium Arsenide, with a two valley bandstructure of the conduction band (see e.g. [11]). The most important effect of this configuration is that electrons in the lower valley have a much smaller effective mass than electrons in the upper valley. Therefore, electrons in the lower (upper) valley are called light (heavy) electrons. We denote their concentration densities by n_1 (n_2). As a consequence their drift velocities $v_1(E)$ and $v_2(E)$ due to the acceleration by an electric field E are of different orders of magnitude, i.e. $v_1(E) \gg v_2(E)$. Therefore, the terms light and fast (heavy and slow) are used simultaneously for the electrons in the lower (upper) valley. The functions v_1 and v_2 are odd, monotone increasing functions of the electric field. For small fields they are essentially linear and they saturate at values v_1^s and v_2^s as $E \to \infty$ (see Figure 2).

The intervalley scattering of electrons between these two states at a certain point within the semiconductor depends mainly on the strength of the electric field at this point (see [9]). We use a simple phenomenological description of this process suitable for the macroscopic drift-diffusion model written in scaled variables.

$$
(1) \qquad \varepsilon \frac{\partial E}{\partial x} = n_1 + n_2 - 1
$$

$$
\frac{\partial n_1}{\partial t} = \frac{\partial}{\partial x}\left(\varepsilon\delta \frac{\partial n_1}{\partial x} - v_1(E)n_1 \right) - \frac{\alpha(E)n_1 - n_2}{\lambda^2}
$$

$$
\frac{\partial n_2}{\partial t} = \frac{\partial}{\partial x}\left(\varepsilon\delta \frac{\partial n_2}{\partial x} - v_2(E)n_2 \right) + \frac{\alpha(E)n_1 - n_2}{\lambda^2}
$$

* Institut für Angewandte und Numerische Mathematik, TU-Wien, Austria. The work of the second author has been supported by the Fonds zur Förderung der wissenschaftlichen Forschung, Austria and by the Institute for Mathematics and its Applications with funds provided by the National Science Foundation.

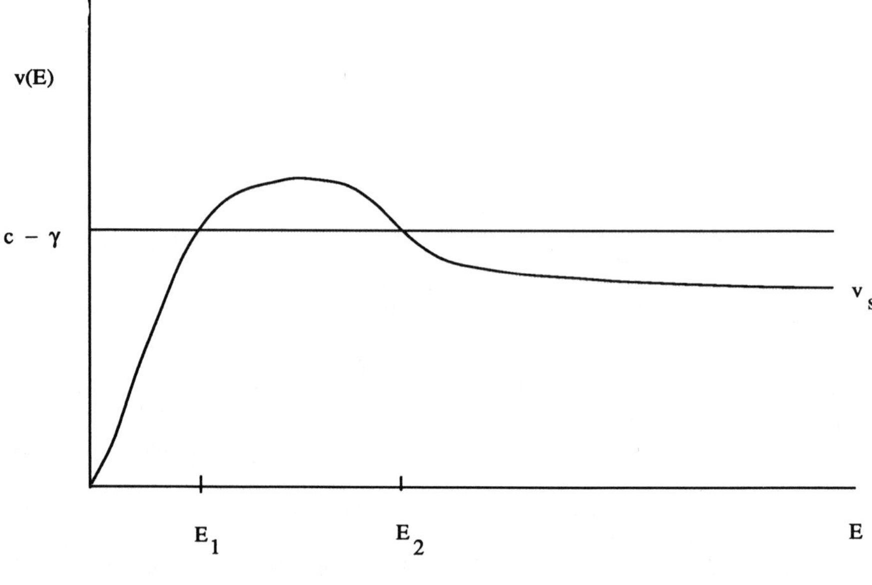

FIG. 1.

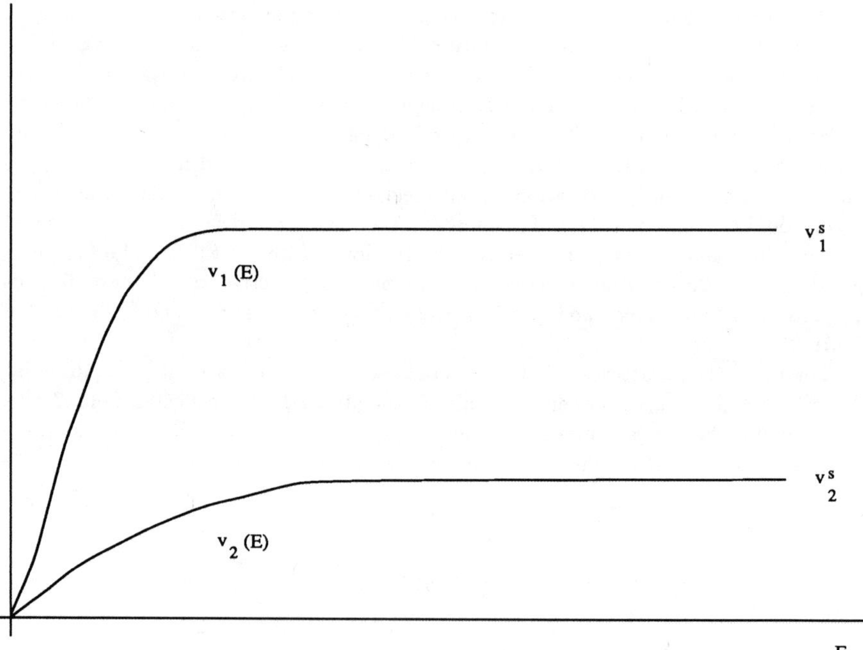

FIG. 2.

We denote time by t, the spatial variable is x. The first equation in (1) is Poisson's equation which determines the (negative) electric field E for a given space charge $\rho := n_1 + n_2 - 1$. The constant one in the expression for ρ represents a uniform background charge density called doping profile. The two other equations are continuity equations for the concentration densities of fast and slow electrons. The dimensionless constants ε and δ are given by

$$(2) \qquad \varepsilon = \frac{\varepsilon_s E_0}{q C_0 L}, \quad \delta = \frac{q D C_0}{v_0 \varepsilon_s E_0},$$

where C_0, E_0, v_0, and L are reference values for the doping profile, the electric field, the electron velocity, and the length of the device. The parameters q, ε_s, and D are the elementary charge, the permittivity of the semiconductor material, and the diffusion constant which we assume to be the same for slow and fast electrons. For typical values of these parameters ε and δ are both small. We would like to make the following comments on the form of the source term $S := (\alpha(E) n_1 - n_2)/\lambda^2$ which models the intervalley transfer: (i) the source term S appears with different signs in the continuity equations since the total electron concentration $n := n_1 + n_2$ remains unchanged by the scattering process; (ii) the parameter λ^2 is very small, because the intervalley transfer occurs on a much faster time scale than the other dynamics; (iii) by transforming system (1) to the fast time scale $\tau := t/\lambda^2$ and by keeping only the leading order terms we obtain the following system of linear differential equations for n_1 and n_2 with the variables x and $E(x)$ acting as parameters

$$(3) \qquad \begin{aligned} \frac{\partial n_1}{\partial \tau} &= -\alpha(E) n_1 + n_2 \\ \frac{\partial n_2}{\partial \tau} &= \alpha(E) n_1 - n_2. \end{aligned}$$

All solutions of the linear equation (3) converge exponentially to the stationary solution given by

$$(4) \qquad n_1 = \frac{n}{1 + \alpha(E)}, \quad n_2 = \frac{n\,\alpha(E)}{1 + \alpha(E)}$$

where $n := n_1 + n_2$ denotes the total electron concentration which does not change during this simplified fast time scale process. This motivates the assumptions : (A_1) α is a monotone increasing function of the field strength, i.e. $\alpha(E) = \alpha(|E|)$, and (A_2) $\alpha(0) = 0$. A particular simple choice for the function $\alpha(E)$ is an even power of E which has been used in [6]. We expect from the simple dynamics of the fast time scale process (3) that the distribution of the electrons between the two valleys for the full singularly perturbed problem is – at least approximately – described by equations (4), i.e. for low fields the majority of the electrons is in the lower valley whereas for high fields most of the electrons are in the upper valley.

By eliminating the fast time scale process, i.e. by assuming the validity of equations (4) we obtain formally the reduced problem corresponding to system (1). By adding the two continuity equations we obtain an equation for the total electron concentration $n := n_1 + n_2$

$$(5) \qquad \begin{aligned} \varepsilon \frac{\partial E}{\partial x} &= n - 1 \\ \frac{\partial n}{\partial t} &= \frac{\partial}{\partial x}\left(\varepsilon\delta \frac{\partial n}{\partial x} - v(E)n \right). \end{aligned}$$

Thus we are left with a single continuity equation where the averraged electron velocity $v(E)$ is given by

$$(6) \qquad v(E) := \frac{v_1(E) + \alpha(E)v_2(E)}{1 + \alpha(E)}.$$

Due to our assumptions on the electron velocities v_1, v_2 and the function α the velocity coefficient $v(E)$ exhibits negative differential resistance (see Figure 1). For small values of the field the velocity is almost linear $v(E) \sim v_1(E)$, as the field increases the proportion of electrons in the upper (slower) valley increases causing a decrease of the velocity for large electric field. For even larger fields, i.e. if most of the electrons are in the upper valley, we have $v(E) \sim v_2(E)$. We do not bother to give precise conditions on the coefficients v_1, v_2, and α which guarantee this behavior we simply assume that we have a configuration with this qualitative behavior.

The above arguments suggest to rewrite system (1) in terms of the slow variable $n := n_1 + n_2$ and the fast variable n_1

$$(7) \qquad \varepsilon \frac{\partial E}{\partial x} = n - 1$$
$$\frac{\partial n}{\partial t} = \frac{\partial}{\partial x}\left(\varepsilon\delta \frac{\partial n}{\partial x} - v_1(E)n_1 - v_2(E)(n - n_1) \right)$$
$$\frac{\partial n_1}{\partial t} = \frac{\partial}{\partial x}\left(\varepsilon\delta \frac{\partial n_1}{\partial x} - v_1(E)n_1 \right) + \frac{n - (1 + \alpha(E))n_1}{\lambda^2}.$$

One expects - at least heuristically - that for λ sufficiently small the solution of system (7) is close to the solution of the reduced problem (5).

The Gunn effect has been thoroughly investigated by physicists and device engineers as documented in the monographs [2], [10]. More recently some of the more mathematical aspects of the interesting dynamics associated with the Gunn effect have been analysed in [12], [13]. It is the purpose of this paper to present some of these results in a unified form. Our main tool are singular perturbation methods based on the smallness of the parameters λ, ε, and δ. In Section 2 we consider the Gunn effect in the framework of traveling waves in an infinite long device. We prove the existence of traveling wave solutions of system (1) close to the well known traveling wave solutions of system (5). In the rest of the paper we consider system (5) in a finite geometry. In Section 3 we present new results on bifurcating steady state and time periodic solutions. These results prove analytically that system (5) supports periodically oscillating solutions. In Section 4 we give an asymptotic expansion of a steady moving dipole layer away from the boundaries.

2. Traveling waves. A first explanation of the Gunn effect by traveling wave solutions of system (5) is classical. A mathematical analysis of the existence and stability of these waves is given in [12]. In this section we will report on the existence of traveling wave solutions of system (1). Traveling wave solutions are solutions depending on the single variable $s := x - ct$. By transforming system (7) to the variable s and by integrating the equation for n once we obtain the following four-dimensional nonlinear system of singularly perturbed ordinary differential equations

$$(1) \qquad \varepsilon \dot{E} = (n - 1)$$
$$\varepsilon\delta \dot{n} = v_1(E)u + v_2(E)(n - u) - cn + \gamma$$

$$\lambda \dot{u} = w$$
$$\lambda \varepsilon \delta \dot{w} = \lambda(v_1(E) - c)w + \lambda^2 v_1'(E)u(n-1) + (1 + \alpha(E))u - n .$$

where we have used the substitutions $u := n_1$, $w := \lambda \dot{n}_1$ and γ is an arbitrary integration constant. We consider λ as the smallest parameter, i.e. the last two equations are singularly perturbed with respect to λ for fixed ε and δ. Physically significant traveling wave solutions correspond to homoclinic or heteroclinic orbits connecting fixed points of equation (1). As a first step we construct the homoclinic or heteroclinic orbits of the reduced problem

$$
\begin{array}{rcl}
\varepsilon \dot{E} &=& n - 1 \\
\varepsilon \delta \dot{n} &=& (v(E) - c)\,n + \gamma \\
0 &=& w \\
0 &=& (1 + \alpha(E))\,u - n
\end{array}
\tag{2}
$$

obtained by setting $\lambda = 0$ in system (1) where the function $v(E)$ is given by equation (6). Obviously, the first two equations correspond to the traveling wave problem for system (5). We restrict ourselves to the case of waves traveling to the right, i.e. $c > 0$. However, all the traveling waves exist for $c < 0$ as well under the obvious changes. The fixed points of (1) are given by $p_i := (E_i, 1, 1/(1 + \alpha(E_i), 0)$, $i = 1, 2$ where the values E_i are the solutions of the equation $v(E) = c - \gamma$. Let v_{max} denote the local maximum of $v(E)$ and let v_s denote the saturation velocity. For $c - \gamma \in (v_s, v_{max})$ there exist two fixed points which we assume to be ordered as in Figure 1. An easy calculation shows that the fixed point p_1 is a saddle for all choices of the parameters. The fixed point p_2 is a center for $\gamma = 0$ and a stable or unstable node or focus depending on the parameters c, γ and δ. We are not more specific on that since p_2 is not important in the following. The substitution $n = e^z$ transforms the first two equations of the reduced problem (2) to

$$
\begin{array}{rcl}
\varepsilon \dot{E} &=& e^z - 1 \\
\varepsilon \delta \dot{z} &=& v(E) - c + \gamma e^{-z}
\end{array}
\tag{3}
$$

which is a Hamiltonian system for $\gamma = 0$.

THEOREM 2.1. *For $\gamma = 0$ equation (2) has an an orbit $\omega(c)$ homoclinic to the point $p_1(c)$ for all $c \in (v_s, v_{max})$. There exists a family of periodic orbits around the center p_2. Furthermore, the electron concentration n is strictly positive along these orbits. In the case $\gamma \neq 0$ no closed orbits exist.*

Proof: By using E as independent variable we obtain the differential equation

$$
\delta \frac{d}{dE} n = \frac{(v(E) - c)n}{n - 1}
\tag{4}
$$

By integrating this equation starting at the point p_1 we obtain the equation

$$
n(E) - \ln(n(E)) - 1 = \frac{1}{\delta} \int_{E_1}^{E} (v(e) - c)\, de
\tag{5}
$$

which describes the stable and unstable manifolds of p_1. The value $n = 1$ is the only zero of the function $n - \ln n - 1$. The unstable manifold of p_1 crosses the line $n = 1$ at a maximal value of the field E_{max} which is determined by the 'equal area rule'

$$
\int_{E_1}^{E_{max}} (v(e) - c)\, de = 0,
$$

and returns to the fixed point forming the homoclinic orbit. For $c \to v_{max}-$ the homoclinic loop $\omega(c)$ shrinks to the corresponding fixed point p_1, for $c \to v_s+$ the homoclinic loop $\omega(c)$ becomes unbounded, i.e. $E_{max} \to \infty$ holds. The existence of the family of periodic orbits inside the homoclinic loop is proved in the same way.

In the case $\gamma \neq 0$ no closed orbits exist because the divergence of the vector-field (3) is strictly positive resp. negative for $\gamma < 0$ resp. $\gamma > 0$. $\square$

The traveling waves given by Theorem (2.1) have the form of a single high field pulse corresponding to a dipole layer of the of the electron concentration. By using methods from dynamical systems theory [4], [14] the following theorem is proved in [13]

THEOREM 2.2. *Assume that* $\alpha : \mathbf{R} \to \mathbf{R}$ *is* C^r, $r > 2$. *Then for* $[c_-, c_+] \subset (v_s, v_{max})$ *there exist* $\lambda_1 > 0$ *and a* C^{r-1}*-function* $\gamma : [c_-, c_+] \times [0, \lambda_1) \to \mathbf{R}$ *such that the singularly perturbed problem (1) has an orbit* $\omega_\lambda(c)$ *homoclinic to the fixed point* p_1 *for* $\gamma = \gamma(c, \lambda)$, $c \in [c_-, c_+]$, *and* $\lambda \in (0, \lambda_1)$. *The electron concentrations* n, u, *and* $n - u$ *are strictly positive along all the homoclinic orbits* $\omega_\lambda(c)$.

We conclude from Theorem 2.2 that all homoclinic orbits of the reduced problem (2) are slightly perturbed to homoclinic orbits of the singularly perturbed problem (1) for small λ. Since the parameter γ is just an integration constant in system (1) we obtain the existence of traveling wave solutions of system (1) in an interval of possible wavespeeds. A unique traveling wave solution is determined by the value of $E(\pm\infty)$ which determines the wavespeed c by the equation $v(E(\pm\infty)) = c - \gamma(c, \lambda)$ and vice versa. We will show in Section 4 that for system (5) on a finite domain a unique wave is determined by appropriate boundary conditions. Our analysis shows that - in the context of the traveling wave description of the Gunn-effect - the reduced problem (5) is a valid zeroth order approximation of the more sophisticated two-valley model (1).

3. Bifurcation analysis. We consider equations (5) on a finite domain which is the interval $[0, 1]$ because of our scaling. We need to specify boundary conditions. One boundary condition is given by

$$(1) \qquad \int_0^1 E(x, t)\, dx = U$$

where U is the applied bias. It is well known that effects at the contacts are crucial for the mode of operation of a Gunn diode (see e.g. [10]). However, we consider just two simple types of boundary conditions to be able to obtain analytical results. The first type of boundary conditions which we consider is vanishing space charge at the boundaries

$$(2) \qquad n(0, t) = n(1, t) = 1.$$

The other type is to prescribe the electric field at the boundary

$$(3) \qquad E(0, t) = E_0, \quad E(1, t) = E_1.$$

Actually we will assume that $E_0 = E_1 = U$ holds. Under this assumption equation (5) has the trivial stationary solution

$$(4) \qquad n \equiv 1, \quad E \equiv U$$

for both types of boundary conditions. In the following we will linearize system (5) at the stationary solution (4) and analyse the spectrum of the linerized operator. We will show that the trivial stationary solution looses its stability and that bifurcations of nontrivial stationary solutions of system (5), (1), and (2) and of periodic solutions of system (5), (1), and (3) occur for appropriate values of the parameters U, ε, and δ. We will use that $\varepsilon << 1$ resp. $\varepsilon\delta << 1$ hold, therefore, we repeat the asymptotic dependence of these nondimensional parameters on the physical parameters: $L \to \infty$ implies $\varepsilon \to 0$ and $\delta = const.$, $C \to \infty$ implies $\varepsilon \to 0$, $\delta \to \infty$, $\varepsilon\delta = const.$, $D \to 0$ implies $\varepsilon = const.$, $\delta \to 0$. We regard L and C as the main parameters which specify a certain device whose operation mode is determined by the bifurcation parameter U. We are somewhat hesitant to use the limit $\delta \to 0$ which is extensively used in [1] because this limit can not be realized by varying physical parameters of a device.

THEOREM 3.1. *For $U < E_T$ the solution (4) of system (5), (1), and (2) is stable for all values of the parameters ε, δ. For sufficiently small ε and $\varepsilon\delta = O(1)$ there exist $U_1(\varepsilon), U_2(\varepsilon)$ such that the trivial solution is stable for $U < U_1$ or $U > U_2$ and unstable for $U_1 < U < U_2$. At $U = U_1$ a nontrivial solution bifurcates which is unstable for $U < U_1$ and stable for $U > U_1$. At $U = U_2$ a nontrivial solution bifurcates which is stable for $U < U_2$ and unstable for $U > U_2$.*

Proof : We set $E = U + e$ and $n = 1 + w$ and obtain the following eigenvalue problem for the linearization of system (5), (1)

$$(5) \qquad \varepsilon\delta w_{xx} - v(U)w_x - \frac{v'(U)}{\varepsilon}w = \mu w$$

subject to the boundary conditions $w(0) = w(1) = 0$. The corresponding simple eigenvalues are given by

$$(6) \qquad \mu_k = -\frac{v'(U)}{\varepsilon} - \frac{v^2(U)}{4\varepsilon\delta} - k^2\pi^2\varepsilon\delta, \quad k = 1, 2, \dots$$

In the case $U < E_T$ the largest eigenvalue μ_1 is always negative since $v'(U) > 0$ holds which proves the first part of the theorem. For sufficiently small ε and $\varepsilon\delta = O(1)$ there exist $E_T < U_1(\varepsilon) < U_2(\varepsilon)$ such that the largest eigenvalue μ_1 is negative for $U < U_1$ or $U > U_2$, and is positive for $U_1 < U < U_2$. Since the eigenvalues and eigenfunctions can be computed explicitly it is easy to verify that at $U = U_1$ resp. $U = U_2$ the conditions for the theorem on bifurcation from a simple zero eigenvalue (see [3]) are satisfied from which the existence and stability properties of the bifurcating nontrivial solutions follow. $\square$

We conclude from the above remarks on the asymptotic dependence of the dimensionless parameters ε, δ on the physical parameters that the assumptions of Theorem 3.1 are satisfied in a device of constant length for high doping profile. We remark that nontrivial stationary solutions bifurcate from the eigenvalues μ_k, $k > 1$ as well, however these solutions are unstable. These analytical results and our numerical experiments indicate that system (5) with boundary conditions (1), (2) has no time periodic solutions, i.e. these boundary conditions are not able to describe the Gunn effect. Now we consider the second type of boundary conditions (1), (3) where we assume additionally that $E_0 = E_1 = U$ holds.

THEOREM 3.2. *For $U < E_T$ the solution (4) of system (5), (1), and (3) is stable for small values of the parameter $\varepsilon\delta$. For sufficiently small ε there exist $U_1(\varepsilon), U_2(\varepsilon)$*

such that the trivial solution is stable for $U < U_1$ or $U > U_2$ and unstable for $U_1 < U < U_2$. At $U = U_1$ and $U = U_2$ branches of periodic solutions bifurcate from the trivial solution by a Hopf bifurcation. Due to the algebraic complexity of the problem we do not compute the direction of the bifurcating branches. However, numerical experiments indicate that both Hopf bifurcations are supercritical, i.e. the bifurcating periodic solutions exist for $U > U_1$ resp. $U < U_2$ and are stable.

Proof: The proof is an application of the Hopf bifurcation, i.e. the periodic solutions are generated as a pair of purely imaginary eigenvalues of the linearization crosses the imaginary axis with nonzero speed as U passes through U_1 resp. U_2. We shall only verify this eigenvalue condition and omit the proof that the semiflow generated by system (5), (1), and (3) has the smoothness and compactness properties necessary for the Hopf bifurcation for an evolution equation in a Banach space (see [7]). We set $E = U + e$, $n = 1 + w$ and obtain the eigenvalue problem of the linearization of system (5), (1), and (3)

$$(7) \qquad\qquad\qquad \varepsilon e_x - w = 0$$

$$\varepsilon \delta w_{xx} - v(U) w_x - \frac{v'(U)}{\varepsilon} w = \mu w$$

subject to the boundary conditions

$$(8) \qquad\qquad e(0) = 0, \quad e(1) = 0, \quad \int_0^1 e(x)\, dx = 0.$$

By using $e(x) = \int_0^x w(s)\, ds$ we obtain nonlocal conditions on w

$$(9) \qquad\qquad \int_0^1 w(x)\, dx = 0, \quad \int_0^1 \int_0^x w(s)\, ds dx = 0.$$

If we set

$$\eta = \frac{\mu}{v(U)} + \frac{v'(U)}{\varepsilon v(U)}, \qquad \beta = \frac{\varepsilon \delta}{v(U)}$$

and use the equation for w to simplify the boundary condition (9) the above eigenvalue problem is (for $\eta \neq 0$) equivalent to

$$\beta w_{xx} - w_x - \eta w = 0$$

with the coupled boundary conditions

$$-\beta w_x(1) + (1 + \beta) w(1) - \beta w(0) = 0, \quad \beta w_x(0) - (1 - \beta) w(0) - \beta w(1) = 0.$$

A long calculation gives the eigenvalue equation

$$(10) \qquad D(\eta, \beta) = \exp\left(\frac{2\eta}{1 + \sqrt{1 + 4\beta\eta}}\right)(\eta - \sqrt{1 + 4\beta\eta}) + \sqrt{1 + 4\beta\eta} -$$

$$\exp\left(-\frac{1 + \sqrt{1 + 4\beta\eta}}{2\beta}\right)(\eta + \sqrt{1 + 4\beta\eta}) + e^{-1/\beta}\sqrt{1 + 4\beta\eta} = 0.$$

For $\beta = 0$ the eigenvalue equation reduces to

$$e^{\eta}(\eta - 1) + 1 = 0.$$

Note that all zeros (except $\eta = 0$) have negative real parts and that $\eta = 0$ corresponds to no eigenvalue of the original problem. The zero with the largest real part is $\eta_1 \sim -1.774 \pm 5.202i$. Thus the eigenvalue of the original problem with the largest real part is asymptotically given by

$$(11) \qquad \mu = -\frac{v'(U)}{\varepsilon} + \eta_1 v(U) + O(\varepsilon\delta).$$

For $U < E_T$ all eigenvalues have negative real parts for $\varepsilon\delta$ small and the trivial solution is stable. For sufficiently small ε the first term in equation (11) dominates. It is easy to see that there exist $E_T < U_1 < U_2$ such that the eigenvalue with the largest real part crosses the imaginary axis at $U = U_1$ resp. U_2 with nonzero speed. Thus the eigenvalue conditions for the occurence of a Hopf bifurcation are satisfied. $\square$

The conditions of the theorem are satisfied in an sufficiently long device with arbitrary doping profile. The current oscillations corresponding to the periodic solutions which we obtain by the Hopf bifurcation are small amplitude sinusoidal oscillations still far away from the spikelike oscillations observed in the Gunn effect. However, our analysis proves that periodic solutions exist. We conjecture based on numerical experiments that the amplitude of the oscillations grows as the applied voltage is encreased and become more spikelike as ε is decreased.

Another interesting result of our stability- and bifurcation analysis is that the trivial solution loses its stability at a certain voltage $U_1 > E_T$ due to the negative sign of $v'(U)$. However, because of velocity saturation $v'(U) \to 0$ as $U \to \infty$ and consequently the trivial solution regains its stability again at the applied voltage U_2. Since a realistic field-velocity curve is - unlike Figure 1 - rather flat in the region of negative differential resistance the length of interval (U_1, U_2) on which the instabilities and oscillations are observed could be rather small for a moderate value of ε. This could explain some of the experiments[1] reported in [10] on the existence or nonexistence of instabilities as the applied voltage is varied in samples of different length.

It is an interesting mathematical problem whether the branches of bifurcating solutions given in Theorems 3,4 which start at U_1 and U_2 are connected. This global problem is not answered by our local results.

4. Dipole waves on a finite domain. In this section we will show that the travelling wave solutions given by Theorem 1 are also relevant in a finite geometry as long as the pulse stays away from the boundaries. We will give a matched asymptotic expansion of a steady moving dipole wave for small values of ε. By setting $\varepsilon = 0$ in system (5) we obtain the outer problem

$$(1) \qquad n = 1, \quad \frac{\partial}{\partial x} v(E) = 0$$

which implies that the outer solution is $\bar{n} = 1$ and $\bar{E}$ is constant since we look for a single pulse.

The approximate solution has to satisfy the boundary conditions (1), (2) or (3). The outer solution $\bar{n} = 1$ satisfies condition (2) therefore no boundary layers occur in this case. If the field is prescribed at both endpoints boundary layers of a standard

---

[1] We are grateful to H.L.Grubin for pointing out this fact.

type occur which have no impact on the following analysis. The crucial condition is the integral condition (1) which would imply $\bar{E} = U$. However, this trivial solution is unstable for $U > E_T$. Thus we look for an approximate solution

$$E = \bar{E} + \hat{E}$$

which is essentially the traveling wave solution from Theorem 1. The layer term $\hat{E}$ depends on a stretched moving variable $\xi = (x-ct)/\varepsilon^\alpha$ and should decay exponentially for $\xi \to \pm\infty$ which is only possible for $\bar{E} < E_T$. The moving internal layer term $\hat{E}$ has to satisfy

$$(2) \qquad \bar{E} + \int_0^1 \hat{E}\, dx = U\,.$$

The traveling wave equation (2) suggests that $\alpha = 1$ gives the correct layer variable ξ. However, in that case the contribution of $\hat{E}$ in the above integral is just $O(\varepsilon)$ because the width of the layer is $O(\varepsilon)$ and $E = O(1)$. This difficulty is resolved by choosing the speed c close to the saturation velocity v_s which makes the layer wider and the maximal value of the field higher (see the proof of Theorem 1). We follow the asymptotic expansion given in [8]

$$(3) \qquad \bar{E} = E_0 + \sqrt{\varepsilon}E_1 + O(\varepsilon), \qquad \hat{E} = \frac{e(\xi)}{\sqrt{\varepsilon}} + O(1)$$

with $\xi = (x - ct)/\sqrt{\varepsilon}$. The value E_0 is the solution of $v(E) = v_s$ and E_1, c have to be determined in the solution process. In the following we use the notation $E_\infty = E_0 + \sqrt{\varepsilon}E_1$. By inserting the expansion (3) into system (5) we obtain the layer equation

$$(4) \qquad \begin{aligned} e' &= n - 1 \\ \sqrt{\varepsilon}\delta n' &= \left(v(E_\infty + \frac{e}{\sqrt{\varepsilon}}) - v(E_\infty) \right) n \end{aligned}$$

where we have used that $c = v(E_\infty)$ is necessary for the existence of a decaying solution. A similar analysis as in the proof of Theorem 1 shows that system (4) has an orbit homoclinic to the fixed point $(0,1)$ which gives us the moving internal layer. However, system (4) still depends on ε therefore a more careful analysis is necessary, which reveals a double layer structure and enables us to determine E_1 and hence the wavespeed c.

As long as $e = O(1)$ holds system (4) is asymptotically

$$(5) \qquad \begin{aligned} e' &= n - 1 \\ \sqrt{\varepsilon}\delta n' &= \left(v_s - v(E_0) + \sqrt{\varepsilon}v'(E_0)E_1 \right) n + O(\varepsilon), \end{aligned}$$

which gives $\delta n' = -v'(E_0)E_1 n$. We integrate this equation to obtain

$$(6) \qquad n(\xi) = e^{-v'(E_0)E_1\xi/\delta}$$

for ξ in an interval which will be determined later. Our choice of the integration constant in equation (6) implies $n(0) = 1$. To obtain the interval of validity of the above equation we use the equivalent of formula (5) for system (4)

$$(7) \qquad n(e) - \ln(n(e)) - 1 = \frac{1}{\delta} \int_{E_\infty}^{E_\infty + e/\sqrt{\varepsilon}} (v(y) - c)\, dy\,.$$

If we assume that the convergence of $v(y) \to v_s$ for $y \to \infty$ is sufficiently fast and by using $c = v(E_0) + \sqrt{\varepsilon}v'(E_0)E_1 + O(\varepsilon)$ we obtain asymptotically

$$(8) \qquad n(e) - \ln(n(e)) - 1 \sim \frac{1}{\delta} \int_{E_0}^{\infty} (v(y) - v_s)\, dy \; - \frac{v'(E_0)E_1 e}{\delta} = A - BE_1 e \,.$$

From this equation we obtain

$$(9) \qquad e = \frac{A - n + \ln(n) + 1}{BE_1}$$

which is valid for positive e, i.e. for n between the zeros $0 < n_2 < n < n_1$ of the right hand side. Thus, equation (6) and n_1, n_2 determine $\xi_1 < 0 < \xi_2$ such that n, e are given asymptotically by equations (6), (9) for $\xi_1 < \xi < \xi_2$. At ξ_1 the electron concentration jumps from one to n_1, at ξ_2 the electron concentration jumps from n_2 to one. These jumps in the electron concentrations correspond to corners in the electric field. These discontinuities can be smoothed out by additional layer terms on the faster scale

$$\zeta = \frac{\xi - \xi_i}{\sqrt{\varepsilon}}, \quad i = 1, 2.$$

We compute that to leading order

$$\int_0^1 \hat{E}\, dx = \int_{\xi_1}^{\xi_2} e(\xi)\, d\xi = K/E_1^2$$

holds where K is given by

$$K = B^{-2}\left(n_2 - n_1 - \ln(\frac{n_2}{n_1}\left(\frac{1}{2}\ln(n_1 n_2) + A + 1\right)\right) .$$

From formula (2) we obtain

$$(10) \qquad E_1 = \sqrt{K/(U - E_0)} \,.$$

We conclude that the traveling wave solutions given in Theorem 1 are essential in a description of the Gunn effect in a finite geometry. Our analysis reveals the double layer structure of the moving dipole layer and its sensitive dependence on the applied voltage for small ε. The speed of the dipole wave is always close to the saturation velocity. Obviously, our description of the dipole layer can be only valid as long as the layer does not interact with the boundaries. The difficult problem of the generation of the dipole wave at one end of the device and of a dipole leaving the device at the other end has to be analysed to obtain a full understanding of the Gunn effect. Interesting results on this problem have been obtained in [1] in the limit[2] $\delta \to 0$.

REFERENCES

[1] L.L.Bonilla, Solitary waves in semiconductors with finite geometry and the Gunn effect, SIAM J.Appl.Math 51, Nr.3, (1991).
[2] B.G.Bosch, R.W.H.Engelmann, *Gunn Effect Electronics*, Halsted Press, (1975).

---

[2] A detailed asymptotic analysis of a moving dipole layer in that limit has been presented by L.V.Kalachev at this conference.

[3] M.Crandall and P.Rabinowitz, Bifurcation, perturbation of simple eigenvalues and linearized stability, Arch. Rat. Mech. Anal. 52, (1972), pp.161-181.

[4] N.Fenichel, Geometric singular perturbation theory, Journal of Differential Equations 31, (1979), pp. 53-98.

[5] J.B.Gunn, Microwave oscillations of current in III-V semiconductors, Solid State Comm.1, (1963), pp.88-91.

[6] H.Kroemer, Nonlinear space-charge domain dynamics in a semiconductor with negative differential mobility, IEEE Trans. Electron Devices, Vol. Ed.-13, (1966), pp.27-40.

[7] J.E.Marsden and M.McCracken, *The Hopf Bifurcation and Its Applications*, Springer, New York (1976).

[8] P.A.Markowich, C.A.Ringhofer, and C.Schmeiser, *Semiconductor Equations* Springer, New York (1990).

[9] B.K.Ridley, The inhibition of negative resistance dipole waves and domains in n-GaAs, IEEE Trans. Electron Devices, Vol. Ed.-13, (1966), pp.41-43.

[10] M.P.Shaw, H.L.Grubin, and P.R.Solomon, *The Gunn-Hilsum Effect*, Academic Press, New York (1979).

[11] S.M.Sze, *Physics of semiconductor devices*, 2nd. Ed., Wiley, New York (1981).

[12] P.Szmolyan, Traveling waves in GaAs-semiconductors, Physica D 39, (1989), pp.393-404.

[13] P.Szmolyan, Analysis of a singularly perturbed traveling wave problem, IMA-preprint 649 (1990), to appear SIAM J.Appl.Math.

[14] P.Szmolyan, Transversal heteroclinic and homoclinic orbits in singular perturbation problems, J.Differential Equations 92, Nr.2, (1991), pp.252-281.

SOME EXAMPLES OF SINGULAR PERTURBATION PROBLEMS IN DEVICE MODELING

MICHAEL J. WARD*, LUIS REYNA AND F. ODEH†

1. Introduction. In this article we give a brief survey of some recent asymptotic results obtained for three singular perturbation problems arising in device modeling. The emphasis will be to obtain some rather explicit representations of the solutions and the input-output characteristics for these three 'typical' problems.

In §2 we show how the above asymptotic method can be used to analyze the two-dimensional equilibrium potential near the source and drain for a simplified MOSFET structure under strong inversion conditions. The analysis relies on an explicit solution to a free boundary problem found in [2]. We will then show how to obtain an analytical expression for the current-voltage relation near equilibrium of a MOSFET structure having $1 - 10\mu$ long channels and constant channel doping levels on the order of 10^{16} cm^{-3}.

In §3 we describe some recent work in the construction of multiple steady state solutions for a one-dimensional PNPN structure. We also briefly comment on some of the difficulties associated with including the effect of recombination into the analysis for such a structure. Finally, in §4 we introduce a high field scaling and analysis of the drift-diffusion model for a strongly forward biased PN or PNPN structure.

2. An application of a free boundary problem in MOSFET modeling. In this section we analyze the equilibrium potential distribution in the two-dimensional regions near the source and drain for a MOSFET structure under strong inversion conditions. The equilibrium potential is used to derive a composite expansion for the mobile charge, which includes the charge contributions from the gate, source, and drain regions. Using the mobile charge we will derive the current-voltage relation, valid for small source-drain biases, for a moderately short-channel MOSFET. This current-voltage relation will be compared to the corresponding result for an infinitely long-channel. The details of the analysis can be found in [12].

Introducing non-dimensional variables as in [12], the equilibrium potential w and electron quasi-Fermi potential ϕ for an n-channel MOSFET structure, with constant channel doping, satisfy

$$(2.1a) \qquad \tilde{\nabla}^2 w = \frac{1}{\lambda} \left(e^{(w-\phi)\log\lambda} - e^{-w\log\lambda} \right) + 1 .$$

$$(2.1b) \qquad \tilde{\nabla}^2 \phi + \log\lambda\, \tilde{\nabla}\phi \cdot \tilde{\nabla}(w - \phi) = 0 .$$

Here we have defined $\tilde{\nabla} \equiv (\partial_x, \epsilon\partial_y)$ where $\epsilon = (\lambda^{-1}\log\lambda)^{1/2} L_d/L$ is the aspect ratio given in terms of the normalized channel doping, λ, channel length, L, and intrinsic

*Courant Institute of Mathematical Sciences, 251 Mercer St., New York, New York 10012

†Dept. of Mathematical Sciences, I.B.M. Thomas Watson Research Center, Yorktown Heights, N.Y. 10598.

Debye length, L_d. Typically $\lambda \approx 10^6$ and, for silicon, $L_d \approx 33\mu$. The system (2.1) is to be solved in the rectangular region $0 \leq x \leq x^*$, $0 \leq y \leq 1$ representing a simplified MOSFET structure. Here $x = 0$ corresponds to the oxide-semiconductor interface, $y = 0$ the source, and $y = 1$ the drain.

Assuming a very thin oxide, the boundary conditions for (2.1) are

$$(2.2a) \qquad \phi(x,0) = 0\,, \quad \phi(x,1) = \frac{v_d}{\log \lambda}\,, \quad \phi_x(0,y) = \phi_x(x^*,y) = 0\,,$$

$$(2.2b)\ w(x,0) = w_{bi}\,, \quad w(x,1) = w_{bi} + \frac{v_d}{\log \lambda}\,, \quad -w_x(0,y) + \gamma\, w(0,y) = \gamma\, \frac{v_g}{\log \lambda}\,.$$

Here $w_{bi} \equiv \log \lambda_+ / \log \lambda$ where $\lambda_+ \approx 10^8$ is the normalized doping level of the source and drain wells. The constant γ in (2.2b) is related to the oxide thickness, t_{ox}, and the permitivities, ξ_i, ξ_s, of the oxide and semiconductor by $\gamma = (\log \lambda)^{1/2} \lambda^{-1/2} c_{ox}$, where $c_{ox} = \xi_i L_d / t_{ox} \xi_s$. The parameters v_d, v_g are the drain and gate bias, respectively. To complete the formulation, the source-drain current I_d is defined in terms of a dimensional constant I_c by

$$(2.3) \qquad I_d = I_c \left(\frac{\log \lambda}{\lambda} \right)^{1/2} \int_0^{x^*} \log \lambda\, \frac{\partial \phi}{\partial y}\, e^{(w-\phi)\log \lambda}\, dx\,.$$

In equilibrium, which occurs when $v_d = 0$, it follows from (2.1b), (2.2a) and (2.3) that $\phi \equiv 0$ and $I_d = 0$. In this case we now analyze (2.1a), setting $\phi \equiv 0$, in the long channel limit $\epsilon \ll 1$. The large doping limit $\lambda \gg 1$ will then be used to simplify some of the resulting equations.

The structure of the solution to (2.1a) (with $\phi \equiv 0$) in the limit $\epsilon \ll 1$ is of boundary layer type. In the gate-controlled 'outer' region, located away from the source and drain, the potential is one-dimensional and is obtained by setting $\epsilon = 0$ in (2.1a). This 'outer' problem for the potential was analyzed in [11] using the method of matched asymptotic expansions in the large doping limit $\lambda \gg 1$.

In the source and drain regions, near $y = 0$ and $y = 1$, respectively, the equilibrium potential is fully two-dimensional. By symmetry we need only analyze the drain region. Introducing the stretched variables $\eta = \epsilon^{-1}(1-y)$, $\xi = x$ and $u(\xi, \eta) \equiv w(\xi, 1 - \epsilon\eta)$, we obtain the following 'inner' problem from (2.1a):

$$(2.4) \qquad \begin{aligned} u_{\xi\xi} + u_{\eta\eta} &= \frac{1}{\lambda} \left(e^{u \log \lambda} - e^{-u \log \lambda} \right) + 1\,, \qquad \xi, \eta \geq 0\,, \\ u(\xi, 0) &= \frac{\log \lambda_+}{\log \lambda}\,, \qquad -u_\xi(0, \eta) + \gamma\, u_d(\eta) = \gamma\, \frac{v_g}{\log \lambda}\,. \end{aligned}$$

Here $u_d(0, \eta) \equiv u(0, \eta)$ is the surface potential in the drain region. For simplicity we have assumed that the depth of the n-well, x^*, is sufficiently large so that the condition that u tends to a one-dimensional solution as $\xi \to \infty$ can be imposed. When $\lambda \approx 10^6$, this condition requires that the depth of the n-well be no smaller than about $.45\mu$ (see [12] for details).

We now solve (2.4) in the limit $\lambda \gg 1$. The structure of the solution to (2.4) in this limit is shown in Fig. 1. Since we have assumed that the gate voltage v_g

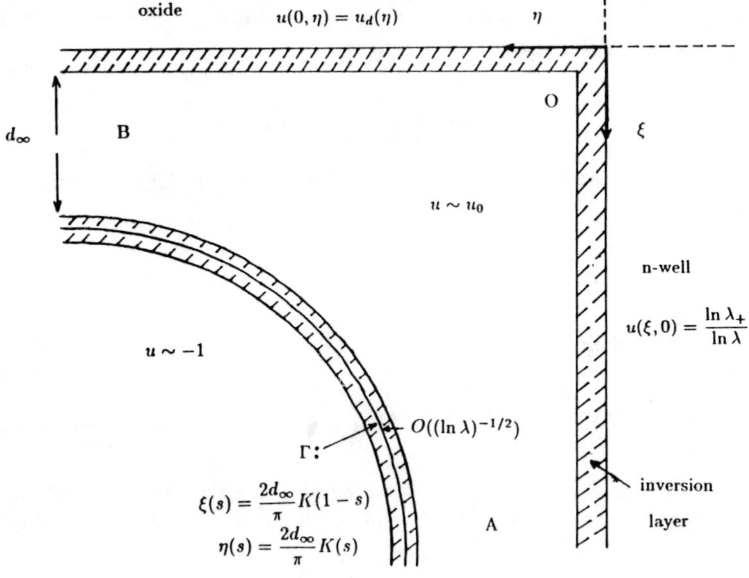

Figure 1

is sufficiently large, then there are thin inversion layers near $\xi = 0$ and $\eta = 0$ in which the term $\lambda^{-1} e^{u \log \lambda}$ dominates the space charge density. In each of these boundary layers the potential can be determined from the solution to an ordinary differential equation in the direction perpendicular to the coordinate axes. However, the asymptotic potential in each of these layers is determined only up to an unknown 'constant' of integration. In each of the inversion layers, this unknown 'constant' can depend on the variable parallel to that layer.

Away from these thin quasi one-dimensional inversion layers, it is shown in [12] that the asymptotic potential $u \sim u_0$ satisfies the following free boundary problem:

$$(2.5) \qquad \begin{aligned} \nabla^2 u_0 &= 1, \qquad \xi, \eta \geq 0, \\ u_0(0,\eta) = u_0(\xi,0) &= 1 + \frac{\log(\log \lambda)}{\log \lambda}, \qquad u_0 \left|_\Gamma = -1, \quad u_{0n} \right|_\Gamma = 0. \end{aligned}$$

Here u_{0n} denotes the normal derivative of u_0. The problem (2.5) is a two-dimensional version of the well-known depletion approximation in which the potential is patched for C^1 continuity to the bulk solution $u \equiv -1$ along some unknown curve Γ. In actuality, a more refined analysis shows that the unknown curve Γ is located inside a thin transition layer of width $O((\log \lambda)^{-1/2})$ near the bulk in which the terms $\lambda^{-1} e^{-u \log \lambda}$ and $+1$ in (2.4) are of the same order of magnitude. By analyzing this layer we find, that to leading order in $O(1/\log \lambda)$, the curve Γ can be obtained from (2.5).

The free boundary problem (2.5), which also arises in a fluid mechanical context, was solved up to a quadrature in [2] using a complex variable method. In [12], unknown 'constants' in the inversion layers were found by matching the inversion layers to this free boundary solution. Next, the equilibrium mobile charge in the drain region $Q_d(\eta)$ defined by

$$(2.6) \qquad Q_d(\eta) \equiv \int_0^{x^*} e^{u \log \lambda}\, d\xi$$

was computed asymptotically in the limit $\lambda \gg 1$ using Laplace's method. This yielded

$$(2.7) \qquad Q_d(\eta) \sim \frac{\lambda^{[1+u_d(\eta)]/2}}{(\log \lambda)^{1/2}} \left[(1 + c_b^2)^{1/2} - c_b \right].$$

Here $c_b \equiv c_b(\eta)$ is defined parametrically by

$$(2.8)$$
$$c_b(\eta(s)) = \frac{\sqrt{2}\, d_\infty}{\pi \sqrt{s}} \lambda^{[1-u_d(\eta)]/2} (\log \lambda)^{1/2} K\left(\frac{s-1}{s} \right), \qquad \eta(s) = \frac{2\, d_\infty}{\pi \sqrt{s}} K\left(1/s \right).$$

Here $K(m)$ denotes the complete elliptic integral of the first kind of modulus $m^{1/2}$ and d_∞, which is shown in Fig. 1, is given by $d_\infty \equiv \sqrt{2}(2 + \log(\log \lambda)/\log \lambda)^{1/2}$. Thus (2.7), (2.8) expresses the charge in the drain region in terms of the as yet unknown surface potential $u_d(\eta)$. Satisfying the mixed boundary condition for the potential in (2.2b), the surface potential is to be found from the following nonlinear algebraic equation:

$$(2.9) \qquad \frac{v_g}{\log \lambda} - u_d(\eta) - \sqrt{2}\gamma^{-1}(\log \lambda)^{-1/2} \lambda^{[u_d(\eta)-1]/2} \left(1 + c_b^2(\eta) \right)^{1/2} = 0.$$

Eliminating $u_d(\eta)$ in (2.7) by using (2.8) and (2.9) then determines the drain-controlled charge Q_d as a function of η. The details of this analysis can be found in [12].

Finally, composite expansions for the equilibrium mobile charge and the surface potential, denoted by $Q_c(y)$ and $u_c(y)$, respectively, which are valid along the entire channel, are given by

$$(2.10) \quad Q_c(y) \equiv Q_d(\epsilon^{-1}(1 - y)), \qquad u_c(y) \equiv u_d(\epsilon^{-1}(1 - y)), \qquad \text{for } \frac{1}{2} \le y \le 1.$$

Using the symmetry about the mid-channel line, we then have $Q_c(y) = Q_c(1 - y)$ and $u_c(y) = u_c(1 - y)$ on $0 \le y \le 1/2$. Away from the source and drain regions near $y = 0$ and $y = 1$, each having width $O(\epsilon)$, we obtain $Q_c(y) \sim Q_d(\infty)$, which is the long-channel result derived in [11].

To derive the current-voltage relation in the ohmic regime, we first linearize (2.3) for small drain biases by replacing $e^{(w-\phi)\log \lambda}$ by $e^{w \log \lambda}$. A similar linearization in (2.1a) shows that the potential for $v_d \ll 1$ is given by the equilibrium potential. Next, we use Laplace's method for $\log \lambda \gg 1$ to express (2.3) in terms of the composite expansion for the equilibrium mobile charge $Q_c(y)$ defined in (2.10). This yields

$$(2.11) \qquad \frac{I_d}{I_c} \left(\frac{\lambda}{\log \lambda} \right)^{1/2} \sim \phi_y(0, y) \log \lambda\, Q_c(y).$$

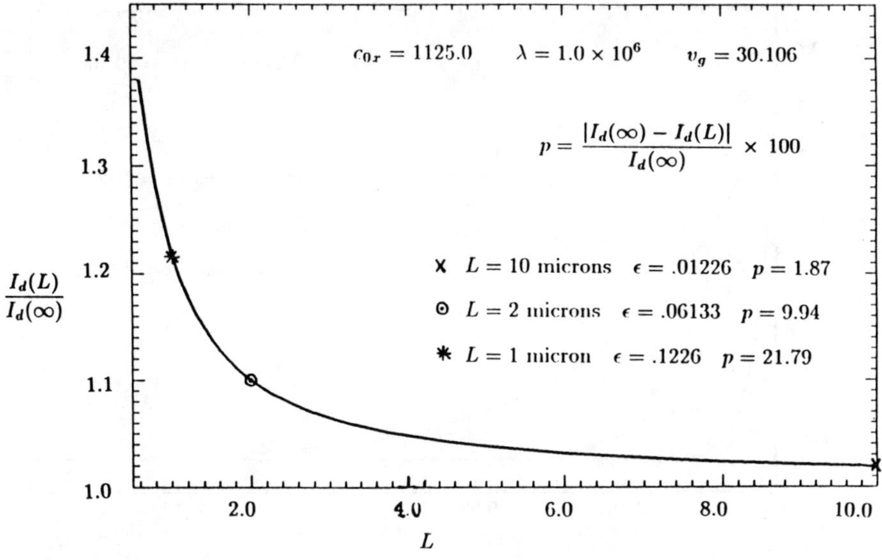

Figure 2

Integrating (2.11) with respect to y and using (2.2a), we obtain the following expression for the current as a function of the channel length:

(2.12)
$$\frac{I_d(L)}{I_c} = v_d \left(\frac{\log \lambda}{\lambda} \right)^{1/2} \left(2\epsilon \int_0^{(2\epsilon)^{-1}} \frac{1}{Q_d(\eta)} \, d\eta \right)^{-1}, \qquad \epsilon = \left(\frac{\log \lambda}{\lambda} \right)^{1/2} \frac{L_d}{L}.$$

The long-channel result is recovered by letting $L \to \infty$ ($\epsilon \to 0$) in (2.12), which yields $I_d(\infty)/I_c = v_d(\log \lambda/\lambda)^{1/2} Q_d(\infty)$.

Taking the parameter values $L_d = 33\mu$, $\lambda = 10^6$, $\gamma = 4.18$ and $v_g = 30.106$, which ensures that an inversion layer exists near the oxide boundary, we now compute $I_d(L)/I_c$ numerically for various channel lengths. The results are shown in Fig. 2. From this figure we observe that the long channel result $I_d(\infty)$ underestimates the actual current $I_d(L)$ by about 2% for a 10μ long device. This error rises to roughly 10% for a 2μ device and to 20% for a 1μ device. Clearly, however, the predictions given by (2.12) must become inaccurate when the channel becomes too short. To qualitatively assess the range of validity of (2.12), in Fig. 3 we plot the mobile charge $Q_c(y)$ on the interval $[0, 1/2]$ for several values of L. From this figure we note that even for a 1.25μ device, for which $\epsilon \approx .10$, there is a region near $y = 1/2$ where the mobile charge can be approximated by the one-dimensional, gate-controlled result. Thus we predict that the asymptotic result (2.12) should be capable of describing the ohmic regime of the current-voltage relation for $1 - 10\mu$ long devices, with inversion layers, and with doping levels on the order of $\lambda \approx 10^6$. Naturally, to properly determine the range of channel lengths for which (2.12) gives accurate results we should compare it with corresponding current-voltage curves obtained from full numerical solutions to the drift-diffusion model. This comparison has yet to be done.

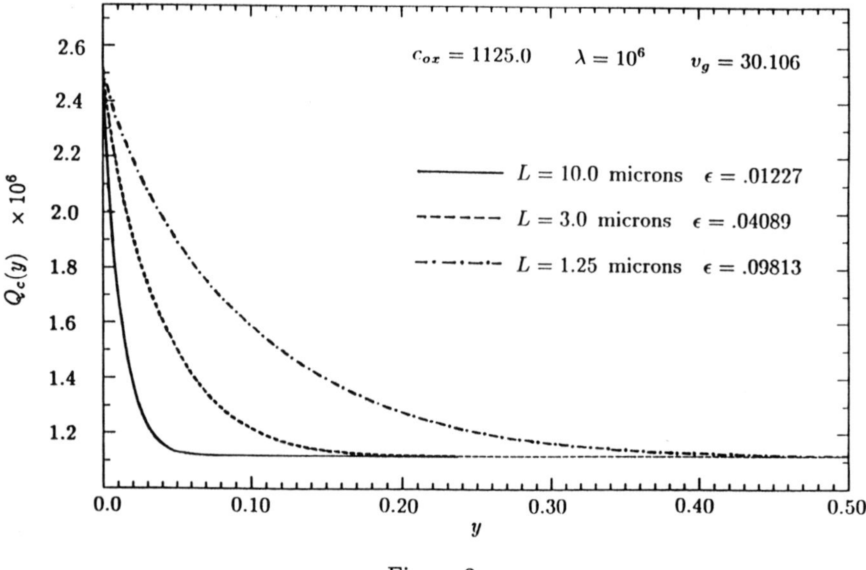

Figure 3

3. Multiple steady state solutions for a PNPN structure.

Multiple steady state solutions, within the context of the drift-diffusion model, can occur for some device geometries. One of the simplest configurations known to exhibit an S-shaped current-voltage curve is a one-dimensional PNPN structure. Numerical evidence supporting the existence of multiple steady state solutions for this structure was given in [5]. More recently, PNPN structures have been analyzed asymptotically in the limit of large doping densities in [7], [9], [10] and [13]. Despite these efforts, the criteria for the occurence of multiple steady state solutions under various recombination and generation terms is still not completely established. In this section we discuss some of the results obtained from the recent asymptotic studies.

Allowing for Shockley-Read-Hall (SRH) recombination, but assuming constant mobilities, the one-dimensional drift-diffusion model on the interval $[0, L]$ is given in a standard notation by

$$(3.1a) \qquad \xi_s \psi'' = q\left(n - p + N_d\, d(x)\right), \qquad E = \psi',$$

$$(3.1b) \qquad J_n = q\mu_n\left(-nE + \frac{k_b T}{q} n'\right),$$

$$(3.1c) \qquad J_p = q\mu_p\left(-pE - \frac{k_b T}{q} p'\right), \qquad J = J_n + J_p, \qquad J' = 0,$$

$$(3.1d) \qquad J_n' = q\, R_s(n, p) \equiv \frac{q\left(np - n_i^2\right)}{\tau_p n + \tau_n p + (\tau_p + \tau_n)n_i}.$$

Assuming perfect Ohmic contacts at $x = 0$ and $x = L_x$, the boundary conditions, for $N_d/n_i \gg 1$, reduce to

$$(3.2a) \quad n - p + N_d\, d(x) = 0 \quad \text{at} \quad x = 0, L_x\,; \quad np = n_i^2 \quad \text{at} \quad x = 0, L_x\,,$$

$$(3.2b) \qquad \psi(L_x) = \frac{k_b T}{q} \log\left(\frac{N_d|d(L_x)|}{n_i}\right) - U\,, \quad \psi(0) = -\frac{k_b T}{q} \log\left(\frac{N_d|d(0)|}{n_i}\right)\,.$$

Here $U > 0$ is the applied voltage.

We now non-dimenionalize $(3.1)-(3.2)$ by introducing the following dimensionless variables in the overbar notation: $n = n_i\bar{n}$, $p = n_i\bar{p}$, $\psi = k_b T\bar{\psi}/q$, $J_n = J_r\bar{J}_n$, $J_p = J_r\bar{J}_p$, $J = J_r\bar{J}$, $\tau_n = \tau_r\bar{\tau}_n$, $\tau_p = \tau_r\bar{\tau}_p$, and $E = k_b T\bar{E}/L_d q$. Lengths are normalized by $y = x/L_d$. Here $L_d \equiv \left(k_b T\xi_s/n_i q^2\right)^{1/2}$, $J_r \equiv q\mu_{p0}n_i k_b T/L_d$ and $\tau_r \equiv \xi_s/q\mu_{p0}n_i$. Introducing this scaling into (3.1), and dropping the overbar notation, we obtain

$$(3.3a) \qquad \psi'' = n - p + \lambda d(y)\,, \qquad E = \psi'\,,$$

$$(3.3b) \qquad J_n/\mu = n' - n\psi'\,,$$

$$(3.3c) \qquad J_n - J = -J_p = p' + p\psi'\,, \qquad J' = 0\,,$$

$$(3.3d) \qquad J_n' = R_s(n,p) \equiv \frac{np - 1}{\tau_p n + \tau_n p + (\tau_p + \tau_n)}\,,$$

on $0 \le y \le L \equiv L_x/L_d$. The constant μ is the ratio of the electron to hole mobility. Similarly, the boundary conditions from (3.2) become

$$(3.4a) \qquad n - p + \lambda d(y) = 0 \quad \text{at} \quad y = 0, L\,; \quad np = 1 \quad \text{at} \quad y = 0, L\,,$$

$$(3.4b) \qquad \psi(L) = \log\left(\lambda|d(L)|\right) - V\,, \quad \psi(0) = -\log\left(\lambda|d(0)|\right)\,.$$

Other dependent variables, used below, are the quasi-Fermi potentials ϕ_n, ϕ_p defined by $\phi_n \equiv \psi - \log n$ and $\phi_p \equiv \psi + \log p$. The key non-dimensional parameters are $\lambda \equiv N_d/n_i \approx 10^5 \gg 1$ and $V \equiv Uq/k_b T$. To complete the formulation, we assume that the doping profile $d(y)$ is a piecewise constant function with three sign alterations:

$$(3.5) \qquad d(y) = (-1)^{i-1}d_i \quad \text{on} \quad (y_{i-1}, y_i) \quad \text{for} \quad i = 1,..,4\,, \qquad \Delta_i = y_i - y_{i-1}\,.$$

Here $d_i > 0$ and we have labelled $y_0 = 0$ and $y_4 = L$. Our goal then is to construct the $J(V)$ curve in the limit $\lambda \gg 1$ and to determine the parameter regime in which the $J(V)$ curve is S-shaped.

In equilibrium $(V = 0)$, the potential is determined from the nonlinear Poisson equation $\psi'' = e^\psi - e^{-\psi} + \lambda d(y)$. In the limit $\lambda \gg 1$, the equilibrium potential is given asymptotically by $\psi \sim (-1)^i \log(\lambda d_i)$ away from the junctions y_i, and has rapid variations on a scale of $(\log \lambda/\lambda)^{1/2}$ near the junctions. Since $\psi'' \approx 0$ away from the junctions, these regions are essentially charge-neutral and are called quasi-neutral regions.

Based on the structure of the equilibrium potential when $\lambda \gg 1$ an asymptotic simplification of (3.3), referred to as the quasi-neutral (QN) formulation, was proposed in [7] to treat the *non-equilibrium* case under the assumption of no recombination. An attempt to extend this proposal to allow for a non-vanishing recombination term was made in [9]. The QN formulation used in [7] and [9] is based on three assumptions: the layers near the junctions are infinitesimally thin, ψ'' is slowly varying away from the junctions so that we can set $\psi'' = 0$ there to a first approximation, and continuity conditions across the junctions can be used to replace the missing ψ'' term.

Using the first two assumptions of the QN formulation it is possible to decouple the electric field from the carrier concentrations on each subinterval (y_{i-1}, y_i). Defining σ by $\sigma = n + p$, it can be shown from (3.3) that on each subinterval we have

$$(3.6a) \quad \sigma'' - \frac{\lambda d}{\sigma^2}\left(\frac{\alpha\sigma\sigma' - \sigma J}{\sigma(1+\alpha) - \alpha\lambda d}\right)\sigma' = R(\sigma)\left((1 + \mu^{-1}) - \frac{\lambda d}{\sigma}(1 - \mu^{-1})\right),$$

$$(3.6b) \quad \psi' = \left(\frac{\alpha\sigma' - J}{\sigma(1+\alpha) - \alpha\lambda d}\right) \equiv -F(\sigma; J).$$

Here $\alpha \equiv (\mu - 1)/2$ and $R(\sigma) \equiv R_s[(\sigma - \lambda d)/2, (\sigma + \lambda d)/2]$. The boundary conditions for σ from (3.4) are, in the limit $\lambda \gg 1$, given by $\sigma(0) = \lambda d_1$ and $\sigma(L) = \lambda d_4$.

The final ingredient needed for the QN formulation is the derivation of appropriate continuity conditions for σ and ψ across the junctions. This is in general no easy task. These continuity conditions must be found by a separate asymptotic analysis similar to that done in [6] for a single forward biased junction under SRH recombination. In [6] a detailed analysis showed that the form of the continuity conditions depended rather sensitively on the magnitude of the current. In addition, since a PNPN structure can have both forward and reverse bias junctions, depending on the parameter values in (3.3), the analysis in [6] must be extended to treat reverse biased junctions. Furthermore, what is needed for (3.6) is a *uniformly* valid form of these continuity conditions applicable to both reverse and forward biased structures under a a wide current range. To our knowledge this has yet to be done. Thus we will first consider the special case of no recombination where these continuity conditions are easier to obtain.

In the absence of recombination, the current densities are constant and so are continuous across the junctions. A second continuity condition, established in [6] and [8] for forward and reverse biased junctions respectively, is that the product np is continuous across the junctions. Thus in terms of σ, the continuity conditions for (3.6a) are

$$(3.7) \qquad [\sigma^2 - \lambda^2 d^2]_i = 0, \qquad \left[\frac{\sigma\sigma' + \sigma J - J\lambda d}{\sigma(1+\alpha) - \alpha\lambda d}\right]_i = 0.$$

Here we have defined $[v]_i$ by $[v]_i \equiv v(y_{i+}) - v(y_{i-})$.

An additional continuity condition is needed for (3.6b). Since the quasi-Fermi potentials ϕ_n, ϕ_p are never greater than the applied bias, it is plausible to assume

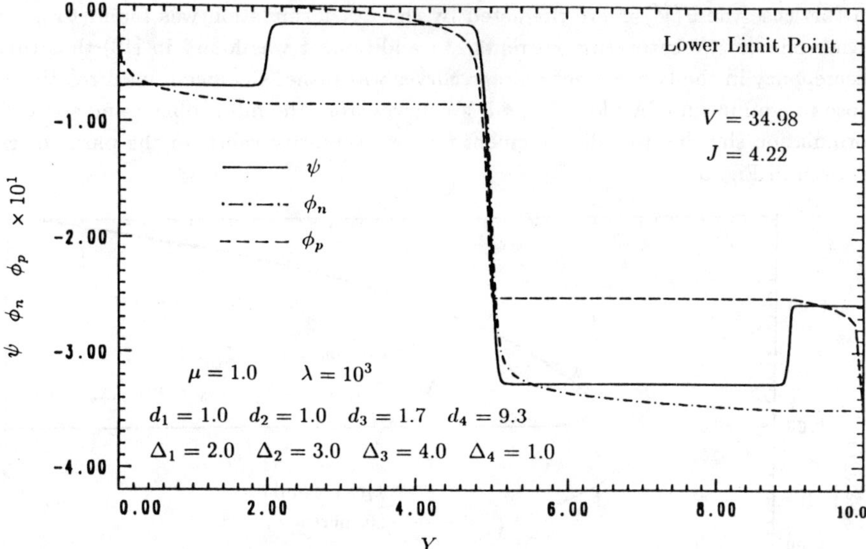

Figure 4

that they are smooth variables across the junctions, at least near equilibrium. Although the quasi-Fermi potentials can vary significantly across a strongly reverse biased junction (see Fig. 4), we will impose the condition, used in [7], [9], [10] and [13], that $[\phi_n]_i = 0$. Notice that since np is continuous across the junctions it then follows that ϕ_p must be continuous as well. With this condition, we can derive from (3.4b) and (3.6b) the following implicit expression for the $J(V)$ curve:

$$(3.8) \qquad V = \int_0^L F(\sigma(\eta); J)\, d\eta + \sum_{i=1}^{3} \big[\log(\sigma + \lambda d)\big]_i + \log(\lambda^2 d_1 d_4).$$

When $R \equiv 0$, (3.6a) can be integrated explicitly as in [7]. Imposing (3.7) then determines $\sigma(y)$ in terms of a nonlinear algebraic system which must be solved numerically at each current level. The $J(V)$ curve is then obtained from (3.8).

The $J(V)$ curve obtained in this way erroneously predicts that the current can saturate as $V \to \infty$. Depending on the parameter values in (3.6), the QN formulation may predict either one or two saturation currents. As shown in [13], the $J(V)$ curve obtained from such a formulation typically agrees well with the $J(V)$ curve obtained from the numerical solution to the full problem (3.3), (3.4) when the full problem has a monotonic $J(V)$ curve. However, there is a small parameter range where the $J(V)$ curve for the full problem is monotonic but the $J(V)$ curve from the QN formulation has two saturation currents. We also mention that although it was shown in [9] that the $J(V)$ curve obtained from the QN formulation can be S-shaped in a very limited parameter range, this S-shaped behavior does not occur for the full model in the same parameter range. Related difficulties with the QN formulation occurred when the full model had an S-shaped current-voltage curve.

398

In this case the $J(V)$ curve predicted by the QN formulation was found to have either one or two saturation currents. In addition, it was found in [13] that this discrepancy in the two current-voltage curves was rather significant even relatively close to equilibrium. A plot of the $J(V)$ curves from the full problem and the QN formulation showing this disagreement for representative values of the parameters is given in Fig. 5.

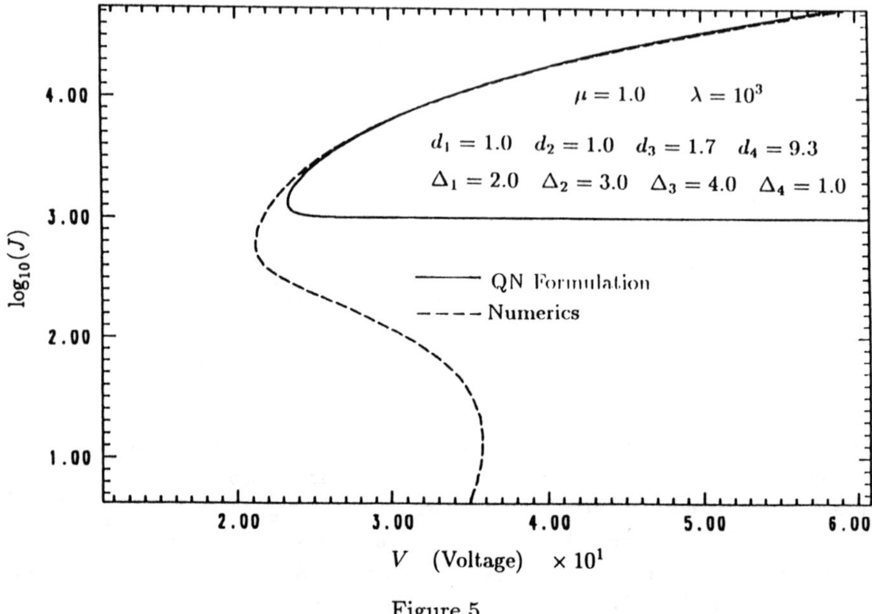

Figure 5

The comparisons above show that at least one of the three assumptions on which the QN formulation is based is invalid. From the numerical solution to (3.3), (3.4) obtained in [13] it is not immediately clear which of these assumptions should be modified. In Fig. 4 we plot the potential and quasi-Fermi potentials near the lower fold point in the case where the full problem has an S-shaped $J(V)$ curve. From this figure we observe that ψ is roughly constant in the quasi-neutral regions and that the layers near the junctions are very thin. Thus the first two assumptions of the QN formulation appear to be warranted. The only assumption of the QN formulation which seems to be in doubt from Fig. 4 is the assumed continuity of the quasi-Fermi potentials across the strongly reverse biased middle junction. Modifying this assumption, however, introduces only small quantitative improvements and does not eliminate the prediction of saturation currents by the QN formulation.

The resolution of this difficulty was given in [13]. A related approach is found in [10]. It was found in [13] that we must abandon the assumption that the layers near the junctions are infinitesimally thin. Instead the width of these layers must be able to respond according to the jump in the potential across each of the layers. As discussed more fully from a physical viewpoint in [13], this modification of the original QN formulation prohibits the carrier concentrations predicted by it from becoming negative and thus current saturation is eliminated. Since the equation (3.6a) for σ remains unchanged, and thus can still be integrated explicitly, it is a relatively easy matter to incorporate this modification into the original QN formulation. This was done in [13].

In Fig. 6 we compare the $J(V)$ curve predicted by the modified QN formulation with the numerical $J(V)$ curve obtained from the full model (3.3) using the same parameter values as in Fig. 5. We now find close agreement over a very wide current range. We remark, however, that the agreement deteriorates somewhat at very large current levels. To understand why, we plot in Fig. 7 the potential and quasi-Fermi potentials at a point on the upper branch of the $J(V)$ curve. From this figure we observe that the potential has significant variations across the device and a boundary layer structure for the carrier concentrations near the contacts has now formed. These two observations signify that the QN formulation is not applicable for arbitrarily large current flow. A different asymptotic approach such as that described in §4 is then needed.

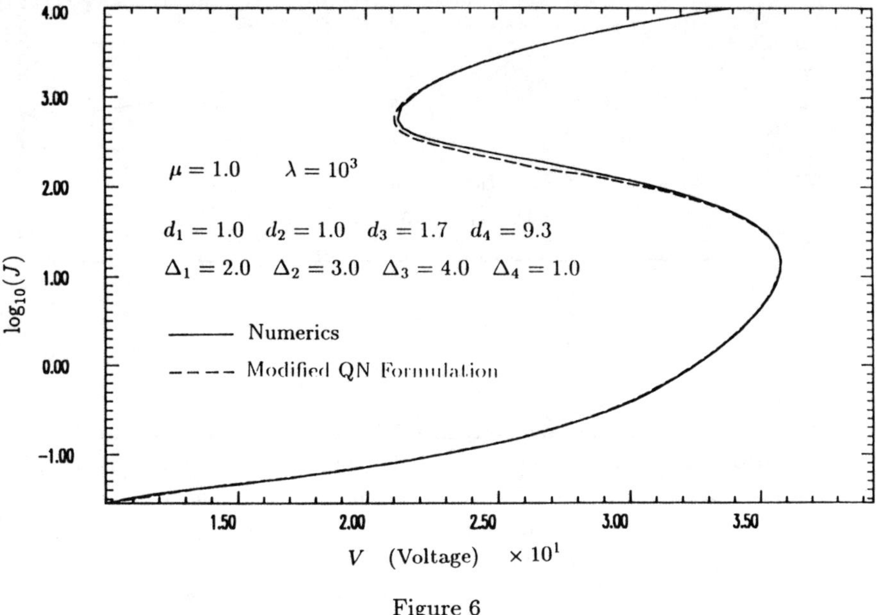

Figure 6

The construction of multiple steady state solutions when $R \equiv 0$ is thus rather well understood. However, this is not the case for the finite lifetime problem and only some partial results are available. In [13] it was shown how to construct the $J(V)$ curve from the original QN formulation (3.6) with continuity conditions (3.7) for the finite lifetime problem. The construction was based on analyzing the $J(V)$ curve for three ranges of current: $J \gg O(\lambda)$, $J = O(\lambda)$, $J << O(\lambda)$. In each of these ranges, (3.6) can be simplified asymptotically and the form of the $J(V)$ curve corresponding to each of these ranges can be found. The results obtained from this analysis were qualitatively similar to corresponding results for the case of no recombination. Namely, it was typically found that when the full model (3.3) had a monotonic $J(V)$ curve it was approximated well by the $J(V)$ curve obtained from the original QN formulation. The reason for this success can be inferred from [6]. From [6] we know that across a forward biased junction the current densities do not have a significant variation expect possibly at very small current levels. Thus, since (3.3) has a monotonic $J(V)$ curve only when the three junctions are forward biased at all current levels, the continuity conditions (3.7) used by the QN formulation must be correct in this case.

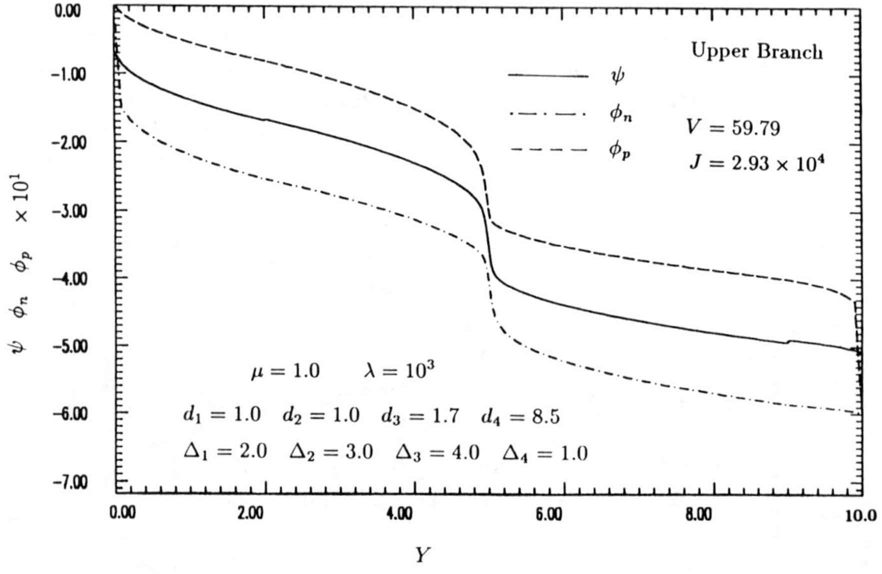

Figure 7

We are presently working on modifying the QN formulation for the finite lifetime problem to determine the correct form of the continuity conditions allowing for both forward and reverse biased junctions. From some preliminary numerical and asymptotic work we now state a plausible conjecture on the qualitative effect of recombination on solution multiplicity for a PNPN structure. *Conjecture*: In the limit of strong recombination $\tau_n, \tau_p \to 0$, the problem (3.3) has a monotonic $J(V)$ curve. Finally, we mention that it would also be interesting to investigate the effect of avalanche multiplication on the solution multiplicity.

4. High field scaling of the drift-diffusion model. In this section we analyze the drift-diffusion equations, allowing for field dependent mobilities and SRH recombination, which model a strongly forward biased PN or PNPN structure. A novel scaling of the drift-diffusion model, in the limit where the applied bias greatly exceeds the thermal voltage, is introduced. The resulting equations are then analyzed by the method of matched asymptotic expansions and the current-voltage relation, when velocity saturation effects become significant, is determined. The details of the analysis can be found in [14].

We consider (3.1), (3.2) for a forward biased PN or PNPN silicon structure. We shall use the following empirical models for the mobilities as a function of the electric field:

$$(4.1) \qquad \mu_n(E) = \frac{\mu_{n0}}{(1 + b_n \, |E/E_c|^m)^{1/m}}, \qquad \mu_p(E) = \frac{\mu_{p0}}{(1 + b_p \, |E/E_c|^m)^{1/m}}.$$

Here $|E_c| \approx 1.0 \times 10^4$ volts/cm is a typical magnitude of the electric field when saturation effects become significant. The material constants μ_{n0} and μ_{p0} are the constant low field mobilities and the dimensionless parameters b_n, b_p and m are of order unity. The doping profile in (3.1a), relevant to either a PN or PNPN structure, is taken to be a piecewise constant function of the form
(4.2)
$$d(x) = (-1)^{i-1} d_i \quad \text{on} \quad (x_{i-1}, x_i) \quad \text{for} \quad i = 1, .., k \qquad k = 2, 4, \qquad (d_i > 0).$$

Choosing U_r to be a typical magnitude of the applied voltage when velocity saturation becomes significant, we introduce the following dimensionless variables in the overbar notation: $n = N_d \bar{n}$, $p = N_d \bar{p}$, $\psi = U_r \bar{\psi}$, $J_n = J_r \bar{J}_n$, $J_p = J_r \bar{J}_p$, $J = J_r \bar{J}$, $\tau_n = \tau_r \bar{\tau}_n$, $\tau_p = \tau_r \bar{\tau}_p$, $\mu_n = \mu_{n0} \bar{\mu}_n$, $\mu_p = \mu_{p0} \bar{\mu}_p$ and $E = U_r \bar{E}/\bar{L}$. Here we have chosen $J_r = q \mu_{p0} N_d U_r / \bar{L}$, $\tau_r = \xi_s / q \mu_{p0} N_d$ and $\bar{L} = \left(\xi_s U_r / q N_d \right)^{1/2}$. Introducing this scaling into (3.1), setting $y = x/\bar{L}$, and dropping the overbar notation, we find

$$(4.3a) \qquad E' = n - p + d(y), \qquad \psi' = E,$$

$$(4.3b) \qquad J_n/\mu = \mu_n(E)(-nE + \epsilon n'),$$

$$(4.3c) \qquad J - J_n = J_p = \mu_p(E)(-pE - \epsilon p'), \qquad J' = 0,$$

$$(4.3d) \qquad J_n' = R_s(n, p, \lambda) \equiv \frac{np - \lambda^{-2}}{\tau_p n + \tau_n p + (\tau_p + \tau_n)\lambda^{-1}},$$

on $0 \le y \le L$. The scaled mobility models appearing in (4.3b,c) are given by

$$(4.4) \qquad \mu_n(E) = \frac{1}{(1 + \beta_n |E|^m)^{1/m}}, \qquad \mu_p(E) = \frac{1}{(1 + \beta_p |E|^m)^{1/m}},$$

and the doping profile $d(y)$ is given by (4.2) where $x_i = y_i \bar{L}$. In terms of these new variables, the boundary conditions (3.2) become

$$(4.5a) \qquad n - p + d(y) = 0 \quad \text{at} \quad y = 0, L; \qquad np = \lambda^{-2} \quad \text{at} \quad y = 0, L,$$

$$(4.5b) \quad \psi(L) = -V + \epsilon \ln(\lambda d_k), \qquad \psi(0) = -\epsilon \ln(\lambda d_1).$$

The current-voltage relation is then determined from

$$(4.6) \qquad \int_0^L E(y)\, dy = -V + \epsilon \ln(d_1 d_k \lambda^2).$$

Since $V > 0$ we will assume for the moment that the device is sufficiently forward biased to ensure that $E(y) < 0$ for $y \in (0, L)$.

The dimensionless parameters ϵ, λ, μ, L, V, β_n, and β_p appearing in (4.3)$-$(4.5) are defined by $\epsilon = k_b T/q U_r$, $\lambda = N_d/n_i$, $\mu = \mu_{n0}/\mu_{p0}$, $L = L_x/\bar{L}$, $V = U/U_r$, $\beta_n = b_n(U_r/|E_c|\bar{L})^m$, and $\beta_p = b_p(U_r/|E_c|\bar{L})^m$. Taking typical parameters for silicon at room temperature and assuming that $L_x \approx 5$ microns, $N_d \approx 10^{15}\text{cm}^{-3}$, and $E_c = 1.0 \times 10^4\text{volts/cm}$, we find upon choosing $U_r = 2$ volts that

$$\beta_n = 2.0, \quad \beta_p = .50, \quad L = 5.9, \quad \epsilon = .015, \quad \lambda = 10^5, \quad \mu = 3.$$

Therefore since $\epsilon \ll 1$ we will construct the solution to (4.3) $-$ (4.5) in the high field limit $\epsilon \to 0$. The analysis will be done for the case of finite carrier lifetimes and we will then simplify the results to the special case of no recombination. For $\epsilon \ll 1$ the structure of the solution to (4.3) is of boundary layer type. Away from the contacts in the 'outer' region the diffusion terms in (4.3b,c) can be neglected. Boundary layers near the contacts at $y = 0, L$ must then be inserted for n and p in order to satisfy (4.5a).

In the outer region, away from the contacts, we expand the solution to (4.3) in a regular expansion in powers of ϵ with leading terms E_0, n_0, p_0, J_{n0} and J_0. From (4.3b,c) we find

$$(4.7) \qquad n_0(y) = \frac{-J_{n0}/\mu}{E_0(y)\,\mu_n[E_0(y)]}, \qquad p_0(y) = -\frac{(J_0 - J_{n0})}{E_0(y)\,\mu_p[E_0(y)]}.$$

Substituting (4.7) into (4.3a,d) we obtain the following coupled system for $E_0(y)$, $J_{n0}(y)$, with J_0 as parameter:

$$(4.8a) \qquad E_0' = -\frac{J_{n0}/\mu}{E_0\mu_n(E_0)} + \frac{J_0 - J_{n0}}{E_0\mu_p(E_0)} + d(y),$$

$$(4.8b) \qquad J_{n0}' = R_s\left[-\frac{J_{n0}/\mu}{E_0\mu_n(E_0)}, \frac{J_{n0} - J_0}{E_0\mu_p(E_0)}, \lambda\right].$$

We now derive boundary conditions for (4.8) by analyzing the boundary layers near the contacts.

Near the cathode at $y = 0$ we let $z = \epsilon^{-1}y$ and define $E_c(z) = E(\epsilon z)$, $n_c(z) = n(\epsilon z)$, $p_c(z) = p(\epsilon z)$ and $J_{nc}(z) = J_n(\epsilon z)$. Expanding E_c, n_c, p_c and J_{nc} in a regular expansion in powers of ϵ, with leading terms E_{c0}, n_{c0}, p_{c0} and J_{nc0}, we obtain from (4.3a,d) that E_{c0} and J_{nc0} are constant in this layer. Then from (4.3b,c) we find

$$(4.9) \quad n_{c0}'(z) - E_{c0}\,n_{c0}(z) = \frac{J_{nc0}}{\mu\,\mu_n(E_{c0})}, \qquad p_{c0}'(z) + E_{c0}\,p_{c0}(z) = \frac{J_{nc0} - J_0}{\mu_p(E_{c0})}.$$

Eliminating exponential growth in the hole concentration, we find that p_{c0} is given by the constant solution $p_{c0} = (J_{nc0} - J_0)/E_{c0}\mu_p(E_{c0})$. Satisfying the boundary condition (4.5a), then gives the following relation between E_{c0} and J_{nc0}:

$$(4.10) \qquad -E_{c0}\,\mu_p(E_{c0}) = \frac{2(J_0 - J_{nc0})}{d_1 + (d_1^2 + 4/\lambda^2)^{1/2}} \sim \frac{J_0 - J_{nc0}}{d_1}, \quad \text{for } \lambda \gg 1.$$

Using the form of μ_p given in (4.4), the unique root to (4.10) with $E_{c0} < 0$ is, for $\lambda \gg 1$,

$$(4.11) \qquad E_{c0} \sim -\frac{(J_0 - J_{nc0})}{d_1}\left[1 - \left(\frac{J_0 - J_{nc0}}{d_1}\right)^m \beta_p\right]^{-1/m},$$

provided that $J_0 - J_{nc0} < d_1 \beta_p^{-1/m}$. With E_{c0} specified in terms of J_{nc0} and J_0, we now determine $n_{c0}(z)$ by solving (4.9), (4.5a). This yields

$$(4.12) \qquad n_{c0}(z) = -\frac{J_{nc0}/\mu}{E_{c0}\,\mu_n(E_{c0})} + \left(\lambda^{-2}d_1 + \frac{J_{nc0}/\mu}{E_{c0}\,\mu_n(E_{c0})}\right)e^{E_{c0}z},$$

which matches directly onto the outer solution given in (4.7). This qualitative result that a boundary layer for the electrons occurs in the cathode region is clearly seen in Fig. 4. Matching the electric field and electron current density in the outer and cathode regions then enforces that $E_0(0) = E_{c0}$ and $J_{n0}(0) = J_{nc0}$.

A very similar analysis can be done for the anode boundary layer near $y = L$. Qualitatively, we find that there is a boundary layer for holes rather than electrons, which is also clearly seen from Fig. 4. In addition, the analysis near the anode provides a relation between $E_0(L)$ and $J_{n0}(L)$, similar in form to (4.11), which is written below in (4.13b).

To determine E_0 and J_{n0} in the outer region, we then solve (4.8), for fixed J_0, subject to the boundary conditions

$$(4.13a) \qquad E_0(0) = -\frac{\left(J_0 - J_{n0}(0)\right)}{d_1}\left[1 - \left(\frac{J_0 - J_{n0}(0)}{d_1}\right)^m \beta_p\right]^{-1/m},$$

$$(4.13b) \qquad E_0(L) = -\frac{J_{n0}(L)}{\mu d_k}\left[1 - \left(\frac{J_{n0}(L)}{\mu d_k}\right)^m \beta_n\right]^{-1/m}.$$

This system is easily solved by a shooting method for each fixed J_0. An Euler continuation in J_0 is then combined with the shooting method to determine $E_0(y)$ and $J_{n0}(y)$ at each current level. The integral constraint (4.6), with E replaced by E_0, determines the $J(V)$ curve. Finally, composite expansions for the electron and hole concentrations, determined from $E_0(y)$ and $J_{n0}(y)$, are given by

$$(4.14a) \quad n(y) \sim \frac{-J_{n0}(y)/\mu}{E_0(y)\,\mu_n[E_0(y)]} + \left(\frac{1}{\lambda^2 d_1} + \frac{J_{n0}(0)/\mu}{E_0(0)\,\mu_n[E_0(0)]}\right)e^{E_0(0)y/\epsilon},$$

$$(4.14b) \quad p(y) \sim \frac{-(J_0 - J_{n0}(y))}{E_0(y)\,\mu_p[E_0(y)]} + \left(\frac{1}{\lambda^2 d_k} + \frac{(J_0 - J_{n0}(L))}{E_0(L)\,\mu_p[E_0(L)]}\right)e^{E_0(L)(L-y)/\epsilon}.$$

In the special case of no recombination, $R_s = 0$, we find from (4.8b) that J_{n0} is constant. In this case (4.8a) can be integrated once and, upon enforcing the continuity of E_0 at $y = y_i$, we find

$$(4.15) \quad \Delta_i = \int_{E_0(y_{i-1})}^{E_0(y_i)} \left[-\frac{J_{n0}/\mu}{s\,\mu_n(s)} + \frac{(J_0 - J_{n0})}{s\,\mu_p(s)} + (-1)^{i-1}d_i\right]^{-1} ds\,, \quad i = 1,..,k,$$

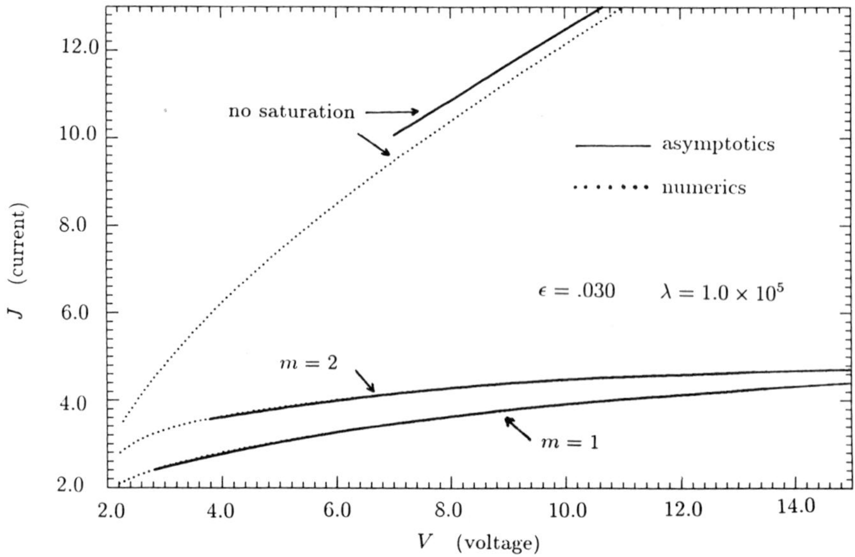

Figure 8

Here $\Delta_i = y_i - y_{i-1}$ and we have labelled $y_0 = 0$ and $y_k = L$. With $E(0)$, $E(L)$ given in (4.13), the nonlinear algebraic system (4.15) is solved numerically for $E_0(y_1), .., E_0(y_{k-1})$ and J_{n0} at each current level J_0 satisfying $J_0 < d_1 \beta_p^{-1/m} + \mu d_k \beta_n^{-1/m}$.

We now give an example of our asymptotic results applied to a strongly forward biased PN junction under the assumption of no recombination. The parameter values we take are

$$(4.16) \qquad \begin{aligned} \mu &= 3.0\,, \quad \beta_n = 1.0\,, \quad \beta_p = .25\,, \quad L = 5\,, \quad \lambda = 10^5\,, \quad \epsilon = .030 \\ d_1 &= 1.0\,, \quad d_2 = 1.0\,, \quad \Delta_1 = 2.5\,, \quad \Delta_2 = 2.5\,. \end{aligned}$$

In Fig. 8 we compare our asymptotic current-voltage curve obtained from (4.6), (4.13) and (4.15) with the corresponding current-voltage relation obtained from a full numerical solution to (4.3). In this figure we have also shown the form of the current-voltage relation when saturation effects are neglected, obtained by setting $\beta_n = \beta_p = 0$ in (4.4). Clear agreement between the asymptotic and numerical results is seen when $m = 1$ and $m = 2$. We remark that the results for the case of no saturation agree more favorably at slightly higher voltage levels. In Fig. 9 we compare the asymptotic carrier concentrations obtained from the composite expansion (4.14), at one current level, with those obtained from the full numerical solution to (4.3), again showing clear agreement. Further examples comparing asymptotics and numerics for a PNPN structure and for a PN structure with recombination can be found in [14].

We now remark on three limitations of the asymptotic analysis presented above. The analysis presented above applies only when $E_0(y) < 0$ for $y \in [0, L]$. Our

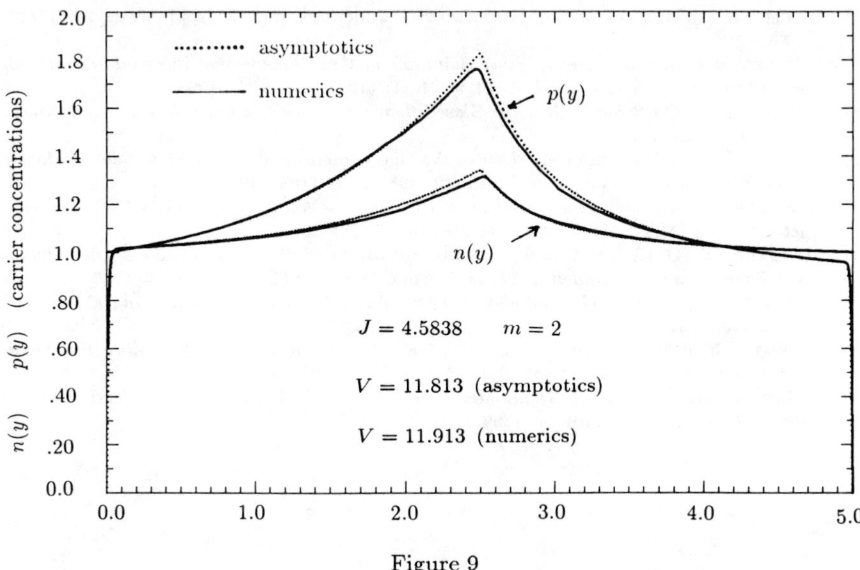

Figure 9

analysis breaks down when the electric field vanishes at some point inside the domain for then (4.7) predicts infinite carrier concentrations. The inequality $E_0 < 0$ will certainly be violated if the total current J_0 becomes too small or the device length L becomes too long. In particular, our analysis breaks down near equilibrium, where it is known that the diffusion terms in (4.3) cannot be neglected. The resulting abrupt termination of the asymptotic current-voltage curve as J is decreased can be seen in Fig. 8. In addition, even when the inequality $E_0(y) < 0$ is satisfied we observe, from (4.7), (4.8a) and Fig. 9, that the asymptotic carrier concentrations are not differentiable at $y = y_i$. Finally, as discussed in [14], this high field analysis does not apply to strongly reverse biased structures for which E_0 has an opposite sign. An analysis relevant to this case was done in [8].

We are currently working on extending this high field asymptotic approach to analyze the hydrodynamic model of semiconductor devices describing an N^+NN^+ structure.

<h2 style="text-align:center">REFERENCES</h2>

[1]   A. GNUDI, F. ODEH, M. RUDAN, *Non-Local Effects in Small Semiconductor Devices*, European Transactions on Telecommunications, 3, 1 (1990), pp. 307-312.

[2]   S. HOWISON, J. KING, *Explicit Solutions to Six Free-Boundary Problems in Fluid Flow and Diffusion*, IMA J. Appl. Math, 42 (1989), pp. 155-176.

[3]   P. MARKOWICH, *The Stationary Semiconductor Device Equations*, Springer-Verlag, Vienna-New York, 1986.

[4]   P. MARKOWICH, C. RINGHOFER, C. SCHMEISER, *The Semiconductor Equations*, Springer-Verlag, Vienna-New York, 1990.

[5]   M. MOCK, *An Example of Nonuniqueness of Stationary Solutions in Semiconductor*, Device Models, COMPEL (1982), pp. 165–174.

[6]   C. PLEASE, *An Analysis of Semiconductor P-N Junctions*, IMA J. Appl. Math, 28 (1982), pp. 301–318.

[7]   I. RUBINSTEIN, *Multiple Steady State Solutions in One-Dimensional Electrodiffusion with Local Electroneutrality*, SIAM J. Appl. Math, 47 (1987), pp. 1076–1093.

[8]   C. SCHMEISER, *On Strongly Reverse Biased Semiconductor Diodes*, SIAM J. Appl. Math, 49 (1989), pp. 1734–1748.

[9]   H. STEINRÜCK, *A Bifurcation Analysis of the One-Dimensional Steady State Semiconductor Device Equations*, SIAM J. Appl. Math, 49 (1989), pp. 1102–1121.

[10]   H. STEINRÜCK, *An Asymptotic Analysis of the Current-Voltage Curve of a PNPN Semiconductor Device*, IMA J. Appl. Math, 43 (1990), pp. 243–261.

[11]   M. WARD, F. ODEH, D. COHEN, *Asymptotic Methods for Metal Oxide Semiconductor Field-Effect Transistor Modeling*, SIAM J. Appl. Math, 50 (1990), pp. 1099–1126.

[12]   M. WARD, *Singular Perturbations and a Free Boundary Problem in the Modeling of Field-Effect Transistors*, SIAM J. Appl. Math., 52 (1992), pp. 112–139.

[13]   M. WARD, L. REYNA, F. ODEH, *Multiple Steady State Solutions in a Multi-Junction Semiconductor Device*, SIAM J. Appl. Math, 51 (1991), pp. 90–123.

[14]   L. REYNA, M. WARD, *Drift Dominated Current Flow in Forward Biased Semiconductor Devices*, IMA J. Appl. Math., 48 (1992), pp. 1–21.